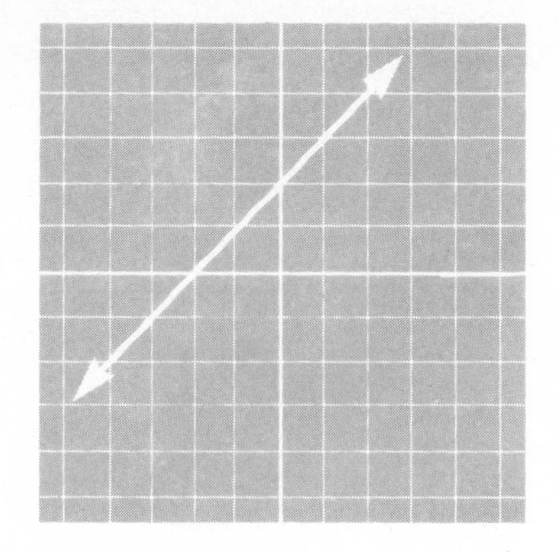

Introductory Algebra

SECOND EDITION

D. Franklin Wright
Cerritos College

Bill D. New
Cerritos College

ALLYN AND BACON, INC.

Boston · London · Sydney · Toronto

Series Editor, Carol Nolan-Fish
Editorial Production Services, Barbara A. Willette
Copy Editor, Margaret Shaffer
Cover Administrator, Linda Dickinson
Cover Designer, Christy Rosso

Library of Congress Cataloging-in-Publication Data

Wright, D. Franklin.
Introductory algebra.

Includes index.
1. Algebra. I. New, Bill D. II. Title.
QA152.2.W74 1986 512.9 85-11181
ISBN 0-205-08512-1

Printed in the United States of America.
10 9 8 7 6 5 4 3 2 1 90 89 88 87 86 85

$$\text{slope} = \frac{\text{rise}}{\text{run}} = \frac{y_2 - y_1}{x_2 - x_1} = \frac{y_1 - y_2}{x_1 - x_2}$$

$Ax + By = C$ The **standard form** of a **linear equation**

$y = mx + b$ The **slope-intercept form** for the equation of a line

$y - y_1 = m(x - x_1)$ The **point-slope form** for the equation of a line

Properties of Exponents

If a and b are nonzero numbers and m and n are integers, then

Property 1 $a^m \cdot a^n = a^{m+n}$

Property 2 $a^0 = 1$

Property 3 $a^{-n} = \frac{1}{a^n}$

Property 4 $\frac{a^n}{a^m} = a^{n-m}$

Property 5 $(ab)^n = a^n b^n$

Property 6 $\left(\frac{a}{b}\right)^n = \frac{a^n}{b^n}$

Property 7 $(a^m)^n = a^{mn}$

General Quadratic Equation

$$ax^2 + bx + c = 0 \qquad a \neq 0$$

Quadratic Formula

$$x = \frac{-b \pm \sqrt{b^2 - 4ac}}{2a}$$

INTRODUCTORY ALGEBRA

Contents

3 Solving First-Degree Equations 99

4 Exponents and Polynomials 151

5 Factoring Polynomials 195

6 Rational Expressions 235

7 Applications with First-Degree Equations and Inequalities 279

8 Graphing Linear Equations 319

Preface

THE PURPOSE

Introductory Algebra has been written to provide a smooth transition from arithmetic to the more abstract skills and reasoning abilities developed in a beginning algebra course. This second edition has taken advantage of feedback from users of the first edition. Several sections have been rewritten and reorganized.

We have assumed only that students have some basic skills with arithmetic and no previous knowledge of algebra. Chapter 1 now contains not only an arithmetic review but also an early introduction to solving equations, use of decimal numbers and fractions in equations, and use of geometric formulas, both to serve as reference for later applications and to illustrate the use of algebra in a familiar setting. We stress the similarities between arithmetic and algebra to give the students confidence in their ability to learn algebra.

The text is traditional, with very little use of sets and no reference to set properties or operations. Whenever feasible, information is presented in list form for easy reference. We have included many more examples in this edition and have improved the descriptions and discussions of the examples. There are more diagrams for the word problems. Many sections have a few extra problems for practice with calculators.

Two of the most important topics in algebra—solving equations and solving word problems—have been treated with special emphasis. In Chapter 3 we have presented skills for solving first-degree equations in a step-by-step approach over three sections. These skills are used throughout the text, particularly in Chapters 5, 7, 9, and 11. Chapters 1, 2, 3, 5, 6, 7, 9, and 11 each contain at least one section on word problems. There are several optional sections entitled "Additional Applications" that are designed to show applications of algebra in real-life situations such as physics, nursing, chemistry, and agriculture. The students see and manipulate some formulas that they might otherwise see only in specialized programs.

Each chapter now has a Chapter Test as well as a Chapter Summary of ideas and an extensive Chapter Review. A special feature for student interest is a computer program written in BASIC with a printout of results at the end of each chapter, showing how to use the computer to deal with a topic covered in the chapter.

THE EXERCISES

More than 4000 carefully selected and graded exercises proceed from relatively easy problems to more difficult ones. There is particular emphasis on solving equations and solving word problems.

SPECIAL FEATURES

1. Each chapter has
   "Did You Know?" historical commentaries
   Chapter Summary of key ideas and terms
   Chapter Review for extra practice
   Chapter Test
   "Using the Computer" for special interest
2. Answers to the odd-numbered exercises and to all the Chapter Review and Chapter Test questions are in the back of the book.
3. There are over 440 numbered examples completely worked out with detailed analysis. There are also practice quiz problems and many examples in the discussions.
4. Key ideas, procedures, and rules are in list form for easy identification and reference.
5. The format allows for easy identification of the various parts: text material, examples, practice quizzes, and exercises.
6. Additional Applications sections show real-life applications that students may not think involve algebra. The problems are designed to be easy so as to stimulate interest and to help answer the question that confronts every algebra teacher: "What good is algebra?"

THE CONTENT

Chapter 1 (Whole Numbers and Fractions) is a review of arithmetic and an introduction to algebra. Variables and exponents are introduced and are sued in the review of basic topics such as fractions, decimals and percents, least common multiple, and order of operations.

We also have included an introduction to equations and work with geometric formulas. Most of these ideas are used throughout the text in word problems and at higher skill levels.

Chapter 2 (Signed Numbers) develops the basic skills of adding, subtracting, multiplying, and dividing with positive and negative numbers. Absolute value is defined in terms of distance from zero; we have left the more abstract definition for Intermediate Algebra. Decimals and fractions are included in the exercises to allow for more varied and interesting exercises and word problems in Chapter 3.

In Chapter 3 (Solving First-Degree Equations), step-by-step techniques for solving first-degree equations are developed over three sections.

Chapter 4 (Exponents and Polynomials), Chapter 5 (Factoring Polynomials), and Chapter 6 (Rational Expressions) cover the standard skills of adding, subtracting, multiplying, and dividing with polynomials and exponents. We have reorganized Chapter 4 so that all the basic formulas for exponents (including negative exponents) are introduced in Sections 4.1 and 4.2. For more practice and interest, scientific notation now appears in Section 4.2 instead of in the Appendix as in the first edition.

Chapter 7 (Applications with First-Degree Equations and Inequalities) starts with a review of solving first-degree equations, including equations with variable denominators. Then first-degree inequalities are discussed along with graphing intervals of real numbers. Three sections of applications allow the students to develop their reasoning abilities and equation-solving skills more thoroughly and to see algebra in a "useful" light.

Chapter 8 (Graphing Linear Equations) discusses the graphs of straight lines in a plane and devotes an entire section to each of the different algebraic forms for equations of straight lines. Linear inequalities are now included here (as an optional section) instead of in the Appendix.

Chapter 9 (Systems of Linear Equations) shows how to solve systems of linear equations graphically and algebraically and how to apply these ideas in solving word problems.

Chapter 10 (Real Numbers and Radicals) discusses the real numbers in detail and introduces radical notation with emphasis on square roots. The formula for the distance between two points is developed and illustrated by an immediate application of radicals easily understood by the students.

Chapter 11 (Quadratic Equations) first reminds the students that some quadratic equations can be solved by factoring (as in Chapter 5), then discusses completing the square. With this background the quadratic formula is developed, and a variety of applications are given, include the Pythagorean Theorem, work, distance, and geometry.

We recommend that the topics be covered in the order presented in the text, since most sections assume knowledge of the material in previous sections.

ADDITIONAL AIDS

Supplementary materials include a Study Guide for students who desire additional exercises and explanations and an Instructor's Manual that contains the answers to the even-numbered exercises and sample tests for each chapter.

ACKNOWLEDGMENTS

We want to thank our editor on this second edition, Carol Nolan-Fish, for her support and enthusiasm for the entire project, including our second edition of Intermediate Algebra. Barbara Willette has done another outstanding job on production. The manuscript typing by Pat Wright was terrific as usual and has made everyone's job that much easier. Our students, colleagues at Cerritos College, and the following reviewers have all been helpful with their constructive and critical comments:

Reviewers, Second Edition

Lu Ann Blair, *University of Wisconsin–Parkside*
Duane E. Deal, *Ball State University*
John F. Haldi, *Spokane Community College*
Helen M. Hancock, *Shoreline Community College*
Vivian Heigl, *University of Wisconsin–Parkside*
Linda Holden, *Indiana University*
Carol Johnson, *Blue Jay, California*
Ben G. Matley, *Ventura College*
Doris Nice, *University of Wisconsin–Parkside*
Karen L. Pender, *Chaffey College*
Kathleen C. Stiehl, *University of Wisconsin–Green Bay*
Richard Watkins, *Tidewater Community College*

Accuracy Reviewers, Second Edition

Lois McBride, *Stark Technical College*
Nancy Ann McCullough, *Big Rapids, Michigan*

Reviewers, First Edition

Ignacio Bello, *Hillsborough Community College*
Edward Bouse, *Community College of Denver*
Dean Buzzard, *Charles Stewart Mott Community College*
Mary Connolly, *Indiana University–South Bend*
Elwyn Cutler, *Ferris State College*

Robert Davies, *Cuyahoga Community College*
Rosemary Degnan, *El Paso Community College*
Albert Giambrone, *Sinclair Community College*
George Grisham, *Illinois Central College*
James Hall, *University of Wisconsin*
Nancy Halford, *Rio Hondo College*
George Holloway, *L.A. Valley College*
Carol Johnson, *Golden West College*
Jerry Karl, *Golden West College*
Samuel Sargis, *Modesto Junior College*
Albert Thompson, Jr., *Tidewater Community College*
Dwann Veroda, *El Camino College*
Glorya Welch, *Cerritos College*

The essays entitled "Did You Know?" were written by Deann Christianson, University of the Pacific, Stockton, California. Carol Johnson and Nancy Moore are responsible for the Study Guide and Instructor's Manual. Jae Chung wrote the BASIC Programs at the end of each chapter. Thank you all.

INTRODUCTORY ALGEBRA

The essence of mathematics lies in its freedom.

GEORG CANTOR 1845–1918

DID YOU KNOW?

Almost all of mathematics uses the language of sets to simplify notation and to help in understanding concepts. The theory of sets is one of the few branches of mathematics that was initially developed almost completely by one person, Georg Cantor.

Georg Cantor was born in St. Petersburg, Russia, but spent his adult life in Germany, first as a student at the University of Berlin and later as a professor at the University of Halle. Cantor's research and development of set ideas was met by ridicule and public attacks on his character and work. In 1885, Cantor suffered the first of a series of mental breakdowns probably caused by the attacks on his work by other mathematicians, most notably a former teacher, Leopold Kronecker. It is suggested that Kronecker kept Cantor from becoming a professor at the prestigious University of Berlin.

We are not used to seeing intolerance toward new ideas among the scientific community, and, as you study mathematics, you will probably have difficulty imagining how anyone could have felt hostile toward or threatened by Cantor's ideas. However, his idea of a set with an **infinite** number of elements was thought of as revolutionary by the mathematical establishment.

The problem is that infinite sets have the following curious property: a part may be numerically equal to a whole. For example,

$$\begin{array}{rl} N = \{1, & 2, \; 3, \; 4, \;\; 5, \ldots, \;\; n, \ldots\} \\ & \updownarrow \; \updownarrow \; \updownarrow \; \updownarrow \;\; \updownarrow \qquad\;\; \updownarrow \\ E = \{2, & 4, \; 6, \; 8, \; 10, \ldots, 2n, \ldots\} \end{array}$$

Sets N and E are numerically equal because set E can be put in one-to-one correspondence (matched or counted) with set N. So there are as many even numbers as there are natural numbers! This is contrary to common sense and, in Cantor's time, contrary to usual mathematical assumptions.

Whole Numbers and Fractions

1.1 OPERATIONS WITH WHOLE NUMBERS

The numbers 1, 2, 3, 4, 5, . . . are called the **counting numbers** or **natural numbers.** The three dots indicate that the pattern is to continue without end. Putting 0 with the set of natural numbers gives 0, 1, 2, 3, 4, 5, . . . , called the **whole numbers.** Thus, 0 is a whole number but not a natural number.

In this section, we will review the four basic operations with whole numbers—addition, subtraction, multiplication, and division.

Addition with whole numbers can be indicated by writing the numbers either horizontally with a plus (+) sign between the numbers or in column form with directions to add. For example,

$$25 + 8 + 12 \qquad \text{or} \qquad \text{Add:}\ \begin{array}{r} 25 \\ 8 \\ \underline{12} \end{array}$$

The result of addition is called the **sum.**

$$\underset{\underset{\text{sum}}{\uparrow}}{25 + 8 + 12 = 45} \qquad \text{or} \qquad \begin{array}{r} 25 \\ 8 \\ \underline{12} \\ 45 \leftarrow \text{sum} \end{array}$$

The statement $7 + 6 = 6 + 7$ is called an **equation.** The equal sign (=) means that the number represented on the left is the same as the number represented on the right.

The fact that 7 and 6 can be added in either order to get the same sum, 13, is an example of the **commutative property of addition.** We can illustrate this property more generally with the use of **variables,** that is, symbols or letters that can represent more than one number.

> **Commutative Property of Addition**
>
> If a and b are whole numbers, then
>
> $$a + b = b + a$$

Some examples of the commutative property of addition are

1. $5 + 2 = 2 + 5$
2. $8 + 10 = 10 + 8$
3. $17 + x = x + 17$
   This statement is true for any value of the **variable** x. For example, if $x = 3$, then

$$17 + 3 = 3 + 17$$

since $17 + 3 = 20$ and $3 + 17 = 20$. Or if $x = 5$, then

$$17 + 5 = 5 + 17$$

since $17 + 5 = 22$ and $5 + 17 = 22$.

To add three or more numbers, we use the **associative property of addition,** which allows us to group numbers together differently but still get the same sum. For example,

$$8 + 5 + 9 = (8 + 5) + 9 = 13 + 9 = 22$$

and

$$8 + 5 + 9 = 8 + (5 + 9) = 8 + 14 = 22$$

Associative Property of Addition

If a, b, and c are whole numbers, then

$$a + b + c = (a + b) + c = a + (b + c)$$

Some examples of the associative property of addition are

1. $(3 + 7) + 4 = 3 + (4 + 7)$
   since $(3 + 7) + 4 = (10 + 4) = 14$
   and $3 + (4 + 7) = 3 + (11) = 14$.
2. $(5 + 9) + y = 5 + (9 + y)$
   As an example, we can substitute 18 for y. Then
   $(5 + 9) + 18 = (14) + 18 = 32$
   and $5 + (9 + 18) = 5 + 27 = 32$.
   So, $(5 + 9) + 18 = 5 + (9 + 18)$. In fact, the equation is true for any value of y.

What number would you add to 15 to get 18? The answer is 3, of course. This number is called the **difference** between 15 and 18 and can be found using a reverse addition called **subtraction.** Subtraction is indicated with a minus (−) sign, and

$$18 - 15 = 3 \text{ because } 18 = 15 + 3$$

To subtract large numbers, we must be aware of the place value of each digit in writing a number and of the technique called **borrowing.** For example, 473 − 195 cannot be done in your head because you don't know what to add to 195 to get 473. We can write

$$\begin{array}{rcrcrcr} 473 & = & 400 + 70 + 3 & = & 400 + 60 + 13 & = & 300 + 160 + 13 \\ \underline{-195} & = & \underline{100 + 90 + 5} & = & \underline{100 + 90 + 5} & = & \underline{100 + 90 + 5} \\ & & & & & & 200 + 70 + 8 = 278 \end{array}$$

We "borrowed" 10 from 70, then borrowed 100 from 400. In familiar shorthand,

$$\begin{array}{r} {\scriptstyle 3\ 16\ 1} \\ \not{4}\ \not{7}\ 3 \\ \underline{-1\ 9\ 5} \\ 2\ 7\ 8 \end{array}$$

The commutative and associative properties are **not** true for subtraction. For example,

$$8 - 5 \neq 5 - 8 \quad \text{and} \quad 17 - (6 - 3) \neq (17 - 6) - 3$$

("≠" is read "is not equal to.")

Multiplication is shorthand for repeated addition. For example,

$$613 + 613 + 613 + 613 = 4 \cdot 613 = 2452$$

The repeated number (613) and the number of times it is being used (4) are both called **factors** of the result (2452) which is now called the **product.**

Multiplication can be indicated in any of the following ways:

a. $4 \cdot 613$	b. $4(613)$	c. $(4)613$
d. $(4)(613)$	e. 4×613	f. $\begin{array}{r} 613 \\ \underline{\times 4} \end{array}$

Generally we will avoid types (e) and (f) because the times sign (×) can be confused with the letter x. Also, multiplication can be indicated by writing a number next to a variable or two or more variables next to each other. For example, $7y$ means $7 \cdot y$, and abc means $a \cdot b \cdot c$. Thus, $3a$, ab, and $5xyz$ all indicate multiplication.

Multiplication is both commutative and associative. That is, $3 \cdot 5 = 5 \cdot 3$ and $6 \cdot (4 \cdot 7) = (6 \cdot 4) \cdot 7$.

Commutative Property of Multiplication

If a and b are whole numbers, then

$$a \cdot b = b \cdot a$$

Some examples of the commutative property of multiplication are

1. $4 \cdot 20 = 20 \cdot 4$
2. $3y = y3$
   (Note that $y3$ means the product of y and 3; however, we usually write the number to the left of the variable, such as $3y$ or $7x$ or $18ab$.)
   Substituting $y = 6$ (or any other number), we see
   $3 \cdot 6 = 6 \cdot 3$
   or $18 = 18$
   (Note that the dots are necessary between numbers since 36 does not equal 63.)

Associative Property of Multiplication

If a, b, and c are whole numbers, then

$$a \cdot b \cdot c = (a \cdot b) \cdot c = a \cdot (b \cdot c)$$

Some examples of the associative property of multiplication are

1. $2 \cdot (12 \cdot 5) = (2 \cdot 12) \cdot 5$
   since $2 \cdot (12 \cdot 5) = 2(60) = 120$
   and $(2 \cdot 12) \cdot 5 = (24) \cdot 5 = 120$
2. $8 \cdot x \cdot y = (8x)y = 8(xy)$
   For example, if x is 3 and y is 2, we have
   $(8 \cdot 2) \cdot 3 = (16) \cdot 3 = 48$
   and $8(2 \cdot 3) = 8(6) = 48$.
   So, $(8 \cdot 2) \cdot 3 = 8(2 \cdot 3)$.

One property of numbers that we find extremely useful in algebra combines both addition and multiplication. We can illustrate this property, called the **distributive property of multiplication over addition,** in the following way. If you were to multiply $8(10 + 1)$, you would probably add 10 and 1 first and get $8(10 + 1) = 8(11) = 88$.

We can get the same result as follows:

$$8(10 + 1) = 8 \cdot 10 + 8 \cdot 1 = 80 + 8 = 88$$

The fact that $8(10 + 1) = 8 \cdot 10 + 8 \cdot 1$ illustrates the distributive property.

Distributive Property of Multiplication over Addition

If a, b, and c are whole numbers, then

$$a \cdot (b + c) = a \cdot b + a \cdot c$$

The following geometric figure might help in remembering the distributive property.

a | $b + c$ = a | b + a | c

$$a(b + c) = ab + ac$$

total area of large rectangle = sum of areas of two smaller rectangles

Some examples of the distributive property are

1. $3(5 + 7) = 3 \cdot 5 + 3 \cdot 7 = 15 + 21 = 36$

2. $4(x + 3) = 4 \cdot x + 4 \cdot 3 = 4x + 12$

EXAMPLES

1. Add:

```
 25
 46
193
 50
---
314   sum
```

2. Subtract:

```
742
361
---
381   difference
```

3. Multiply:

$$\begin{array}{r} 82 \\ \underline{34} \\ 328 \\ \underline{246} \\ 2788 \end{array} \quad \text{product}$$

4. Use the distributive property to evaluate $7(3 + 5)$.

$$7(3 + 5) = 7 \cdot 3 + 7 \cdot 5 = 21 + 35 = 56$$

What number would you multiply by 2 to get 14? The answer is 7, of course. This number is called the **quotient** of 14 and 2 and can be found using a reverse multiplication called **division.** Division can be indicated using a divide ($\div$) sign, or a bar (—) called fraction form, or a $\overline{)}$ sign. Thus, we can write

$$14 \div 2 = 7, \qquad \frac{14}{2} = 7, \qquad 2\overset{7}{\overline{)14}}$$

and $14 \div 2 = 7$ because $14 = 2 \cdot 7$.

In general, for nonzero b, $\dfrac{a}{b} = x$ if and only if $a = b \cdot x$.

Division by 0 is Undefined

Consider $\frac{6}{0} = \square$. Whatever $\square$ is, $6 = 0 \cdot \square$. But $0 \cdot \square = 0$ whatever $\square$ represents, so that $6 = 0 \cdot \square = 0$. This is impossible. Next consider $\frac{0}{0} = \square$. Then $0 = 0 \cdot \square$, which is true regardless of what $\square$ represents. This means $\frac{0}{0}$ could be any number, an unacceptable situation. **Thus $\mathbf{\dfrac{a}{0}}$ is undefined for any whole number a.**

Does $14 \div 2 = 2 \div 14$? Does $36 \div (12 \div 3) = (36 \div 12) \div 3$? The answer to both questions is No. In other words, division, like subtraction, is neither commutative nor associative.

Another topic related to operations with whole numbers is **average,** such as a batter's average in baseball or the average value of stocks on the stock market.

To Find the Average of a Collection of Numbers

1. Add all the numbers in the collection.
2. Divide this sum by the number of numbers in the collection.

This quotient is the **average.**

EXAMPLES

5. 120 ÷ 4

$$\begin{array}{r} 30 \\ 4\overline{)120} \\ \underline{12} \\ 00 \\ \underline{00} \end{array}$$ quotient

6. Divide:

$$\begin{array}{r} 46 \\ 7\overline{)322} \\ \underline{28} \\ 42 \\ \underline{42} \end{array}$$ quotient

7. Find the average of the numbers 82, 91, 63, 51, and 48.

$$\begin{array}{r} 82 \\ 91 \\ 63 \\ 51 \\ \underline{48} \\ 335 \end{array} \qquad \begin{array}{r} 67 \\ 5\overline{)335} \\ \underline{30} \\ 35 \\ \underline{35} \end{array}$$ average

Since 5 numbers are added, the sum is divided by 5 to find the average.

EXERCISES 1.1

Add in Exercises 1–10.

1. $\begin{array}{r} 38 \\ 57 \\ \underline{91} \end{array}$ **2.** $\begin{array}{r} 63 \\ 42 \\ \underline{86} \end{array}$ **3.** $\begin{array}{r} 97 \\ 132 \\ 61 \\ \underline{5} \end{array}$ **4.** $\begin{array}{r} 153 \\ 201 \\ 89 \\ \underline{62} \end{array}$ **5.** $\begin{array}{r} 653 \\ 1402 \\ 119 \\ \underline{88} \end{array}$

6. 9 + 11 + 13 + 2 **7.** 15 + 20 + 13 + 9 **8.** 17 + 2 + 12 + 3

9. 7 + 13 + 6 + 18 **10.** 23 + 19 + 8 + 16

Subtract in Exercises 11–20.

11. $\begin{array}{r}36\\ \underline{21}\end{array}$ **12.** $\begin{array}{r}45\\ \underline{14}\end{array}$ **13.** $\begin{array}{r}25\\ \underline{16}\end{array}$ **14.** $\begin{array}{r}63\\ \underline{48}\end{array}$ **15.** $\begin{array}{r}365\\ \underline{79}\end{array}$ **16.** $\begin{array}{r}482\\ \underline{288}\end{array}$

17. $56 - 32$ **18.** $43 - 27$ **19.** $94 - 59$ **20.** $77 - 38$

Multiply in Exercises 21–30.

21. $\begin{array}{r}37\\ \underline{8}\end{array}$ **22.** $\begin{array}{r}748\\ \underline{6}\end{array}$ **23.** $\begin{array}{r}45\\ \underline{27}\end{array}$ **24.** $\begin{array}{r}96\\ \underline{93}\end{array}$ **25.** $\begin{array}{r}723\\ \underline{37}\end{array}$ **26.** $(34)(28)$

27. $(86)(51)$ **28.** $142 \cdot 39$ **29.** $183 \cdot 62$ **30.** $179 \cdot 48$

Divide in Exercises 31–40.

31. $7\overline{)525}$ **32.** $6\overline{)138}$ **33.** $8\overline{)272}$ **34.** $5\overline{)845}$ **35.** $17\overline{)238}$

36. $19\overline{)608}$ **37.** $135 \div 9$ **38.** $336 \div 16$ **39.** $612 \div 18$ **40.** $405 \div 15$

Use the distributive property to evaluate the expressions in Exercises 41–46.

41. $6(3 + 8)$ **42.** $5(7 + 2)$ **43.** $7(8 + 5)$

44. $9(6 + 4)$ **45.** $10(7 + 6)$ **46.** $12(5 + 3)$

Complete the expressions in Exercises 47–56 using the given property.

47. $7 + 3 =$ ______ commutative property for addition

48. $(6 \cdot 9) \cdot 3 =$ ______ associative property for multiplication

49. $19 \cdot 4 =$ ______ commutative property for multiplication

50. $18 + 5 =$ ______ commutative property for addition

51. $6(5 + 8) =$ ______ distributive property

52. $16 + (9 + 11) =$ ______ associative property for addition

53. $2 \cdot (3x) =$ ______ associative property for multiplication

54. $3(x + 5) =$ ______ distributive property

55. $3 + (x + 7) =$ ______ associative property for addition

56. $9(x + 5) =$ ______ distributive property

Name the property used in each of the Exercises 57–66, and show that they are true if, for example, $x = 4$.

57. $6 \cdot x = x \cdot 6$ **58.** $19 + x = x + 19$

59. $8 + (4 + x) = (8 + 4) + x$ **60.** $(2 \cdot 7) \cdot x = 2 \cdot (7 \cdot x)$

61. $5(x + 18) = 5x + 90$ **62.** $(x + 14) + 3 = x + (14 + 3)$

63. $(6 \cdot x) \cdot 9 = 6 \cdot (x \cdot 9)$

64. $11 \cdot x = x \cdot 11$

65. $x + 34 = 34 + x$

66. $3(x + 15) = 3x + 45$

Find the average of each of the sets of numbers in Exercises 67–70.

67. 21, 14, 16

68. 47, 39, 55

69. 27, 18, 25, 30

70. 37, 45, 52, 58

1.2 EXPONENTS AND ORDER OF OPERATIONS

Since $4 \cdot 8 = 32$, 4 and 8 are called **factors** of 32, and 32 is called the **product** of 4 and 8. Similarly, since $9 \cdot 3 = 27$, 9 and 3 are factors of the product 27. If a factor is repeated, such as in $7 \cdot 7 = 49$ and $3 \cdot 3 \cdot 3 \cdot 3 = 81$, there is a shorthand notation using **exponents** to indicate the repetition. An **exponent** is a number that tells how many times a factor occurs in a product. The exponent is written to the right and slightly above the factor, and the factor is called the **base** of the exponent. The product is called a **power** of the factor. Thus, we can write

$7 \cdot 7 = 7^2 = 49$ (exponent: 2; base: 7; power: 49) — 7^2 is read "seven squared" or "seven to the second power."

$6 \cdot 6 = 6^2 = 36$

$2 \cdot 2 \cdot 2 = 2^3 = 8$ — 2^3 is read "two cubed" or "two to the third power."

$10 \cdot 10 \cdot 10 \cdot 10 = 10^4 = 10{,}000$ — 10^4 is read "ten to the fourth power."

$5 \cdot 5 \cdot 3 = 5^2 \cdot 3 = 75$

$a \cdot a \cdot a \cdot a \cdot a = a^5$

> In general,
>
> $$a^n = \underbrace{a \cdot a \cdot a \ldots a}_{n \text{ factors}}$$

A **prime number** is a whole number greater than 1 that has only two factors, itself and 1. For example, 13 is a prime number because it has only two factors, 13 and 1. The number 24 is not prime because it has factors other than 24 and 1, namely, 2, 3, 4, 6, 8, and 12. Whole numbers other than 0 and 1 that are not prime numbers are called **composite** numbers. Thus, 24, 10, and 45 are examples of composite numbers. For reasons related to factoring composite numbers, the two numbers 0 and 1 are neither prime nor composite. The prime numbers less than 50 are

2, 3, 5, 7, 11, 13, 17, 19, 23, 29, 31, 37, 41, 43, and 47

We sometimes want to find all the prime factors (the **prime factorization**) of a composite number. This is particularly helpful in working with fractions and simplifying algebraic expressions.

To Find the Prime Factorization of a Composite Number

1. Find any two factors of the number.
2. Continue to factor each of the factors until all factors are prime numbers.
3. Write the product of all these prime factors. (Use exponents whenever possible.)

EXAMPLES

Find the prime factorizations of each of the following composite numbers.

1. 72 $\quad 72 = 8 \cdot 9 = 2 \cdot 4 \cdot 3 \cdot 3 = 2 \cdot 2 \cdot 2 \cdot 3 \cdot 3 = 2^3 \cdot 3^2$
   Or you might write
   $72 = 12 \cdot 6 = 3 \cdot 4 \cdot 2 \cdot 3 = 3 \cdot 2 \cdot 2 \cdot 2 \cdot 3 = 2^3 \cdot 3^2$
   No matter how you start, the prime factorization will be the same.

2. 60 $\quad 60 = 6 \cdot 10 = 2 \cdot 3 \cdot 2 \cdot 5 = 2^2 \cdot 3 \cdot 5$
   Or you might write
   $60 = 2 \cdot 30 = 2 \cdot 15 \cdot 2 = 2 \cdot 3 \cdot 5 \cdot 2 = 2^2 \cdot 3 \cdot 5$

3. 242 $\quad 242 = 2 \cdot 121 = 2 \cdot 11 \cdot 11 = 2 \cdot 11^2$

4. 230 $\quad 230 = 23 \cdot 10 = 23 \cdot 2 \cdot 5 = 2 \cdot 5 \cdot 23$

Multiples of a number are the products of that number with the counting numbers.

Counting numbers:	1, 2, 3, 4, 5, 6, 7, 8, 9, . . .
Multiples of 6:	6, 12, 18, 24, 30, 36, 42, 48, 54, . . .
Multiples of 8:	8, 16, 24, 32, 40, 48, 56, 64, 72, . . .

For the multiples of 6 and 8, the common multiples are

$$24, 48, 72, 96, 120, \ldots$$

The smallest of these, called the **least common multiple** (**LCM**), is 24. Note that the LCM is **not** $6 \cdot 8 = 48$. In this case, the LCM is 24, and it is smaller than the product of 6 and 8.

The LCM is very useful in working with fractions in arithmetic and algebra. The LCM can be found using prime factorizations. The technique is illustrated in the following example. Note that it gives a number, 180, that is much smaller than the product of 15, 18, and 20 ($15 \cdot 18 \cdot 20 = 5400$). Thus, if you were to add the three fractions $\frac{1}{15} + \frac{1}{18} + \frac{1}{20}$, you would want to use a common denominator of 180, not 5400. This will be discussed again in Section 1.4.

EXAMPLE

5. Find the LCM for the numbers 15, 18, and 20.

   a. Find the prime factorization of each number.

$$15 = 3 \cdot 5$$

$$18 = 2 \cdot 3 \cdot 3 = 2 \cdot 3^2$$

$$20 = 2 \cdot 2 \cdot 5 = 2^2 \cdot 5$$

   b. Form the product $2^2 \cdot 3^2 \cdot 5 = 180 = \text{LCM}$.
   (This product uses each prime factor the greatest number of times that it appears in any one prime factorization.)

To Find the LCM

1. Find the prime factorization of each number.
2. Form the product of all prime factors that appear using each prime factor the greatest number of times that it appears in any one prime factorization.

EXAMPLES

6. Find the LCM for the numbers 27, 15, and 60.

$$\left.\begin{aligned} 27 &= 3 \cdot 3 \cdot 3 = 3^3 \\ 15 &= 3 \cdot 5 \\ 60 &= 2 \cdot 2 \cdot 3 \cdot 5 = 2^2 \cdot 3 \cdot 5 \end{aligned}\right\} \text{LCM} = 2^2 \cdot 3^3 \cdot 5 = 540$$

7. Find the LCM for $4x$, x^2y, $6x^2$, and $18y^3$. (**Hint:** Use each variable as a factor.)

$$\left.\begin{aligned} 4x &= 2^2 \cdot x \\ x^2y &= x^2 \cdot y \\ 6x^2 &= 2 \cdot 3 \cdot x^2 \\ 18y^3 &= 2 \cdot 3^2 \cdot y^3 \end{aligned}\right\} \text{LCM} = 2^2 \cdot 3^2 \cdot x^2 \cdot y^3 = 36x^2y^3$$

In doing each Practice Quiz in this book, cover the answers and uncover them one at a time after you solve each problem.

PRACTICE QUIZ

Questions	Answers
Find the LCM for each set of numbers or expressions.	
1. 4, 10, 12	1. $2^2 \cdot 3 \cdot 5 = 60$
2. $12a^2b$, $18b^3$, $27ab$	2. $2^2 \cdot 3^3 \cdot a^2 \cdot b^3 = 108a^2b^3$

In Section 1.1, we said that division was not associative. So, what is the correct value of the expression $36 \div 12 \div 3$?

Does $\quad 36 \div 12 \div 3 = (36 \div 12) \div 3 = 3 \div 3 = 1$?

Or does $\quad 36 \div 12 \div 3 = 36 \div (12 \div 3) = 36 \div 4 = 9$?

We must agree on one answer.

What is the value of $14 \div 2 + 3 \cdot 2$?

Does $\quad 14 \div 2 + 3 \cdot 2 = 7 + 3 \cdot 2 = 10 \cdot 2 = 20$?

Or does $14 \div 2 + 3 \cdot 2 = 7 + 6 = 13$?

Again, we must agree on one answer.

General agreement has been reached by mathematicians on the following order of operations for evaluating numerical expressions.

Rules for Order of Operations

1. Work within symbols of inclusion (parentheses, brackets, or braces), beginning with the innermost pair.
2. Find any powers indicated by exponents.
3. From left to right, perform any multiplications or divisions in the order they appear.
4. From left to right, perform any additions or subtractions in the order they appear.

EXAMPLES

Simplify the following expressions using the rules for order of operations.

8. $36 \div 12 \div 3 = 3 \div 3$ Divide from left to right.

$= 1$

9. $14 \div 2 + 3 \cdot 2 = 7 + 6$ Divide and multiply before adding.

$= 13$

10. $3^2 - 8 \div 4 = 9 - 8 \div 4$ exponents

$= 9 - 2$ division

$= 7$ subtraction

11. $(7 + 8) \div 5 \cdot 4 + 3 = 15 \div 5 \cdot 4 + 3$ parentheses

$= 3 \cdot 4 + 3$ division

$= 12 + 3$ multiplication

$= 15$ addition

12. $[3(4 + 6) + 14] \div 4 + 7$	$= [3(10) + 14] \div 4 + 7$	parentheses
	$= [30 + 14] \div 4 + 7$	multiplication in brackets
	$= [44] \div 4 + 7$	addition in brackets
	$= 11 + 7$	division
	$= 18$	addition

PRACTICE QUIZ

Questions	Answers
Find the value of each expression.	
1. $2^5 - 8 \cdot 3$	1. 8
2. $(16 \div 2^2 + 6) \div 5 + 5$	2. 7
3. $14 \div 2 + 2 \cdot 6 + 30 \div 3$	3. 29

EXERCISES 1.2

Find each of the products in Exercises 1–10.

1. 3^2 **2.** 4^2 **3.** 2^4 **4.** 3^3 **5.** 8^2 **6.** 9^3

7. 5^4 **8.** 10^3 **9.** 12^2 **10.** 5^3

Write each expression in Exercises 11–20 in exponential form.

11. $5 \cdot 5$ **12.** $2 \cdot 2 \cdot 2 \cdot 2 \cdot 2$ **13.** $7 \cdot 7 \cdot 7 \cdot 7$

14. $a \cdot a \cdot a \cdot a$ **15.** $3 \cdot 3 \cdot a \cdot a \cdot a$ **16.** $2 \cdot 2 \cdot 2 \cdot a \cdot a$

17. $2 \cdot 2 \cdot 5 \cdot 5 \cdot 5 \cdot a \cdot a$ **18.** $3 \cdot a \cdot a \cdot a \cdot b \cdot b$ **19.** $7 \cdot a \cdot a \cdot a \cdot a \cdot b \cdot b \cdot b$

20. $5 \cdot 5 \cdot a \cdot a \cdot b \cdot b \cdot b \cdot b$

Find (a) the factors and (b) the first six multiples of each of the numbers in Exercises 21–25.

21. 6 **22.** 8 **23.** 10 **24.** 12 **25.** 15

Find the prime factorization for each of the numbers in Exercises 26–40

26. 24 **27.** 34 **28.** 28 **29.** 27 **30.** 54

31. 56 **32.** 80 **33.** 120 **34.** 140 **35.** 168

36. 153 **37.** 131 **38.** 189 **39.** 196 **40.** 241

Find the LCM for each set of numbers in Exercises 41–55.

41. 15, 35 **42.** 20, 14 **43.** 24, 20

44. 50, 40 **45.** 24, 15, 10 **46.** 49, 25, 35

47. 15, 26, 39 **48.** x, xy, y **49.** xy, xz, yz

50. xy, x^2y, xy^2 **51.** $3x^2y, 4xy^2, 6xy$ **52.** $8x, 10y, 20xy$

53. $14x^2, 21xy, 35x^2y^2$ **54.** $20xz, 24xy^2, 32yz$ **55.** $60x, 105x^2y, 120xy$

Find the value of each expression in Exercises 56–70 using the rules for order of operations.

56. $3 \cdot 5 - 2 \cdot 6$ **57.** $6 + (12 + 3) \div 5$ **58.** $(21 - 9) \div 3$

59. $18 \div 3 \cdot 6 - 3$ **60.** $3 \cdot 5 + 60 \div 3 \cdot 2$ **61.** $6 \cdot 3 \div 2 + 4 - 2$

62. $(4^2 + 6) \div 11 - 1 \cdot 2$ **63.** $7(4 - 2) \div 7 + 3$

64. $2^3 + 4 \div 2 + 2$ **65.** $3^3 \div 9 + (6 + 4^2) \div 2$

66. $(4^2 - 3^2) \div 7 + [2 + 2(3^2)] \div (2 \cdot 5)$

67. $14 + [11 \cdot 4 - (6 \cdot 3 + 1)]$ **68.** $6 + 3[4 - 2(3 - 1)]$

69. $7 - [4 \cdot 3 - (4 + 3 \cdot 2)]$ **70.** $2[6 + 4(1 + 7)] \div 4$

1.3 MULTIPLICATION AND DIVISION WITH FRACTIONS

A **fraction** is a number that can be written in the form $\frac{a}{b}$ which means $a \div b$. The number a is called the **numerator,** and b is called the **denominator.** Since we cannot divide by 0, the denominator $b \neq 0$.

In this chapter, we will discuss fractions that have whole numbers or whole-number variables for numerator and denominator. In Chapter 6 (Rational Expressions), we will discuss more complicated algebraic fractions. The basic rules we develop here will also apply in Chapter 6, so learn these rules well.

To multiply two fractions, multiply the numerators and multiply the denominators:

$$\frac{a}{b} \cdot \frac{c}{d} = \frac{a \cdot c}{b \cdot d}$$

EXAMPLES

1. $\frac{2}{3} \cdot \frac{7}{5} = \frac{2 \cdot 7}{3 \cdot 5} = \frac{14}{15}$

2. $\frac{1}{8} \cdot \frac{9}{10} = \frac{1 \cdot 9}{8 \cdot 10} = \frac{9}{80}$

The product of 1 with any number is that number, and 1 is called the **multiplicative identity.** That is,

$$\frac{a}{b} \cdot 1 = \frac{a}{b}$$

If $k \neq 0$, then $\frac{k}{k} = 1$, and we have

$$\frac{a}{b} = \frac{a}{b} \cdot 1 = \frac{a}{b} \cdot \frac{k}{k} = \frac{a \cdot k}{b \cdot k}$$

The relationship

$$\frac{a}{b} = \frac{a \cdot k}{b \cdot k} \qquad \text{where } k \neq 0$$

is called the **Fundamental Principle of Fractions.** We can use the Fundamental Principle to build a fraction to **higher terms** (find an equal fraction with a larger denominator) or to reduce to **lower terms** (find an equal fraction with a smaller denominator).

EXAMPLES

3. Raise $\frac{3}{7}$ to higher terms with a denominator of 28.

   Use $k = 4$ since $7 \cdot 4 = 28$.

$$\frac{3}{7} = \frac{3 \cdot \mathbf{4}}{7 \cdot \mathbf{4}} = \frac{12}{28}$$

4. Raise $\frac{5}{8}$ to higher terms with a denominator of $16a$.

   Use $k = 2a$ since $8 \cdot 2a = 16a$.

$$\frac{5}{8} = \frac{5 \cdot \mathbf{2a}}{8 \cdot \mathbf{2a}} = \frac{10a}{16a}$$

To reduce a fraction, factor both the numerator and denominator, then use the Fundamental Principle to "divide out" any common factors. If the numerator and denominator have no common prime factors, the fraction has been **reduced to lowest terms. Finding the prime factorizations of the numerator and denominator before reducing, while not necessary, will help guarantee a fraction is in lowest terms.**

EXAMPLES

Reduce the following fractions to lowest terms.

5. $\dfrac{12}{20} = \dfrac{\cancel{4} \cdot 3}{\cancel{4} \cdot 5} = \dfrac{3}{5}$ or $\dfrac{12}{20} = \dfrac{\cancel{2} \cdot \cancel{2} \cdot 3}{\cancel{2} \cdot \cancel{2} \cdot 5} = \dfrac{3}{5}$

6. $\dfrac{75}{90} = \dfrac{\cancel{15} \cdot 5}{\cancel{15} \cdot 6} = \dfrac{5}{6}$ or $\dfrac{75}{90} = \dfrac{\cancel{3} \cdot 5 \cdot \cancel{5}}{2 \cdot \cancel{5} \cdot 3 \cdot \cancel{3}} = \dfrac{5}{6}$

Factor and reduce before finding the following products.

7. $\dfrac{15}{28} \cdot \dfrac{4}{9} = \dfrac{\cancel{3} \cdot 5 \cdot \cancel{4}}{\cancel{4} \cdot 7 \cdot 3 \cdot \cancel{3}} = \dfrac{5}{21}$

8. $\dfrac{9a}{10b} \cdot \dfrac{25b}{33a^2} = \dfrac{\cancel{3} \cdot 3 \cdot \cancel{a} \cdot \cancel{5} \cdot 5 \cdot \cancel{b}}{2 \cdot \cancel{5} \cdot \cancel{b} \cdot \cancel{3} \cdot 11 \cdot a \cdot \cancel{a}} = \dfrac{15}{22a}$

Note that the number 1 is implied to be a factor even if it is not written. So, if all the prime factors "divide out" in either the numerator or denominator, 1 is still a factor.

9. $\dfrac{4a}{12b} \cdot \dfrac{3b}{7a} = \dfrac{\cancel{4} \cdot \cancel{a} \cdot \cancel{3} \cdot \cancel{b} \cdot 1}{\cancel{4} \cdot \cancel{3} \cdot \cancel{b} \cdot 7 \cdot \cancel{a}}$

$= \dfrac{1}{7}$

Here we write the factor 1 because all other factors have been "divided out."

We could write

$$\frac{4a}{12b} \cdot \frac{3b}{7a} = \frac{4 \cdot a \cdot 3 \cdot b}{4 \cdot 3 \cdot b \cdot 7 \cdot a}$$

$$= \frac{4}{4} \cdot \frac{3}{3} \cdot \frac{a}{a} \cdot \frac{b}{b} \cdot \frac{1}{7}$$

$$= 1 \cdot 1 \cdot 1 \cdot 1 \cdot \frac{1}{7} = \frac{1}{7}$$

If $a \neq 0$ and $b \neq 0$, the **reciprocal** of $\frac{a}{b}$ is $\frac{b}{a}$, and

$$\frac{a}{b} \cdot \frac{b}{a} = 1$$

Now consider the division problem $\frac{2}{3} \div \frac{5}{6}$. Using the definition of division in terms of multiplication, if

$$\frac{2}{3} \div \frac{5}{6} = \square, \text{ then } \frac{2}{3} = \square \cdot \frac{5}{6}$$

Since

$$\frac{2}{3} = \frac{2}{3} \cdot 1 = \frac{2}{3} \cdot \left(\frac{6}{5} \cdot \frac{5}{6}\right) = \boxed{\frac{2}{3} \cdot \frac{6}{5}} \cdot \frac{5}{6}$$

we have

$$\frac{2}{3} \div \frac{5}{6} = \boxed{\frac{2}{3} \cdot \frac{6}{5}}$$

Another approach to understanding division with fractions is to write the indicated division as a fraction.

$$\frac{2}{3} \div \frac{5}{6} = \frac{\frac{2}{3}}{\frac{5}{6}}$$ Write the division as a fraction.

$$= \frac{\frac{2}{3}}{\frac{5}{6}} \cdot \frac{\frac{6}{5}}{\frac{6}{5}}$$ Using the reciprocal of the denominator, multiply by

$$1 = \frac{\frac{6}{5}}{\frac{6}{5}}$$

$$= \frac{\frac{2}{3} \cdot \frac{6}{5}}{\frac{5}{6} \cdot \frac{6}{5}}$$ Simplify.

$$= \frac{\frac{2}{3} \cdot \frac{6}{5}}{1}$$ $$\frac{5}{6} \cdot \frac{6}{5} = 1$$

$$= \frac{2}{3} \cdot \frac{6}{5}$$

In either case, we have

$$\frac{2}{3} \div \frac{5}{6} = \frac{2}{3} \cdot \frac{6}{5} = \frac{2 \cdot 2 \cdot \cancel{3}}{\cancel{3} \cdot 5} = \frac{4}{5}$$

That is, **to divide by a nonzero fraction, multiply by its reciprocal.**

$$\frac{a}{b} \div \frac{c}{d} = \frac{a}{b} \cdot \frac{d}{c}$$

EXAMPLES

Divide and reduce all answers.

10. $\dfrac{3}{4} \div \dfrac{2}{5} = \dfrac{3}{4} \cdot \dfrac{5}{2} = \dfrac{15}{8}$

(In algebra, $\frac{15}{8}$, an improper fraction, is preferred to the mixed number $1\frac{7}{8}$. Improper fractions are perfectly acceptable as long as they are reduced, that is, as long as the numerator and denominator have no common prime factors.)

11. $\dfrac{26}{35} \div \dfrac{39}{20} = \dfrac{26}{35} \cdot \dfrac{20}{39} = \dfrac{2 \cdot \cancel{13} \cdot 2 \cdot 2 \cdot \cancel{5}}{\cancel{5} \cdot 7 \cdot 3 \cdot \cancel{13}} = \dfrac{8}{21}$

Note carefully that in Example 11 we factored and reduced without actually multiplying the numerators and denominators. It would not be wise to multiply first because we would then just have to factor two large numbers. For example,

$$\frac{26}{35} \div \frac{39}{20} = \frac{26}{35} \cdot \frac{20}{39} = \frac{520}{1365}$$

and now we have to factor 520 and 1365. But these were already factored for us since $26 \cdot 20 = 520$ and $35 \cdot 39 = 1365$.

12. $\dfrac{21a}{5b} \div 3a = \dfrac{21a}{5b} \cdot \dfrac{1}{3a} = \dfrac{\cancel{3} \cdot 7 \cdot \cancel{a}}{5 \cdot \cancel{3} \cdot \cancel{a} \cdot b} = \dfrac{7}{5b}$

Note: The reciprocal of $3a$ is $\dfrac{1}{3a}$ since $3a = \dfrac{3a}{1}$.

13. $\dfrac{5}{8} \div \dfrac{3}{4} \cdot \dfrac{6}{25} = \dfrac{5}{8} \cdot \dfrac{4}{3} \cdot \dfrac{6}{25}$

The reciprocal of $\dfrac{3}{4}$ is $\dfrac{4}{3}$.

$= \dfrac{\cancel{5} \cdot \cancel{4} \cdot \cancel{3} \cdot \cancel{2} \cdot 1}{\cancel{4} \cdot \cancel{2} \cdot \cancel{3} \cdot \cancel{5} \cdot 5}$

We need 1 in the numerator.

$= \dfrac{1}{5}$

14. **Division by 0 is undefined.** We know $15 \div 3 = 5$ because $15 = 3 \cdot 5$. Consider $6 \div 0 = x$. Then $6 = 0 \cdot x$. But this is impossible since $0 \cdot x = 0$ for any value of x.

Thus, $18 \div 0$ is undefined and $\frac{1}{4} \div 0$ is undefined. Remember, though, 0 can be divided by a nonzero number.

$$0 \div 8 = 0 \quad \text{since } 0 = 8 \cdot 0$$

$$0 \div \frac{2}{5} = 0 \quad \text{since } 0 = \frac{2}{5} \cdot 0$$

$$\frac{0}{32} = 0 \quad \text{since } 0 = 32 \cdot 0$$

PRACTICE QUIZ

Questions	Answers
Perform the indicated operations and reduce all answers to lowest terms.	
1. $5 \cdot \frac{3}{10}$	1. $\frac{3}{2}$
2. $\frac{3}{4} \div \frac{4}{3}$	2. $\frac{9}{16}$
3. $\frac{36}{27} \div \frac{48}{18}$	3. $\frac{1}{2}$
4. $\frac{7}{3} \div \frac{4}{5} \cdot \frac{3}{20}$	4. $\frac{7}{16}$
5. $\frac{7}{3} \cdot \frac{7}{20} \div \frac{7}{15}$	5. $\frac{7}{4}$

EXERCISES 1.3

Reduce each fraction in Exercises 1–12 to lowest terms.

1. $\frac{15}{18}$ **2.** $\frac{25}{40}$ **3.** $\frac{22}{33}$ **4.** $\frac{18}{45}$ **5.** $\frac{24}{56}$ **6.** $\frac{35}{63}$

7. $\frac{28}{28}$ **8.** $\frac{39}{13}$ **9.** $\frac{54}{36}$ **10.** $\frac{72}{27}$ **11.** $\frac{150}{350}$ **12.** $\frac{72}{108}$

Find the missing numerator by changing each fraction to an equal fraction with the indicated denominator in Exercises 13–24.

13. $\frac{5}{8} = \frac{}{32}$

14. $\frac{5}{6} = \frac{}{48}$

15. $\frac{3}{7} = \frac{}{21}$

16. $\frac{3}{13} = \frac{}{52}$

17. $\frac{0}{7} = \frac{}{35}$

18. $\frac{5}{17} = \frac{}{51}$

19. $\frac{7}{5} = \frac{}{60}$

20. $\frac{16}{3} = \frac{}{12}$

21. $\frac{6}{11} = \frac{}{121a}$

22. $\frac{0}{9} = \frac{}{63b}$

23. $\frac{11}{16} = \frac{}{112x}$

24. $\frac{7}{24} = \frac{}{144x}$

Perform the indicated operations in Exercises 25–60. Reduce each answer to lowest terms.

25. $\frac{3}{8} \cdot \frac{4}{9}$

26. $\frac{4}{5} \cdot \frac{3}{7}$

27. $\frac{5}{16} \cdot \frac{9}{8}$

28. $\frac{10}{7} \cdot \frac{14}{15}$

29. $\frac{2}{3} \cdot \frac{5}{4}$

30. $\frac{8}{5} \cdot \frac{8}{12}$

31. $\frac{11}{9} \cdot \frac{15}{22}$

32. $\frac{13}{7a} \cdot \frac{5a}{26}$

33. $\frac{16x}{7x} \cdot \frac{49}{64}$

34. $\frac{26b}{51a} \cdot \frac{34a}{39}$

35. $\frac{3}{8} \div \frac{3}{4}$

36. $\frac{4}{5} \div \frac{8}{3}$

37. $\frac{5}{4a} \div \frac{1}{5a}$

38. $\frac{2x}{5} \div \frac{2x}{5}$

39. $\frac{7b}{10} \div \frac{3a}{5}$

40. $\frac{40}{9} \div \frac{2}{15}$

41. $0 \div \frac{7}{3}$

42. $5 \div \frac{2}{5}$

43. $\frac{11}{3} \div 5x$

44. $\frac{3}{8a} \div 6c$

45. $\frac{11}{16} \div \frac{11}{16}$

46. $0 \div \frac{5}{9}$

47. $\frac{4}{13} \div 0$

48. $\frac{15}{8} \div \frac{3}{4}$

49. $\frac{3}{5} \cdot \frac{2}{7} \cdot 3$

50. $\frac{8}{3} \cdot \frac{18}{5} \cdot \frac{5}{4}$

51. $\frac{9a}{15} \cdot \frac{10}{3b} \cdot \frac{1}{5a^2}$

52. $\frac{a}{3} \cdot \frac{5}{4} \cdot 5b$

53. $\frac{15}{4} \cdot \frac{5}{6} \cdot \frac{16}{9}$

54. $\frac{7}{5} \cdot \frac{3}{14} \cdot 5$

55. $\frac{13}{7} \cdot \frac{5}{26} \cdot \frac{14}{9}$

56. $2 \cdot \frac{7}{5} \div \frac{7}{12}$

57. $\frac{10}{13} \div \frac{14}{17} \cdot \frac{7x}{10}$

58. $\frac{13}{6x} \div \frac{2}{5} \cdot \frac{3x}{25}$

59. $\frac{7a}{10x} \div \frac{2}{5} \cdot \frac{3x^2}{28}$

60. $\frac{7a}{18} \cdot \frac{9}{28} \div \frac{3x}{10}$

1.4 ADDITION AND SUBTRACTION WITH FRACTIONS

Finding the sum of two or more fractions with the same denominator is similar to adding whole numbers of some particular item. For example, the sum of 5 apples and 6 apples is 11 apples. Similarly, the sum of 5 seventeenths and 6 seventeenths is 11 seventeenths, or $\frac{5}{17} + \frac{6}{17} = \frac{11}{17}$.

To add two fractions $\frac{a}{b}$ and $\frac{c}{b}$ with common denominator b, add the numerators a and c and use the common denominator:

$$\frac{a}{b} + \frac{c}{b} = \frac{a + c}{b}$$

A formal proof of this relationship involves the fact that $\frac{a}{b} = \frac{1}{b} \cdot a$ and the distributive property.

$$\begin{aligned}\frac{a}{b} + \frac{c}{b} &= \frac{1}{b} \cdot a + \frac{1}{b} \cdot c \\ &= \frac{1}{b}(a + c) \qquad \text{distributive property} \\ &= \frac{a + c}{b}\end{aligned}$$

EXAMPLES

1. $\frac{3}{8} + \frac{4}{8} = \frac{3 + 4}{8} = \frac{7}{8}$

2. $\frac{9}{10} + \frac{3}{10} = \frac{9 + 3}{10} = \frac{12}{10} = \frac{\not{2} \cdot 6}{\not{2} \cdot 5} = \frac{6}{5}$

3. $\frac{2}{15} + \frac{3}{15} + \frac{1}{15} + \frac{6}{15} = \frac{2 + 3 + 1 + 6}{15} = \frac{12}{15} = \frac{\not{3} \cdot 4}{\not{3} \cdot 5} = \frac{4}{5}$

If the fractions do not have the same denominators, find the LCM of the denominators. This number will be the least common denominator. Then change each fraction to an equal fraction so that all fractions have the same denominator.

EXAMPLES

4. $\dfrac{7}{15} + \dfrac{5}{18}$

$$\left.\begin{array}{l} 15 = 3 \cdot 5 \\ 18 = 2 \cdot 3 \cdot 3 \end{array}\right\} \text{LCM} = 2 \cdot 3 \cdot 3 \cdot 5 = 90$$

$= 15 \cdot 6$ Denominator 15 needs to be multiplied by 6.

$= 18 \cdot 5$ Denominator 18 needs to be multiplied by 5.

$$\frac{7}{15} + \frac{5}{18} = \frac{7 \cdot 6}{15 \cdot 6} + \frac{5 \cdot 5}{18 \cdot 5} = \frac{42}{90} + \frac{25}{90} = \frac{42 + 25}{90} = \frac{67}{90}$$

5. $\dfrac{1}{4} + \dfrac{3}{8} + \dfrac{3}{10}$

$$\left.\begin{array}{l} 4 = 2 \cdot 2 \\ 8 = 2 \cdot 2 \cdot 2 \\ 10 = 2 \cdot 5 \end{array}\right\} \text{LCM} = 2 \cdot 2 \cdot 2 \cdot 5 = 40$$

$= 4 \cdot 10$

$= 8 \cdot 5$

$= 10 \cdot 4$

$$\frac{1}{4} + \frac{3}{8} + \frac{3}{10} = \frac{1 \cdot 10}{4 \cdot 10} + \frac{3 \cdot 5}{8 \cdot 5} + \frac{3 \cdot 4}{10 \cdot 4} = \frac{10}{40} + \frac{15}{40} + \frac{12}{40} = \frac{37}{40}$$

6. $\dfrac{5}{21a} + \dfrac{5}{28a}$

$$\left.\begin{array}{l} 21a = 3 \cdot 7 \cdot a \\ 28a = 2 \cdot 2 \cdot 7 \cdot a \end{array}\right\} \text{LCM} = 2 \cdot 2 \cdot 3 \cdot 7 \cdot a = 84a$$

$= 21a \cdot 4$

$= 28a \cdot 3$

$$\frac{5}{21a} + \frac{5}{28a} = \frac{5 \cdot 4}{21a \cdot 4} + \frac{5 \cdot 3}{28a \cdot 3} = \frac{20}{84a} + \frac{15}{84a} = \frac{35}{84a} = \frac{\not{7} \cdot 5}{\not{7} \cdot 12a} = \frac{5}{12a}$$

The difference of two fractions with a common denominator is found by subtracting the numerators and using the common denominator.

$$\frac{a}{b} - \frac{c}{b} = \frac{a - c}{b}$$

If the two fractions do not have the same denominator, find equal fractions with the least common denominator, just as with addition.

EXAMPLES

7. $\frac{5}{8} - \frac{1}{8} = \frac{5-1}{8} = \frac{4}{8} = \frac{\cancel{4} \cdot 1}{\cancel{4} \cdot 2} = \frac{1}{2}$

8. $\frac{1}{45} - \frac{1}{72}$

$$\left.\begin{aligned} 45 &= 3 \cdot 3 \cdot 5 \\ 72 &= 2 \cdot 2 \cdot 2 \cdot 3 \cdot 3 \end{aligned}\right\} \quad \text{LCM} = 2 \cdot 2 \cdot 2 \cdot 3 \cdot 3 \cdot 5 = 360$$

$$= 45 \cdot 8$$

$$= 72 \cdot 5$$

$$\frac{1}{45} - \frac{1}{72} = \frac{1 \cdot 8}{45 \cdot 8} - \frac{1 \cdot 5}{72 \cdot 5} = \frac{8}{360} - \frac{5}{360} = \frac{3}{360} = \frac{\cancel{3} \cdot 1}{\cancel{3} \cdot 120} = \frac{1}{120}$$

To evaluate an expression such as $\frac{1}{2} + \frac{3}{8} \div \frac{3}{4}$ that involves more than one operation, we use the same rules for order of operations discussed in Section 1.2.

EXAMPLES

9. $\frac{1}{2} + \frac{3}{8} \div \frac{3}{4} = \frac{1}{2} + \frac{3}{8} \cdot \frac{4}{3}$ Divide.

$= \frac{1}{2} + \frac{\cancel{3} \cdot \cancel{4} \cdot 1}{2 \cdot \cancel{4} \cdot \cancel{3}}$ Reduce.

$= \frac{1}{2} + \frac{1}{2}$ Add.

$= \frac{2}{2} = 1$

10. $\frac{2}{7} \cdot \frac{14}{3} + \frac{5}{8} \cdot \frac{2}{5} = \frac{2 \cdot 14}{7 \cdot 3} + \frac{5 \cdot 2}{8 \cdot 5}$ Multiply.

$= \frac{2 \cdot \cancel{7} \cdot 2}{\cancel{7} \cdot 3} + \frac{\cancel{5} \cdot \cancel{2} \cdot 1}{\cancel{2} \cdot 4 \cdot \cancel{5}}$ Reduce.

$= \frac{4}{3} + \frac{1}{4}$

$= \frac{4 \cdot 4}{3 \cdot 4} + \frac{1 \cdot 3}{4 \cdot 3}$ Common denominator.

$= \frac{16}{12} + \frac{3}{12} = \frac{19}{12}$ Add.

PRACTICE QUIZ

Questions	Answers
Simplify the following expressions.	
1. $\frac{2}{3}+\frac{1}{2}$	1. $\frac{7}{6}$
2. $\frac{2}{2a}-\frac{1}{2a}$	2. $\frac{1}{2a}$
3. $\frac{5}{24}-\frac{7}{36}$	3. $\frac{1}{72}$
4. $\frac{1}{3}\div\frac{1}{2}+\frac{1}{5}\cdot\frac{5}{3}$	4. 1

EXERCISES 1.4

Add or subtract as indicated in Exercises 1–55 from left to right. Reduce all answers.

1. $\frac{3}{7}+\frac{5}{7}$ **2.** $\frac{2}{9}+\frac{5}{9}$ **3.** $\frac{11}{12}+\frac{17}{12}$

4. $\frac{5}{16}+\frac{9}{16}$ **5.** $\frac{7}{15}-\frac{4}{15}$ **6.** $\frac{6}{17}-\frac{5}{17}$

7. $\frac{3}{19}+\frac{8}{19}+\frac{2}{19}$ **8.** $\frac{14}{23}-\frac{9}{23}$ **9.** $\frac{2}{7}+\frac{8}{7}+\frac{4}{7}$

10. $\frac{3}{8}+\frac{1}{8}+\frac{5}{8}$ **11.** $\frac{11}{12}+\frac{5}{12}-\frac{1}{12}$ **12.** $\frac{3}{10}+\frac{7}{10}-\frac{5}{10}$

13. $\frac{5}{9}+\frac{7}{9}-\frac{2}{9}$ **14.** $\frac{5}{5}-\frac{1}{5}+\frac{3}{5}$ **15.** $\frac{5}{8}-\frac{3}{8}-\frac{1}{8}$

16. $\frac{8}{7}-\frac{2}{7}-\frac{4}{7}$ **17.** $\frac{6}{11}+\frac{8}{11}-\frac{3}{11}$ **18.** $\frac{11}{15a}-\frac{7}{15a}+\frac{2}{15a}$

19. $\frac{17}{32a}+\frac{13}{32a}-\frac{5}{32a}$ **20.** $\frac{19}{25x}-\frac{7}{25x}-\frac{9}{25x}$ **21.** $\frac{1}{2}+\frac{1}{3}$

22. $\frac{4}{5}-\frac{2}{3}$ **23.** $\frac{3}{8}-\frac{1}{4}$ **24.** $\frac{5}{7}+\frac{1}{4}$ **25.** $\frac{3}{5}-\frac{1}{4}$

26. $\dfrac{27}{40} - \dfrac{5}{8}$ **27.** $\dfrac{5}{6} + \dfrac{7}{24}$ **28.** $\dfrac{5}{6} + \dfrac{1}{8}$ **29.** $\dfrac{1}{3} + \dfrac{1}{4}$ **30.** $\dfrac{2}{7} - \dfrac{1}{6}$

31. $\dfrac{4}{5} - \dfrac{1}{4}$ **32.** $\dfrac{7}{8} - \dfrac{3}{5}$ **33.** $\dfrac{5}{9a} - \dfrac{1}{6a}$ **34.** $\dfrac{11}{15c} - \dfrac{7}{20c}$

35. $\dfrac{7}{8c} - \dfrac{9}{32c}$ **36.** $\dfrac{1}{3} + \dfrac{1}{4} + \dfrac{5}{6}$ **37.** $\dfrac{2}{5} - \dfrac{1}{3} + \dfrac{9}{10}$

38. $\dfrac{1}{3} + \dfrac{11}{15} - \dfrac{7}{30}$ **39.** $\dfrac{3}{8} + \dfrac{5}{6} + \dfrac{2}{3}$ **40.** $\dfrac{11}{8} + \dfrac{5}{16} + \dfrac{3}{32}$

41. $\dfrac{15}{16} - \dfrac{7}{12} + \dfrac{3}{8}$ **42.** $\dfrac{5}{21} + \dfrac{4}{3} - \dfrac{7}{9}$ **43.** $\dfrac{5}{6} + \dfrac{9}{20} - \dfrac{7}{15}$

44. $\dfrac{5}{8} - \dfrac{2}{7} + \dfrac{1}{2}$ **45.** $\dfrac{4}{5} + \dfrac{7}{8} - \dfrac{7}{10}$ **46.** $\dfrac{4}{25} + \dfrac{7}{20} - \dfrac{3}{10}$

47. $\dfrac{5}{12} + \dfrac{7}{15} - \dfrac{11}{30}$ **48.** $\dfrac{5}{24} + \dfrac{7}{18} - \dfrac{3}{8}$ **49.** $\dfrac{8}{35} + \dfrac{6}{10} - \dfrac{5}{14}$

50. $\dfrac{8}{9} - \dfrac{5}{12} - \dfrac{1}{4}$ **51.** $\dfrac{2}{3x} + \dfrac{1}{4x} - \dfrac{1}{6x}$ **52.** $\dfrac{7}{10y} - \dfrac{1}{5y} + \dfrac{1}{3y}$

53. $\dfrac{11}{24x} - \dfrac{7}{18x} + \dfrac{5}{36x}$ **54.** $\dfrac{7}{6b} - \dfrac{9}{10b} + \dfrac{4}{15b}$ **55.** $\dfrac{1}{28a} + \dfrac{8}{21a} - \dfrac{5}{12a}$

Simplify the expressions in Exercises 56–65 using the rules for order of operations.

56. $\dfrac{3}{8} \cdot \dfrac{4}{5} + \dfrac{1}{15}$ **57.** $\dfrac{1}{3} \div \dfrac{1}{2} - \dfrac{5}{6} \cdot \dfrac{3}{4}$ **58.** $\dfrac{5}{6} \div \dfrac{5}{12} - \dfrac{3}{8}$

59. $\dfrac{2}{5} \cdot \dfrac{3}{8} + \dfrac{1}{5} \cdot \dfrac{3}{4}$ **60.** $\dfrac{3}{4} \div \dfrac{3}{16} - \dfrac{2}{3} \cdot \dfrac{3}{4}$ **61.** $\dfrac{3}{4} + \dfrac{5}{6} \div \dfrac{5}{8} - \dfrac{1}{3}$

62. $\dfrac{5}{7} + \dfrac{1}{2} \cdot \dfrac{2}{3} - \dfrac{1}{12}$ **63.** $\left(\dfrac{7}{8} - \dfrac{7}{16}\right) \cdot \dfrac{6}{7} - \dfrac{1}{4}$ **64.** $\left(\dfrac{7}{10} + \dfrac{3}{5}\right) \div \left(\dfrac{1}{2} + \dfrac{3}{7}\right)$

65. $\left(\dfrac{2}{3} - \dfrac{1}{5}\right) \div \left(\dfrac{5}{7} - \dfrac{1}{3}\right)$

1.5 DECIMALS AND PERCENTS

A **decimal number** (or simply a **decimal**) is a fraction that has a power of ten as its denominator. Thus, $\frac{3}{10}$, $\frac{51}{100}$, $\frac{24}{10}$, and $\frac{3}{1000}$ are all decimal numbers. Decimal numbers can be written with a decimal

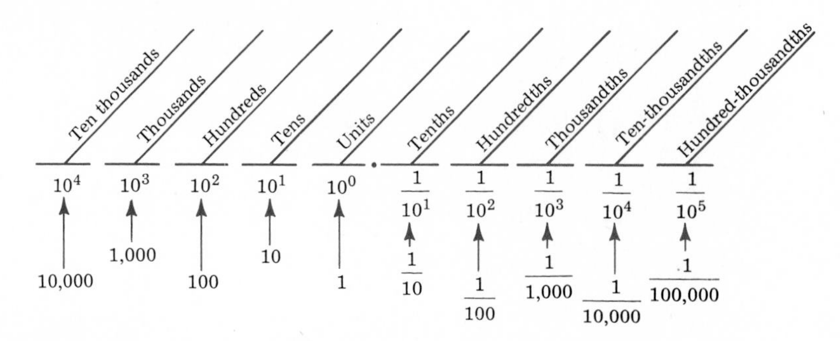

FIGURE 1.1

point and a place-value system that indicates whole numbers to the left of the decimal point and fractions less than 1 to the right of the decimal point. The values of each place are indicated in Figure 1.1.

To add or subtract decimal numbers, line up the decimal points, one under the other, then add or subtract as with whole numbers. Place the decimal point in the answer in line with the other decimal points. This technique guarantees that digits having the same place value will be added together or subtracted from each other.

EXAMPLES

1. Add: $37.498 + 5.63 + 42.781$

$$\begin{array}{r} 37.498 \\ 5.630 \\ 42.781 \\ \hline 85.909 \end{array}$$

2. Subtract: $26.872 - 13.99$

$$\begin{array}{r} 26.872 \\ 13.990 \\ \hline 12.882 \end{array}$$

To find the product of two decimal numbers, multiply as with whole numbers, then place the decimal point so that the number of digits to its right is equal to the sum of the number of digits to the right of the decimal points in the numbers being multiplied.

EXAMPLES

Find the following products.

3.
$$\begin{array}{r} 4.78 \\ 0.3 \\ \hline 1.434 \end{array}$$

4.
$$\begin{array}{r} 16.4 \\ 0.517 \\ \hline 1148 \\ 164 \\ 8\,20 \\ \hline 8.4788 \end{array}$$

Decimals and percents (%) are closely related because "percent" means hundredths. Thus, 70% and 0.70 have the same meaning. Similarly,

$$65\% = 0.65 \qquad \text{and} \qquad 125\% = 1.25$$

To change a decimal to a percent, move the decimal point two places to the right and write the % sign.

To change a percent to a decimal, move the decimal point two places to the left and drop the % sign.

EXAMPLES

5. Change the following decimals to percents.

   a. 0.3 = 30% b. 0.73 = 73% c. 1.4 = 140% d. 0.356 = 35.6%

6. Change the following percents to decimals.

   a. 28% = 0.28 b. 4.3% = 0.043 c. 5% = 0.05 d. 50% = 0.50

In many word problems, such as in finding sales tax, discounts, and sales commissions for work done, we are to find a percent of a number. To find a percent of a number, change the percent to a decimal, then multiply. For example, to find 52% of 85, change 52% to 0.52, then multiply by 85.

$$\begin{array}{r} 85 \\ 0.52 \\ \hline 1\,70 \\ 42\,5 \\ \hline 44.20 \end{array}$$

Thus, 52% of 85 is equal to 44.2. The product, 44.2, is sometimes called a **percentage.**

EXAMPLES

7. Find 40% of 15.

```
  15
0.40
----
6.00
```

40% of 15 is 6.

8. Find 160% of 35.

```
   35
 1.60
-----
21 00
35
-----
56.00
```

160% of 35 is 56.

Note that more than 100% of a number is larger than the number.

Calculators are particularly useful in performing operations with decimal numbers. Remember, though, you must be able to change percents to decimals and know what operation to use.

CALCULATOR EXAMPLES

9. Use a calculator to find the sum:

8.6321 + 7.5476 + 2.143 + 17.8293

Solution: 36.152

10. Use a calculator to find the product:

(14.763)(0.47)(321.6)

Solution: 2231.457

11. Use a calculator to find 18.3% of 210.55.

Solution: 38.53065

EXERCISES 1.5

Change the decimals to percents in Exercises 1–10.

1. 0.37 **2.** 0.91 **3.** 0.723 **4.** 0.625 **5.** 0.032 **6.** 1.37

7. 2.8 **8.** 3.62 **9.** 1.23 **10.** 0.125

In Exercises 11–20, change the percents to decimals.

11. 43% **12.** 69% **13.** 6% **14.** 7.5% **15.** 11.3%

16. 162% **17.** 238% **18.** 9.5% **19.** 200% **20.** 14.5%

Add or subtract as indicated in Exercises 21–35.

21. 6.31 + 87.2 + 63.71

22. 19.62 + 11.15 + 2.8

23. 243.7 + 65.22 + 8.31

24. 29.51 + 17.2 + 72.4

25. 94.32 − 27.21

26. 37.69 − 15.72

27. 420.43 − 156.92

28. 87.0 − 66.31

29. 47.32 + 56.1 + 83.59

30. 11.69 + 73.48 + 23.2

31. 65.13 + 44.81 − 75.9

32. 84.1 + 17.63 − 12.98

33. 147.0 + 79.6 − 85.43

34. 6.49 + 103.81 − 59.62

35. 94.29 + 120.3 − 206.41

Find the indicated products in Exercises 36–45.

36. $\begin{array}{r} 7.4 \\ \underline{0.9} \end{array}$ **37.** $\begin{array}{r} 8.9 \\ \underline{0.07} \end{array}$ **38.** $\begin{array}{r} 60.4 \\ \underline{1.8} \end{array}$ **39.** $\begin{array}{r} 29.6 \\ \underline{5.2} \end{array}$ **40.** $\begin{array}{r} 21.6 \\ \underline{0.83} \end{array}$

41. (4.38)(7.5) **42.** (1.23)(6.2) **43.** (15.83)(0.24) **44.** (43.61)(12.7)

45. (571.6)(20.8)

46. Find 13% of 68.

47. Find 27% of 49.

48. What is 11% of 93?

49. What is 6% of 480?

50. What is 38% of 147?

51. 81% of 76 is equal to ______.

52. 102% of 87 is equal to ______.

53. 112% of 620 is equal to ______.

54. What is 108% of 735?

55. What is 125% of 350?

56. 9.5% of 570 is equal to ______.

57. 10.3% of 986 is equal to ______.

58. Find 24.3% of 85.

59. Find 118.6% of 68.

60. Find 106.5% of 524.

Calculator Problems

Use a calculator to solve Exercises 61–70.

61. 11.869 + 19.407 − 9.8531

62. 27.081 + 13.1152 − 20.539

63. 57.9381 − 22.651 − 15.0932

64. (83.69)(41.982)

65. (0.6534)(142.5)

66. (11.36)(19.83)(2.64)

67. Find 16.67% of 24.3.

68. Find 1.05% of 5.3.

69. What is 106.5% of 40.6?

70. 8.25% of 60.4 is equal to ______.

1.6 WORD PROBLEMS

The word problems in this section involve any or all the ideas presented in the first five sections of Chapter 1. The student must read each problem and, on the basis of personal experience and understanding, decide whether to add, subtract, multiply, divide, find a percent of a number, or perform combinations of operations.

Included in some of the word problems is the idea of a fractional part of a number. This is similar to finding a percent of a number in that we multiply the fraction times the number. For example, to find $\frac{2}{3}$ of 12, we multiply $\frac{2}{3} \cdot 12 = \frac{2}{3} \cdot \frac{12}{1} = \frac{24}{3} = 8$. Thus, 8 is $\frac{2}{3}$ of 12.

EXAMPLES

1. Six monthly payments to purchase a used car were \$150, \$175, \$230, \$200, \$180, and \$259. What was the total of the payments? What was the average monthly payment?

Solution: To find the total, add all six numbers. To find the average monthly payment, divide the sum by 6.

$$\begin{array}{r} \$\ 150 \\ 175 \\ 230 \\ 200 \\ 180 \\ \underline{259} \\ \$1194 \end{array} \text{ total} \qquad \begin{array}{r} \$199 \\ 6\overline{)1194} \\ \underline{6} \\ 59 \\ \underline{54} \\ 54 \\ \underline{54} \end{array} \text{ average monthly payment}$$

2. A bicycle was purchased for $\frac{3}{4}$ of the asking price of $160. If 6% sales tax was added to the purchase price, what was paid for the bicycle?

Solution:

$$\frac{3}{4} \cdot 160 = \frac{3}{\cancel{4}} \cdot \frac{\overset{40}{\cancel{160}}}{1} = \$120 \quad \text{the purchase price}$$

$$\begin{array}{r} \$120 \\ .06 \\ \hline \$7.20 \end{array} \quad \text{sales tax} \qquad \begin{array}{r} \$120.00 \\ 7.20 \\ \hline \$127.20 \end{array} \quad \text{total price including sales tax}$$

EXERCISES 1.6

1. During the first four months of the year, a company showed profits of $7483, $9157, $10,544, and $9280. What was the total profit? What was the average profit per month?
2. For the months January through May, a family's electricity bills were $68.45, $56.83, $49.78, $46.90, and $42.39. Find the average monthly cost of electricity for these months.
3. Sound travels approximately 1080 feet per second. If the sound of thunder is heard 9 seconds after the lightning is seen, how far away was the lightning?
4. In 12 hours of driving, the odometer on a car changed from 36,849 to 37,485. Find the average miles traveled per hour.
5. How tall is a stack of 14 boards if each board is 0.75 inch thick?
6. Five boards, each exactly 2 feet long, are cut from a board 12 feet long. If the waste for each cut is 0.012 foot, what is the length of the remaining piece?
7. How many pieces of wood $\frac{2}{3}$ foot long can be cut from a board 6 feet long?
8. Twelve people in a math class made above 90% on a test. If this was $\frac{3}{8}$ of the entire class, how many people were in the class?
9. Seventy-two boys reported to football practice. If $\frac{2}{3}$ of the boys made the team, how many were on the team?
10. In scale for a blueprint of a building, 1 in. represents 48 ft. If on the drawing the building is $\frac{7}{8}$ in. by $\frac{5}{4}$ in., what will be the actual size of the building?

11. The seventeenth hole at the local golf course is a par 4 hole. Ralph drove his ball 258 yards. If this distance was $\frac{3}{4}$ the length of the hole, how long is the seventeenth hole?

12. Divide the product of $\frac{5}{8}$ and $\frac{9}{10}$ by the product of $\frac{9}{10}$ and $\frac{4}{5}$.

13. Multiply the quotient of $\frac{19}{2}$ and $\frac{13}{4}$ by $\frac{13}{12}$.

14. Find the product of $\frac{11}{4}$ with the quotient of $\frac{8}{5}$ and $\frac{11}{6}$.

15. Find the result of multiplying $\frac{5}{3}$ by $\frac{3}{4}$, then dividing by $\frac{21}{8}$, then dividing by $\frac{10}{7}$.

16. The product of $\frac{9}{16}$ and $\frac{13}{7}$ is added to the quotient of $\frac{10}{3}$ and $\frac{7}{4}$. What is the sum?

17. Find the quotient of the sum of $\frac{4}{5}$ and $\frac{11}{15}$ divided by the difference between $\frac{7}{12}$ and $\frac{4}{15}$.

18. How much space on a shelf will Betty need to hold 3 books $\frac{3}{4}$ in. thick and 4 books $\frac{7}{8}$ in. thick?

19. On a recent trip, June noticed that the odometer on her car read 9710 when she left home and 10,280 when she returned. If 38 gallons of gasoline were used, find her average miles per gallon.

20. On a loan of $6000, Georgia must pay 11% interest. How much interest will she pay in one year?

21. A salesperson works on a commission of 9%. What would be the commission on sales of $8420?

22. A salesman's salary is determined as follows: $300 per month plus a commission of 7% based on the amount of sales exceeding $5000. Find the amount of last month's check if he sold $11,470.

23. To buy a used car, you can pay $3250 cash or you can put down $800 and make 24 monthly payments of $131.50. How much can you save by paying cash?

24. A ski shop is selling out its new skis and its used rental skis. If the new skis are priced at $160.00 per pair and the used ones at $69.50 per pair, what will be the total received if 8 pairs of new skis and 13 pairs of used ones are sold?

25. Two couples went out to dinner. The prices of the dinners were \$5.95, \$7.25, \$7.95, and \$7.95. If a sales tax of 6% is added to the bill, what will be the total bill? If they leave a tip of 15% of the total bill, what is the amount of the tip? (Round off to the nearest cent.)

Calculator Problems

26. Five bars of iron weigh 5.75, 6.375, 4.5625, 5.125, and 7.5 pounds. Find the total weight and the average weight.

27. A piece of wood $7\frac{11}{16}$ (or 7.6875) inches long is cut from a board 18 inches long. If the cut itself wastes $\frac{1}{8}$ (or 0.125) inch, how long is the remaining piece?

28. Four pieces of ribbon, each measuring $3\frac{5}{8}$ (or 3.625) inches, are cut from a piece of ribbon 20 inches long. How long is the remaining piece?

29. How many pieces of wire, each $2\frac{3}{8}$ (or 2.375) inches long, can be cut from a piece of wire 22 inches long?

30. Natalie works part time. For one three-day period, she worked 3.25, 4.5, and 3.75 hours. She is paid \$5.65 per hour. How much did she earn? If 12% of her wages is held out for taxes, how much tax did she pay?

1.7 VARIABLES AND SOLUTIONS TO EQUATIONS

A **variable** is a symbol or letter used to represent more than one number. We can use letters such as a, b, x, y, and z or other symbols from other alphabets such as α (alpha), β (beta), and γ (gamma) from the Greek alphabet. Variables are used to represent unknown numbers in **algebraic expressions.** An **algebraic expression** is an expression that indicates the sum, product, difference, or quotient of constants and variables. For example,

$$x + y, \qquad 3a - 7, \qquad 2(x + 3), \qquad \text{and} \qquad 8a^2 + 3b^2 - 5$$

are all algebraic expressions.

If an expression contains variables, then the value of that expression depends on the numbers substituted for the variables. **Any value substituted for a variable must be substituted for every occurrence of that variable in an expression.**

For example, using the rules for order of operations, evaluate the following expression:

$$2x + 3 + 5x \qquad \text{for } x = 4 \text{ and again for } x = 3$$

a. If $x = 4$, then

$$2x + 3 + 5x = 2 \cdot 4 + 3 + 5 \cdot 4$$
$$= 8 + 3 + 20 = 31$$

b. If $x = 3$, then

$$2x + 3 + 5x = 2 \cdot 3 + 3 + 5 \cdot 3$$
$$= 6 + 3 + 15 = 24$$

EXAMPLES

1. Evaluate the expression

$3a + 5 - 2a \qquad$ for $a = 2$

Solution:

$$3a + 5 - 2a = 3 \cdot 2 + 5 - 2 \cdot 2$$
$$= 6 + 5 - 4$$
$$= 11 - 4 = 7$$

2. Evaluate the expression

$3a^2 - 7a \qquad$ for $a = \frac{5}{2}$

Solution:

$$3a^2 - 7a = 3 \cdot \left(\frac{5}{2}\right)^2 - 7 \cdot \frac{5}{2}$$
$$= 3 \cdot \frac{25}{4} - \frac{35}{2}$$
$$= \frac{75}{4} - \frac{70}{4} = \frac{5}{4}$$

An **equation** states that two algebraic expressions are equal. That is, both expressions represent the same numbers. All of the following are examples of equations:

$$6 + 7 = 10 + 3, \qquad x + 3 = 11, \qquad \text{and} \qquad 2y - 5 = y + 7$$

We will discuss techniques for solving equations throughout the text. In this section, you will be asked only to verify that a certain number is a **solution** to an equation.

Any value for a variable that gives a true statement when that value is substituted for the variable is a **solution to the equation.** For example,

a. $x = 5$ **is** a solution to $3x + 4 = 19$
since $3 \cdot 5 + 4 = 19$ is true.

b. $x = 3$ **is not** a solution to $3x + 4 = 19$
since $3 \cdot 3 + 4 = 9 + 4 = 13$ and $13 \neq 19$.

EXAMPLES

Determine whether or not the given number is a solution to the given equation by substituting it into the equation.

3. $5x = 20;\ x = 4$
Substitute 4 for x:

$$5x = 20$$

$$5 \cdot 4 = 20 \qquad \text{true}$$

4 is a solution of the equation.

4. $3q - 8 = 13;\ q = 7$
Substitute 7 for q:

$$3q - 8 = 13$$

$$3 \cdot 7 - 8 = 13$$

$$21 - 8 = 13$$

$$13 = 13 \qquad \text{true}$$

7 is a solution of the equation.

5. $5a + 4 = 6 + 3a;\ a = 2$
Substitute 2 for a:

$$5a + 4 = 6 + 3a$$

$$5 \cdot 2 + 4 = 6 + 3 \cdot 2$$

$$10 + 4 = 6 + 6$$

$$14 = 12 \qquad \text{false } (14 \neq 12)$$

2 is **not** a solution of the equation.

6. $5a + 4 = 6 + 3a$; $a = 1$
   Substitute 1 for a:

$$5a + 4 = 6 + 3a$$

$$5 \cdot 1 + 4 = 6 + 3 \cdot 1$$

$$5 + 4 = 6 + 3$$

$$9 = 9 \qquad \text{true}$$

   1 is a solution of the equation.

CALCULATOR EXAMPLE

7. Use a calculator to verify that $y = 3.75$ is a solution to $4.1y - 1.6 = 13.775$.

   Solution: $4.1(3.75) - 1.6 = 15.375 - 1.6 = 13.775$

EXERCISES 1.7

Evaluate the expressions in Exercises 1–20 for the given value of the variable.

1. $x + 7$; $x = 2$

2. $x + 11$; $x = 0$

3. $8 - x$; $x = 4$

4. $17 - x$; $x = 6$

5. $2x - 1$; $x = 1$

6. $3y + 5$; $y = 5$

7. $5y + 1$; $y = 0$

8. $8x + 3$; $x = 3$

9. $11 - 2x$; $x = 5$

10. $14 - 5x$; $x = 2$

11. $3y + 4 - y$; $y = 3$

12. $x + 5 + 2x$; $x = 1$

13. $4x + x + 3$; $x = 6$

14. $8 - 3y + y$; $y = 4$

15. $x^2 + 3$; $x = 6$

16. $3x^2 + x$; $x = 0$

17. $5y^2 - 3y + 1$; $y = 2$

18. $2x^2 + 4x + 4$; $x = 3$

19. $4x^2 + x + 7$; $x = 0$

20. $x^2 + 9x + 4$; $x = 5$

Evaluate the expressions in Exercises 21–26 if (a) $x = \frac{1}{2}$ and (b) $x = \frac{1}{3}$.

21. $x + 4$

22. $x + 2$

23. $2x + 5$

24. $3x + 6$

25. $3x - \frac{1}{4}$

26. $4x - \frac{1}{6}$

Evaluate the expressions in Exercises 27–32 if (a) $x = 1.4$ and (b) $x = 0.25$.

27. $2x + 3$

28. $4x + 2$

29. $5x + 8$

30. $7x + 4$

31. $8x - 1.3$

32. $9x - 2.03$

Determine whether or not the given number is a solution to the given equation in Exercises 33–52.

33. $3x = 18;\ x = 6$

34. $8x = 32;\ x = 4$

35. $5x + 1 = 17;\ x = 3$

36. $4x - 3 = 11;\ x = 4$

37. $4y - 1 = 17;\ y = 4$

38. $8 - 3y = 2;\ y = 2$

39. $15 - 2x = 5;\ x = 5$

40. $23 - 6y = 5;\ y = 3$

41. $7y + 3 = 2y + 8;\ y = 1$

42. $5x + 2 = 4x - 3;\ x = 5$

43. $6 - x = 4x - 9;\ x = 3$

44. $11 - 3x = 2x + 1;\ x = 2$

45. $2x + 3 = 8x;\ x = \frac{1}{2}$

46. $6 - 3x = 6x + 3;\ x = \frac{1}{3}$

47. $9y = y + 2;\ y = \frac{1}{4}$

48. $6y + 4 = 5y + 4;\ y = \frac{1}{5}$

49. $4x + 6 = 6.8;\ x = 0.2$

50. $5y + 3 = 2.8;\ y = 0.5$

51. $3x + 4 = 7x;\ x = 1.4$

52. $2x + 3 = 4x - 2;\ x = 2.5$

Calculator Problems

Use a calculator to determine whether or not the given number in Exercises 53–58 is a solution to the given equation.

53. $6.2x + 1.57 = 15.52;\ x = 2.25$

54. $5.3y + 3.261 = y + 9.496;\ y = 1.45$

55. $4.2y - 1.573 = 2.3y + 2.9;\ y = 2.36$

56. $1.3x + 4.614 = 2.4x + 0.6;\ x = 3.62$

57. $0.3x + 8.537 = 3.1x + 2.713;\ x = 2.08$

58. $0.5x + 3.172 = 2x - 2.012;\ x = 3.456$

1.8 FORMULAS IN GEOMETRY

A **formula** is an equation that represents a general relationship between two or more quantities or measurements. Several variables may appear in a formula, and, if values are known for all but one of these variables, the remaining value can be found. Formulas are useful in mathematics, economics, chemistry, medicine, physics, and many other fields of study. In this section, we will discuss some formulas related to simple geometric figures such as rectangles, circles, and triangles.

Perimeter

The **perimeter** of a geometric figure is the total distance around the figure. Perimeters are measured in units of length such as inches, feet, yards, miles, centimeters, and meters. The formulas for the perimeters of various geometric figures are shown in Figure 1.2.

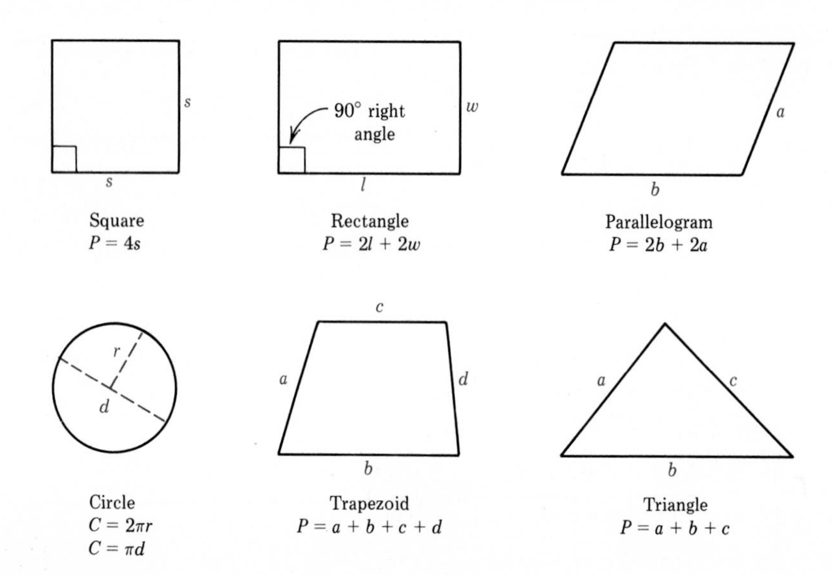

FIGURE 1.2

Some special comments about circles:

1. The perimeter of a circle is called its **circumference.**
2. The distance from the center to a point on the circle is called its **radius.**
3. The distance from one point on a circle to another point on the circle measured through its center is called its **diameter.** The diameter is twice the radius.

4. **Pi** (π) is the symbol used for the constant 3.1415926535.... For our purposes, we will use $\pi = 3.14$, but you should understand that this is only an approximation.

EXAMPLES

1. Find the perimeter of a rectangle with a length of 10 feet and a width of 8 feet.

Solution:

a. Sketch the figure.

b. $P = 2l + 2w$

$P = 2 \cdot 10 + 2 \cdot 8$

$= 20 + 16 = 36$ ft

The perimeter is 36 feet.

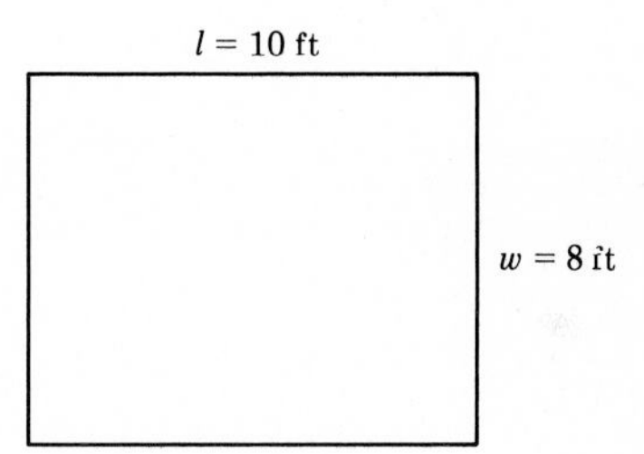

2. Find the circumference of a circle with a diameter of 3 centimeters.

Solution:

a. Sketch the figure.

b. $C = \pi d$

$C = 3.14(3)$

$= 9.42$ cm

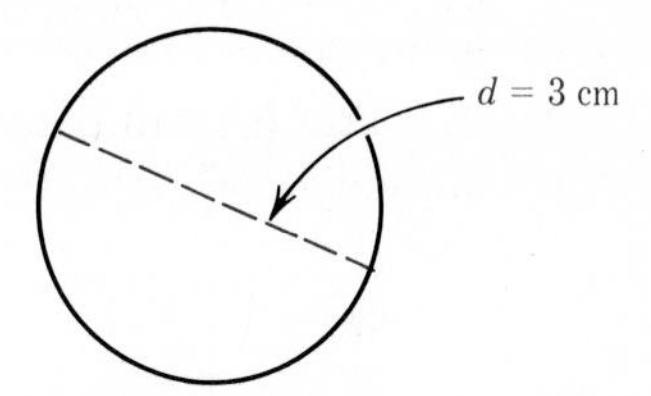

The circumference is approximately 9.42 centimeters.

3. Find the perimeter of the triangle with sides as labeled in the given figure.

Solution:

$P = a + b + c$

$P = 3 + 6.2 + 8.1$

$= 17.3$ in.

The perimeter is 17.3 inches.

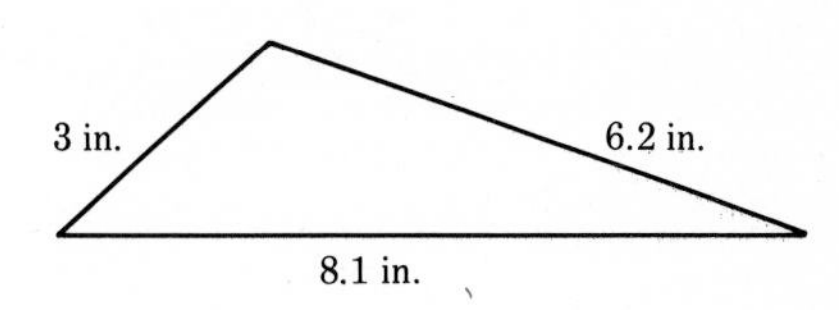

Area

Area is a measure of the interior, or enclosure, of a surface. Area is measured in square units such as square feet, square inches, square meters, and square miles. For example, the area enclosed by a square

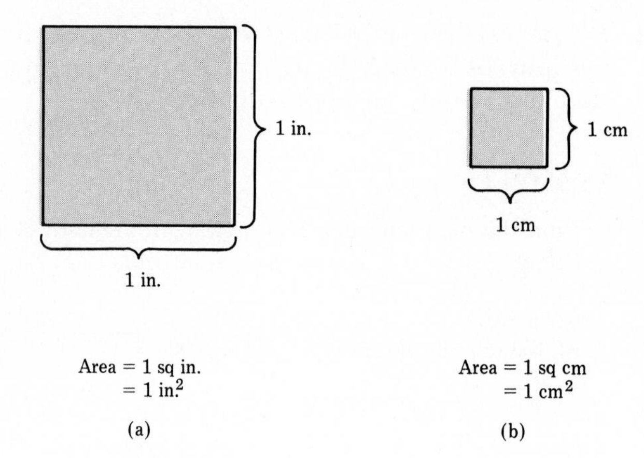

FIGURE 1.3

of 1 inch on each side is 1 sq in. [see Figure 1.3(a)], while the area enclosed by a square of 1 centimeter on each side is 1 sq cm [see Figure 1.3(b)].

The formulas for finding the areas of several geometric figures are shown in Figure 1.4.

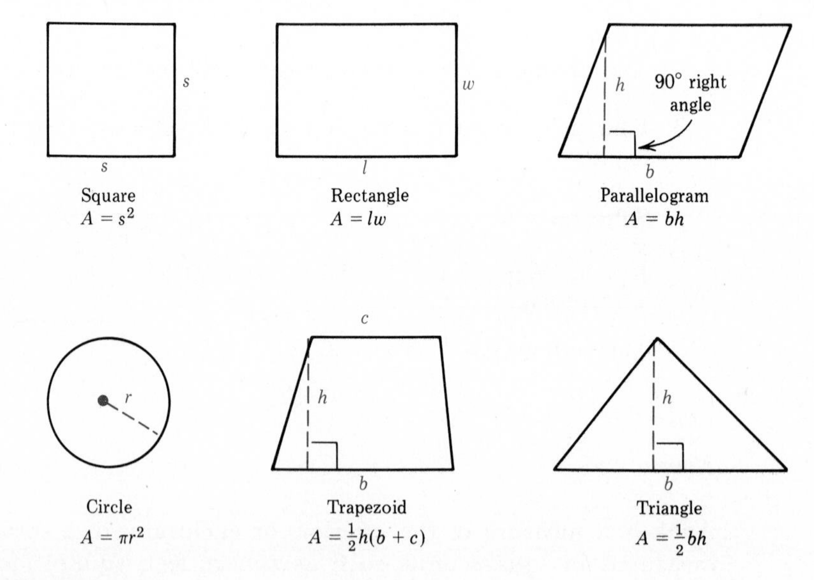

FIGURE 1.4

EXAMPLES

4. Find the area of a triangle with a height of 3 centimeters and a base of 4 centimeters.

Solution:

a. Sketch the figure.

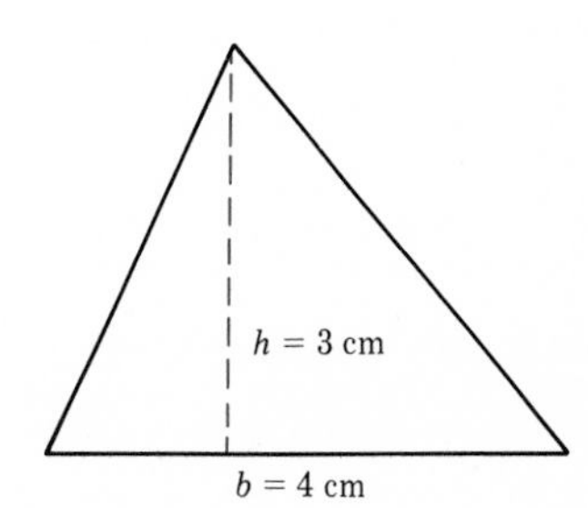

b. $A = \frac{1}{2} \cdot b \cdot h$

$A = \frac{1}{2} \cdot 3 \cdot 4$

$= \frac{1}{2} \cdot 12 = 6$ sq cm or 6 cm^2

The area is 6 square centimeters.

5. Find the area of a circle with a radius of 6 inches.

Solution:

a. Sketch the figure.

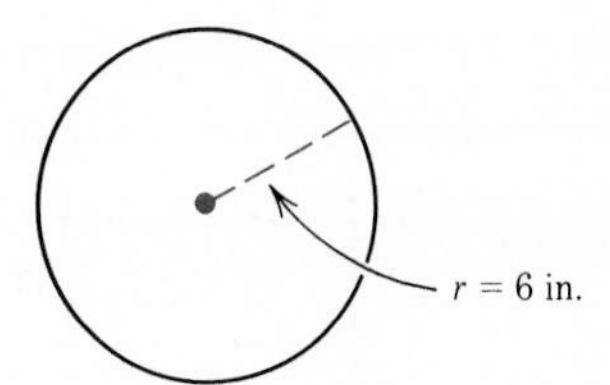

b. $A = \pi r^2$

$A = 3.14(6^2)$

$= 3.14(36)$

$= 113.04$ sq in. or 113.04 in.^2

The area is approximately 113.04 square inches.

Volume

Volume is a measure of the space enclosed by a three-dimensional figure. Volume is measured in cubic units such as cubic inches, cubic centimeters, and cubic feet. For example, the volume enclosed by a cube of 1 inch on each edge is 1 cu in. [see Figure 1.5(a) on page 44], while the volume enclosed by a cube of 1 centimeter on each edge is 1 cu cm [see Figure 1.5(b) on page 44].

The formulas for finding the volumes of some geometric solids are given in Figure 1.6 on page 44.

EXAMPLES

6. Find the volume of a right circular cylinder with a radius of 2 centimeters and a height of 5 centimeters. (Solution is on page 45.)

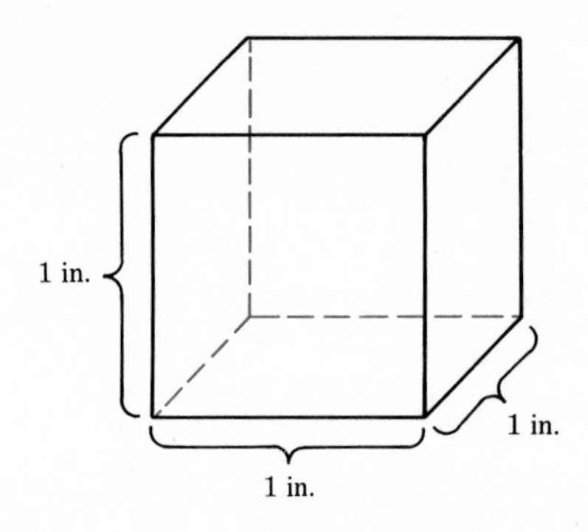

Volume = 1 cu in.
= 1 in^3

(a)

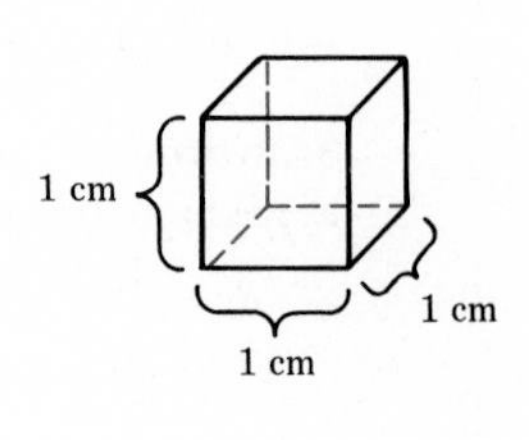

Volume = 1 cu cm
= 1 cm^3

(b)

FIGURE 1.5

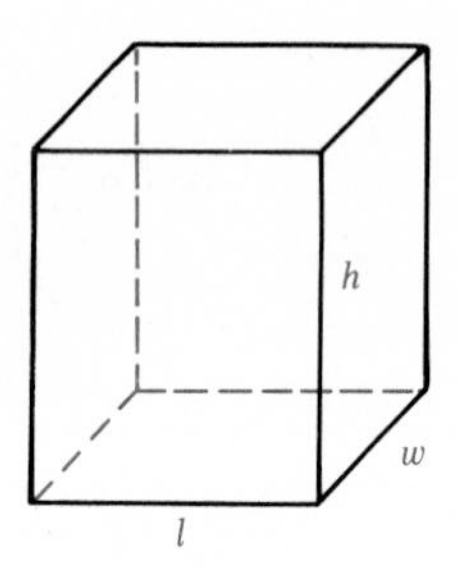

Rectangular solid
$V = lwh$

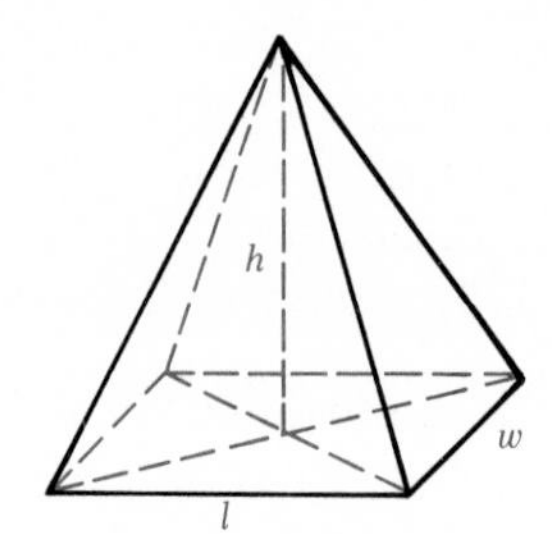

Rectangular pyramid
$V = \frac{1}{3}lwh$

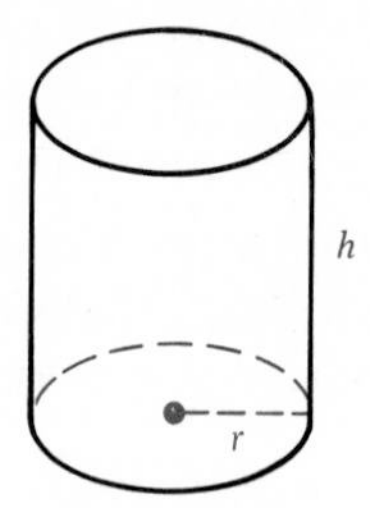

Right circular cylinder
$V = \pi r^2 h$

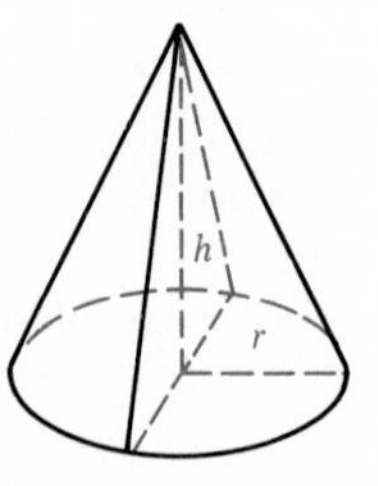

Right circular cone
$V = \frac{1}{3}\pi r^2 h$

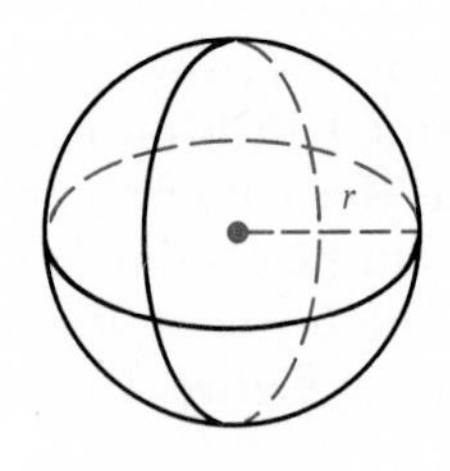

Sphere
$V = \frac{4}{3}\pi r^3$

FIGURE 1.6

Solution:

a. Sketch the figure.

b. $V = \pi r^2 h$

$V = 3.14(2^2) \cdot 5$

$= 3.14(4) \cdot 5$

$= 3.14(20)$

$= 62.80 \text{ cm}^3$

The volume is approximately 62.80 cubic centimeters.

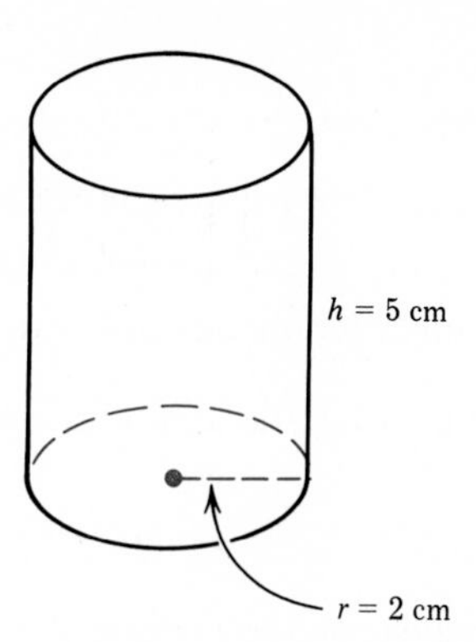

7. Find the volume of a sphere with a radius of 4 feet.

Solution:

a. Sketch the figure.

b. $V = \frac{4}{3}\pi r^3$

$V = \frac{4}{3}(3.14) \cdot 4^3$

$= \frac{4(3.14) \cdot 64}{3}$

$= \frac{803.84}{3} \text{ ft}^3$ or about 267.95 ft^3

$r = 4$ ft

The volume is approximately $\frac{803.84}{3}$ cubic feet.

EXERCISES 1.8

In Exercises 1–15, select the answer from the right-hand column that correctly matches the statement.

1. The formula for the perimeter of a square ______ **a.** $V = \frac{1}{3} lwh$

2. The formula for the circumference of a circle ______ **b.** $A = \frac{1}{2} h(b + c)$

3. The formula for the perimeter of a triangle ______ **c.** $A = s^2$

4. The formula for the area of a rectangle ______

5. The formula for the area of a square ______

6. The formula for the area of a trapezoid ______

7. The formula for the area of a triangle ______

8. The formula for the area of a parallelogram ______

9. The formula for the perimeter of a rectangle ______

10. The formula for the volume of a rectangular pyramid ______

11. The formula for the volume of a rectangular solid ______

12. The formula for the area of a circle ______

13. The formula for the volume of a right circular cylinder ______

14. The formula for the volume of a sphere ______

15. The formula for the volume of a right circular cone ______

d. $A = \frac{1}{2}bh$

e. $P = 4s$

f. $A = lw$

g. $P = 2l + 2w$

h. $V = \frac{1}{3}\pi r^2 h$

i. $V = \frac{4}{3}\pi r^3$

j. $C = 2\pi r$

k. $P = a + b + c$

l. $V = lwh$

m. $V = \pi r^2 h$

n. $A = \pi r^2$

o. $A = bh$

Find (a) the perimeter and (b) the area of each of the figures given in Exercises 16–21.

16.

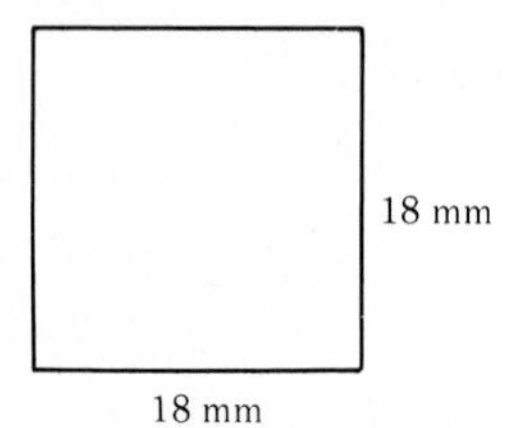

$P =$ ______

$A =$ ______

17.

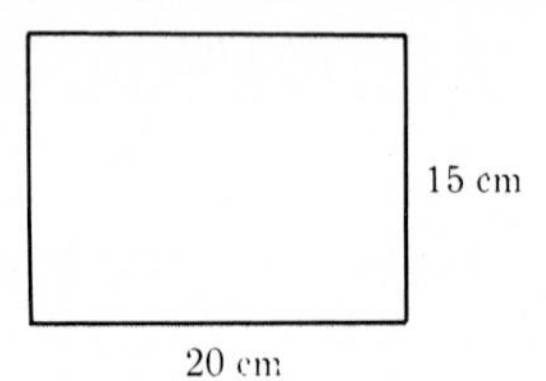

$P =$ ______

$A =$ ______

18.

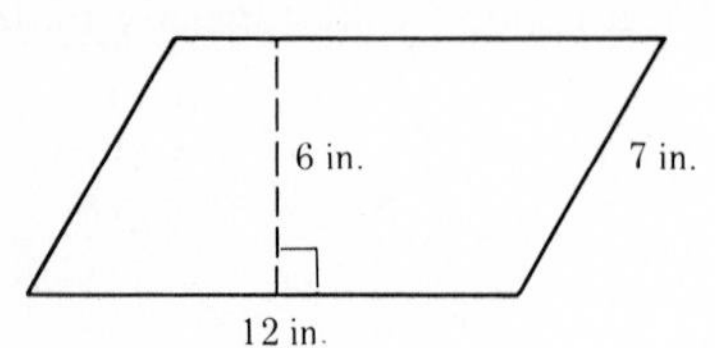

$P =$ ______

$A =$ ______

19.

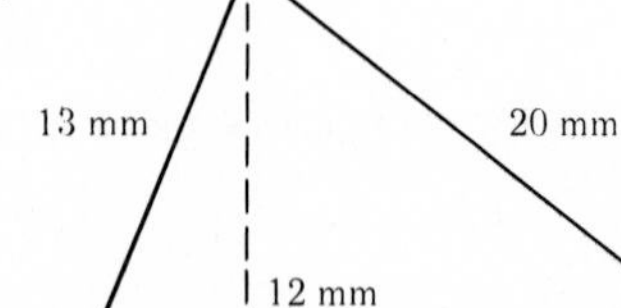

$P =$ ______

$A =$ ______

20.

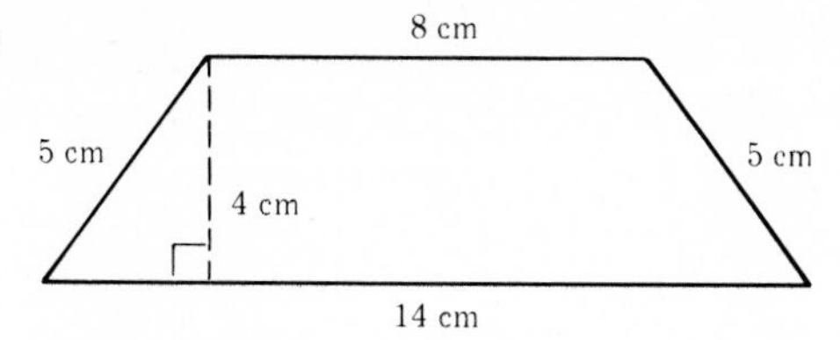

$P =$ ______

$A =$ ______

21.

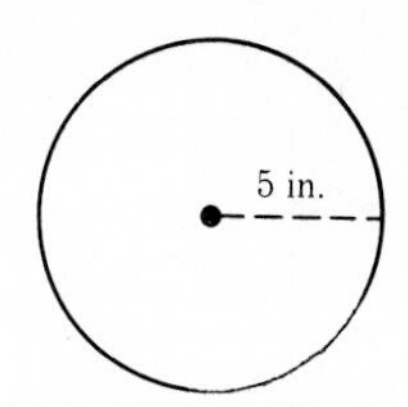

$C =$ ______

$A =$ ______

Find the volume of each figure in Exercises 22–26.

22.

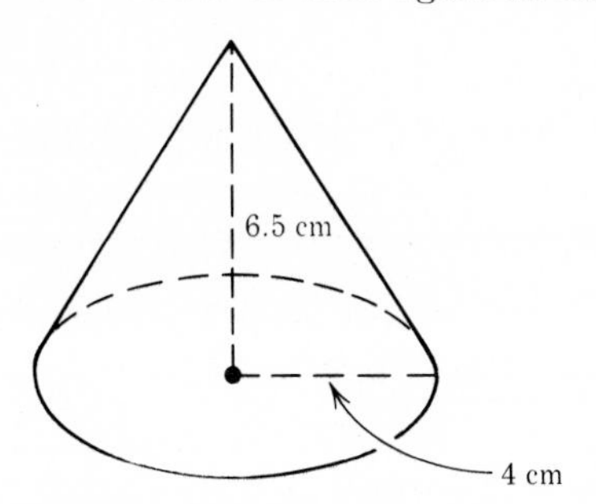

$V =$ ______

23.

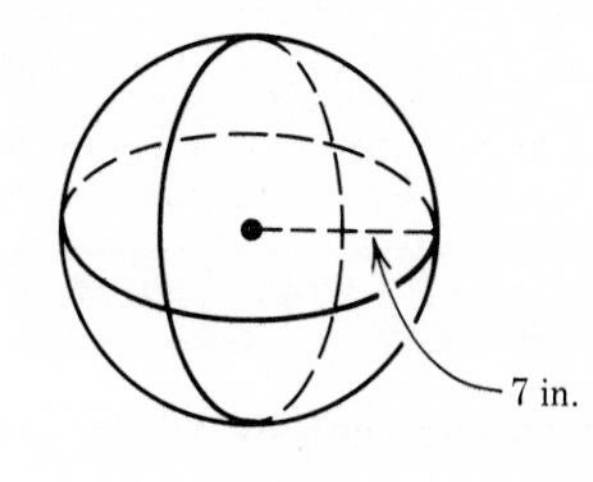

$V =$ ______

24.

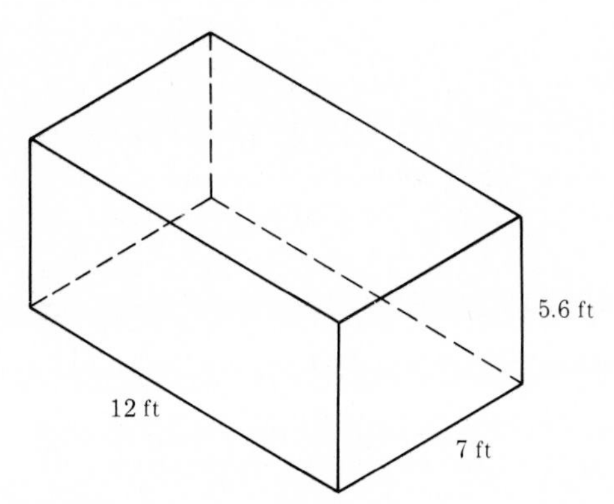

$V =$ ______

25.

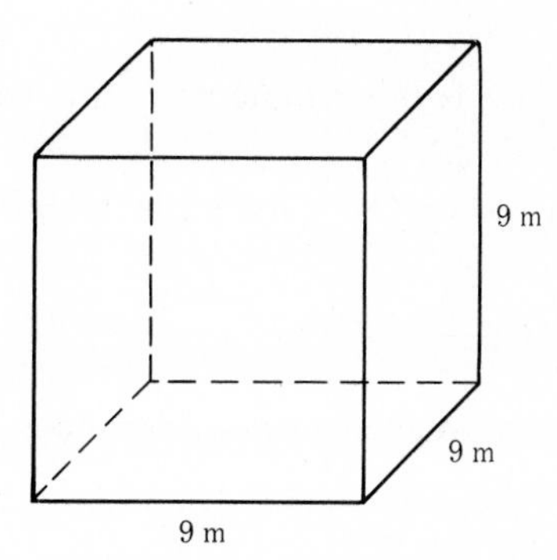

$V =$ ______

26.

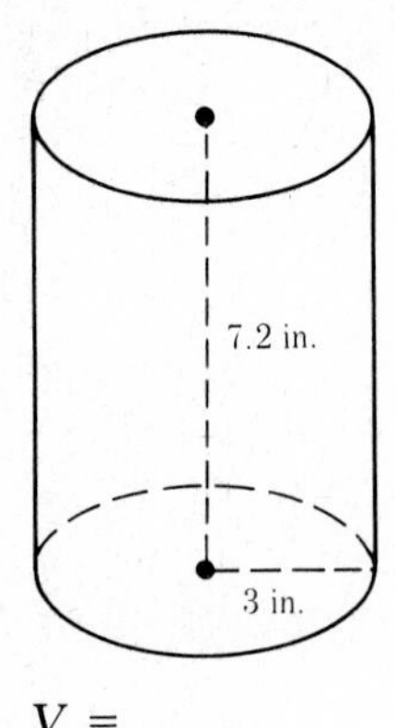

V = ______

Solve the problems in Exercises 27–38.

27. What is the perimeter of a rectangle whose length is 17 in. and width is 11 in.?

28. Find the area of a square whose sides are 9 ft long.

29. The base of a triangle is 14 cm and the height is 9 cm. Find the area.

30. Find the circumference of a circle with a radius of 8 m ($\pi = 3.14$).

31. The radius of the base of a cylindrical tank is 14 ft. If the tank is 10 ft high, find the volume.

32. The sides of a triangle are 6.2 m, 8.6 m, and 9.4 m. Find the perimeter.

33. What is the volume of a cube whose edge is 5 ft?

34. A rectangular garden plot is 60 ft long and 42 ft wide. Find the area.

35. A parallelogram has a base of 20 cm and a height of 13.6 cm. Find the area.

36. A rectangular box is 18 in. long, 10.3 in. wide, and 8 in. high. Find the volume.

37. The diameter of the base of a right circular cone is 15 in. If the height of the cone is 9 in., find the volume.

38. Find the volume of a sphere whose radius is 10 cm.

Find the perimeter and area for the figures in Exercises 39 and 40.

39.

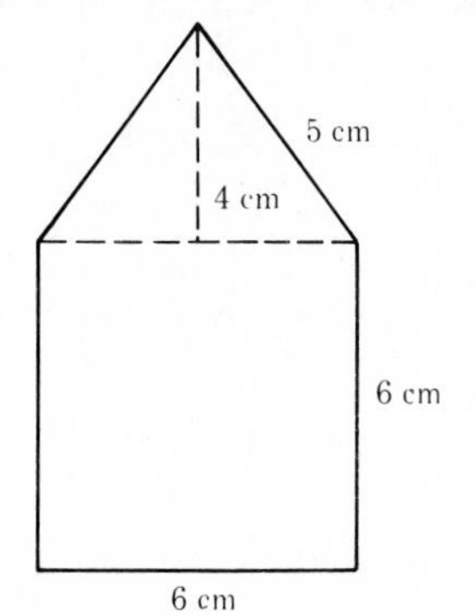

$P =$ ______

$A =$ ______

40.

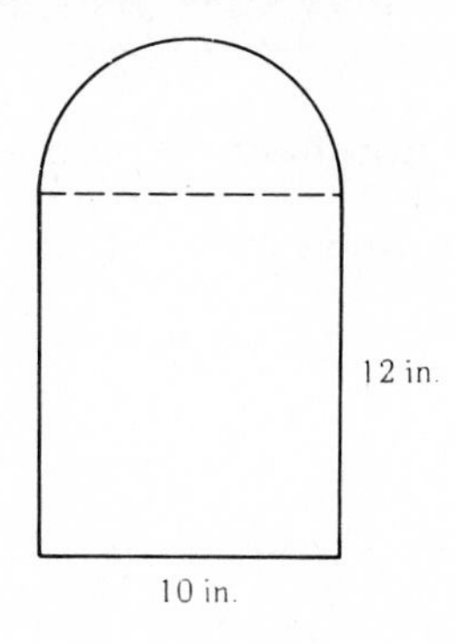

$P =$ ______

$A =$ ______

CHAPTER 1 SUMMARY

The **counting numbers** (or natural numbers) are 1, 2, 3, 4, 5, ...

The **whole numbers** are 0, 1, 2, 3, 4, 5, ...

A **variable** is a symbol or letter that can represent more than one number.

Commutative Property of Addition

If a and b are whole numbers, then

$$a + b = b + a$$

Associative Property of Addition

If a, b, and c are whole numbers, then

$$a + b + c = (a + b) + c = a + (b + c)$$

Commutative Property of Multiplication

If a and b are whole numbers, then

$$a \cdot b = b \cdot a$$

Associative Property of Multiplication

If a, b, and c are whole numbers, then

$$a \cdot b \cdot c = (a \cdot b) \cdot c = a \cdot (b \cdot c)$$

Distributive Property of Multiplication over Addition

If a, b, and c are whole numbers, then

$$a \cdot (b + c) = a \cdot b + a \cdot c$$

In general, for nonzero b, $\frac{a}{b} = x$ if and only if $a = b \cdot x$.

Division by 0 is Undefined.

To Find the Average of a Collection of Numbers

1. Add all the numbers in the collection.
2. Divide this sum by the number of numbers in the collection.

This quotient is the **average.**

An **exponent** is a number that tells how many times a factor occurs in a product. The factor is the **base** of the exponent, and the product is the **power.**

To Find the Prime Factorization of a Composite Number

1. Find any two factors of the number.
2. Continue to factor each of the factors until all factors are prime numbers.
3. Write the product of all these prime factors. (Use exponents whenever possible.)

To Find the LCM

1. Find the prime factorization of each number.
2. Form the product of all prime factors that appear using each prime factor the greatest number of times that it appears in any one prime factorization.

Rules for Order of Operations

1. Work within symbols of inclusion (parentheses, brackets, or braces), beginning with the innermost pair.
2. Find any powers indicated by exponents.
3. From left to right, perform any multiplications or divisions in the order they appear.
4. From left to right, perform any additions or subtractions in the order they appear.

The following properties of fractions are true if no denominator is 0:

$$\frac{a}{b} \cdot \frac{c}{d} = \frac{a \cdot c}{b \cdot d}$$

$$\frac{a}{b} = \frac{a \cdot k}{b \cdot k} \qquad \text{where } k \neq 0$$

$$\frac{a}{b} \div \frac{c}{d} = \frac{a}{b} \cdot \frac{d}{c}$$

$$\frac{a}{b} + \frac{c}{b} = \frac{a + c}{b}$$

$$\frac{a}{b} - \frac{c}{b} = \frac{a - c}{b}$$

The **reciprocal** of $\frac{a}{b}$ is $\frac{b}{a}$; and $\frac{a}{b} \cdot \frac{b}{a} = 1$.

A **decimal number** (or simply a **decimal**) is a fraction that has a power of ten as its denominator.

To change a decimal to a percent, move the decimal point two places to the right and write the % sign.

To change a percent to a decimal, move the decimal point two places to the left and drop the % sign.

An **equation** states that two algebraic expressions are equal.

A **formula** is an equation that represents a general relationship between two or more quantities or measurements.

The **perimeter** of a geometric figure is the total distance around the figure.

Area is a measure of the interior, or enclosure, of a surface.

Volume is a measure of the space enclosed by a three-dimensional figure.

Formulas in Geometry:

Perimeter	**Area**
$P = 4s$ (square)	$A = s^2$ (square)
$P = 2l + 2w$ (rectangle)	$A = lw$ (rectangle)
$P = 2a + 2b$ (parallelogram)	$A = bh$ (parallelogram)
$P = a + b + c$ (triangle)	$A = \frac{1}{2}bh$ (triangle)
$C = 2\pi r$ (circle) $C = \pi d$ (circle)	$A = \pi r^2$ (circle)
$P = a + b + c + d$ (trapezoid)	$A = \frac{1}{2}h(b + c)$ (trapezoid)

Volume

$V = lwh$ (rectangular solid)

$V = \frac{1}{3}lwh$ (rectangular pyramid)

$V = \pi r^2 h$ (right circular cylinder)

$V = \frac{1}{3}\pi r^2 h$ (right circular cone)

$V = \frac{4}{3}\pi r^3$ (sphere)

CHAPTER 1 REVIEW

Perform the indicated operations in Exercises 1–12.

1. $18 + 7 + 8 + 3$ **2.** $9 + 16 + 11 + 4$ **3.** $26 - 17$

4. $33 - 18$

5. $(14)(9)$

6. $(27)(17)$

7. $273 \div 7$

8. $744 \div 6$

9. Add: $\begin{array}{r} 75 \\ 34 \\ 608 \\ 123 \\ \hline \end{array}$

10. Subtract: $\begin{array}{r} 738 \\ 542 \\ \hline \end{array}$

11. Multiply: $\begin{array}{r} 42 \\ 79 \\ \hline \end{array}$

12. Divide: $17\overline{)408}$

Use the distributive property to evaluate Exercises 13 and 14.

13. $7(3 + 6)$

14. $8(9 + 7)$

Find the average of each of the sets of numbers in Exercises 15 and 16.

15. 17, 14, 20

16. 18, 22, 27, 37

Find the indicated products in Exercises 17 and 18.

17. 3^4

18. 11^2

Write Exercises 19 and 20 in exponential form.

19. $2 \cdot 2 \cdot 2 \cdot a \cdot a \cdot b$

20. $3 \cdot 3 \cdot 5 \cdot 5 \cdot a \cdot b \cdot b \cdot b$

Find the prime factorization for Exercises 21 and 22.

21. 60

22. 153

Find the LCM for each set of numbers in Exercises 23 and 24.

23. 18, 27, 36

24. $12x$, $9xy$, $24xy^2$

Find the value of each expression in Exercises 25 and 26 using the rules for order of operations.

25. $(13 \cdot 5 - 5) \div 2 \cdot 3$

26. $4[6 - (4 \cdot 5 - 2 \cdot 9) - 1]$

Reduce each fraction in Exercises 27 and 28 to lowest terms.

27. $\frac{20}{44}$

28. $\frac{36}{54}$

Find the missing numerator by changing each fraction in Exercises 29 and 30 to an equal fraction with the indicated denominator.

29. $\frac{5}{12} = \frac{\quad}{72}$

30. $\frac{4}{5} = \frac{\quad}{65}$

Name the property used in each of the Exercises 31–33.

31. $x + 17 = 17 + x$

32. $(y + 11) + 4 = y + (11 + 4)$

33. $6(x + 7) = 6x + 42$

Perform the indicated operations in Exercises 34–41. Reduce each answer to lowest terms.

34. $\frac{3}{8} + \frac{7}{10}$

35. $\frac{4}{9} - \frac{5}{12}$

36. $\frac{4}{5y} + \frac{2}{3y}$

37. $\frac{7}{4a} - \frac{9}{6a}$

38. $\frac{20}{6} \cdot \frac{9}{4}$

39. $\frac{11}{9} \div \frac{22}{15}$

40. $\frac{7b}{15} \div \frac{2b}{6}$

41. $\frac{8x}{21} \cdot \frac{7}{12x}$

Simplify Exercises 42–45 using the rules for order of operations.

42. $\frac{7}{18} + \frac{5}{24} - \frac{3}{8}$

43. $\frac{3}{5} \div \frac{8}{5} \cdot \frac{8}{3}$

44. $\frac{1}{3} + \frac{2}{5} \cdot \frac{5}{8} \div \frac{3}{2} - \frac{1}{2}$

45. $\left(\frac{3}{4} - \frac{1}{3}\right) \div \left(\frac{2}{3} + \frac{2}{7}\right)$

Change the decimals in Exercises 46 and 47 to percents.

46. 0.73

47. 0.065

Change the percents in Exercises 48 and 49 to decimals.

48. 7%

49. 12.5%

Perform the indicated operations in Exercises 50–57.

50. $15.8 + 9.1 + 7.63$

51. $19.31 - 14.62$

52. $24.3 + 6.81 - 16.51$

53. $(25.3)(2.3)$

54. $(14.7)(0.36)$

55. Find 18% of 60.

56. What is 106% of 55?

57. 11.5% of 380 is equal to ______.

Evaluate the expressions in Exercises 58 and 59 if $x = 5$.

58. $3x + 2 - x$

59. $x^2 + 2x$

Evaluate the expressions in Exercises 60 and 61 if $x = \frac{1}{4}$.

60. $2x + 3$

61. $3x - \frac{1}{2}$

Evaluate the expressions in Exercises 62 and 63 if $x = 1.2$.

62. $4x - 3.1$

63. $5x - 2.34$

Determine whether or not the given number is a solution to the given equation in Exercises 64–67.

64. $5x + 3 = 10$; $x = 2$

65. $7x - 8 = 13$; $x = 3$

66. $4x + 1 = 2x + 2$; $x = \frac{1}{2}$

67. $3x + 7 = 11.5$; $x = 1.5$

68. A rectangle is 23 ft long and 15 ft wide. Find the perimeter.

69. Find the area of a circle with a radius of 9 meters.

70. Find the volume of a rectangular box that is 11 in. long, 9 in. wide, and 7 in. high.

71. Lucia is making punch for a party. She is making 5 gallons. The recipe calls for $\frac{2}{3}$ cup of sugar per gallon of punch. How many cups of sugar does she need?

72. Ozzie is building some bookshelves. He needs 3 pieces 25 in. long, 1 piece 19 in. long, 2 pieces 64 in. long, and 3 pieces 38 in. long. The wood costs $0.75 per 12 in. How much will the wood cost?

73. Julie grew corn for a school experiment. She planted 6 seeds. At the end of 7 weeks, she measured each cornstalk. The heights were 18.6, 20.4, 23.5, 19.7, 22.1, and 20.5 in. What was the average height of Julie's plants?

74. Last year, a rancher sold 72 head of cattle. It cost him $17,280 to raise and feed these cattle. He wants to make a profit of $125 per head. What price per head must he receive for his cattle?

75. Matt's Little League team is selling magazine subscriptions to earn extra money. If they receive 15% of sales, how much will they earn if they sell $453.40 worth of subscriptions?

CHAPTER 1 TEST

1. Add: $32 + 688 + 1013 + 61$

2. Subtract: $453 - 87$

3. $437 \div 19$

4. Identify the property used: $8x + 20 = 4(2x + 5)$.

5. Find the average of the numbers: 34, 61, 72, 45.

6. Write in exponential form: $2 \cdot 3 \cdot 3 \cdot x \cdot y \cdot y \cdot y$.

7. Find the prime factorization of 315.

8. Find the LCM of $10xy^2$, $18x^2y$, and $15xy$.

9. Find the value of $4[5^2 - (6 + 5 \cdot 3) \div 7]$.

10. Reduce to lowest terms: $\frac{72}{156}$.

11. Find the missing numerator: $\frac{5}{9} = \frac{\quad}{108}$.

Perform the indicated operations and reduce to lowest terms.

12. $\frac{7}{12} \cdot \frac{9}{28} \div \frac{5}{16}$

13. $\frac{4}{15} + \frac{1}{3} - \frac{3}{10}$

14. $\frac{7}{8} - \frac{1}{3} \div \frac{5}{6} + \frac{1}{4}$

15. $28.63 + 7.9 - 15.47$

16. Find the product: $(31.6)(0.43)$.

17. Find 12.5% of 340.

18. Evaluate the expression $4x + 6 - x^2$ if $x = 3$.

19. Find the area of a rectangle 12 ft long and $8\frac{1}{2}$ ft wide.

20. Jaime bought a pair of skis priced at $240. He paid 20% down in cash. The balance was paid off in 6 equal payments. How much was each payment?

```
10 REM
20 REM This program computes the average and sum of given
30 REM numbers.
40 REM
50 PRINT "How many numbers do you have ";
60 INPUT HOWMANY
70 DIM RLIST(HOWMANY)
80 PRINT "Type the numbers you have."
90 PRINT "After you type in each number, hit the return key."
100 REM
110 REM   Get the numbers from key-boards.
120 REM
130 FOR COUNT = 1 TO HOWMANY      ' Loop for reading each number.
140 INPUT RLIST(COUNT)            ' Get a number from key board.
150 NEXT COUNT                    ' End loop.
160 REM
170 REM   Compute the sum and average.
180 REM
190 LET SUM = 0
200 FOR COUNT = 1 TO HOWMANY      ' Loop for adding each number.
210 LET SUM = SUM + RLIST(COUNT)  ' Add each number.
220 NEXT COUNT                    ' End loop.
230 LET AVERAGE = SUM / HOWMANY   ' Compute the average.
240 PRINT "The sum is "SUM        ' Display the result.
250 PRINT "The average is "AVERAGE
260 END                           ' End program.
```

```
How many numbers do you have  10
Type the numbers you have.
After you type in each number, hit the return key.
 1
 2
 3
 4
 5
 6
 7
 8
 9
 10
The sum is  55
The average is  5.5
```

How can it be that mathematics, being after all a product of human thought independent of experience, is so admirably adapted to the objects of reality.

ALBERT EINSTEIN 1879–1955

DID YOU KNOW?

Arithmetic operations defined on the set of positive integers, negative integers, and zero are studied in this chapter. The integer zero will be shown to have interesting properties under the operations of addition, subtraction, multiplication, and division.

Curiously, zero was not recognized as a number by early Greek mathematicians. When Hindu scientists developed the place-value numeration system we currently use, the zero symbol was initially a place holder but not a number. The spread of Islam transmitted the Hindu number system to Europe where it became known as the Hindu-Arabic system and replaced Roman numerals. The word **zero** comes from the Hindu word meaning "void," which was translated into Arabic as *sifr* and later into Latin as *zephirum*, hence the derivation of our English words *zero* and *cipher*.

Almost all of the operational properties of zero were known to the Hindus. However, the Hindu mathematician, Bhaskara the Learned (1114–1185?), asserted that a number divided by zero was zero, or possibly infinite. Bhaskara did not seem to understand the role of zero as a divisor, since division by zero is undefined and hence an impossible operation in mathematics.

Albert Einstein, in his development of a proof that the universe was stable and unchangeable in time, divided both sides of one of his intermediate equations by a complicated expression which under certain circumstances could become zero. When the expression became zero, Einstein's proof did not hold and the possibility of a pulsating, expanding, or contracting universe had to be considered. This error was pointed out to Einstein and he was forced to withdraw his proof that the universe was stable. The moral of this story is that, although zero seems like a "harmless" number, its operational properties are different from those of the positive or negative integers.

2 Signed Numbers

2.1 NUMBER LINES

The closest thing to a picture of a set of numbers is its graph on a number line. We generally will use horizontal and vertical lines for number lines. Choose some point on a line and label it with the number 0 (Figure 2.1).

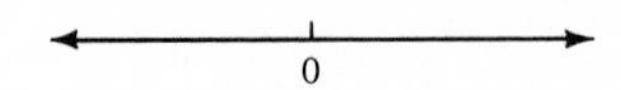

FIGURE 2.1

Now choose another point on the line to the right of 0 and label it with the number 1 (Figure 2.2).

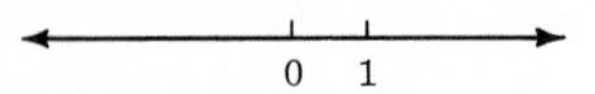

FIGURE 2.2

We now have a number line. Points corresponding to all the whole numbers are determined. The point corresponding to 2 is the same distance from 1 as 1 is from 0, 3 from 2, 4 from 3, and so on (Figure 2.3).

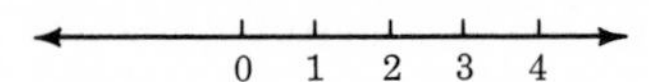

FIGURE 2.3

The **graph** of a number is the point that corresponds to the number, and the number is called the **coordinate** of the point. We will follow the convention of using the terms "number" and "point" interchangeably. Thus, a point can be called "seven" or "two." The graph of 7 is indicated by marking the point corresponding to 7 with a large dot (Figure 2.4).

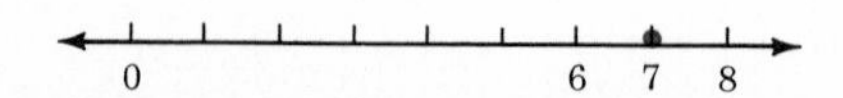

FIGURE 2.4

The graph of the set $A = \{2, 4, 6\}$ is shown in Figure 2.5.

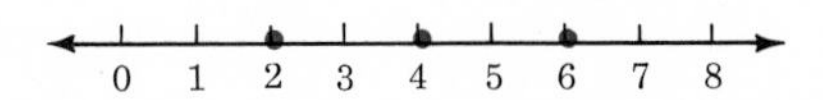

FIGURE 2.5

On a horizontal number line, the point one unit to the left of 0 is the **opposite of 1.** It is called **negative 1** and is symbolized -1. Similarly, the point two units to the left of 0 is the opposite of 2, called negative 2, and symbolized -2, and so on (Figure 2.6).

The opposite of 1 is -1;
the opposite of 2 is -2;
the opposite of 3 is -3;
and so on.

The opposite of -1 is $-(-1) = +1$;
the opposite of -2 is $-(-2) = +2$;
the opposite of -3 is $-(-3) = +3$;
and so on.

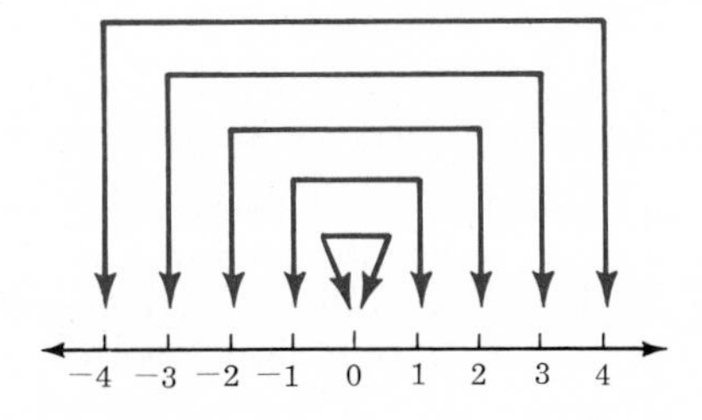

FIGURE 2.6

The set of numbers consisting of the whole numbers and their opposites is called the set of **integers.** The natural numbers are called **positive integers.** Their opposites are called **negative integers. Zero is its own opposite and is neither positive nor negative** (Figure 2.7). Note that the opposite of a positive integer is a negative integer, and the opposite of a negative integer is a positive integer.

- Integers: $\ldots, -3, -2, -1, 0, 1, 2, 3, \ldots$
- Positive integers: $1, 2, 3, 4, 5, \ldots$
- Negative integers: $\ldots, -4, -3, -2, -1$

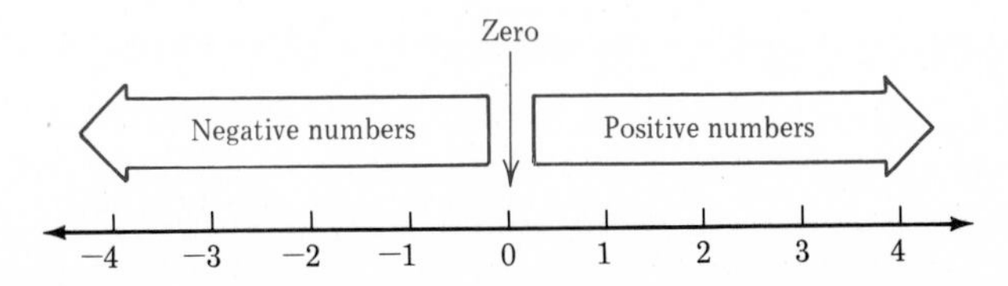

FIGURE 2.7

EXAMPLES

1. Find the opposite of 7.

   Solution: -7

2. Find the opposite of -3.

   Solution: $-(-3)$ or $+3$

3. Graph the set $\{-3, -1, 1, 3\}$.

   Solution:

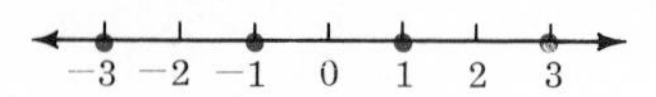

4. Graph the set $\{\ldots, -5, -4, -3\}$.

   Solution:

. . .
−6 −5 −4 −3 −2 −1 0

The integers are not the only numbers that can be represented on a number line. Fractions and decimal numbers such as $\frac{1}{2}$, $\frac{3}{4}$, $-\frac{4}{3}$, and 1.2 can also be represented (Figure 2.8).

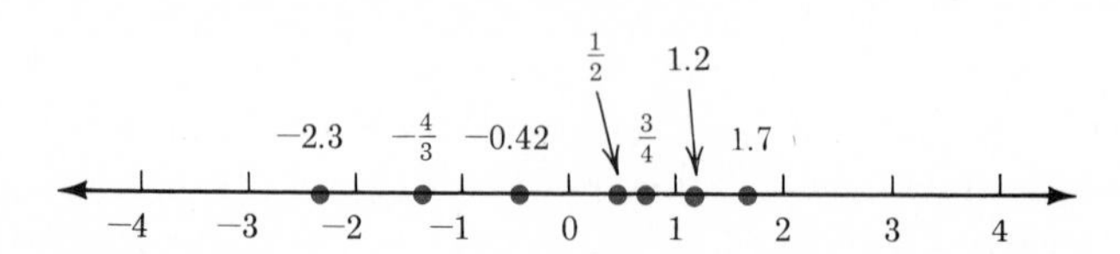

FIGURE 2.8

Numbers that can be written as fractions with integers as numerator and denominator are called **rational numbers.** They include positive and negative decimal numbers and the integers themselves. For example, the following are rational numbers:

$$1.3 = \frac{13}{10} \quad \text{and} \quad 5 = \frac{5}{1} \quad \text{and} \quad -4 = \frac{-4}{1}$$

All rational numbers have corresponding points on a number line.

- Rational numbers $\left\{ \begin{array}{l} \text{numbers that can be written as} \\ \dfrac{a}{b} \text{ where } a \text{ and } b \text{ are integers, } b \neq 0 \end{array} \right.$

Other numbers on a number line such as $\sqrt{2}$, $\sqrt{3}$, π, and $\sqrt[3]{5}$ are called **irrational numbers.** These numbers will be discussed in some detail in Chapter 10 and in later courses in mathematics. All the rational and irrational numbers, positive and negative, can be referred to as **signed numbers.**

On a horizontal number line, **smaller numbers are always to the left of larger numbers.** Each number is smaller than any number to its right and larger than any number to its left. Two symbols used to indicate order are

$<$, read "is less than"
and $>$, read "is greater than."

−2 −1 0 1 2 3 4 5 6 7

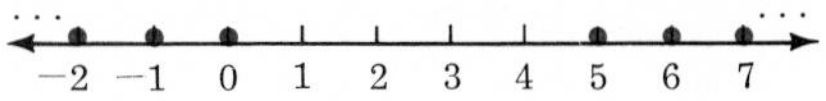

FIGURE 2.9

Using the number line in Figure 2.9, you can see the following relationships:

	Using	*OR*	*Using*
$\frac{1}{4} < \frac{9}{8}$	$\frac{1}{4}$ is less than $\frac{9}{8}$	$\frac{9}{8} > \frac{1}{4}$	$\frac{9}{8}$ is greater than $\frac{1}{4}$
$0 < 3$	0 is less than 3	$3 > 0$	3 is greater than 0
$-2 < 1$	−2 is less than 1	$1 > -2$	1 is greater than −2
$-7 < -4$	−7 is less than −4	$-4 > -7$	−4 is greater than −7

Two other symbols commonly used are

$\leq$, read "is less than or equal to"
and $\geq$, read "is greater than or equal to."

For example, $5 \geq -10$ is true since 5 is greater than −10. Also, $5 \geq 5$ is true since 5 does equal 5.

Table of Symbols

$=$	is equal to	$\neq$	is not equal to
$<$	is less than	$>$	is greater than
$\leq$	is less than or equal to	$\geq$	is greater than or equal to

EXAMPLES

5. Determine whether each of the following statements is true or false:

a. $7 < 15$ True, since 7 is less than 15.
b. $3 > -1$ True, since 3 is greater than -1.
c. $4 \geq -4$ True, since 4 is greater than -4.
d. $2.7 \leq 2.7$ True, since 2.7 is equal to 2.7.
e. $-5 < -6$ False, since -5 is greater than -6.

6. Graph the set of numbers $\{-\frac{3}{4}, 0, 1, 1.5, 3\}$.

Solution:

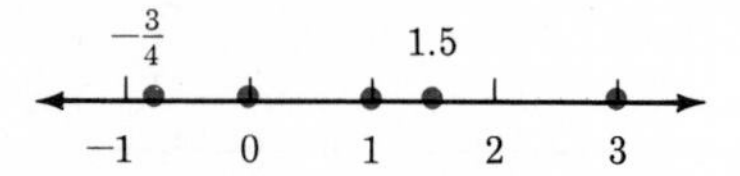

7. Graph all natural numbers less than or equal to 3.

Solution:

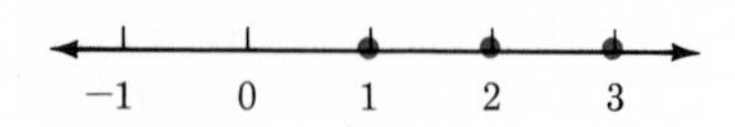

(**Note:** Remember that the natural numbers are 1, 2, 3, 4, . . .)

8. Graph all integers less than 1.

Solution:

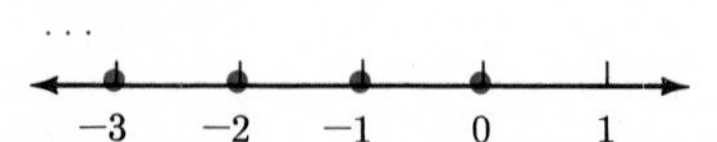

(**Note:** The three dots above the graph line indicate that the pattern of graphing is to continue without end.)

EXERCISES 2.1

Fill in the blank in Exercises 1–15 with the appropriate symbol: <, >, or =.

1. 6 _____ 4

2. −3 _____ 1

3. −2 _____ −4

4. 5 _____ −(−5)

5. $\frac{1}{3}$ _____ $\frac{1}{2}$

6. $-\frac{2}{3}$ _____ $\frac{1}{8}$

7. $-\frac{2}{8}$ _____ $-\frac{1}{4}$

8. −8 _____ 0

9. 1.6 _____ 2.3

10. $-\frac{3}{4}$ ______ -1 **11.** $\frac{9}{16}$ ______ $\frac{3}{4}$ **12.** $-\frac{1}{2}$ ______ $-\frac{1}{3}$

13. -2.3 ______ $-2\frac{3}{10}$ **14.** 5.6 ______ $-(-4.7)$ **15.** $-\frac{4}{3}$ ______ $-\left(-\frac{1}{3}\right)$

Determine whether each statement in Exercises 16–35 is true or false.

16. $0 = -0$ **17.** $-22 < -16$ **18.** $-9 > -8.5$

19. $11 = -(-11)$ **20.** $-17 \leq -17$ **21.** $-6 < -8$

22. $4.7 \geq 3.5$ **23.** $-\frac{1}{3} \leq 0$ **24.** $-8.1 < -8.1$

25. $-7.3 \leq -8.6$ **26.** $\frac{3}{5} > \frac{1}{4}$ **27.** $-2.3 < 1$

28. $-9 > -7.69$ **29.** $4.6 > 4.1$ **30.** $0 > -\frac{2}{5}$

31. $4 + 3 < 2 + 5$ **32.** $4 - 2 < 5 - 1$ **33.** $14.3 > 8.1 + 5.9$

34. $13.6 - 7.8 > 2.3 + 1.5$ **35.** $6.3 + 5.2 \geq 12.0 - 0.5$

In Exercises 36–55, graph each set of numbers on a number line.

36. $\{1, 2, 5, 6\}$ **37.** $\{-3, -2, 0, 1\}$ **38.** $\{2, -3, 1, 0, -1\}$

39. $\{-2, -1, 4, -3\}$ **40.** $\left\{0, -1, \frac{5}{4}, 3, 1\right\}$ **41.** $\left\{-2, -1, -\frac{1}{3}, 2\right\}$

42. $\left\{-\frac{3}{4}, 0, 2, 3.6\right\}$ **43.** $\left\{-3.4, -2, 0.5, 1, \frac{5}{3}\right\}$

44. $\left\{-\frac{7}{2}, -1.5, 1, \frac{4}{3}, 2\right\}$ **45.** $\left\{-4, -\frac{7}{3}, -1, 0.2, \frac{5}{2}\right\}$

46. all natural numbers less than 4

47. all natural numbers less than 7

48. all positive integers less than or equal to 3

49. all negative integers greater than or equal to -3

50. all integers less than -6

51. all integers greater than or equal to -1

52. all whole numbers less than 8

53. all whole numbers less than or equal to 4.3

54. all negative integers greater than or equal to -5.7

55. all integers greater than or equal to -1.6

2.2 ABSOLUTE VALUE

In working with the number line in Section 2.1, you may have noticed that any integer and its opposite lie the same number of units from 0 on the number line. For example, both $+7$ and -7 are seven units from 0 (Figure 2.10). The $+$ and $-$ signs indicate direction and the 7 indicates distance.

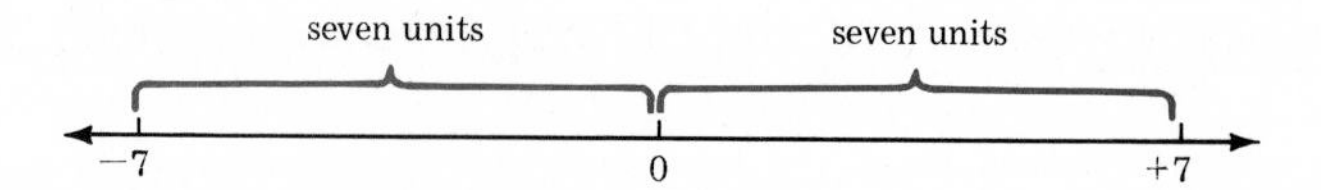

FIGURE 2.10

The distance a number is from 0 on a number line is called its **absolute value** and is symbolized by two vertical bars, $|\ \ |$. Thus, $|+7| = 7$ and $|-7| = 7$. Similarly,

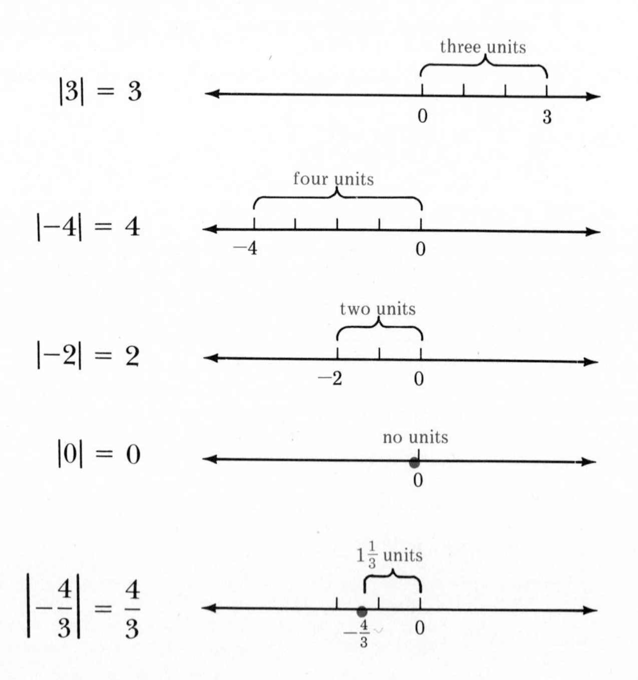

Definition: The **absolute value** of a number is its distance from 0. The absolute value of a number is never negative.

EXAMPLES

1. $|6.3| = 6.3$

2. $|-5.1| = 5.1$

3. If $|x| = 7$, what are the possible values for x?

   Solution: $x = 7$ or $x = -7$ since $|7| = 7$ and $|-7| = 7$.

4. If $|x| = 1.35$, what are the possible values for x?

   Solution: $x = 1.35$ or $x = -1.35$ since $|1.35| = 1.35$ and $|-1.35| = 1.35$.

5. True or False: $|-4| \leq 4$

   Solution: True, since $|-4| = 4$ and $4 = 4$.

6. True or False: $|-5\frac{1}{2}| > 5\frac{1}{2}$

   Solution: False, since $|-5\frac{1}{2}| = 5\frac{1}{2}$.

7. If $|x| = -3$, what are the possible values for x?

   Solution: There are no values of x for which $|x| = -3$. The absolute value can never be negative.

EXERCISES 2.2

In Exercises 1–10, graph the sets on a number line.

1. $\{-5, 3, 2, |-1|, |4|\}$

2. $\{-7, |-3|, |-2|, 0, 1\}$

3. $\{0, -1, |-5|, |2|, -3\}$

4. $\{-2, |-6|, 5, |-2|, 7\}$

5. $\{|-8|, 0, 8, |6|, -8\}$

6. $\{-3, 1, |-3|, |3|, -2\}$

7. $\left\{-2, \left|-\frac{3}{2}\right|, -1, 2, |2.7|\right\}$

8. $\left\{\left|-\frac{2}{3}\right|, -1, |-2|, |-1.5|\right\}$

9. $\left\{0, \left|-\frac{4}{3}\right|, -2.5, |2|, -2.1\right\}$

10. $\left\{-1.6, |-2.5|, \frac{3}{2}, |2.5|, 3\right\}$

In Exercises 11–20, determine whether the statements are true or false.

11. $|-5| = |5|$

12. $|-8| \geq 4$

13. $|-6| \geq 6$

14. $|-7| < |7|$

15. $|-1.9| < 2$

16. $|-1.6| < |-2.1|$

17. $\left|-\frac{5}{2}\right| < 2$

18. $\frac{2}{3} < |-1|$

19. $3 > \left|-\frac{4}{3}\right|$

20. $|-3.4| < 0$

List the numbers that satisfy the conditions stated in Exercises 21–40.

21. $|x| = 4$

22. $|y| = 6$

23. $|x| = 9$

24. $13 = |x|$

25. $0 = |y|$

26. $-2 = |x|$

27. $|x| = -3$

28. $|x| = 3.5$

29. $|x| = 4.7$

30. $|y| = |-8|$

31. $|-12| = |y|$

32. $|x| = \frac{5}{2}$

33. $|x| = \frac{4}{7}$

34. $|x| = \frac{4}{3}$

35. $|x| = \left|-\frac{5}{4}\right|$

36. $|y| = -\frac{5}{8}$

37. $|y| = -2.7$

38. $|x| = 4.16$

39. $|x| = -\frac{3}{7}$

40. $|x| = 11.624$

2.3 ADDITION WITH SIGNED NUMBERS

Picture a straight line in an open field and numbers marked on a number line. An archer stands at 0 and shoots an arrow to +3, then stands at 3 and shoots the arrow 5 more units in the positive direction (to the right). Where will the arrow land? (See Figure 2.11.)

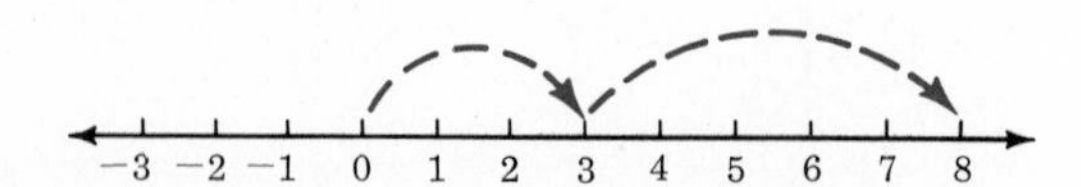

FIGURE 2.11

Naturally, you have figured that the answer is +8. What you have done is add the two positive numbers, +3 and +5.

$$(+3) + (+5) = +8 \qquad \text{or} \qquad 3 + 5 = 8$$

Suppose another archer shot arrows in the same manner as the first but in the opposite direction. Where would his second arrow light? The arrow lights at −8. You have just added −3 and −5 (Figure 2.12).

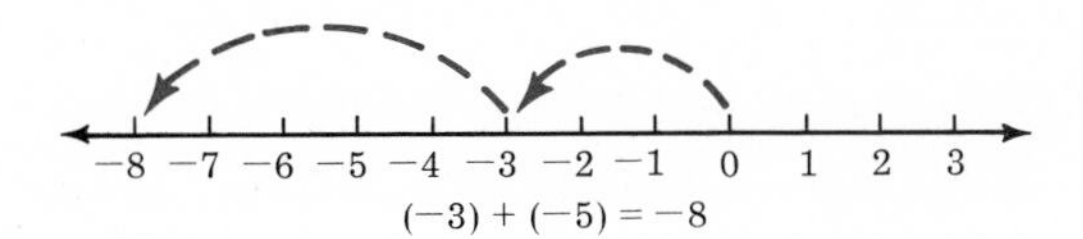

FIGURE 2.12

If an arrow is shot to +3, and then the archer goes to +3 and turns around and shoots an arrow 5 units in the opposite direction, where will the arrow stick? Would you believe at −2? (See Figure 2.13.)

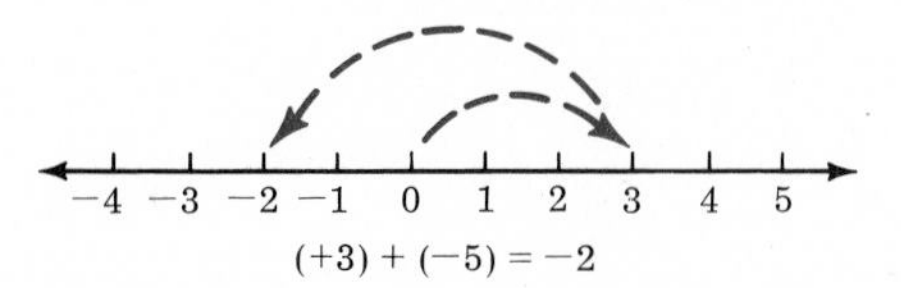

FIGURE 2.13

For our final archer, we observe a first shot to −3. Then, after going to −3, he turns around and shoots 5 units in the opposite direction. Where is the arrow? It is at +2 (Figure 2.14).

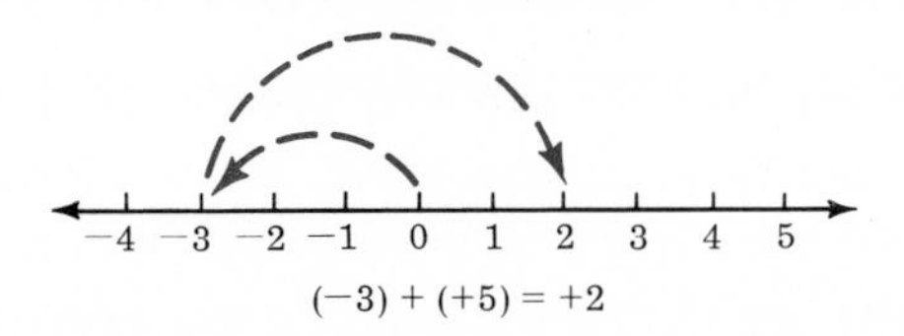

FIGURE 2.14

In summary,

1.	$(+3)$	$+$	$(+5)$	$=$	$+8$
	↑	↑	↑		↑
	positive	plus	positive	is	positive
2.	(-3)	$+$	(-5)	$=$	-8
	↑	↑	↑		↑
	negative	plus	negative	is	negative
3.	$(+3)$	$+$	(-5)	$=$	-2
	↑	↑	↑		↑
	positive	plus	negative	is	negative
4.	(-3)	$+$	$(+5)$	$=$	$+2$
	↑	↑	↑		↑
	negative	plus	positive	is	positive

PRACTICE QUIZ

Questions	**Answers**
Find each sum.	
1. $(-14) + (-6) =$	1. -20
2. $(+16) + (-10) =$	2. $+6$
3. $(-12) + (+8) =$	3. -4
4. $11 + 7 =$	4. 18
5. $(-13) + (+13) =$	5. 0
6. $(-11) + (-8) =$	6. -19
7. $(+6) + (-7) =$	7. -1
8. $(+100) + (-100) =$	8. 0
9. $(-5.2) + (+16.3) =$	9. 11.1
10. $9.7 + 4.1 =$	10. 13.8
11. $(-4\frac{1}{2}) + (-3\frac{1}{4}) =$	11. $-7\frac{3}{4}$
12. $7.8 + (-7.8) =$	12. 0

You probably did quite well and understand how to add numbers. The rules can be written out in the following rather formal manner.

Rules for Adding Signed Numbers

1. To add two numbers with like signs, add their absolute values and use the common sign:

$$(+7) + (+3) = +(|+7| + |+3|) = +(7 + 3) = +10$$
$$(-7) + (-3) = -(|-7| + |-3|) = -(7 + 3) = -10$$

2. To add two numbers with unlike signs, subtract their absolute values (the smaller from the larger) and use the sign of the number with the larger absolute value:

$$(-12) + (+10) = -(|-12| - |+10|) = -(12 - 10) = -2$$
$$(+12) + (-10) = +(|+12| - |-10|) = +(12 - 10) = +2$$
$$(-15) + (+15) = (|-15| - |+15|) = (15 - 15) = 0$$

Since equations in algebra are almost always written horizontally, you should become used to working with sums written horizontally. However, there are situations (as in long division) where sums (and differences) are written vertically with one number directly under another.

EXAMPLES

Find each sum.

1.	2.	3.	4.
-10	-4	-5	-10.5
7	6	-8	$+3.2$
-3	-15	-9	$+6.8$
	-13	-22	-0.5

Questions 5 and 8 on the Practice Quiz illustrate an important relationship between a number and its opposite.

Definition: The **opposite** of a number is called its **additive inverse.** The sum of a number and its additive inverse is 0. Symbolically, for any number x,

$$x + (-x) = 0$$

EXAMPLES

Find the additive inverse (opposite) of each number.

5. 3

Solution: The additive inverse of 3 is -3.

$$3 + (-3) = 0$$

6. -7

Solution: The additive inverse of -7 is $-(-7) = +7$.

$$(-7) + (+7) = 0$$

7. -4.8

Solution: The additive inverse of -4.8 is $-(-4.8) = +4.8$.

$$(-4.8) + (+4.8) = 0$$

8. 0

Solution: The additive inverse of 0 is $-0 = 0$.

$$(0) + (0) = 0$$

Now that we know how to add positive and negative numbers, we can determine whether or not a particular signed number satisfies an equation.

EXAMPLES

Determine whether or not the given number is a solution to the given equation by substituting and adding.

9. $x + 5 = -2;\ x = -7$

Solution: $(-7) + 5 = -2$ is true, so -7 is a solution.

10. $y + (-4) = -6.3;\ y = -2.3$

Solution: $(-2.3) + (-4) = -6.3$ is true, so -2.3 is a solution.

11. $14 + z = -3;\ z = -11$

Solution: $14 + (-11) = -3$ is false since $14 + (-11) = +3$. So, -11 is *not* a solution.

EXERCISES 2.3

Find the additive inverse for the numbers given in Exercises 1–10.

1. 11 **2.** 17 **3.** -6 **4.** -23 **5.** 4.7

6. -3.4 **7.** 0 **8.** $-\frac{2}{3}$ **9.** $-\frac{5}{16}$ **10.** -2.57

Find the sum in Exercises 11–56.

11. $4 + 9$ **12.** $8 + (-3)$ **13.** $(-9) + 5$

14. $(-7) + (-3)$ **15.** $(-9) + 9$ **16.** $2 + (-8)$

17. $11 + (-6)$ **18.** $(-12) + 3$ **19.** $-18 + 5$

20. $26 + (-26)$ **21.** $-5 + |-3|$ **22.** $11 + |-2|$

23. $(-2) + (-8)$ **24.** $10 + (-3)$ **25.** $17 + (-17)$

26. $(-7) + 20$ **27.** $21 + (-4)$ **28.** $(-5) + (-3)$

29. $(-12) + (-7)$ **30.** $26 + (-26)$ **31.** $-4\frac{2}{3} + (-5\frac{1}{6})$

32. $(-6\frac{3}{5}) + (-8\frac{7}{10})$ **33.** $9\frac{5}{16} + (-12\frac{3}{4})$ **34.** $-12\frac{3}{4} + 9\frac{5}{8}$

35. $38.5 + (-16.48)$ **36.** $(-20.3) + (-11.81)$

37. $(-33.62) + (-21.9)$ **38.** $(-21.6) + 18.5$

39. $-3 + 4 + (-8)$ **40.** $(-9) + (-6) + 5$

41. $(-9) + (-2) + (-5)$ **42.** $(-21) + 6 + 15$

43. $-13 + (-1) + (-12)$ **44.** $-19 + (-2) + (-4)$

45. $27 + (-14) + (-13)$ **46.** $-33 + 29 + 2$

47. $-43 + (-16) + 27$ **48.** $-68 + (-3) + 42$

49. $-38 + 49 + (-6)$ **50.** $102 + (-93) + (-6)$

51. $\begin{array}{r} -21 \\ \underline{-62} \end{array}$ **52.** $\begin{array}{r} -12 \\ \underline{17} \end{array}$ **53.** $\begin{array}{r} -15 \\ 8 \\ \underline{19} \end{array}$ **54.** $\begin{array}{r} -7 \\ 23 \\ \underline{-9} \end{array}$

55. $\begin{array}{r} -163 \\ 204 \\ \underline{-73} \end{array}$ **56.** $\begin{array}{r} -93 \\ -87 \\ \underline{147} \end{array}$

Determine whether or not the number given in each of the Exercises 57–65 is a solution to the given equation by substituting and then evaluating.

57. $x + 4 = 2;\ x = -2$ **58.** $x - 7 = -10;\ x = -3$

59. $-10 + x = -14;\ x = -4$ **60.** $2x + 9 = 7;\ x = 1$

61. $17 + x = 10;\ x = -7$ **62.** $2x - 1 = 9;\ x = 5$

63. $3x - 12 = 6;\ x = 2$ **64.** $x + 3.5 = 2.8;\ x = -1.7$

65. $x + \frac{3}{4} = -\frac{1}{4};\ x = -1$

Choose the response that correctly completes the sentence in Exercises 66–70.

66. If x is a positive number and y is a negative number, then $x + y$ is (never, sometimes, always) a negative number.

67. If x and y are numbers, then $x + y$ is (never, sometimes, always) equal to 0.

68. If x and y are positive numbers, then $x + y$ is (never, sometimes, always) equal to 0.

69. If y is a number, then $y + (-y)$ is (never, sometimes, always) equal to 0.

70. If x and y are negative numbers, then $x + y$ is (never, sometimes, always) a positive number.

Calculator Problems

Use a calculator to evaluate Exercises 71–73.

71. $47.832 + (-29.572) + 66.919$

72. $56.473 + (-41.031) + (-28.638)$

73. $(-16.945) + (-27.302) + (-53.467)$

2.4 SUBTRACTION WITH SIGNED NUMBERS

In Section 2.3, we added numbers such as $5 + (-2) = 3$ and $26 + (-9) = 17$. Note that in each case, subtraction will give the same answers; that is, $5 - 2 = 3$ and $26 - 9 = 17$. Thus, it seems that subtraction and addition are closely related—and, in fact, they are.

From arithmetic, $5 - 2$ is asking, "What number **added** to 2 gives 5?" That is,

$$5 - 2 = 3 \quad \text{because} \quad 5 = 2 + 3$$

$$26 - 9 = 17 \quad \text{because} \quad 26 = 9 + 17$$

What do we mean by $4 - (-1)$? Do we mean, "What number added to -1 gives 4?" Precisely. Thus,

$$4 - (-1) = 5 \quad \text{because} \quad 4 = (-1) + 5$$

$$4 - (-2) = 6 \quad \text{because} \quad 4 = (-2) + 6$$

$$4 - (-3) = 7 \quad \text{because} \quad 4 = (-3) + 7$$

But note the following results:

$$4 + 1 = 5$$

$$4 + 2 = 6$$

$$4 + 3 = 7$$

The following relationship between subtraction and addition becomes the basis for subtraction with all numbers:

$$4 - (-1) = 4 + (+1) = 5$$

$$4 - (-2) = 4 + (+2) = 6$$

$$4 - (-3) = 4 + (+3) = 7$$

EXAMPLES

1. $(-1) - (-4) = (-1) + (+4) = +3$
2. $(-1) - (-5) = (-1) + (+5) = +4$
3. $(-1) - (-6) = (-1) + (+6) = +5$
4. $(-10) - (-2) = (-10) + (+2) = -8$
5. $(-10) - (-5.7) = (-10) + (+5.7) = -4.3$

PRACTICE QUIZ

Questions	Answers
1. $(-4) - (-1) = (-4) + (+1) =$	1. -3
2. $(-4) - (-2) = (-4) + (+2) =$	2. -2
3. $(-4) - (-3) = (-4) + (+3) =$	3. -1
4. $(-4) - (-4) = (-4) + (+4) =$	4. 0
5. $(-3) - (-8) = (-3) + (+8) =$	5. 5
6. $(-3) - (+8) = (-3) + (-8) =$	6. -11
7. $(25.4) - (46.7) = 25.4 + (-46.7) =$	7. -21.3
8. $(13\frac{3}{4}) - (8\frac{1}{2}) = (13\frac{3}{4}) + (-8\frac{1}{2}) =$	8. $5\frac{1}{4}$
9. $14 - (-5) = 14 + (5) =$	9. 19
10. $14 - (5) = 14 + (-5) =$	10. 9

You may have noticed that in subtraction, the **opposite** of the number being subtracted is **added.** For example, we could write

$$(-1) - (-4) = (-1) + [-(-4)] = (-1) + (+4) = +3$$

or

$$(-10) - (-3) = (-10) + [-(-3)] = (-10) + (+3) = -7$$

Definition: For any numbers a and b,

$$a - b = a + (-b)$$

This definition translates as, "To subtract b from a, **add** the **opposite** of b to a." In practice, the notation $a - b$ is thought of as addition of signed numbers. That is, since $a - b = a + (-b)$, we think of the plus sign, +, as being present in $a - b$. **In fact, an expression such as 4 − 19 can be read "four plus a negative nineteen."** We have

$$4 - 19 = 4 + (-19) = -15$$
$$-25 - 30 = -25 + (-30) = -55$$
$$-3 - 17 = -3 + (-17) = -20$$
$$24 - 11 = 24 + (-11) = 13$$

Generally, the second form is omitted and we go directly to the answer by computing the sum mentally.

$$4 - 19 = -15$$
$$-25 - 30 = -55$$
$$-3 - 17 = -20$$
$$24 - 11 = 13$$

The numbers may also be written vertically, that is, one underneath the other. In this case, the sign of the number being subtracted (the bottom number) is changed and addition is performed. This format is used in long division, as discussed in Chapter 6.

EXAMPLES

6. **Subtract** **(Add)**

$$\begin{array}{r} 43 \\ \underline{-25} \end{array} \qquad \begin{array}{r} 43 \\ \underline{+25} \\ 68 \end{array}$$

7. **Subtract** **(Add)**

$$\begin{array}{r} -38 \\ \underline{+11} \end{array} \qquad \begin{array}{r} -38 \\ \underline{-11} \\ -49 \end{array}$$

8. **Subtract** **(Add)**

$$\begin{array}{r} -73 \\ \underline{-32} \end{array} \qquad \begin{array}{r} -73 \\ \underline{+32} \\ -41 \end{array}$$

9. **Subtract** **(Add)**

$$\begin{array}{r} 17.6 \\ \underline{69.3} \end{array} \qquad \begin{array}{r} 17.6 \\ \underline{-69.3} \\ -51.7 \end{array}$$

Now, using subtraction as well as addition, we can determine whether or not a number is a solution to an equation of a slightly more complex nature.

EXAMPLES

Determine whether or not the given number is a solution to the given equation by substituting and then evaluating.

10. $x - (-5) = 6;\ x = 1$

Solution: $1 - (-5) = 1 + (+5) = 6,$ so 1 is a solution.

11. $5 - y = 7;\ y = -2$

Solution: $5 - (-2) = 5 + (+2) = 7,$ so -2 is a solution.

12. $z - 14 = -3;\ z = 10$

Solution: $10 - 14 = -4$ and $-4 \neq -3,$ so 10 is **not** a solution.

13. $6 - 8 = x - (-3);\ x = -5$

Solution: $6 - 8 = -5 - (-3)$

$$6 - 8 = -2 \quad \text{and} \quad -5 - (-3) = -5 + (+3) = -2$$

so -5 is a solution.

EXERCISES 2.4

Simplify the expressions in Exercises 1–12.

1. $8 - 3$

2. $5 - 7$

3. $-4 - 6$

4. $3 - (-4)$

5. $5 - (-7)$

6. $-18 - 17$

7. $-8 - (-11)$

8. $0 - (-12)$

9. $-4\frac{5}{8} - 2\frac{1}{4}$

10. $8\frac{2}{3} - 7\frac{1}{4}$

11. $16.34 - (-8.4)$

12. $(14.71) - 23.8$

Subtract the bottom number from the top number in Exercises 13–24.

13. $\begin{array}{r} 27 \\ \underline{42} \end{array}$

14. $\begin{array}{r} 19 \\ \underline{26} \end{array}$

15. $\begin{array}{r} 23 \\ \underline{-7} \end{array}$

16. $\begin{array}{r} 41 \\ \underline{-8} \end{array}$

17. $\begin{array}{r} -21 \\ \underline{36} \end{array}$

18. $\begin{array}{r} -47 \\ \underline{13} \end{array}$ **19.** $\begin{array}{r} -17 \\ \underline{-17} \end{array}$ **20.** $\begin{array}{r} 14\frac{3}{5} \\ \underline{17\frac{9}{10}} \end{array}$ **21.** $\begin{array}{r} -7\frac{3}{4} \\ \underline{18\frac{1}{8}} \end{array}$ **22.** $\begin{array}{r} -11\frac{3}{8} \\ \underline{-5\frac{5}{6}} \end{array}$

23. $\begin{array}{r} -41.62 \\ \underline{-10.58} \end{array}$ **24.** $\begin{array}{r} 16.4 \\ \underline{-5.83} \end{array}$

Perform the indicated operations in Exercises 25–36.

25. $-6 + (-4) - 5$

26. $-7 - (-2) + 6$

27. $6 + (-3) + (-4)$

28. $-3 + (-7) + 2$

29. $-5 - 2 - (-4)$

30. $-8 - 5 - (-3)$

31. $-2 - 2 + 11$

32. $-3 - (-3) + (-6)$

33. $-\frac{2}{3} + \frac{4}{5} - \left(-\frac{1}{2}\right)$

34. $\frac{4}{3} - \frac{1}{4} - \frac{5}{6}$

35. $9.37 - 16.42 - (8.21)$

36. $-11.63 + 5.83 - 7.29$

Fill in the blank in Exercises 37–50 with the proper symbol: <, >, or =.

37. $-6 + (-2)$ ______ $3 + (-8)$

38. $-4 - (-3)$ ______ $-4 + (-3)$

39. $5 - 8$ ______ $8 - 5$

40. $7 - (-3)$ ______ $-3 - 7$

41. $11 + (-3)$ ______ $11 - 3$

42. $0 - 6$ ______ $0 - (-6)$

43. $-8 - (-8)$ ______ $-14 - 13$

44. $-7 - (-3)$ ______ $4 - 9$

45. $0 - \left(-\frac{1}{8}\right)$ ______ $0 - \frac{1}{8}$

46. $-\frac{4}{5} - \left(-\frac{1}{2}\right)$ ______ $\frac{3}{5} - \frac{3}{10}$

47. $-\left(-\frac{7}{8}\right)$ ______ $\frac{1}{2} - \left(-\frac{5}{8}\right)$

48. $6.5 - 4.3$ ______ $9.9 - (-1.3)$

49. $-15.71 - 8.46$ ______ $-(10.07 + 14.1)$

50. $7.25 - 21.62$ ______ $-13.31 - 2.53$

Determine whether or not the number given in each of the Exercises 51–60 is a solution to the given equation by substituting and then evaluating.

51. $x + 5 = -3;\ x = -8$

52. $x - 6 = -9;\ x = -3$

53. $15 - y = 17;\ y = -2$

54. $11 - x = 8;\ x = 3$

55. $2x - 3 = -5;\ x = 4$

56. $3y - 2 = 10;\ y = 4$

57. $-9 - 3x = 6;\ x = 5$

58. $4x + 13 = 3;\ x = 2$

59. $5x - 2 = 3;\ x = 1$

60. $-18 - 4y = -30;\ y = 3$

Calculator Problems

Use a calculator to solve Exercises 61–63.

61. $14.685 - 22.753 + 8.33$

62. $-21.5832 + 15.614 - 9.591$

63. $27.681 - 14.117 - (-6.841)$

2.5 MULTIPLICATION AND DIVISION WITH SIGNED NUMBERS

Multiplication with whole numbers is shorthand for repeated addition. That is,

$$7 + 7 + 7 + 7 + 7 = 5 \cdot 7 = 35$$

and

$$\underbrace{7 + 7 + 7 + \cdots + 7}_{105\ 7\text{'s}} = 105 \cdot 7 = 735$$

Similarly, multiplication with signed numbers can be considered shorthand for repeated addition. For example,

$$(-6) + (-6) + (-6) = 3(-6) = -18$$

$$(-2) + (-2) + (-2) + (-2) + (-2) = 5(-2) = -10$$

Repeated addition with a negative number results in a product of a positive number and a negative number. Since the sum of negative numbers is negative, the product of a positive number and a negative number will be negative. In fact, the product of any positive number with a negative number will be negative.

EXAMPLES

1. $5(-3) = (-3) + (-3) + (-3) + (-3) + (-3) = -15$
2. $7(-10) = -70$
3. $42(-1) = -42$
4. $3.1(-5) = -15.5$
5. $+\frac{1}{4}\left(-\frac{1}{2}\right) = -\frac{1}{8}$

The product of two negative numbers does not relate to repeated addition. The following discussion is based on intuition, and no formal proof of the results will be given here. Notice the pattern of the results and see if you can supply the missing products.

$$4(-7) = -28 \qquad 0(-7) = 0$$

$$3(-7) = -21 \qquad -1(-7) = ?$$

$$2(-7) = -14 \qquad -2(-7) = ?$$

$$1(-7) = -7 \qquad -3(-7) = ?$$

Did you get the following answers (noting that each product is 7 more than the previous product)?

$$-1(-7) = +7$$

$$-2(-7) = +14$$

$$-3(-7) = +21$$

You were correct if you did. Although one example does not prove that a procedure is correct, our intuition is good this time. The product of two negative numbers is positive. Again, the rule can be extended to include any signed numbers.

EXAMPLES

6. $(-4)(-9) = +36$
7. $-7(-5) = +35$
8. $-2.1(-6) = +12.6$
9. $\left(-\frac{1}{3}\right)\left(-\frac{5}{8}\right) = +\frac{5}{24}$

What happens if a number is multiplied by 0? For example, $3(0) = 0 + 0 + 0 = 0$. In fact, multiplication by 0 always gives a product of 0.

EXAMPLES

10. $6 \cdot 0 = 0$
11. $(3.7) \cdot 0 = 0$
12. $0 \cdot 8\frac{1}{2} = 0$

The rules for multiplication can be summarized as follows:

Rules for Multiplying Signed Numbers

1. The product of two positive numbers is positive.
2. The product of two negative numbers is positive.
3. The product of a positive number and a negative number is negative.
4. The product of 0 with any number is 0.

The rules can be stated more abstractly:

If a and b are positive numbers,
1. $a \cdot b = ab$
2. $a(-b) = -ab$
3. $(-a)(-b) = ab$
4. $a \cdot 0 = 0$

PRACTICE QUIZ

Questions	Answers
Find the following products.	
1. $5(-3) =$	1. -15
2. $-6(-4) =$	2. 24
3. $-8(4) =$	3. -32
4. $-13(0) =$	4. 0
5. $-9.1(-2) =$	5. 18.2
6. $3(-20.6) =$	6. -61.8
7. $\left(+\frac{3}{4}\right)\left(-\frac{2}{15}\right) =$	7. $-\frac{1}{10}$

The rules for multiplication lead directly to the rules for division since division is defined in terms of multiplication.

Definition: For any numbers a and b where $b \neq 0$,

$$\frac{a}{b} = x \quad \text{means} \quad a = b \cdot x$$

If a is any number, then $\frac{a}{0}$ is **undefined** (see Section 1.1).

EXAMPLES

13. $\dfrac{36}{9} = 4$ because $36 = 9 \cdot 4$.

14. $\dfrac{-36}{9} = -4$ because $-36 = 9(-4)$.

15. $\dfrac{36}{-9} = -4$ because $36 = -9(-4)$.

16. $\dfrac{-36}{-9} = +4$ because $-36 = -9(4)$.

The rules for division can be stated as follows:

Rules for Dividing Signed Numbers

1. The quotient of two positive numbers is positive.
2. The quotient of two negative numbers is positive.
3. The quotient of a positive number and a negative number is negative.

The rules can be stated more abstractly:

If a and b are positive numbers,

1. $\dfrac{a}{b} = \dfrac{a}{b}$
2. $\dfrac{-a}{-b} = \dfrac{a}{b}$
3. $\dfrac{-a}{b} = -\dfrac{a}{b}$ and $\dfrac{a}{-b} = -\dfrac{a}{b}$

The quotient of two integers may not always be another integer. Just as with whole numbers, the quotient may be a fraction. We will discuss these ideas more thoroughly in Chapter 6.

PRACTICE QUIZ

Questions	**Answers**
Find the quotients.	
1. $\frac{-30}{10}$	1. -3
2. $\frac{40}{-10}$	2. -4
3. $\frac{-20}{-10}$	3. 2
4. $\frac{-7}{0}$	4. undefined
5. $\frac{0}{13}$	5. 0

Now that we have all the rules for addition, subtraction, multiplication, and division with signed numbers, we can discuss the solutions to equations that involve any or all of these operations.

EXAMPLES

Determine whether or not the given number is a solution to the given equation by substituting and then evaluating.

17. $7x = -21$; $x = -3$

Solution: $7(-3) = -21$, so -3 is a solution.

18. $-8y = 56$; $y = -7$

Solution: $-8(-7) = +56$, so -7 is a solution.

19. $\frac{y}{-4} = -10$; $y = -40$

Solution: $\frac{-40}{-4} = +10$ and $+10 \neq -10$, so -40 is **not** a solution.

20. $-5x + 7 = -3$; $x = 2$

Solution: $-5(2) + 7 = -10 + 7 = -3$, so 2 is a solution.

EXERCISES 2.5

Find the product in Exercises 1–20.

1. $4 \cdot (-3)$ **2.** $(-5) \cdot 6$ **3.** $(-8)(-7)$ **4.** $12 \cdot 4$

5. $19 \cdot 3$ **6.** $(-11)(-2)$ **7.** $(-14)(-4)$ **8.** $(-3.2)(7)$

9. $(5.4)(-6)$ **10.** $(-1.1)(-0.6)$ **11.** $(-1.3)(-2.1)$

12. $\left(\frac{1}{2}\right)\left(-\frac{1}{3}\right)$ **13.** $\left(-\frac{2}{5}\right)\left(\frac{3}{2}\right)$ **14.** $\left(-\frac{1}{8}\right)\left(-\frac{4}{3}\right)$

15. $\left(-\frac{7}{6}\right)\left(-\frac{9}{14}\right)$ **16.** $(-6)(-3)(-9)$ **17.** $-8 \cdot 4 \cdot 9$

18. $-3 \cdot 2 \cdot (-3)$ **19.** $(-7)(-16) \cdot 0$ **20.** $(-9) \cdot 11 \cdot 4$

Find the quotient in Exercises 21–35.

21. $\frac{-8}{-2}$ **22.** $\frac{-20}{10}$ **23.** $\frac{-30}{5}$ **24.** $\frac{-26}{-13}$

25. $\frac{39}{-13}$ **26.** $\frac{-51}{3}$ **27.** $\frac{-91}{-7}$ **28.** $\frac{0}{6}$

29. $\frac{0}{-7}$ **30.** $\frac{-3}{0}$ **31.** $\frac{16}{0}$ **32.** $\frac{-3.4}{2}$

33. $\frac{4.24}{-4}$ **34.** $\left(-\frac{2}{3}\right) \div \frac{3}{4}$ **35.** $\left(-\frac{1}{2}\right) \div \left(-\frac{7}{8}\right)$

Correctly complete the sentences in Exercises 36–45 with *positive, negative,* 0, or *undefined.*

36. If x is a positive number, then $x(-x)$ is a ____________ number.

37. If x is a negative number, then $x(-x)$ is a ____________ number.

38. If x is a negative number and y is a negative number, then xy is a ____________ number.

39. If x is a positive number and y is a negative number, then xy is a ____________ number.

40. If x and y are positive numbers, then $\frac{x}{y}$ is a ____________ number.

41. If x is a negative number and y is a natural number, then $\frac{x}{y}$ is a ____________ number, if such a number exists.

42. If x and y are negative numbers, then $\frac{x}{y}$ is a ____________ number.

43. If x is a number, then $x \cdot 0$ is ____________.

44. If x is a nonzero number, then $\frac{0}{x}$ is ____________.

45. If x is a number, then $\frac{x}{0}$ is ____________.

Determine whether the equations in Exercises 46–55 are true or false.

46. $(-4) \cdot (6) = 3 \cdot 8$

47. $(-7) \cdot (-9) = 3 \cdot 21$

48. $(-12)(6) = 9(-8)$

49. $(-6)(9) = (18)(-3)$

50. $6(-3) = (-14) + (-4)$

51. $7 + 8 = (-10) + (-5)$

52. $-7 + 0 = (-7) \cdot (0)$

53. $17 + (-3) = (16) + (-4)$

54. $-4(6 + 3) = (-24) + (-12)$

55. $14 + 6 = -2[(-7) + (-3)]$

Determine whether or not the number given in each of the Exercises 56–65 is a solution to the given equation by substituting and then evaluating.

56. $11x = -55;\ x = -5$

57. $-7x = 84;\ x = -12$

58. $\frac{x}{7} = -6;\ x = 42$

59. $\frac{y}{-6} = 12;\ x = -72$

60. $\frac{y}{10} = -9;\ y = -90$

61. $-9x = -72;\ x = -8$

62. $5x + 3 = 18;\ x = 3$

63. $-3x + 7 = -8;\ x = 5$

64. $4x - 3 = -23;\ x = -5$

65. $7x - 6 = -34;\ x = -4$

Calculator Problems

Use a calculator to solve Exercises 66–70.

66. $(2.73)(-0.241)(-1.8)$

67. $(-4.613)(-0.45)(-1.66)$

68. $(5.314)(-1.7)(24)$

69. $(-77.459) \div 29$

70. $(-62.234) \div (-37)$

2.6 WORD PROBLEMS

The word problems in this section involve simple applications of the sums and differences of integers. Subtraction is used, for example, to find the change in values between two readings on a thermometer or the change between two distances or two altitudes. To calculate the change, including direction (positive or negative), **subtract the beginning value from the end value.**

EXAMPLES

1. On a winter day, the temperature dropped from 35°F at noon to 6°F below zero (−6°F) at 7 P.M. What was the change in temperature?

Solution:

$$\underbrace{\text{end value}}_{-6^\circ} - \underbrace{\text{beginning value}}_{(35)^\circ} = -6^\circ + (-35^\circ) = -41^\circ$$

2. A jet pilot flew her plane from an altitude of 30,000 ft to an altitude of 12,000 ft. What was the change in altitude?

Solution:

$$\underbrace{\text{end value}}_{12{,}000} - \underbrace{\text{beginning value}}_{30{,}000} = -18{,}000 \text{ ft}$$

3. Sue weighed 130 lb when she started to diet. The first week she lost 7 lb, the second week she gained 2 lb, and the third week she lost 5 lb. What was her weight after 3 weeks of dieting?

Solution:

$$\begin{aligned} 130 + (-7) + (+2) + (-5) &= 123 + (+2) + (-5) \\ &= 125 + (-5) \\ &= 120 \text{ lb} \end{aligned}$$

EXERCISES 2.6

Write each problem as a sum or difference, then simplify.

1. 23 lb lost, 13 lb gained, 6 lb lost

2. $24 earned, $17 spent, $2 earned

3. 4° rise, 3° rise, 9° drop

4. $10 withdrawal, $25 deposit, $18 deposit, $9 withdrawal

5. $47 earned, $22 spent, $8 earned, $45 spent

6. $20 won, $42 lost, $58 won, $11 lost

7. 14° rise, 6° drop, 11° rise, 15° drop

8. $53 withdrawal, $8 withdrawal, $48 deposit, $17 withdrawal

9. $187 profit, $241 loss, $82 profit, $26 profit

10. Snap-O Mousetrap stock rose $3 Monday, rose $6 Tuesday, and dropped $7 Wednesday. What was the net change in price of the stock?

11. In the first quarter of a recent football game, Fumbles A. Lott carried the ball six times with the following results: a gain of 6 yd, a gain of 3 yd, a loss of 4 yd, a gain of 2 yd, no gain or loss, and a loss of 3 yd. What was his net yardage for the first quarter?

12. Beginning with 7° above zero, the temperature rose 4°, then dropped 2°, and then dropped 6°. Find the final temperature.

13. Starting at the third floor, an elevator went down 1 floor, up 3 floors, up 7 floors, and then down 4 floors. Find the final location of the elevator.

14. Bill lost 2 lb the first week of his diet, lost 6 lb the second week, gained 1 lb the third week, lost 4 lb the fourth week, and gained 3 lb the fifth week. What was the total loss or gain? If Bill weighed 223 lb at the time he began his diet, what was his weight after 5 weeks of dieting?

15. Jeff works Friday, Saturday, and Sunday in a restaurant. His salary is $10 a night plus tips. Friday night he received $5 in tips but spent $2 for food. Saturday he bought a new shirt for $8, spent $3 for food, but received $12 in tips. Sunday he spent $3 for food and received $9 in tips. How much money did he have left after three days of work?

16. What should be added to -5 to get a sum of 17?

17. What should be added to -10 to get a sum of -23?

18. What should be added to 39 to get a sum of 16?

19. What should be added to 24.6 to get a sum of 13.8?

20. What should be added to -37.3 to get a sum of -54.7?

21. What should be added to $15\frac{7}{8}$ to get a sum of 18?

22. What should be added to $-12\frac{3}{10}$ to get a sum of $15\frac{1}{2}$?

23. From the sum of -4 and -17, subtract the sum of -12 and 6.

24. From the sum of 11 and −13, subtract the sum of 19 and −8.

25. From −47, subtract the result of 32 decreased by −7.

26. From 53, subtract the result of 29 diminished by −3.

27. Mr. Adams received a bank statement indicating that he was overdrawn by $63. How much must he deposit to bring his balance to $157?

28. A campus sorority sold tickets to a pancake breakfast. The expenses totaled $87.50. If they realized a profit of $192.80, how much money did they receive from ticket sales?

29. At 2:00 P.M. the temperature was 77°F. At 8:00 P.M. the temperature was 58°F. What was the change in temperature?

30. The temperature at 5:30 A.M. was 37°F below zero; at noon, the temperature was 29°F above zero. What was the change in temperature?

31. Lotsa-Flavor Chewing Gum stock opened on Monday at $47 per share and closed Friday at $39 per share. Find the change in price of the stock.

32. The Greek mathematician and scientist Aristotle lived from 384 B.C. to 322 B.C. How long did he live?

33. If you travel from the top of Mt. Whitney, elevation 14,495 ft, to the floor of Death Valley, elevation 282 ft below sea level, what is the change in elevation?

34. A submarine, submerged 280 ft below the surface of the sea, fired a rocket that reached an altitude of 30,000 ft. What was the change in altitude?

35. A man who was born in 87 B.C. died in 35 B.C. How old was he when he died?

36. At the end of the first round of a golf tournament, Beth was 4 under par. At the end of the second round, she was 7 over par. How much over par did she shoot in the second round?

37. In a four-day golf tournament, Joe scored 2 over par the first day, 3 under par the second day, 3 under par the third day, and 1 under par the fourth day. How much over or under par was he for the tournament?

38. Widget Inc. stock closed Monday at $7\frac{3}{4}$. Tuesday it went up $\frac{5}{8}$ point, and Wednesday it went down $\frac{3}{8}$ point. What was the closing price?

39. On Monday, Alpha Corp. stock opened at $29\frac{3}{8}$ and closed at $31\frac{1}{4}$. What was the change in the price?

40. Claudia had a balance of $307.86 in her checking account. She wrote checks of $23.68, $42.50, and $17.43. She made a deposit of $35.80. What was the balance of her account?

CHAPTER 2 SUMMARY

The **graph** of a number is the point that corresponds to the number, and the number is called the **coordinate** of the point.

The set of numbers consisting of the whole numbers and their opposites is called the set of **integers.**

- Integers: $\ldots, -3, -2, -1, 0, 1, 2, 3, \ldots$
- Positive integers: $1, 2, 3, 4, 5, \ldots$
- Negative integers: $\ldots, -4, -3, -2, -1$

Zero is its own opposite and is neither positive nor negative.

On a horizontal number line, **smaller numbers are always to the left of larger numbers.**

- Rational numbers $\begin{cases} \text{numbers that can be written as } \frac{a}{b} \\ \text{where } a \text{ and } b \text{ are integers, } b \neq 0 \end{cases}$

Table of Symbols

$=$	is equal to	$\neq$	is not equal to
$<$	is less than	$>$	is greater than
$\leq$	is less than or equal to	$\geq$	is greater than or equal to

Definition: The **absolute value** of a number is its distance from 0. The absolute value of a number is never negative.

Rules for Adding Signed Numbers

1. To add two numbers with like signs, add their absolute values and use the common sign.
2. To add two numbers with unlike signs, subtract their absolute values (the smaller from the larger) and use the sign of the number with the larger absolute value.

Definition: The **opposite** of a number is called its **additive inverse.** The sum of a number and its additive inverse is 0. Symbolically, for any number x,

$$x + (-x) = 0$$

Definition: For any numbers a and b,

$$a - b = a + (-b)$$

Rules for Multiplying Signed Numbers

1. The product of two positive numbers is positive.
2. The product of two negative numbers is positive.
3. The product of a positive number and a negative number is negative.
4. The product of 0 with any number is 0.

Rules for Dividing Signed Numbers

1. The quotient of two positive numbers is positive.
2. The quotient of two negative numbers is positive.
3. The quotient of a positive number and a negative number is negative.

If a is any number, then $\frac{a}{0}$ is **undefined**.

Definition: For any numbers a and b where $b \neq 0$,

$$\frac{a}{b} = x \text{ means } a = b \cdot x$$

CHAPTER 2 REVIEW

Fill in the blanks in Exercises 1–6 with the proper symbol: <, >, or =.

1. -3 ______ -2 **2.** -6 ______ $-(6)$ **3.** 2 ______ -2

4. $-(-5)$ ______ 1 **5.** $-\frac{2}{3}$ ______ $-\frac{3}{4}$ **6.** 1.8 ______ -0.3

Graph each of the sets of numbers in Exercises 7–16 on the number line.

7. $\{-3, 2, 1, 5\}$ **8.** $\{-6, 0, 3, 1, -2\}$ **9.** $\{-4, 3, |-6|, 0, 1\}$

10. $\{|-3|, -3, 2, -1\}$ **11.** $\left\{-2, -\frac{2}{3}, 1, \left|-\frac{7}{3}\right|, 4\right\}$

12. $\{-2.1, -0.8, |-1.6|, 2.7\}$

13. all integers greater than -4

14. all negative integers greater than -8

15. all positive integers less than or equal to 6

16. all integers less than -8

List the numbers that satisfy the conditions stated in Exercises 17–22.

17. $|x| = 6$ **18.** $|x| = 10$ **19.** $|y| = \frac{3}{5}$

20. $|y| = 2.9$ **21.** $|x| = -3.1$ **22.** $|y| = 1.724$

Perform the indicated operations in Exercises 23–40.

23. $-13 + 12 + (-7)$ **24.** $-17 + (-3) - (-8)$

25. $-23 - (-8)$ **26.** $7.3 + (-5.5)$

27. $-\frac{4}{5} + \frac{2}{3}$ **28.** $-\frac{4}{3} + \frac{1}{4} - \frac{5}{6}$

29. $8.64 - 10.21 + 0.39$ **30.** $(-19)(6)$

31. $(-23)(-11)$ **32.** $(-8)(-5)4$ **33.** $\left(\frac{3}{8}\right)\left(-\frac{4}{5}\right)$

34. $(2.6)(-1.5)$ **35.** $-28 \div (-7)$ **36.** $11 \div 0$

37. $98 \div (-14)$ **38.** $0 \div \frac{5}{16}$ **39.** $-13.6 \div 2$

40. $-\frac{5}{7} \div \left(-\frac{3}{14}\right)$

Choose the response that correctly completes the sentences in Exercises 41–45.

41. If x is a positive integer and y is a negative integer, then $x + y$ is (never, sometimes, always) a positive integer.

42. If x is a positive integer and y is a negative integer, then $x \cdot y$ is (never, sometimes, always) a positive integer.

43. If x is an integer and y is an integer, then $x - y$ is (never, sometimes, always) equal to $x + (-y)$.

44. If x is an integer and y is an integer, then $x + y$ is (never, sometimes, always) equal to 0.

45. If x is an integer, y is an integer, and $y \neq 0$, then $x \div y$ is (never, sometimes, always) an integer.

Determine whether or not the number given in Exercises 46–55 is a solution to the given equation.

46. $2x - 3 = 9;\ x = 6$

47. $5x + 7 = -3;\ x = -2$

48. $8 - x = 13;\ x = -5$

49. $7x - 1 = 20;\ x = -3$

50. $\frac{y}{8} + 1 = -1;\ y = -16$

51. $\frac{y}{4} + 8 = 12;\ y = 16$

52. $4x - 5 = -2;\ y = \frac{3}{4}$

53. $3x + 4 = 2;\ x = -\frac{2}{3}$

54. $2x + 1.7 = -3.5;\ x = -2.6$

55. $-3x - 4.1 = 8;\ x = 1.3$

Solve the word problems in Exercises 56–60.

56. Stock in the Go-Fly-A-Kite Company opened Monday at $11 and rose $2; Tuesday it dropped $4; Wednesday it rose $3; and Thursday it dropped $2. What was the closing price on Thursday?

57. Ralph started the month of January with $117. The first week he earned $21 and spent $9. The second week he earned $13 and spent $45. The third week he earned $19 and spent $16. The fourth week he earned $18 and spent $30. What is his balance after 4 weeks?

58. From the sum of -12 and -9, subtract the product of -3 and 8.

59. Subtract $8\frac{5}{8}$ from the sum of $6\frac{1}{4}$ and $3\frac{9}{16}$.

60. Subtract the product of 4.07 and -3 from 5.28.

CHAPTER 2 TEST

1. Fill in the blanks with the proper symbol: <, >, or =.

a. -4 ______ -2 **b.** $-(-2)$ ______ 0

2. Graph the following set of numbers on the number line:

$$\left\{-2, -0.4, |-1|, \frac{7}{3}, 3.1\right\}.$$

3. Graph the following set of numbers on the number line: all integers less than or equal to 2.

4. List the numbers that satisfy the condition $|y| = 7$.

5. List the numbers that satisfy the condition $|x| = \frac{7}{8}$.

6. Add: $-15 + 45 + (-17)$.

7. Find the sum: $17.04 + (-10.36)$.

8. Find the sum: $-\frac{2}{3} + \frac{5}{8} + \frac{1}{6}$.

9. Find the difference: $-28 - (-47)$.

10. Perform the indicated operations: $-9 - (-6) + 14$.

11. Find the product: $(-27)(-6)$.

12. Divide: $64 \div (-16)$

13. Multiply: $(4.12)(-6.1)$.

14. Divide: $-\frac{15}{8} \div \frac{5}{6}$

15. If x is a positive number and y is a negative number, then $x - y$ is (never, sometimes, always) a negative number.

16. If x is a positive number and y is a negative number, then $y \div x$ is (never, sometimes, always) a negative number.

17. Is $x = 5$ a solution to $3x - 9 = 6$?

18. Is $y = -1.5$ a solution to $-2x + 5 = 8$?

19. From the sum of -12 and -18, subtract the product of -6 and -5.

20. This morning, Thuy had \$27.63. During the day she earned \$10.56, spent \$14.29, and spent \$8.43. How much money does she have tonight?

```
10  REM
20  REM This program will show you addition, subtraction,
30  REM multiplication and division given two numbers.
40  '*****************************************************
50  '*  SUM         : sum of given two numbers           *
60  '*  DIFFERENCE  : difference of given two numbers    *
70  '*  PRODUCT     : product of given two numbers       *
80  '*  QUOTIENT    : quotient of given two numbers      *
90  '*****************************************************
100 PRINT
110 INPUT "Type in two numbers :",NUM1,NUM2
120 LET SUM = NUM1 + NUM2
130 LET DIFFERENCE = NUM1 - NUM2
140 LET PRODUCT = NUM1 * NUM2
150 LET QUOTIENT = NUM1 / NUM2
160 PRINT "The sum is        ";SUM
170 PRINT "The difference is ";DIFFERENCE
180 PRINT "The product is    ";PRODUCT
190 PRINT "The quotient is   ";QUOTIENT
200 END
```

```
Type in two numbers : 5 , 10
The sum is          15
The difference is -5
The product is      50
The quotient is     .5
```

```
Type in two numbers : 12 , 3
The sum is          15
The difference is  9
The product is      36
The quotient is     4
```

```
Type in two numbers : 5 , 9
The sum is          14
The difference is -4
The product is      45
The quotient is     .555556
```

Algebra is generous, she often gives more than is asked of her.

JEAN le ROND d'ALEMBERT 1717?–83

DID YOU KNOW?

Traditionally, algebra has been defined to mean generalized arithmetic where letters represent numbers. For example, $3 + 3 + 3 + 3 = 4 \cdot 3$ is a special case of the more general algebraic statement that $x + x + x + x = 4 \cdot x$. The name *algebra* comes from an Arabic word, *al-jabr*, which means "to restore."

A famous Moslem ruler, Harun al-Rashid, the caliph made famous in *Tales of the Arabian Nights*, and his son Al-Mamun, brought to their Baghdad court many famous Moslem scholars. One of these scholars was the mathematician, Mohammed ibn-Musa al-Khowarizmi, who wrote a text (c. A.D. 800) entitled *ihm al-jabr w' al-muqabal*. The text included instructions for solving equations by adding terms to both sides of the equation, thus "restoring" equality. The abbreviated title of Al-Khowarizmi's text, *Al-jabr*, became our word for equation solving and operations on letters standing for numbers, **algebra.**

Al-Khowarizmi's algebra was brought to western Europe through Moorish Spain in a Latin translation done by Robert of Chester (c. A.D. 1140). Al-Khowarizmi's name may have sounded familiar to you. It eventually was translated as "algorithm," which came to mean any series of steps used to solve a problem. Thus we speak of the division algorithm used to divide one number by another. Al-Khowarizmi is known as the father of algebra, just as Euclid is known as the father of geometry.

One of the hallmarks of algebra is its use of specialized notation that enables complicated statements to be expressed using compact notation. Al-Khowarizmi's algebra used only a few symbols, with most statements written out in words. He did not even use symbols for numbers! **The development of algebraic symbols and notation occurred over the next 1000 years.**

3 Solving First-Degree Equations

3.1 SIMPLIFYING AND EVALUATING EXPRESSIONS

An expression that involves only multiplication and/or division with constants and/or variables is called a **term. A single constant or variable is also a term.** Examples of terms are

$$3x, \quad 16, \quad -7y, \quad a, \quad 14x^2, \quad \frac{-x}{y^3}$$

Expressions such as

$$3 + x, \quad 5y - 7x, \quad 3x + 4x, \quad \text{and} \quad 5x^2 - 3x^2 + 2x$$

are the algebraic sums of terms. **Like terms** (or **similar terms**) are terms that are constants or terms that contain the same variables that are of the same power.

Like Terms	**Unlike Terms**	
$-2x$ and $3x$	$5x$ and $4x^2$	(x is not of the same power in both terms.)
7 and -10		
$4a^2$ and $-9a^2$	-8 and $6y$	(The term -8 does not have the variable y in it.)
$5x^2y^3$ and $-2x^2y^3$		
$\frac{5}{8}x$ and $-\frac{2}{3}x$	$3c^2d$ and $7c^2$	(There is no d in the second term.)

The numerical part of a term is called the **coefficient** of the variable or variables in the term. For example,

a. in the term $8x$, 8 is the coefficient of x;

b. in the term $-5y^3z$, -5 is the coefficient of y^3z.

If no coefficient is written next to a variable, the coefficient is understood to be 1. Thus,

$$x = 1 \cdot x, \quad y^3 = 1 \cdot y^3, \quad \text{and} \quad -b = -1 \cdot b$$

By applying the distributive property (discussed in Section 1.1), we can **combine like terms.** The distributive property states that

$$a(b + c) = ab + ac$$

or

$$ab + ac = a(b + c)$$

or

$$ba + ca = (b + c)a$$

This last form is particularly useful when b and c are numerical coefficients. For example,

a. $3x + 5x = (3 + 5)x$ By the distributive property.
$= 8x$ Add the coefficients.

b. $3x^2 - 5x^2 = (3 - 5)x^2$ By the distributive property.
$= -2x^2$ Add the coefficients algebraically.

In (a) the like terms $3x$ and $5x$ are **combined** and the result is $8x$.

In (b) the like terms $3x^2$ and $-5x^2$ are combined and the result is $-2x^2$.

EXAMPLES

Combine like terms whenever possible.

1. $8x + 10x = (8 + 10)x$ By the distributive property.
$= 18x$

2. $6y - 2y = (6 - 2)y$
$= 4y$

3. $4(x - 7) + 5(x + 1) = 4x - 28 + 5x + 5$ Use the distributive property twice.
$= 4x + 5x - 28 + 5$
$= 9x - 23$ Combine like terms.

4. $2x^2 + 3a + x^2 - a = 2x^2 + x^2 + 3a - a$ **Note:** $+x^2 = +1x^2$, $-a = -1a$
$= (2 + 1)x^2 + (3 - 1)a$
$= 3x^2 + 2a$

5. $\dfrac{x + 3x}{2} + 5x = \dfrac{4x}{2} + 5x = \dfrac{4}{2} \cdot x + 5x$ A fraction bar is a symbol of inclusion, like parentheses. So combine like terms in the numerator first.
$= 2x + 5x$
$= 7x$

In most cases, if an expression is to be evaluated, like terms should be combined first and then the resulting expression evaluated. Before combining terms and evaluating expressions, we make special mention of one particular situation involving negative numbers. Is there a difference between $(-7)^2$ and -7^2, or are they the same?

In the expression $(-7)^2$, the base is -7, and -7 is to be squared:

$$(-7)^2 = (-7)(-7) = +49$$

However, for -7^2, the base is 7 and the rules for order of operations say to square 7 first:

$$-7^2 = -(7^2) = -49$$

or

$$-7^2 = -1 \cdot 7^2 = -1 \cdot 49 = -49$$

Suppose we want to evaluate $-x^2$ for $x = 3$ and for $x = -4$. We have

$$-x^2 = -3^2 = -9 \qquad \text{for } x = 3$$

and

$$-x^2 = -(-4)^2 = -16 \qquad \text{for } x = -4$$

EXAMPLES

Simplify each expression by combining like terms; then evaluate the resulting expression using the given values for the variables. Refer to the Rules for Order of Operations in Section 1.2.

6. $2x + 5 + 7x$; $x = -3$

Solution:

$$\begin{aligned} 2x + 5 + 7x &= 2x + 7x + 5 \\ &= 9x + 5 \end{aligned}$$

$$\begin{aligned} 9x + 5 &= 9(-3) + 5 \\ &= -27 + 5 \\ &= -22 \end{aligned}$$

7. $3ab - 4ab + 6a - a$; $a = 2$, $b = -1$

Solution:

$$3ab - 4ab + 6a - a = -ab + 5a$$

$$\begin{aligned} -ab + 5a &= -1(2)(-1) + 5(2) \\ &= 2 + 10 \\ &= 12 \end{aligned}$$

8. $\dfrac{5x + 3x}{4} + 2(x + 1);\ x = 5$

Solution:

$$\begin{aligned}\frac{5x + 3x}{4} + 2(x + 1) &= \frac{8x}{4} + 2x + 2\\ &= 2x + 2x + 2\\ &= 4x + 2\end{aligned}$$

$$\begin{aligned}4x + 2 &= 4 \cdot 5 + 2\\ &= 20 + 2\\ &= 22\end{aligned}$$

PRACTICE QUIZ

Questions	**Answers**
Simplify the following expressions by combining like terms.	
1. $-2x - 5x$	1. $-7x$
2. $12y + 6 - y + 10$	2. $11y + 16$
3. $5(x - 1) + 4x$	3. $9x - 5$
4. $2b^2 - a + b^2 + a$	4. $3b^2$
Simplify the expression, then evaluate the resulting expression if $x = 3$ and $y = -2$.	
5. $2(x + 3y) + 4(x - y)$	5. $6x + 2y;\ 14$

EXERCISES 3.1

In each of the following expressions, combine like terms.

1. $8x + 7x$

2. $3y + 8y$

3. $5x - 2x$

4. $7x + (-3x)$

5. $6y^2 - y^2$

6. $16z^2 - 5z^2$

7. $23x^2 - 11x^2$

8. $18x^2 + 7x^2$

9. $4x + 2 + 3x$

10. $3x - 1 + x$

11. $2x - 3y - x$

12. $x + y + x - 2y$

13. $-2x^2 + 5y + 6x^2 - 2y$

14. $4a + 2a - 3b - a$

15. $3(x + 1) - x$

16. $2(x - 4) + x + 1$

17. $5(x - y) + 2x - 3y$

18. $4x - 3y + 2(x + 2y)$

19. $3(2x + y) + 2(x - y)$

20. $4(x + 5y) + 3(2x - 7y)$

21. $2x + 3x^2 - 3x - x^2$

22. $2y^2 + 4y - y^2 - 3y$

23. $2(x^2 + 3x) + 4(-x^2 + 2x)$

24. $3x^2 + 2x - 5 - x^2 + x - 4$

25. $3x^2 + 4xy - 5xy + y^2$

26. $2x^2 - 5xy + 11xy + 3$

27. $\dfrac{x + 5x}{6} + x$

28. $2y - \dfrac{2y + 3y}{5}$

29. $y - \dfrac{2y + 4y}{3}$

30. $z - \dfrac{3z + 5z}{4}$

31. $\dfrac{-3x + 5x}{-2} + x$

32. $\dfrac{-4x - 2x}{3} - 4x$

33. $\dfrac{4x - 2x}{2} + 3(x + 2x)$

34. $5(2x + 2) - \dfrac{3(2x + x)}{9}$

35. $\dfrac{3(5x - x)}{6} - \dfrac{4(2x - x)}{4}$

Evaluate the expressions in Exercises 36–45 if $x = 1$, $y = -3$, and $z = 2$.

36. $3xy^2$

37. x^2yz^2

38. $x^2(y + z)$

39. $xy - (x - y)$

40. $y(3x^2 + z) - 4z$

41. $x^2 + 3x - 4$

42. $\dfrac{4y^2 - z}{10z + y}$

43. $\dfrac{z - 4y}{z^2 - y}$

44. $\dfrac{2y^2 + 3y - 4}{1 - z - z^2}$

45. $\dfrac{3x^2 - 4x + 1}{y^2 + 3y - 4}$

Simplify the expressions in Exercises 46–60 by combining like terms. Then evaluate the resulting expression using the given value.

46. $3x + 4 - x$; $x = 6$

47. $5x - 7 + 8x$; $x = 4$

48. $3(x - 1) + 2(x + 2)$; $x = -5$

49. $4(y + 3) + 5(y - 2)$; $y = -3$

50. $-2x + 5y + 6x - 2y$; $x = 1$, $y = -2$

51. $4x + 2x - 3y - x$; $x = 2$, $y = -1$

52. $4x - 3y + 2(x + 2y)$; $x = -3$, $y = 1$

53. $5(x + y) + 2x - 3y$; $x = -2$, $y = 2$

54. $3xy + 5x - xy + x;\ x = 2,\ y = -1$

55. $7y - 2xy + 4xy - y;\ x = 4,\ y = 3$

56. $\dfrac{3x + 5x}{-2} + x;\ x = 5$

57. $\dfrac{-4x - 2x}{3} + 7x;\ x = -6$

58. $\dfrac{2(4x - x)}{3} - \dfrac{3(6x - x)}{3};\ x = -4$

59. $\dfrac{8(5x + 2x)}{7} - \dfrac{2(3x + x)}{4};\ x = 3$

60. $\dfrac{6(5x - x)}{8} + \dfrac{5(x + 5x)}{3};\ x = 5$

3.2 SOLVING EQUATIONS OF THE FORM $x + b = c$

If an equation contains a variable, the value (or values) that gives a true statement when substituted for the variable is called the **solution** to the equation. In Chapters 1 and 2 you were asked to show that some particular value was or was not a solution to an equation. However, you were not asked **to solve the equation** or **to find a solution** to the equation yourself.

Definition: If a, b, and c are constants and $a \neq 0$, then a **first-degree equation** is an equation that can be written in the form

$$ax + b = c$$

Note: The variable x is understood to have an exponent of 1 ($x = x^1$), thus the term **first-degree.** Also, the variable x is arbitrary. Any other variable such as y or z will serve just as well as x.
All the following equations are first-degree equations:

$$2x + 3 = 7, \qquad 5y + 6 = -13,$$

$$-7z - 1 = 6, \qquad x + 5 = 3x - 5$$

To develop a technique for solving first-degree equations, we need the **addition property of equality.**

Addition Property of Equality

If the same algebraic expression is added to both sides of an equation, the new equation has the same solutions as the original equation. Symbolically, if A, B, and C are algebraic expressions, then the equations

$$A = B$$

and

$$A + C = B + C$$

have the same solutions.

The objective of solving first-degree equations is to get the variable by itself on one side of the equation and any constants on the other side. The following procedures will accomplish this.

Procedures for Solving an Equation That Simplifies to the Form $x + b = c$

1. Combine any like terms on each side of the equation.
2. Use the addition property of equality and add the opposite of one of the constant terms to both sides so that all constants are on one side.
3. Use the addition property of equality and add the opposite of one of the variable terms so that all variables are on the other side.
4. Check your answer by substituting it into the original equation.

EXAMPLES

Solve each of the following equations.

1. $x + 7 = 12$

Solution:

$x + 7 + (-7) = 12 + (-7)$ Add (-7), the opposite of $+7$, to both sides.

$x = 5$ Simplify.

Check:

$$x + 7 = 12$$

$$5 + 7 = 12 \quad \text{Substitute } x = 5.$$

$$12 = 12 \quad \text{True statement.}$$

2. $y - 8 = -2$

Solution:

$$y - 8 + 8 = -2 + 8$$ Add $(+8)$, the opposite of -8, to both sides.

$$y = 6$$ Simplify.

Check:

$$y - 8 = -2$$

$$6 - 8 = -2 \quad \text{Substitute } y = 6.$$

$$-2 = -2 \quad \text{True statement.}$$

3. $4x + 1 - x = -13 + 2x + 5$

Solution:

$$3x + 1 = 2x - 8$$ Combine like terms on each side.

$$3x + 1 - 1 = 2x - 8 - 1$$ Add (-1), the opposite of $+1$, to both sides.

$$3x = 2x - 9$$ Simplify.

$$3x + (-2x) = 2x - 9 + (-2x)$$ Add $(-2x)$, the opposite of $+2x$, to both sides.

$$x = -9$$ Simplify.

Check:

$$4x + 1 - x = -13 + 2x + 5$$

$$4(-9) + 1 - (-9) = -13 + 2(-9) + 5 \quad \text{Substitute } x = -9.$$

$$-36 + 1 + 9 = -13 - 18 + 5 \quad \text{Simplify.}$$

$$-26 = -26 \quad \text{True statement.}$$

4. $6z + 4 = 7z$

Solution:

$$6z + 4 - 6z = 7z - 6z$$ Add $(-6z)$, the opposite of $6z$, to both sides.

$$4 = z$$ Simplify.

Note: The variable can be on the right side as well as the left side.

Check:

$$6z + 4 = 7z$$

$$6 \cdot 4 + 4 = 7 \cdot 4 \quad \text{Substitute } z = 4.$$

$$24 + 4 = 28 \quad \text{Simplify.}$$

$$28 = 28 \quad \text{True statement.}$$

5. $x - \frac{2}{5} = \frac{3}{10}$

Solution:

$$x - \frac{2}{5} + \frac{2}{5} = \frac{3}{10} + \frac{2}{5}$$ Add $\left(\frac{2}{5}\right)$, the opposite of $-\frac{2}{5}$, to both sides.

$$x = \frac{3}{10} + \frac{4}{10}$$ Simplify. (The common denominator is 10.)

$$x = \frac{7}{10}$$ Simplify.

Check:

$$x - \frac{2}{5} = \frac{3}{10}$$

$$\frac{7}{10} - \frac{2}{5} = \frac{3}{10}$$ Substitute $x = \frac{7}{10}$.

$$\frac{7}{10} - \frac{4}{10} = \frac{3}{10}$$ Simplify.

$$\frac{3}{10} = \frac{3}{10}$$ True statement.

6. $7y - 4.3 = 6y + 1.88$

Solution:

$$7y - 4.3 - 6y = 6y + 1.88 - 6y$$ Add $(-6y)$, the opposite of $6y$, to both sides.

$$y - 4.3 = 1.88$$ Simplify.

$$y - 4.3 + 4.3 = 1.88 + 4.3$$ Add (4.3), the opposite of -4.3, to both sides.

$$y = 6.18$$ Simplify.

Check:

$7y - 4.3 = 6y + 1.88$	
$7(6.18) - 4.3 = 6(6.18) + 1.88$	Substitute $y = 6.18$.
$43.26 - 4.3 = 37.08 + 1.88$	Simplify.
$38.96 = 38.96$	True statement.

PRACTICE QUIZ

Questions	Answers
Solve the following equations.	
1. $x + 5 = -16$	1. $x = -21$
2. $5x - 2 = 4x - 2$	2. $x = 0$
3. $7y - 1.5 = 6y + 3.2$	3. $y = 4.7$

EXERCISES 3.2

Solve the equations in Exercises 1–44.

1. $x - 4 = 2$

2. $y - 5 = 1$

3. $x + 7 = 3$

4. $x + 14 = 23$

5. $y - 2 = -6$

6. $y + 5 = 5$

7. $y + 16 = -2$

8. $x + 3 = -7$

9. $x - 12 = 22$

10. $y - 9 = -4$

11. $x + \frac{1}{3} = \frac{5}{6}$

12. $x - \frac{2}{3} = \frac{7}{6}$

13. $y - \frac{2}{5} = \frac{1}{3}$

14. $y + \frac{3}{4} = \frac{1}{8}$

15. $x + 0.23 = 0.47$

16. $x + 3.6 = 2.4$

17. $x - 7.2 = 3.8$

18. $y - 14.6 = -16.3$

19. $2x + 5 = x - 3$

20. $5x = 4x - 1$

21. $3x + 4 = 4x$

22. $x - 9 = 2x - 3$

23. $4 + x = 2x - 8$

24. $6x - 2 = 5x + 8$

25. $9x - 2 = 8x - 11$

26. $7x + 1 = 6x - 5$

27. $8x + 6 = 7x + 6$

28. $13x - 4 = 12x + 21$

29. $4x = 3x + \frac{1}{4}$

30. $7x + \frac{2}{3} = 6x + \frac{1}{6}$

31. $3x + \frac{1}{5} = 2x - \frac{1}{10}$

32. $10x - \frac{1}{2} = 9x - \frac{9}{10}$

33. $2x - 6.5 = x - 2.3$

34. $5x + 3.7 = 4x + 1.8$

35. $9x + 0.86 = 8x + 1.7$

36. $12x - 0.63 = 11x - 2.51$

37. $3x + 5 - x = 2 + x$

38. $x - 8 + 4x = 7 + 4x$

39. $2 + 4x - 6 = 8 + 3x$

40. $2x + 9 - x = 4x + 2 - 2x$

41. $6x - 2 + 3x = 4x + 8 + 4x$

42. $-2x - 9 + 5x = 4x + 2$

43. $9x + 11 = -3 + 8x + 14$

44. $17 + x + 11x = 15x + 3 - 2x$

The profit, P, is equal to the revenue, R, minus the cost, C ($P = R - C$).

45. Find the revenue if the profit is \$684.50 and the cost is \$8329.00.

46. Find the cost if the profit is \$93.25 and the revenue is \$865.90.

The perimeter, P, of a triangle is equal to the sum of the sides, a, b, and c ($P = a + b + c$).

47. Two sides of a triangle measure 43 cm and 26 cm. Find the length of the third side if the perimeter is 98 cm.

48. One side of a triangle is $11\frac{1}{2}$ in. long. A second side is $8\frac{3}{4}$ in. long. If the perimeter is 24 in., find the length of the third side.

The selling price, S, of an item is equal to the sum of the cost, C, and the markup, M ($S = C + M$).

49. The selling price of an item that cost \$9.80 is \$13.50. Find the markup.

50. The selling price of a jacket is \$39.90. If the markup is \$16.70, find the cost.

Calculator Problems

Use a calculator to solve Exercises 51–53.

51. $y + 32.861 = -17.892$

52. $17.61x + 27.059 = 9.845 + 16.61x$

53. $14.38y - 8.65 + 9.73y = 17.437 + 23.11y$

3.3 SOLVING EQUATIONS OF THE FORM ax = c

You may recall that the **reciprocal** of $\frac{3}{4}$ is $\frac{4}{3}$ and that $\frac{3}{4} \cdot \frac{4}{3} = 1$. In general, if $a \neq 0$, the reciprocal of $\frac{a}{b}$ is $\frac{b}{a}$, and $\frac{a}{b} \cdot \frac{b}{a} = 1$. (See Section 1.3.) We can use this relationship, along with the associative property of multiplication, to help in solving first-degree equations. First, we will simplify some algebraic expressions.

EXAMPLES

Simplify each expression using your knowledge of reciprocals and the associative property of multiplication.

1. $\frac{1}{4}(4x)$

Solution:

$$\frac{1}{4}(4x) = \left(\frac{1}{4} \cdot 4\right)x \qquad \text{by the associative property of multiplication}$$

$$= 1 \cdot x \qquad \frac{1}{4} \cdot 4 = \frac{1}{4} \cdot \frac{4}{1} = 1$$

$$= x$$

2. $\frac{2}{3}\left(\frac{3}{2}x\right)$

Solution:

$$\frac{2}{3}\left(\frac{3}{2}x\right) = \left(\frac{2}{3} \cdot \frac{3}{2}\right)x \qquad \text{by the associative property of multiplication}$$

$$= 1 \cdot x \qquad \frac{2}{3} \cdot \frac{3}{2} = 1$$

$$= x$$

3. $-\frac{6}{5}\left(-\frac{5}{6}y\right)$

Solution:

$$-\frac{6}{5}\left(-\frac{5}{6}y\right) = \left(-\frac{6}{5}\right)\left(-\frac{5}{6}\right)y$$

$$= 1 \cdot y \qquad \left(-\frac{6}{5}\right)\left(-\frac{5}{6}\right) = 1$$

$$= y$$

4. $\frac{2}{5}\left(\frac{5z}{2}\right)$

Solution: Note that $\frac{5z}{2} = \frac{5}{2}z$.

$$\begin{aligned} \frac{2}{5}\left(\frac{5z}{2}\right) &= \frac{2}{5}\left(\frac{5}{2}z\right) \\ &= \left(\frac{2}{5} \cdot \frac{5}{2}\right)z \\ &= 1 \cdot z \\ &= z \end{aligned}$$

In Section 3.2, each time we solved an equation, the coefficient of the variable was 1 after we simplified and used the addition property of equality. However, this one property is not sufficient to help us solve equations such as

$$3x = -15 \quad \text{or} \quad -7x = 21 \quad \text{or} \quad \frac{2}{3}x = 8.6$$

We need the **multiplication property of equality.**

Multiplication Property of Equality

If both sides of an equation are multiplied by the same **nonzero** algebraic expression, the new equation has the same solutions as the original equation. Symbolically, if A, B, and C are algebraic expressions, then the equations

$$A = B$$

and

$$AC = BC \quad \text{where } C \neq 0$$

have the same solutions.

Remember that the objective is to get the variable by itself on one side of the equation. That is, we want the variable to have 1 as its coefficient. The following procedures will accomplish this.

Procedures for Solving an Equation That Simplifies to the Form $ax = c$

1. Combine any like terms on each side of the equation.
2. Use the multiplication property of equality and multiply both sides of the equation by the reciprocal of the coefficient of the variable.
3. Check your answer by substituting it into the original equation.

EXAMPLES

Solve each of the following equations.

5. $5x = 20$

Solution:

$\frac{1}{5} \cdot (5x) = \frac{1}{5} \cdot 20$ Multiply both sides by $\left(\frac{1}{5}\right)$, the reciprocal of 5.

$\left(\frac{1}{5} \cdot 5\right)x = \frac{1}{5} \cdot \frac{20}{1}$ By the associative property of multiplication.

$1x = 4$ $\frac{1}{5} \cdot 5 = 1$

$x = 4$

Check:

$5x = 20$

$5 \cdot 4 = 20$ Substitute $x = 4$.

$20 = 20$ True statement.

Multiplying by the reciprocal of the coefficient is the same as **dividing** by the coefficient itself. So, we can multiply both sides by $\frac{1}{5}$, as we did, or we can divide both sides by 5:

$5x = 20$

$\frac{5x}{5} = \frac{20}{5}$ Divide both sides by 5.

$x = 4$ Simplify.

6. $-7y = 21$

Solution:

$$\frac{-7y}{-7} = \frac{21}{-7}$$ Divide both sides by -7.

$$\frac{-7y}{-7} = \frac{21}{-7}$$ Simplify.

$$y = -3$$

Check:

$$-7y = 21$$

$$-7(-3) = 21$$ Substitute $y = -3$.

$$21 = 21$$ True statement.

7. $5x + 4x = 3 - 21$

Solution:

$$5x + 4x = 3 - 21$$

$$9x = -18$$ Combine like terms.

$$\frac{9x}{9} = \frac{-18}{9}$$ Divide both sides by 9.

$$x = -2$$ Simplify.

Check:

$$5x + 4x = 3 - 21$$

$$5(-2) + 4(-2) = 3 - 21$$ Substitute $x = -2$.

$$-10 - 8 = -18$$ Simplify.

$$-18 = -18$$ True statement.

8. $1.3x = 9.1$

Solution:

$$10(1.3x) = 10(9.1)$$ Multiply both sides by 10 so that there will be no decimals.

$$13x = 91$$ Simplify.

$$\frac{1}{13} \cdot 13x = \frac{1}{13} \cdot 91$$ Multiply both sides by $\frac{1}{13}$.

$$x = 7$$ Simplify.

Check:

$$1.3x = 9.1$$

$$1.3(7) = 9.1 \qquad \text{Substitute } x = 7.$$

$$9.1 = 9.1 \qquad \text{True statement.}$$

When decimal coefficients or constants are involved, you might want to use a calculator. We could simply divide both sides by 1.3 and use a calculator:

$$1.3x = 9.1$$

$$\frac{\cancel{1.3}x}{\cancel{1.3}} = \frac{9.1}{1.3} = 7.0 \qquad \text{Using a calculator or just pencil and paper.}$$

$$x = 7.0$$

9. $\dfrac{4x}{5} = \dfrac{3}{10}$ This could be written $\dfrac{4}{5}x = \dfrac{3}{10}$ since $\dfrac{4}{5}x$ is the same as $\dfrac{4x}{5}$.

Solution:

$$\frac{5}{4} \cdot \frac{4}{5}x = \frac{5}{4} \cdot \frac{3}{10} \qquad \text{Multiply both sides by } \frac{5}{4}.$$

$$1x = \frac{\overset{1}{\cancel{5}}}{4} \cdot \frac{3}{\underset{2}{\cancel{10}}} \qquad \text{Simplify.}$$

$$x = \frac{3}{8}$$

Check:

$$\frac{4}{5}x = \frac{3}{10}$$

$$\frac{4}{5} \cdot \frac{3}{8} = \frac{3}{10} \qquad \text{Substitute } x = \frac{3}{8}.$$

$$\frac{3}{10} = \frac{3}{10} \qquad \text{True statement.}$$

10. $8x = -2.4$

Solution:

$$\frac{8x}{8} = \frac{-2.4}{8} \qquad \text{Divide both sides by 8.}$$

$$x = -0.3 \qquad \text{Simplify.}$$

Check:

$$8x = -2.4$$

$$8(-0.3) = -2.4 \qquad \text{Substitute } x = -0.3.$$

$$-2.4 = -2.4 \qquad \text{True statement.}$$

PRACTICE QUIZ

Questions	**Answers**
Solve the following equations.	
1. $4x = -20$	1. $x = -5$
2. $\frac{3}{5}y = 33$	2. $y = 55$
3. $1.7z + 2.4z = 8.2$	3. $z = 2$
4. $3x = 7$	4. $x = \frac{7}{3}$
5. $6x = -1.8$	5. $x = -0.3$

EXERCISES 3.3

In Exercises 1–6, simplify each expression using your knowledge of reciprocals and the associative property of multiplication.

1. $\frac{5}{7}\left(\frac{7}{5}x\right)$ **2.** $\frac{4}{3}\left(\frac{3}{4}x\right)$ **3.** $-\frac{1}{5}(-5x)$

4. $6\left(\frac{1}{6}x\right)$ **5.** $-\frac{16}{7}\left(-\frac{7}{16}y\right)$ **6.** $-9\left(-\frac{1}{9}y\right)$

In Exercises 7–12, find the number that makes the equation true.

7. $(\quad)(8x) = x$ **8.** $(\quad)\left(\frac{7}{9}y\right) = y$ **9.** $(\quad)\left(\frac{1}{10}y\right) = y$

10. $(\quad)\left(-\frac{5}{8}x\right) = x$ **11.** $(\quad)\left(-\frac{9}{4}x\right) = x$ **12.** $(\quad)\left(-\frac{1}{7}y\right) = y$

Solve the equations in Exercises 13–54.

13. $3x = 12$ **14.** $2y = 10$ **15.** $4x = -28$

16. $-8x = 24$

17. $-9y = -54$

18. $-12 = 6x$

19. $20 - 65 = 2x + 3x$

20. $10x - 2x = 36 - 100$

21. $4x - 7x = 42$

22. $15 + 12 = x + 8x$

23. $\frac{1}{3}x = -5$

24. $-\frac{1}{5}x = 5$

25. $-\frac{1}{8}x = 2$

26. $\frac{3}{4}x = 15$

27. $\frac{2}{3}y = 12$

28. $\frac{2}{5}x = 4$

29. $-\frac{2}{7}y = 6$

30. $-\frac{5}{3}x = 10$

31. $-\frac{1}{8}x = 12 - 8$

32. $\frac{5}{6}y = 17 + 13$

33. $\frac{5}{3}x + \frac{2}{3}x = 21 + 7$

34. $\frac{5}{2}x + \frac{4}{2}x = 15 - 33$

35. $5x = 2$

36. $4x = -3$

37. $-2x = -9$

38. $-3x = 7$

39. $-8x = 12$

40. $7x = 15$

41. $\frac{2}{3}x = \frac{1}{2}$

42. $\frac{3}{4}x = \frac{9}{10}$

43. $-\frac{5}{3}x = \frac{5}{9}$

44. $\frac{1}{6}x = \frac{3}{4}$

45. $-\frac{5}{7}x = -\frac{10}{21}$

46. $\frac{3}{8}y = -\frac{9}{16}$

47. $3.2y = 6.4$

48. $0.2x = 1.6$

49. $1.2x - 2.8x = 4.8$

50. $6.4y - 2.0y = 1.2 - 10.0$

51. $9x - 5x = 2.3 - 5.9$

52. $1.5x - 6.5x = 4.6 - 30.1$

53. $\frac{1}{4}x = 2.5$

54. $\frac{1}{3}x = -3.6$

The distance traveled, d, is equal to the product of the rate, r, and the time, t ($d = rt$).

55. How long will it take a truck to travel 350 mi if it travels at an average rate of 50 miles per hour?

56. How long will it take a train traveling at 40 miles per hour to go 140 mi?

The interest, I, is the product of the money invested (principal), p, the rate, r, and the time, t ($I = prt$).

57. The interest on \$3000 invested at 9% is \$810. Find the time.

58. A savings account pays interest at a rate of 12%. How much must be invested to earn \$900 interest in 5 years?

The volume, v, of a pyramid with a rectangular base is equal to one-third the product of the length, l, the width, w, and the height, h $\left(v = \frac{1}{3} lwh\right)$.

59. Find the height of a pyramid with a rectangular base having a volume of 48 cu in., a length of 6 in., and a width of 4 in.

60. The volume of a pyramid with a rectangular base is 60 cu cm. The height is 9 cm and the length is 5 cm. Find the width.

Calculator Problems

Use a calculator to solve Exercises 61–63.

61. $2.637x = 648.702$

62. $-0.3057y = 316.7052$

63. $0.5178y = -257.8644$

3.4 SOLVING EQUATIONS OF THE FORM $ax + b = c$

Now we are ready to apply all the properties and techniques that we have used in Sections 3.1–3.3 to solving first-degree equations of the form $ax + b = c$.

To Solve a First-Degree Equation

1. Simplify each side of the equation by removing any grouping symbols and combining like terms.
2. Use the addition property of equality to add the opposites of constants and/or variables so that variables are on one side and constants on the other.
3. Use the multiplication property of equality to multiply both sides by the reciprocal of the coefficient of the variable (or divide both sides by the coefficient).
4. Check your answer by substituting it into the original equation.

Study each of the following examples carefully. Remember that **the objective is to get the variable on one side by itself.** Note that **the equations are written one under the other.** Do not write several equations on one line or set one equation equal to another equation.

EXAMPLES

Solve each of the following equations.

1. $7 - x = 12$

Solution:

$-7 + 7 - x = -7 + 12$	Add -7 to both sides.
$-x = 5$	Simplify.
$-1 \cdot x = 5$	$-x = -1 \cdot x$
$\dfrac{-1 \cdot x}{-1} = \dfrac{5}{-1}$	Divide both sides by -1. (We could have multiplied both sides by -1.)
$x = -5$	

Check:

$7 - x = 12$	
$7 - (-5) = 12$	Substitute $x = -5$.
$7 + 5 = 12$	
$12 = 12$	True statement.

The solution is -5.

2. $-5x - 1 = -11$

Solution:

$-5x - 1 + 1 = -11 + 1$	Add $+1$ to both sides.
$-5x = -10$	Simplify.
$-\dfrac{1}{5}(-5x) = -\dfrac{1}{5}(-10)$	Multiply both sides by $-\dfrac{1}{5}$.
$x = 2$	Simplify.

Check:

$-5x - 1 = -11$	
$-5 \cdot 2 - 1 = -11$	Substitute $x = 2$.
$-10 - 1 = -11$	
$-11 = -11$	True statement.

The solution is 2.

3. $2(y - 7) = 4(y + 1) - 26$

Solution:

$2y - 14 = 4y + 4 - 26$	Use the distributive property.
$2y - 14 = 4y - 22$	Combine like terms.
$2y - 14 + 22 = 4y - 22 + 22$	Add +22 to both sides. Here we will put the variables on the right side to get a positive coefficient of y.
$2y + 8 = 4y$	Simplify.
$-2y + 2y + 8 = -2y + 4y$	Add $-2y$ to both sides.
$8 = 2y$	Simplify. (Note that the coefficient of y is +2.)
$\frac{8}{2} = \frac{2y}{2}$	Divide both sides by 2.
$4 = y$	Simplify.

The number 4 does check, so the solution is 4.

4. $16.53 - 18.2z = 7.43$

Solution:

$100(16.53 - 18.2z) = 100(7.43)$	Multiply both sides by 100 to clear the decimals. This will give us integer constants and coefficients.
$100(16.53) - 100(18.2z) = 100(7.43)$	Use the distributive property.
$1653 - 1820z = 743$	Simplify.
$-1820z = 743 - 1653$	Add -1653 to both sides.
$-1820z = -910$	Simplify.
$\frac{-1820z}{-1820} = \frac{-910}{-1820}$	Divide both sides by -1820.
$z = 0.5$	
or $z = \frac{1}{2}$	

Check:

$$16.53 - 18.2z = 7.43$$

$$16.53 - 18.2(0.5) = 7.43 \quad \text{Substitute } z = 0.5.$$

$$16.53 - 9.1 = 7.43 \quad \text{Simplify.}$$

$$7.43 = 7.43 \quad \text{True statement.}$$

The number 0.5 checks and it is the solution.

5. $\frac{2z}{5} + 2 = 6$

Solution:

$$\frac{2z}{5} + 2 - 2 = 6 - 2 \quad \text{Add } -2 \text{ to both sides.}$$

$$\frac{2z}{5} = 4 \quad \text{Simplify.}$$

$$\frac{5}{2} \cdot \frac{2z}{5} = \frac{5}{2} \cdot 4 \quad \text{Multiply both sides by } \frac{5}{2}.$$

$$z = 10 \quad \text{Simplify.}$$

Check:

$$\frac{2z}{5} + 2 = 6$$

$$\frac{2 \cdot 10}{5} + 2 = 6 \quad \text{Substitute } z = 10.$$

$$4 + 2 = 6 \quad \text{Simplify.}$$

$$6 = 6 \quad \text{True statement.}$$

The number 10 checks, and it is the solution.

Checking can be quite time-consuming and need not be done for every problem. This is particularly important on exams. You should check only if you have time after the entire exam is completed.

PRACTICE QUIZ

Questions	Answers
Solve the following equations.	
1. $3x + 4 = -2$	1. $x = -2$

2. $x + 14 - 6x = 2x - 7$	2. $x = 3$
3. $\frac{2y}{3} - 4 = 8$	3. $y = 18$
4. $6.4z + 2.1 = 3.1z - 1.2$	4. $z = -1$

Example 4 illustrated how to multiply terms with decimals so that you get integer constants and coefficients. You may choose instead to use your calculator. Example 6 shows what you should write down while you have the calculator do the arithmetic calculations.

CALCULATOR EXAMPLE

6. Solve $16.53 - 18.2z = 7.43$.

Solution:

$$-18.2z = -9.10 \qquad \text{Add } -16.53 \text{ to both sides.}$$

$$\frac{-18.2z}{-18.2} = \frac{-9.10}{-18.2} \qquad \text{Divide both sides by } -18.2.$$

$$z = 0.5$$

EXERCISES 3.4

Solve the following equations.

1. $x + 8 = -3$ **2.** $x + 14 = 12$ **3.** $-6x = -12$

4. $5x = 20$ **5.** $3x - 12 = 18$ **6.** $5x - 7 = 18$

7. $7x - 3 = -17$ **8.** $3x + 7 = -2$ **9.** $5x + 6 = 26$

10. $6x - 5 = -11$ **11.** $-3x + 7 = 1$ **12.** $9 - 4x = -7$

13. $5x + 6 = 7$ **14.** $3x + 11 = 4$ **15.** $-2y + 4 = -1$

16. $4x + 3 = 8$ **17.** $3y + 1.6 = 7$ **18.** $7x - 3.4 = 11.3$

19. $9x + 4.7 = -3.4$ **20.** $-5x + 2.9 = 9.9$ **21.** $\frac{x}{3} + 1 = 7$

22. $\frac{x}{5} + 2 = -3$ **23.** $\frac{3x}{5} - 1 = 2$ **24.** $\frac{2x}{7} - 3 = 5$

25. $\frac{2x}{3} - 4 = 8$

26. $\frac{5x}{4} + 1 = 11$

27. $\frac{3}{4}x + 4 = 5$

28. $\frac{4}{5}x - 2 = 1$

29. $-\frac{3}{2}x - 3 = 2$

30. $-\frac{8}{3}x + 1 = 5$

31. $3x + 18 = 7x - 6$

32. $3x + 2 = x - 8$

33. $5x - 3 = 2x + 6$

34. $x - 7 = 5x + 9$

35. $4x + 3 - x = x - 9$

36. $5x + 13 = x - 8 - 3x$

37. $4x + 3 = 2x + 4$

38. $5y - 2 = 2y + 5$

39. $8 - 3x = 4x + 10$

40. $14 - 2x = 8x - 1$

41. $\frac{2x}{3} + 1 = x - 6$

42. $\frac{4x}{5} + 2 = 2x - 4$

43. $\frac{3x}{2} + 1 = x - 1$

44. $\frac{7x}{3} + 2 = x + 6$

45. $0.23x + 0.18 = 0.08x - 0.27$

46. $0.9x + 3 = 0.4x + 1.5$

47. $1.1x - 4 = 0.5x + 2$

48. $0.67x + 3 = 2.7 + 0.63x$

49. $2(x - 3) = 5x + 9$

50. $3(x - 1) = 4x + 6$

51. $3(2x - 3) = 4x + 5$

52. $5(3x - 7) = 6x + 1$

53. $4(3 - 2x) = 2(x - 4)$

54. $-3(x + 5) = 6(x + 2)$

55. $2(x + 3) = 4(x + 5) - 7$

56. $7(2x - 1) = 5(x + 6) - 13$

57. $8(3x + 5) - 9 = 9(x - 2) + 14$

The perimeter, P, of a rectangle is the sum of twice the length, l, plus twice the width, w $(P = 2l + 2w)$.

58. Eighty-four feet of fencing is needed to enclose a small garden plot. If the plot is 18 ft wide, find the length.

59. A rectangle is 36 cm long and has a perimeter of 85 cm. Find the width.

The area, A, of a trapezoid is the product of $\frac{1}{2}$ the height, h, times the sum of the lengths, a and b, of the two bases $\left[A = \frac{1}{2}h(a + b)\right]$.

60. The area of a trapezoid is 108 sq cm. The height is 12 cm, and the length of one base is 8 cm. Find the length of the other base.

61. The lengths of the bases of a trapezoid are 10 in. and 15 in. Find the height if the area is 225 sq in.

When purchasing an item on the installment plan, you find the total cost, C, by multiplying the monthly payment, p, by the number of months, t, and adding the product to the down payment, d ($C = pt + d$).

62. A refrigerator costs \$857.60 if purchased on the installment plan. If the monthly payments are \$42.50 and the down payment is \$92.60, how long will it take to pay for the refrigerator?

63. A used automobile will cost \$3250 if purchased on an installment plan. If the monthly payments will be \$115 for 24 months, what will be the down payment?

Calculator Problems

Use a calculator to solve Exercises 64–66. (Round off answers to the nearest hundredth.)

64. $0.17x - 23.014 = 1.35x + 36.234$

65. $48.512 - 1.63x = 2.58x + 87.635$

66. $327.93 + 22.62x = 17.76x + 463.52x$

3.5 WRITING ALGEBRAIC EXPRESSIONS

Algebra is a language of mathematicians, and in order to understand mathematics, you must understand the language. We want to be able to change English phrases into their "algebraic" equivalents, and vice versa. So, if a problem is stated in English, we can translate the phrases into algebraic symbols and proceed to solve the problem according to the rules developed for algebra.

The following examples illustrate how certain key words can be translated into algebraic symbols.

EXAMPLES

English Phrase	***Algebraic Expression***
1. 3 **multiplied by** the number represented by x the **product** of 3 and x 3 **times** x	$3x$
2. a number **added to** 3 the **sum** of z and 3 z **plus** 3	$z + 3$
3. two **times** the quantity found by **adding** a number to 1 **twice** the **sum** of x and 1 the **product** of 2 with the **sum** of x and 1	$2(x + 1)$

4. **twice** x **plus** 1 two **times** x **increased by** 1 one **more than** the **product** of 2 and a number	$2x + 1$
5. the **product** of two numbers x **times** y **multiply** x and y	xy
6. the **difference** between 5 **times** a number and 2 **times** the same number the **product** of 5 and a number **minus** the **product** of 2 and that number the **difference** between $5x$ and $2x$	$5x - 2x$

Certain words, such as those in boldface type in the previous examples, are the keys to the operations. Learn to look for these words and those from the following list.

Addition	***Subtraction***	***Multiplication***	***Division***
add	subtract	multiply	divide
sum	difference	product	quotient
plus	minus	times	
more than	less than	twice	
increased by	decreased by		

PRACTICE QUIZ

Questions	**Answers**
Change the following phrases to algebraic expressions.	
1. 7 less than a number	1. $x - 7$
2. twice the product of two unknown numbers	2. $2ab$
3. the quotient of y and 5	3. $\frac{y}{5}$
4. an unknown amount less than 10	4. $10 - x$
5. 14 more than 3 times a number	5. $3y + 14$

6. the product of 5 with the difference of 2 and x	6. $5(2 - x)$
7. four less than the product of 2 with x minus 3	7. $2(x - 3) - 4$
8. the sum of the product of 5 with a number and the product of 3 with that number	8. $5x + 3x$

Special mention should be made of the words **quotient** and **difference**. As illustrated in Problems 3 and 6 in the Practice Quiz, the division and subtraction are done with the values in the order they are given in the problem. For example, the difference between 3 and 5 is $3 - 5 = -2$, while the difference between 5 and 3 is $5 - 3 = 2$. Similarly, the quotient between y and 5 is $y \div 5$ or $\frac{y}{5}$.

EXERCISES 3.5

Write out in words what each of the expressions in Exercises 1–12 means.

1. $4x$

2. $x + 6$

3. $2x + 1$

4. $4x - 7$

5. $7x - 5.3$

6. $3.2(x + 2.5)$

7. $-2(x - 8)$

8. $10(x + 4)$

9. $5(2x + 3)$

10. $3(4x - 5)$

11. $6(x - 1) + \frac{2}{3}$

12. $9(x + 3) - \frac{8}{15}$

Write out each pair of expressions in Exercises 13–16 in words. Notice the difference.

13. $3x + 7$; $3(x + 7)$

14. $4x - 1$; $4(x - 1)$

15. $7x - 3$; $7(x - 3)$

16. $5(x + 6)$; $5x + 6$

Write the algebraic expression described by each of the word phrases in Exercises 17–35. Choose your own variable.

17. 6 added to a number

18. 7 more than a number

19. 4 less than a number

20. a number decreased by 13

21. 5 less than 3 times a number

22. the difference between twice a number and 10

23. the difference between x and 3, all divided by 7

24. 9 times the sum of a number and 2

25. 3 times the difference between a number and 8

26. 13 less than the product of 4 with the sum of a number and 1

27. 5 subtracted from three times a number

28. the sum of twice a number and four times the number

29. 8 minus twice a number all added to three times the number

30. the sum of a number and 9 more than the number

31. 4 more than the product of 8 with the difference between a number and 6

32. twenty decreased by the sum of twice a number and 4.8

33. the difference between three times a number and 9, all decreased by five times the number

34. the sum of a number and three times itself, all increased by 8

35. the sum of a number and 5, all multiplied by 4, then decreased by twice the number

Write the algebraic expression described by each of the word phrases in Exercises 36–45.

36. the cost of x pounds of candy at $4.95 a pound

37. the annual interest on x dollars if the rate is 11% per year

38. the number of days in t weeks and 3 days

39. the number of minutes in h hours and 20 minutes

40. the points scored by a football team on n touchdowns (6 points each) and 1 field goal (3 points)

41. the cost of renting a car for one day and driving m miles if the rate is $20 per day plus 20 cents per mile

42. the cost of purchasing a fishing rod and reel if the rod costs x dollars and the reel costs $8 more than twice the cost of the rod

43. a salesperson's weekly salary if he receives $250 as his base plus 9% of the weekly sales of x dollars

44. the selling price of an item that costs c dollars if the markup is 20% of the cost

45. the perimeter of a rectangle if the width is w centimeters and the length is 3 cm less than twice the width

3.6 WORD PROBLEMS

In Section 3.5 we discussed translating English phrases into algebraic expressions. The phrase "8 added to twice a number" translates algebraically to $2x + 8$. How do you translate "4 more than a number"? If you said $x + 4$, you are correct. Now, the object is to translate an entire sentence into an equation and then to solve the equation. The two phrases above might be involved in a sentence such as the following:

"If **8 is added to twice a number,** the result is **4 more than the number.**"

Algebraically,

$$2x + 8 = x + 4 \qquad \text{"The result is" translates as } =.$$

Solving, we have

$$\begin{aligned} 2x + 8 &= x + 4 \\ 2x + 8 - x &= x + 4 - x \\ x + 8 &= 4 \\ x + 8 - 8 &= 4 - 8 \\ x &= -4 \end{aligned}$$

In this section, the word problems will be simply exercises in translating sentences into equations and solving these equations. More sophisticated "application" problems will be discussed in later chapters. Such problems will involve geometric formulas, distance, interest, work, inequalities, and mixture.

EXAMPLES

1. Three times the sum of a number and 5 is equal to twice the number plus 5. Find the number.

Solution: Let x = the unknown number.

$$\underbrace{\text{3 times the sum of a number and 5}}_{3(x + 5)} \quad \underbrace{\text{is equal to}}_{=} \quad \underbrace{\text{twice the number plus 5}}_{2x + 5}$$

$$\begin{aligned} 3x + 15 &= 2x + 5 \\ 3x + 15 - 2x &= 2x + 5 - 2x \\ x + 15 &= 5 \\ x + 15 - 15 &= 5 - 15 \\ x &= -10 \end{aligned}$$

The number is -10.

2. If a number is decreased by 36 and the result is 76 less than twice the number, what is the number?

Solution: Let n = the unknown number.

$$\underbrace{n - 36}_{\text{a number decreased by 36}} \underbrace{=}_{\text{the result is}} \underbrace{2n - 76}_{\text{76 less than twice the number}}$$

$$\begin{aligned} n - 36 - n &= 2n - 76 - n \\ -36 &= n - 76 \\ -36 + 76 &= n - 76 + 76 \\ 40 &= n \end{aligned}$$

The number is 40.

3. Joe pays \$300 per month to rent an apartment. If this is $\frac{2}{5}$ of his monthly income, what is his monthly income?

Solution: Let x = Joe's monthly income.

$$\underbrace{\frac{2}{5}x}_{\frac{2}{5}\text{ of monthly income}} \underbrace{=}_{\text{is}} \underbrace{300}_{\$300}$$

$$\begin{aligned} \frac{5}{2} \cdot \frac{2}{5}x &= \frac{5}{2} \cdot \frac{300}{1} \\ x &= 750 \end{aligned}$$

Joe's monthly income is \$750.

4. A textbook is on sale for 25% off the original price. If you pay \$22.50 for the book, what was the original price?

Solution: Let p = the original price.

$$\underbrace{p}_{\text{original price}} - \underbrace{0.25p}_{\text{discount}} \underbrace{=}_{\text{is}} \underbrace{22.50}_{\text{what you pay}}$$

$$\begin{aligned} 100(p - 0.25p) &= 100(22.50) \quad \text{Multiply both sides by 100.} \\ 100p - 25p &= 2250 \\ 75p &= 2250 \\ \frac{75p}{75} &= \frac{2250}{75} \\ p &= 30 \end{aligned}$$

The original price was \$30.00.

Even integers are integers that are divisible by 2. The even integers are

$$E = \{\ldots, -6, -4, -2, 0, 2, 4, 6, \ldots\}$$

Odd integers are integers that are not even. The odd integers are

$$0 = \{\ldots, -5, -3, -1, 1, 3, 5, \ldots\}$$

Consecutive integers are two integers that differ by 1. That is, the second integer is 1 more than the first integer. For example, 21 and 22 are consecutive integers. -14 and -13 are consecutive integers. In general, if n is one integer, then $n + 1$ is the next consecutive integer.

An example of three consecutive integers is 51, 52, 53. Another example is -9, -8, -7. If n is one integer, then $n + 1$ is the next consecutive integer and $n + 2$ is the third consecutive integer.

Consecutive even integers are even integers that differ by 2; that is, the second integer is 2 more than the first. For example, 36 and 38 are two consecutive even integers. Also, -12, -10 and -8 are three consecutive even integers. If n is an even integer, then $n + 2$ is the next consecutive even integer and $n + 4$ is the third consecutive even integer.

Consecutive odd integers are odd integers that differ by 2; again, the second integer is 2 more than the first. For example, -15 and -13 are two consecutive odd integers. Also, 17, 19, and 21 are three consecutive odd integers. If n is an odd integer, then $n + 2$ is the next consecutive odd integer and $n + 4$ is the third consecutive odd integer.

EXAMPLES

Consecutive Integers

5. Find three consecutive integers such that the sum of the first and third is 76 less than three times the second.

Solution: Let

$$\begin{aligned}
n &= \text{the first integer} \\
n + 1 &= \text{the second integer} \\
n + 2 &= \text{the third integer} \\
n + (n + 2) &= 3(n + 1) - 76 \\
2n + 2 &= 3n + 3 - 76 \\
2n + 2 &= 3n - 73 \\
2n + 2 \mathbf{+ 73 - 2n} &= 3n - 73 \mathbf{+ 73 - 2n} \\
75 &= n \\
76 &= n + 1 \\
77 &= n + 2
\end{aligned}$$

The three consecutive integers are 75, 76, and 77.

6. Three consecutive odd integers are such that their sum is -3. What are the integers?

Solution: Let

$$n = \text{the first odd integer}$$
$$n + 2 = \text{the second odd integer}$$
$$n + 4 = \text{the third odd integer}$$
$$n + (n + 2) + (n + 4) = -3$$
$$3n + 6 = -3$$
$$3n = -9$$
$$n = -3$$
$$n + 2 = -1$$
$$n + 4 = +1$$

The three consecutive odd integers are -3, -1, and $+1$.

EXERCISES 3.6

Write an equation for each problem and then solve it.

1. If 7 is added to a number, the result is 43. Find the number.

2. A number decreased by 10 is 23. What is the number?

3. A number subtracted from 12 is equal to twice the number. Find the number.

4. Five less than a number is equal to 13 decreased by the number. Find the number.

5. Three less than twice a number is equal to the number. What is the number?

6. Thirty-six is 4 more than twice a certain number. Find the number.

7. Fifteen decreased by twice a number is 27. Find the number.

8. Seven times a certain number is equal to the sum of twice the number and 35. What is the number?

9. The difference between twice a number and 5 is equal to the number increased by 8. Find the number.

10. Fourteen more than three times a number is equal to 6 decreased by the number. Find the number.

11. Two added to the quotient of a number and 7 is equal to negative three. What is the number?

12. The quotient of twice a number and 5 is equal to the number increased by 6. What is the number?

13. Three times the sum of a number and 4 is equal to -9. Find the number.

14. Four times the difference between a number and 5 is equal to the number increased by 4. Find the number.

15. When 17 is added to six times a number, the result is equal to 1 plus twice the number. What is the number?

16. If the sum of twice a number and 5 is divided by 11, the result is equal to the difference between 4 and the number. Find the number.

17. Twice a number increased by three times the number is equal to 4 times the sum of the number and 3. Find the number.

18. Find two consecutive integers whose sum is 53.

19. Find three consecutive integers whose sum is 69.

20. The sum of three consecutive integers is 207. What are the three integers?

21. The sum of two consecutive odd integers is 60. What are the integers?

22. Find two consecutive integers such that twice the first plus three times the second equals 83.

23. The sum of three consecutive even integers is -156. Find the integers.

24. Find three consecutive even integers such that the first plus twice the second is 54 less than four times the third.

25. One integer is 5 more than another. The sum of the two integers is 171. Find the integers.

26. One integer is three times another. If their difference is 72, find the two integers.

27. One integer is 3 more than twice a second integer. Their sum is 114. Find the two integers.

28. The sum of two integers is 57. One of the integers is 7 less than three times the other integer. Find the two integers.

29. The sum of three integers is 39. The second integer is 5 less than the first, and the third integer is twice the first. What are the three integers?

30. The sum of three integers is 49. The first integer is 7 more than twice the second integer, and the third integer is 6 more than the second integer. Find the three integers.

31. The width of a room is $\frac{2}{3}$ of the length. If the room is 18 ft wide, find the length.

32. The length of a field is 40 yards greater than twice the width. If the field is 400 yards long, find the width.

33. Last week, Ralph earned \$37 more than he did this week. If he earned \$248 this week, how much did he earn last week?

34. Susan made a score of 86 on her math test. If this is 7 points less than her score on the last test, what was her score on the last test?

35. Ann pays a monthly rent of \$260. This represents $\frac{2}{7}$ of her monthly income. What is her monthly income?

36. Bob paid \$6770 for a new car. This is $\frac{1}{3}$ of his yearly salary. Find his yearly salary.

37. A pair of shoes is on sale for 15% off the original price. If you pay \$51 for the shoes, what was the original price?

38. Two-ninths of the people who came for tryouts for the football team will not make the team. If 42 people make the team, how many came for tryouts?

39. A girl is 22 years younger than her mother. The sum of their ages is 44. Find the age of each.

40. A man is 5 years older than three times his son's age. If the total of their ages is 53 years, find the age of each.

41. Matt bought a fish for his aquarium. The total cost was the marked price plus sales tax of 6% of the marked price. If the total cost was \$9.01, find the marked price.

42. A furniture store sells a desk for \$350. The price is determined by adding the cost and the markup. If the markup is 25% of the cost, find the cost.

43. A math student bought a calculator and textbook for a total cost of \$37.95. If the book cost \$6.75 more than the calculator, find the cost of each.

44. A woman bought a skirt and blouse for \$67.40. The skirt cost \$12.50 more than the blouse. Find the cost of each.

45. Lucy bought two boxes of golf balls. She gave the salesperson \$40 and received \$10.50 in change. What was the cost of the golf balls per box?

3.7 SOLVING FOR ANY TERM IN A FORMULA

Formulas are general rules or principles stated mathematically. In Section 1.8, we discussed several formulas related to geometric figures. There are many formulas in fields of study such as business, economics, medicine, physics, and chemistry as well as mathematics.

Some of these formulas and their meanings are shown here.

Formula	***Meaning***
1. $I = prt$	The simple interest (I) earned by investing money is equal to the product of the principal (p) times the rate of interest (r) times the time (t) in years.
2. $C = \frac{5}{9}(F - 32)$	Temperature in degrees Celsius (C) equals $\frac{5}{9}$ the difference between the Fahrenheit temperature (F) and 32.
3. $d = rt$	The distance traveled (d) equals the product of the rate of speed (r) and the time (t).
4. $P = 2l + 2w$	The perimeter (P) of a rectangle is equal to twice the length (l) plus twice the width (w).
5. $L = 2\pi rh$	The lateral surface area of a cylinder (L) is equal to 2π times the radius (r) of the base and the height (h).
6. $\text{IQ} = \frac{100M}{C}$	Intelligence quotient (IQ) is calculated by multiplying 100 times mental age (M) as measured by some test and dividing by chronological age (C).
7. $\alpha + \beta + \gamma = 180$	The sum of the angles (α, β, and γ) of a triangle is 180°. **Note:** α, β, and γ are the Greek letters alpha, beta, and gamma, respectively.

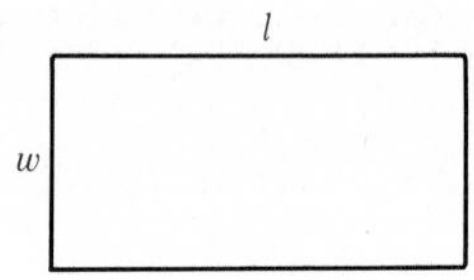

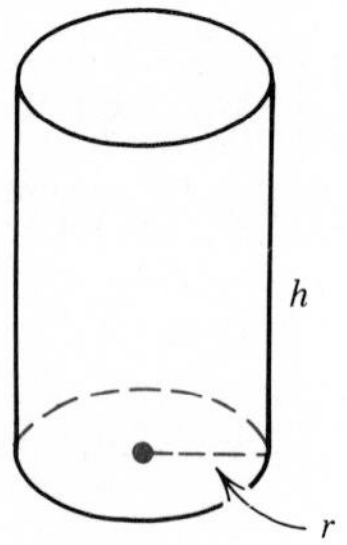

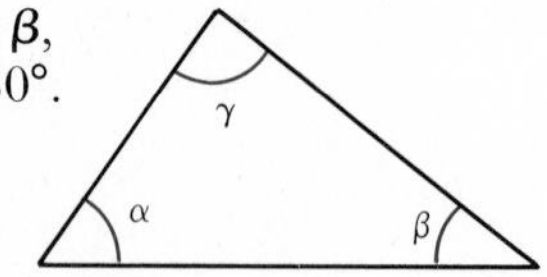

If you know values for all but one variable in a formula, you can substitute those values and find the value of the unknown variable by using the techniques for solving equations discussed in this chapter.

EXAMPLE

1. Given the formula $C = \frac{5}{9}(F - 32)$, find (a) C if $F = 212°$ and (b) F if $C = 20°$.

Solutions:

a. $F = 212°$, so substitute 212 for F in the formula:

$$C = \frac{5}{9}(212 - 32) = \frac{5}{9}(180) = 100$$

That is, 212°F is the same as 100°C. Water will boil at 212°F at sea level. This means that, if the temperature is measured in degrees Celsius instead of degrees Fahrenheit, then water will boil at 100°C at sea level.

b. $C = 20°$, so substitute 20 for C in the formula:

$$20 = \frac{5}{9}(F - 32) \qquad \text{Now solve for } F.$$

$$\frac{9}{5} \cdot 20 = \frac{9}{5} \cdot \frac{5}{9}(F - 32) \qquad \text{Multiply both sides by } \frac{9}{5}.$$

$$36 = F - 32 \qquad \text{Simplify.}$$

$$68 = F \qquad \text{Add 32 to both sides.}$$

That is, a temperature of 20°C is the same as a comfortable spring day temperature of 68°F.

We say that the formula $d = rt$ is "solved for" d in terms of r and t. Similarly, the formula $A = \frac{1}{2}bh$ is solved for A in terms of b and h, and the formula $P = S - C$ (profit is equal to selling price minus cost) is solved for P in terms of S and C. Many times we want to use a certain formula in another form. We want the formula "solved for" some variable other than the one given in terms of the remaining variables. Study the following examples carefully.

EXAMPLES

2. Given $d = rt$, solve for t in terms of d and r. We want to represent the time in terms of distance and rate. We will use this concept later in word problems.

Solution:

$$d = rt$$ Treat r and d as if they were constants.

$$\frac{d}{r} = \frac{rt}{r}$$ Divide both sides by r.

$$\frac{d}{r} = t$$ Simplify.

3. Given $P = a + b + c$, solve for a in terms of P, b, and c. This would be a convenient form for the case in which we know the perimeter and two sides of a triangle and want to find the third side.

Solution:

$$P = a + b + c$$ Treat P, b, and c as if they were constants.

$$P - b - c = a + b + c - b - c$$ Add $-b - c$ to both sides.

$$P - b - c = a$$ Simplify.

4. Given $C = \frac{5}{9}(F - 32)$ as in Example 1(b), solve for F in terms of C. This would give us a formula for finding Fahrenheit temperature given a Celsius temperature value.

Solution:

$$C = \frac{5}{9}(F - 32)$$ Treat C as a constant.

$$\frac{9}{5} \cdot C = \frac{9}{5} \cdot \frac{5}{9}(F - 32)$$ Multiply both sides by $\frac{9}{5}$, as in Example 1(b).

$$\frac{9}{5}C = F - 32$$ Simplify.

$$\frac{9}{5}C + 32 = F$$ Add 32 to both sides.

Thus,

$$F = \frac{9}{5}C + 32 \text{ is solved for } F$$

$$C = \frac{5}{9}(F - 32) \text{ is solved for } C$$

These are two forms of the same formula.

5. Suppose you are given $2x + 4y = 10$. (a) Solve first for x in terms of y. (b) Then solve for y in terms of x. This equation is typical of the algebraic equations that we will discuss in Chapter 8.

Solutions:

a. Solving for x yields

$$2x + 4y = 10$$

$$2x + 4y - 4y = 10 - 4y$$ Subtract $4y$ from both sides. (This is the same as adding $-4y$.)

$$2x = 10 - 4y$$ Simplify.

$$\frac{2x}{2} = \frac{10 - 4y}{2}$$ Divide both sides by 2.

$$x = \frac{10}{2} - \frac{4y}{2}$$ Simplify.

$$x = 5 - 2y$$

b. Solving for y yields

$$2x + 4y = 10$$

$$2x + 4y - 2x = 10 - 2x$$ Subtract $2x$ from both sides.

$$4y = 10 - 2x$$ Simplify.

$$\frac{4y}{4} = \frac{10 - 2x}{4}$$ Divide both sides by 4.

$$y = \frac{10}{4} - \frac{2x}{4}$$ Simplify.

$$y = \frac{5}{2} - \frac{1}{2}x$$

or we can write

$$y = \frac{5 - x}{2}$$ Both forms are correct.

6. Given $3x - y = 15$, solve (a) for x in terms of y and (b) for y in terms of x.

Solutions:

a. Solving for x, we have

$$3x - y = 15$$

$$3x = 15 + y \qquad \text{Add } y \text{ to both sides.}$$

$$\frac{1}{3}(3x) = \frac{1}{3}(15 + y) \qquad \text{Multiply both sides by } \frac{1}{3} \text{ (or divide both sides by 3).}$$

$$x = \frac{1}{3} \cdot 15 + \frac{1}{3} \cdot y \qquad \text{Simplify using the distributive property.}$$

$$x = 5 + \frac{1}{3}y$$

b. Solving for y, we have

$$3x - y = 15$$

$$-y = 15 - 3x \qquad \text{Subtract } 3x \text{ from both sides.}$$

$$-1(-y) = -1(15 - 3x) \qquad \text{Multiply both sides by } -1 \text{ (or divide both sides by } -1\text{).}$$

$$y = -15 + 3x \qquad \text{Simplify using the distributive property.}$$

EXERCISES 3.7

Solve for the indicated variable.

1. $P = a + b + c$; solve for b.

2. $p = 3s$; solve for s.

3. $f = ma$; solve for m.

4. $C = \pi d$; solve for d.

5. $A = lw$; solve for w.

6. $P = R - C$; solve for C.

7. $R = n \cdot p$; solve for n.

8. $v = k + gt$; solve for k.

9. $I = A - p$; solve for p.

10. $L = 2\pi rh$; solve for h.

11. $A = \dfrac{m + n}{2}$; solve for m.

12. $W = RI^2t$; solve for R.

13. $p = 4s$; solve for s.

14. $C = 2\pi r$; solve for r.

15. $d = rt$; solve for t.

16. $p = a + 2b$; solve for a.

17. $I = prt$; solve for t.

18. $R = \dfrac{E}{I}$; solve for E.

19. $p = a + 2b$; solve for b.

20. $c^2 = a^2 + b^2$; solve for b^2.

21. $\alpha + \beta + \gamma = 180$; solve for β.

22. $A = \frac{h}{2}(a + b)$; solve for h.

23. $y = mx + b$; solve for x.

24. $V = lwh$; solve for h.

25. $A = 4\pi r^2$; solve for r^2.

26. $V = \pi r^2 h$; solve for h.

27. $\text{IQ} = \frac{100M}{C}$; solve for M.

28. $A = \frac{R}{2L}$; solve for R.

29. $V = \frac{1}{3}\pi r^2 h$; solve for h.

30. $A = \frac{1}{2}bh$; solve for b.

31. $R = \frac{E}{I}$; solve for I.

32. $\text{IQ} = \frac{100M}{C}$; solve for C.

33. $A = \frac{R}{2L}$; solve for L.

34. $K = \frac{mv^2}{2g}$; solve for g.

35. $v = k + gt$; solve for t.

36. $L = 2\pi rh$; solve for h.

37. $S = 2\pi rh + 2\pi r^2$; solve for h.

38. $A = \frac{h}{2}(a + b)$; solve for a.

39. $S = \frac{a}{1 - r}$; solve for r.

40. $K = \frac{mv^2}{2g}$; solve for v^2.

41. $F = \frac{9}{5}C + 32$; solve for C.

42. $a = p + prt$; solve for t.

43. $K = \frac{mv^2}{2g}$; solve for m.

44. $3x + y = 7$; solve for y.

45. $x - y = 3$; solve for y.

46. $x + y = 4$; solve for y.

47. $2x + y = 8$; solve for y.

48. $3x - y = 14$; solve for y.

49. $x + 2y = 5$; solve for x.

50. $-x + 5y = 6$; solve for x.

51. $4x - 3y = 9$; solve for y.

52. $2x - 5y + 8 = 0$; solve for x.

53. $3x + 8y - 4 = 0$; solve for x.

54. $6x - y = -3$; solve for y.

55. $1.2x + 1.5y = 3$; solve for y.

56. $2x + 1.7y = 5.1$; solve for y.

3.8 ADDITIONAL APPLICATIONS (OPTIONAL)

We have discussed solving equations and working with formulas throughout Chapter 3. In this section, we will present more formulas from a variety of real-life situations, with complete descriptions of the meanings of these formulas. The objectives are simply to give you more practice with formulas and illustrate the wide use of formulas. The given information is to be substituted into the appropriate formula and the resulting equation solved for the unknown variable.

EXAMPLE

1. The lifting force, F, exerted on an airplane wing is found by multiplying some constant, k, times the area, A, of the wing's surface and times the square of the plane's velocity, v. The formula is $F = kAv^2$. Find the force on a plane's wing of area 120 sq ft if k is $\frac{4}{3}$ and the plane is traveling 80 miles per hour.

Solution: We know that $k = \frac{4}{3}$, $A = 120$, and $v = 80$. Substitution gives

$$F = \frac{4}{3} \cdot 120 \cdot 80^2$$

$$= \frac{4}{\cancel{3}} \cdot \overset{40}{\cancel{120}} \cdot 6400$$

$$= 160 \cdot 6400$$

$$F = 1{,}024{,}000 \text{ lb}$$

(The force is measured in pounds.)

EXERCISES 3.8

In the following problems, read the descriptive information carefully and then substitute the values given in the problem for the corresponding variables in the formulas. Evaluate the resulting expression for the unknown variable.

If an object is shot upward with an initial velocity v_0 feet per second, the velocity, v, in feet per second is given by the formula $v = v_0 - 32t$, where t is time in seconds.

1. Find the velocity at the end of 3 seconds if the initial velocity is 144 ft per second.

2. Find the initial velocity of an object if the velocity after 4 seconds is 48 ft per second.

3. An object projected upward with an initial velocity of 106 ft per second has a velocity of 42 ft per second. How many seconds have passed?

In nursing, one procedure for determining the dosage for a child is

$$\text{child's dosage} = \frac{\text{age of child in years}}{\text{age of child} + 12} \cdot \text{adult dosage}$$

4. If the adult dosage of a drug is 20 milliliters, how much should a 3-year-old child receive?

5. If the adult dosage of a drug is 340 milligrams, how much should a 5-year-old child receive?

The amount of money due from investing P dollars is given by the formula $A = P + Prt$, where r is the rate expressed as a decimal and t is the time in years.

6. Find the amount due if \$1000 is invested at 6% for 2 years.

7. How long will it take an investment of \$600, at a rate of 5%, to be worth \$750?

The number, N, of rafters in a roof or studs in a wall can be found by the formula $N = \frac{l}{d} + 1$, where l is the length of the roof or wall and d is the center-to-center distance from one rafter or stud to the next. l and d must be in the same units.

8. How many rafters will be needed to build a roof 26 ft long if they are placed 2 ft on center?

9. A wall has studs placed 16 in. on center. If the wall is 20 ft long, how many studs are in the wall?

10. How long is a wall if it requires 22 studs placed 16 in. on center?

The total cost, C, of producing x items can be found by the formula $C = ax + k$, where a is the cost per item and k is the fixed costs (rents, utilities, and so on).

11. Find the cost of producing 30 items if each costs \$15 and the fixed costs are \$580.

12. It costs \$1097.50 to produce 80 dolls per week. If each doll costs \$9.50 to produce, find the fixed costs.

13. It costs a company \$3.60 to produce a calculator. Last week the total costs were \$1308. If the fixed costs are \$480 weekly, how many calculators were produced last week?

Many items decrease in value as time passes. This decrease in value is called **depreciation.** One type of depreciation is called **linear depreciation.** The value, V, of an item after t years is given by $V = C - Crt$, where C is the original cost and r is the rate of depreciation expressed as a decimal.

14. If you buy a car for \$6000 and depreciate it linearly at a rate of 10% per year, what will be its value after 6 years?

15. A contractor buys a piece of heavy equipment for \$20,000. If it cost \$25,000 new and is 4 years old, find the rate of depreciation.

In Exercises 16–25, (a) create a formula suggested by the stated relationships and (b) solve as indicated with the given information.

16. The perimeter of a square is the product of 4 with the length of a side. Find the length of the sides if the perimeter is 64 meters.

17. The circumference of a circle is the product of 2π and the radius. Find the radius if the circumference is 34π centimeters.

18. The interest on a loan is the product of the amount borrowed (principal) times the rate times the time of the loan. Find the interest on $1100 at 9% borrowed for 3 years.

19. The perimeter of an isosceles triangle (one with 2 sides the same length) is the sum of the base and twice the length of one of the two equal sides. If the perimeter is 72 and the base is 28, what is the length of each of the equal sides?

20. The Celsius temperature can be found by the product of $\frac{5}{9}$ with the difference of the Fahrenheit temperature and 32. Find the Celsius temperature, C, if the Fahrenheit temperature is 77 degrees.

21. The area of a circle is the product of π and the square of the radius. If a rotating sprinkler sprays water a distance of 7 meters, how many square meters will it cover in one revolution?

22. The IQ of a person is 100 times the quotient of the mental age and the chronological age. Find the IQ of someone $10\frac{1}{2}$ years old if the mental age is 12 years.

23. In electricity, the resistance of an electrical circuit measured in ohms is the quotient of the volts and the intensity measured in amperes. What is the resistance of a 120-volt circuit with an intensity of 20 amperes?

24. The volume of a cylinder is the product of π, radius squared, and height. Find the height of a cylinder whose volume is 252π cubic centimeters if the radius is 6 centimeters.

25. The area of a trapezoid is the product of $\frac{1}{2}$ the height times the sum of the two bases. If the area of a trapezoid is 108 square centimeters, the height is 12 centimeters, and the length of one base is 8 centimeters, find the length of the other base.

CHAPTER 3 SUMMARY

An expression that involves only multiplication and/or division with constants and/or variables is called a **term.** A single constant or variable is also a term.

Like terms (or **similar terms**) are terms that are constants or terms that contain the same variables that are of the same power.

A value that gives a true statement when substituted for the variable in an equation is called a **solution** to the equation.

Definition: If a, b, and c are constants and $a \neq 0$, then a **first-degree equation** is an equation that can be written in the form

$$ax + b = c$$

Addition Property of Equality

If the same algebraic expression is added to both sides of an equation, the new equation has the same solutions as the original equation. Symbolically, if A, B, and C are algebraic expressions, then the equations

$$A = B$$

and

$$A + C = B + C$$

have the same solutions.

Multiplication Property of Equality

If both sides of an equation are multiplied by the same nonzero algebraic expression, the new equation has the same solutions as the original solution. Symbolically, if A, B, and C are algebraic expressions, then the equations

$$A = B$$

and

$$AC = BC \quad \text{where } C \neq 0$$

have the same solutions.

To Solve a First-Degree Equation

1. Simplify each side of the equation by removing any grouping symbols and combining like terms.
2. Use the addition property of equality to add the opposites of constants and/or variables so that variables are on one side and constants on the other.
3. Use the multiplication property of equality to multiply both sides by the reciprocal of the coefficient of the variable (or divide both sides by the coefficient).
4. Check your answer by substituting it into the original equation.

List of Key Words:

Addition	*Subtraction*	*Multiplication*	*Division*
add	subtract	multiply	divide
sum	difference	product	quotient
plus	minus	times	
more than	less than	twice	
increased by	decreased by		

Consecutive integers are two integers that differ by 1.

Consecutive even integers are even integers that differ by 2.

Consecutive odd integers are odd integers that differ by 2.

CHAPTER 3 REVIEW

Combine like terms in Exercises 1–5.

1. $3x + 2 + 4x - 1$

2. $-4(x + 3) + 2x$

3. $x + \dfrac{x - 5x}{4}$

4. $\dfrac{2(x + 3x)}{4} + 2x$

5. $2(x^2 - 3x) + 5(2x^2 + 7x)$

Evaluate Exercises 6–10 if $a = 2$, $b = -1$, and $c = 5$.

6. $4a^2 - 3a + 2$

7. $3a - 2ab + c$

8. $-b^2 - 7b + 5$

9. $a(4b^2 + 3bc)$

10. $\frac{3a^3 - b}{c} + ab^2$

Solve the equations in Exercises 11–25.

11. $x + 3.2 = 1.7$

12. $x - \frac{1}{2} = \frac{3}{4}$

13. $8x = 12$

14. $5x = 20.5$

15. $7x + 4 = -17$

16. $9x - 11 = x + 5$

17. $4x + 3 = 39 - 2x$

18. $5(1 - 2x) = 3x + 57$

19. $2(x + 6) = 6(x + 4)$

20. $5(2x + 3) = 3(x - 4) - 1$

21. $\frac{2}{3}x + 5 = 7$

22. $\frac{4}{5}x - 3 = 2$

23. $-\frac{3}{4}x + 2 = -6$

24. $0.5x + 3 = 0.3x - 5$

25. $4.4 + 0.6x = 1.2 - 0.2x$

Write out in words each expression in Exercises 26–30.

26. $3x - 1$

27. $4 - 7y$

28. $5(y + 1)$

29. $2(4n - 1)$

30. $\frac{4}{x + 7}$

Write an algebraic expression described by each of the phrases in Exercises 31–35.

31. 4 added to three times a number

32. the difference between 9 and twice a number

33. 11 times the difference between a number and 4

34. the number of hours in x days and 5 hours

35. the number of points scored by a basketball team on x field goals (2 points each) and 17 free throws (1 point each)

Solve for each indicated variable in Exercises 36–40.

36. $E = mc^2$; solve for m.

37. $F = k \cdot \frac{ab}{d^2}$; solve for b.

38. $V = \frac{1}{3}\pi r^2 h$; solve for h.

39. $2x + 7y = 8$; solve for y.

40. $5x + 2y = -3$; solve for x.

In Exercises 41–50, set up an equation for each word problem and solve.

41. The perimeter of a rectangle is 240 in. If the length is 85 in., find the width.

42. A car will cost $6400 if purchased on an installment plan. If the monthly payments are $150 and the down payment is $1000, how long will it take to pay for the car?

43. If three times a certain number is decreased by 9, the result is 30. Find the number.

44. If twice a certain number is increased by 3, the result is 8 less than three times the number. Find the number.

45. Three times a certain number is 10 more than twice the number. Find the number.

46. The sum of two consecutive odd integers is 84. Find the integers.

47. Find two consecutive integers such that twice the first subtracted from 5 is equal to the second plus 19.

48. Sally bought a softball and bat. The bat cost $9.35 more than the ball. If the total cost was $16.25, find the cost of each.

Optional

49. If you buy a car for $7200 and depreciate it linearly at a rate of 8% per year, what will be its value after 7 years?

50. The volume of a right circular cone is one-third of the product of π, radius squared, and height. Find the height of a cone whose volume is 147π cubic inches and whose radius is 7 in.

CHAPTER 3 TEST

Combine like terms in Exercises 1 and 2.

1. $6(x^2 - 2x) - (-2x^2 + 5x)$

2. $\dfrac{3(3x - x)}{6} + 2(2x - 5)$

Evaluate Exercises 3 and 4 if $x = -3$, $y = 2$, and $z = -1$.

3. $2x + 3z - xy$

4. $3x^2 - 2xz - y^2$

Solve the equations in Exercises 5–9.

5. $\dfrac{5}{3}x + 1 = -4$

6. $4x - 5 - x = 2x + 5 - x$

7. $8(3 + x) = -4(2x - 6)$

8. $\dfrac{3}{4}x + \dfrac{1}{2} = x - 3$

9. $0.7x + 2 = 0.4x + 8$

10. Write out in words: $4(x - 2)$.

11. Write out in words: $3(x + 4) - 7$.

12. Write an algebraic expression described by the following phrase: 3 times the sum of twice a number and 5.

13. Write an algebraic expression described by the following phrase: the number of quarts in x gallons and 3 quarts (1 gallon = 4 quarts).

Solve for each indicated variable in Exercises 14–16.

14. $S = 2\pi rh$; solve for h.

15. $N = mrt + p$; solve for m.

16. $5x - 3y + 7 = 0$; solve for y.

17. One number is 5 more than twice another. Their sum is -22. Find the numbers.

18. Find two consecutive integers such that twice the first added to three times the second is equal to 83.

19. A rectangle is 37 in. long and has a perimeter of 122 in. Find the width of the rectangle.

20. Quang had a hamburger and chocolate shake for lunch. The hamburger cost 90¢ more than the chocolate shake. If the total cost was \$2.80, how much did each cost?

USING THE COMPUTER

```
10  REM  This program will give you a menu and allow you to
20  REM  choose one of three geometric figures then find the
30  REM  area and perimeter of that figure.
40  REM  *************************************************
50  REM  * L : length   W : width of a rectangle          *
60  REM  * RADIUS : radius of circle                      *
70  REM  * BASE, HEIGHT : from right triangle             *
80  REM  *************************************************
90  PRINT "          MENU"
100 PRINT "1) Area of rectangle &
110 PRINT "   Perimeter of rectangle
120 PRINT "2) Area of circle &
130 PRINT "   Perimeter of circle
140 PRINT "3) Area of right triangle &
150 PRINT "   Perimeter of right triangle
160 PRINT "4) Quit"
170 PRINT
180 INPUT "Type in your choice 1,2,3 or 4 :",CHOICE$
190 IF CHOICE$="4" THEN GOTO 250
200 IF CHOICE$="1" THEN GOSUB 260 : GOTO 90
210 IF CHOICE$="2" THEN GOSUB 340 : GOTO 90
220 IF CHOICE$="3" THEN GOSUB 420 : GOTO 90
230 PRINT "Please, type in the menu number between 1 and 4"
240 GOTO 90
250 END
260 REM subroutine for rectangle
270 INPUT "Type in the length of rectangle:",L
280 INPUT "Type in the width of rectangle :",W
290 LET PERIMETER = 2*W + 2*L
300 LET AREA = L*W
310 GOSUB 500
320 PRINT
330 RETURN
340 REM subroutine for circle
350 LET PI = 3.14159
360 INPUT "Type in the radius of circle :",RADIUS
370 LET AREA = PI*RADIUS^2
380 LET PERIMETER = 2*PI*RADIUS
390 GOSUB 500
400 PRINT
410 RETURN
420 REM subroutine for right triangle
430 INPUT "Type in the base of right triangle :",BASE
440 INPUT "Type in the height of right triangle :",HEIGHT
450 LET AREA = .5*BASE*HEIGHT
460 LET PERIMETER = BASE + HEIGHT + (BASE^2 + HEIGHT^2)^.5
470 GOSUB 500
480 PRINT
490 RETURN
500 REM subroutine for area
510 PRINT "The area is "AREA
520 PRINT "The perimeter is "PERIMETER
530 RETURN
```

```
          MENU
1) Area of rectangle &
   Perimeter of rectangle
2) Area of circle &
   Perimeter of circle
3) Area of right triangle &
   Perimeter of right triangle
4) Quit

Type in your choice 1,2,3 or 4 :1
Type in the length of rectangle: 2
Type in the width of rectangle : 3
The area is  6
The perimeter is  10

          MENU
1) Area of rectangle &
   Perimeter of rectangle
2) Area of circle &
   Perimeter of circle
3) Area of right triangle &
   Perimeter of right triangle
4) Quit

Type in your choice 1,2,3 or 4 :2
Type in the radius of circle : 2
The area is  12.5664
The perimeter is  12.5664

          MENU
1) Area of rectangle &
   Perimeter of rectangle
2) Area of circle &
   Perimeter of circle
3) Area of right triangle &
   Perimeter of right triangle
4) Quit

Type in your choice 1,2,3 or 4 :3
Type in the base of right triangle : 3
Type in the height of right triangle : 4
The area is  6
The perimeter is  12

          MENU
1) Area of rectangle &
   Perimeter of rectangle
2) Area of circle &
   Perimeter of circle
3) Area of right triangle &
   Perimeter of right triangle
4) Quit

Type in your choice 1,2,3 or 4 :4
```

Mathematics is the queen of the sciences and arithmetic the queen of mathematics.

KARL F. GAUSS 1777–1855

DID YOU KNOW?

One of the most difficult problems for students in beginning algebra is to become comfortable with the idea that letters or symbols can be manipulated just like numbers in arithmetic. These symbols may be the cause of "math anxiety."

A great deal of publicity has recently been given to the concept that a large number of people suffer from math anxiety, a painful uneasiness caused by mathematical symbols or a problem-solving situation. Persons affected by math anxiety find it almost impossible to learn mathematics, or they may be able to learn but be unable to apply their knowledge or do well on tests. Persons suffering from math anxiety often develop math avoidance, and they avoid careers, majors, or classes that will require mathematics courses or skills. The sociologist Lucy Sells has determined that mathematics is a critical filter in the job market. Persons who lack quantitative skills are channeled into high-unemployment, low-paying, nontechnical areas.

What causes math anxiety? Researchers are investigating the following hypotheses:

1. a lack of skills which leads to lack of confidence and, therefore, to anxiety;
2. an attitude that mathematics is not useful or significant to society;
3. career goals that seem to preclude mathematics;
4. a self-concept that differs radically from the stereotype of a mathematician;
5. perceptions that parents, peers, or teachers have low expectations for the person in mathematics;
6. social conditioning to avoid mathematics (a particular problem for women).

We hope that you are finding your present experience with algebra successful and that the skills you are acquiring now will enable you to approach mathematical problems with confidence.

4 Exponents and Polynomials

4.1 EXPONENTS

In Section 1.2, an **exponent** was defined as a number that tells how many times a factor occurs in a product. This definition is valid only if the exponents are natural (counting) numbers. In this section, we will develop four properties of exponents that will help in simplifying algebraic expressions and expand your understanding of exponents to include the exponent 0 and negative exponents. (**Note:** Fractional exponents are an important topic in algebra but are not discussed in this text. They are a major topic in intermediate algebra.)

We know that

$$6^2 = 6 \cdot 6 = 36$$

and $$6^3 = 6 \cdot 6 \cdot 6 = 216$$

Also, we know that

$$x^3 = x \cdot x \cdot x$$

and $$x^5 = x \cdot x \cdot x \cdot x \cdot x$$

If we want to multiply $6^2 \cdot 6^3$ or $x^3 \cdot x^5$, we could write down all the factors as follows:

$$6^2 \cdot 6^3 = (6 \cdot 6) \cdot (6 \cdot 6 \cdot 6) = 6^5$$

and $$x^3 \cdot x^5 = (x \cdot x \cdot x) \cdot (x \cdot x \cdot x \cdot x \cdot x) = x^8$$

What do you think would be a simplified form for the product $3^4 \cdot 3^3$? You were right if you thought 3^7. That is, $3^4 \cdot 3^3 = 3^7$.

The preceding discussion, along with the basic concept of whole number exponents, leads to the following property of exponents.

> **Property 1 of Exponents**
>
> If a is a nonzero number and m and n are integers, then
>
> $$a^m \cdot a^n = a^{m+n}$$

If a variable or constant has no exponent written, it is understood to be 1. For example,

$$y = y^1, \qquad 7 = 7^1, \qquad \text{and} \qquad a = a^1$$

EXAMPLES

Use property 1 of exponents to simplify the following expressions.

1. $x^2 \cdot x^4$

   Solution: $x^2 \cdot x^4 = x^{2+4} = x^6$

2. $y \cdot y^6$

   Solution: $y \cdot y^6 = y^1 \cdot y^6 = y^{1+6} = y^7$

3. $4^2 \cdot 4$

   Solution: $4^2 \cdot 4 = 4^{2+1} = 4^3 = 64$

4. $2^3 \cdot 2^2$

   Solution: $2^3 \cdot 2^2 = 2^{3+2} = 2^5 = 32$

5. $(-2)^4(-2)^3$

   Solution: $(-2)^4(-2)^3 = (-2)^{4+3} = (-2)^7 = -128$

6. $2y^2 \cdot 3y^9$

   Solution: $2y^2 \cdot 3y^9 = 2 \cdot 3 \cdot y^2 \cdot y^9 = 6y^{2+9} = 6y^{11}$

7. $3 \cdot 2^4 \cdot 2$

   Solution: $3 \cdot 2^4 \cdot 2 = 3 \cdot 2^{4+1} = 3 \cdot 2^5 = 3 \cdot 32 = 96$

8. $5 \cdot 3 \cdot 3^2$

   Solution: $5 \cdot 3 \cdot 3^2 = 5 \cdot 3^{1+2} = 5 \cdot 3^3 = 5 \cdot 27 = 135$

Property 1 is true if m and n are **integers**. So we need to discuss the meaning of 0 as an exponent and negative exponents.

Study the following patterns of numbers. What do you think are the missing values for 2^0, 3^0, and 5^0?

$2^5 = 32$	$3^5 = 243$	$5^5 = 3125$
$2^4 = 16$	$3^4 = 81$	$5^4 = 625$
$2^3 = 8$	$3^3 = 27$	$5^3 = 125$
$2^2 = 4$	$3^2 = 9$	$5^2 = 25$
$2^1 = 2$	$3^1 = 3$	$5^1 = 5$
$2^0 = ?$	$3^0 = ?$	$5^0 = ?$

Notice that in the column of powers of 2, each number is $\frac{1}{2}$ of the preceding number. Since $\frac{1}{2} \cdot 2 = 1$, a reasonable guess is that $2^0 = 1$. Similarly, in the column of powers of 3, $\frac{1}{3} \cdot 3 = 1$, so $3^0 = 1$ seems reasonable. Also, $\frac{1}{5} \cdot 5 = 1$, so $5^0 = 1$ fits the pattern. In fact, the missing values are $2^0 = 1$, $3^0 = 1$, and $5^0 = 1$.

Another approach to understanding 0 as an exponent involves property 1. Consider

$$2^0 \cdot 2^3 = 2^{0+3} = 2^3 \quad \text{using property 1}$$

and

$$1 \cdot 2^3 = 2^3$$

So, $2^0 \cdot 2^3 = 1 \cdot 2^3$ and $2^0 = 1$. Similarly,

$$7^0 \cdot 7^2 = 7^{0+2} = 7^2$$

and

$$1 \cdot 7^2 = 7^2$$

So, $7^0 \cdot 7^2 = 1 \cdot 7^2$ and $7^0 = 1$.

This discussion leads directly to property 2.

Property 2 of Exponents

If a is a nonzero number, then

$$a^0 = 1$$

The expression 0^0 is undefined.

EXAMPLES

Simplify the following expressions using property 2.

9. 10^0

   Solution: $10^0 = 1$

10. $x^0 \cdot x^3$

   Solution: $x^0 \cdot x^3 = x^{0+3} = x^3$

11. 49^0

   Solution: $49^0 = 1$

12. $(-6)^0$

   Solution: $(-6)^0 = 1$

Again, looking at a pattern of powers, we can understand the meaning of negative exponents, as in 2^{-1}, 2^{-2}, 3^{-1}, and 3^{-2}. What do you think the missing values are in the following patterns? Notice that in the column of powers of 2, each number is $\frac{1}{2}$ of the preceding number. In the column of powers of 3, each number is $\frac{1}{3}$ of the preceding number.

$2^3 = 8$	$3^3 = 27$
$2^2 = 4$	$3^2 = 9$
$2^1 = 2$	$3^1 = 3$
$2^0 = 1$	$3^0 = 1$
$2^{-1} = ?$	$3^{-1} = ?$
$2^{-2} = ?$	$3^{-2} = ?$

The values are $2^{-1} = \frac{1}{2}$ and $2^{-2} = \frac{1}{2^2} = \frac{1}{4}$. Also, $3^{-1} = \frac{1}{3}$ and $3^{-2} = \frac{1}{3^2} = \frac{1}{9}$. These lead to property 3.

Property 3 of Exponents

If a is a nonzero number and n is an integer, then

$$a^{-n} = \frac{1}{a^n}$$

EXAMPLES

Simplify each expression so that it contains only positive exponents.

13. 5^{-1}

Solution: $5^{-1} = \frac{1}{5^1} = \frac{1}{5}$ using property 3

14. x^{-3}

Solution: $x^{-3} = \frac{1}{x^3}$ using property 3

15. $2^{-5} \cdot 2^8$

Solution: $2^{-5} \cdot 2^8 = 2^{-5+8} = 2^3 = 8$ using property 1

16. $x^{-9} \cdot x^7$

Solution: $x^{-9} \cdot x^7 = x^{-9+7} = x^{-2} = \frac{1}{x^2}$

(Here, property 1 is used first, then property 3.)

Special note: A simplified expression will not contain negative exponents. (In later courses in algebra, this rule will be adjusted to allow for negative exponents in the numerator of a fraction.)

Now we are ready to consider an expression that involves division as well as exponents. For example,

$$\frac{3^3}{3^5} = \frac{\cancel{3} \cdot \cancel{3} \cdot \cancel{3} \cdot 1}{\cancel{3} \cdot \cancel{3} \cdot \cancel{3} \cdot 3 \cdot 3} = \frac{1}{3^2} = \frac{1}{9} \quad \text{or} \quad \frac{3^3}{3^5} = 3^{3-5} = 3^{-2} = \frac{1}{3^2} = \frac{1}{9}$$

$$\frac{5^4}{5^2} = \frac{\cancel{5} \cdot \cancel{5} \cdot 5 \cdot 5}{\cancel{5} \cdot \cancel{5} \cdot 1} = \frac{5^2}{1} = 25 \quad \text{or} \quad \frac{5^4}{5^2} = 5^{4-2} = 5^2 = 25$$

$$\frac{2^2}{2^7} = \frac{\cancel{2} \cdot \cancel{2} \cdot 1}{\cancel{2} \cdot \cancel{2} \cdot 2 \cdot 2 \cdot 2 \cdot 2 \cdot 2} = \frac{1}{2^5} = \frac{1}{32} \quad \text{or} \quad \frac{2^2}{2^7} = 2^{2-7} = 2^{-5} = \frac{1}{2^5} = \frac{1}{32}$$

Thus, we are led to property 4.

Property 4 of Exponents

If a is a nonzero number and m and n are integers, then

$$\frac{a^m}{a^n} = a^{m-n}$$

EXAMPLES

Simplify each expression using the appropriate property of exponents. Remember, a simplified expression does not contain negative exponents.

17. $\dfrac{x^6}{x}$

Solution: $\dfrac{x^6}{x} = x^{6-1} = x^5$

18. $\dfrac{x^6}{x^{-1}}$

Solution: $\dfrac{x^6}{x^{-1}} = x^{6-(-1)} = x^{6+1} = x^7$

19. $\dfrac{y^{-2}}{y^{-7}}$

Solution: $\dfrac{y^{-2}}{y^{-7}} = y^{-2-(-7)} = y^{-2+7} = y^5$

20. $\dfrac{10^{-8}}{10^{-2}}$

Solution: $\dfrac{10^{-8}}{10^{-2}} = 10^{-8-(-2)} = 10^{-8+2} = 10^{-6}$

$$= \frac{1}{10^6} \text{ or } \frac{1}{1{,}000{,}000}$$

21. $\dfrac{x^{10} \cdot x^2}{x^{15}}$

Solution: $\dfrac{x^{10} \cdot x^2}{x^{15}} = \dfrac{x^{10+2}}{x^{15}} = \dfrac{x^{12}}{x^{15}} = x^{12-15} = x^{-3} = \dfrac{1}{x^3}$

22. $\dfrac{2^{-5} \cdot 2^8}{2^3}$

Solution: $\dfrac{2^{-5} \cdot 2^8}{2^3} = \dfrac{2^{-5+8}}{2^3} = \dfrac{2^3}{2^3} = 2^{3-3} = 2^0 = 1$

Of special interest are expressions with negative exponents in the denominator, such as $\dfrac{1}{a^{-n}}$. We can write

$$\frac{1}{a^{-n}} = \frac{a^0}{a^{-n}} = a^{0-(-n)} = a^{0+n} = a^n$$

or

$$\frac{1}{a^{-n}} = \frac{1}{\dfrac{1}{a^n}} = 1 \cdot \frac{a^n}{1} = a^n$$

Thus,

$$a^{-n} = \frac{1}{a^n} \quad \text{and} \quad \frac{1}{a^{-n}} = a^n$$

In effect, the sign of an exponent is changed whenever a term is moved from numerator to denominator, or vice versa.

PRACTICE QUIZ

Questions	Answers
Simplify each expression.	
1. $2^3 \cdot 2^4$	1. $2^7 = 128$
2. $\frac{2^3}{2^4}$	2. $\frac{1}{2}$
3. $\frac{x^0 \cdot x^7 \cdot x^{-3}}{x^{-2}}$	3. x^6
4. $\frac{10^{-8} \cdot 10^2}{10^{-7}}$	4. 10

EXERCISES 4.1

Simplify each expression so that it has only positive exponents. Assume that all variables represent nonzero numbers.

1. $3^2 \cdot 3$ **2.** $7^2 \cdot 7^3$ **3.** $8^3 \cdot 8^0$ **4.** 3^{-1}

5. 4^{-2} **6.** $(-5)^{-2}$ **7.** 6^{-3} **8.** $(-2)^4 \cdot (-2)^0$

9. $(-4)^3 \cdot (-4)^0$ **10.** $3 \cdot 2^3$ **11.** $6 \cdot 3^2$ **12.** $-4 \cdot 5^3$

13. $-2 \cdot 3^3$ **14.** $3 \cdot 2^{-3}$ **15.** $4 \cdot 3^{-2}$ **16.** $-3 \cdot 5^{-2}$

17. $-5 \cdot 2^{-2}$ **18.** $x^2 \cdot x^3$ **19.** $x^3 \cdot x$ **20.** $y^2 \cdot y^0$

21. $y^3 \cdot y^8$ **22.** x^{-3} **23.** y^{-2} **24.** $2x^{-1}$

25. $5y^{-4}$ **26.** $x \cdot x^{-1}$ **27.** $x \cdot x^{-3}$ **28.** $x^0 \cdot x^{-2}$

29. $y^5 \cdot y^{-2}$ **30.** $y^4 \cdot y^{-6}$ **31.** $\frac{3^4}{3^2}$ **32.** $\frac{7^3}{7}$

33. $\frac{9^5}{9^2}$ **34.** $\frac{10^3}{10^4}$ **35.** $\frac{10^2}{10^5}$ **36.** $\frac{2^3}{2^6}$

37. $\frac{x^4}{x^2}$ **38.** $\frac{x^5}{x^3}$ **39.** $\frac{x^3}{x}$ **40.** $\frac{y^6}{y^4}$

41. $\frac{x^7}{x^3}$ **42.** $\frac{x^8}{x^3}$ **43.** $\frac{x^{-2}}{x^2}$ **44.** $\frac{x^{-3}}{x}$

45. $\frac{x^4}{x^{-2}}$ **46.** $\frac{x^5}{x^{-1}}$ **47.** $\frac{x^{-3}}{x^{-5}}$ **48.** $\frac{x^4}{x^{-1}}$

49. $\dfrac{y^{-2}}{y^{-4}}$ **50.** $\dfrac{y^{3}}{y^{-3}}$ **51.** $3x^3 \cdot x^0$ **52.** $3y \cdot y^4$

53. $(5x^3)(2x^2)$ **54.** $(3x^2)(3x)$ **55.** $(4x^3)(9x^0)$ **56.** $(5x^2)(3x^4)$

57. $(-2x^2)(7x^3)$ **58.** $(3y^3)(-6y^2)$ **59.** $(-4x^5)(3x)$ **60.** $(6y^4)(5y^5)$

61. $\dfrac{8y^3}{2y^2}$ **62.** $\dfrac{12x^4}{3x}$ **63.** $\dfrac{9y^5}{3y^3}$ **64.** $\dfrac{-10x^5}{2x}$

65. $\dfrac{-8y^4}{4y^2}$ **66.** $\dfrac{12x^6}{-3x^3}$ **67.** $\dfrac{21x^4}{-3x^2}$ **68.** $\dfrac{10 \cdot 10^3}{10^{-3}}$

69. $\dfrac{10^4 \cdot 10^{-3}}{10^{-2}}$ **70.** $\dfrac{10 \cdot 10^{-1}}{10^2}$ **71.** $\dfrac{10^0 \cdot 10^3}{10^5}$ **72.** $\dfrac{x^2 \cdot x^{-3}}{x^4}$

73. $\dfrac{x^4 \cdot x^{-2}}{x^{-3}}$ **74.** $\dfrac{y^5 \cdot y^0}{y^{-2}}$ **75.** $\dfrac{y^6 \cdot y^{-2}}{y^4}$

4.2 MORE ON EXPONENTS AND SCIENTIFIC NOTATION

In Section 4.1, we discussed the following four properties of exponents.

> **For a Nonzero Real Number *a* and Integers *m* and *n***
>
> 1. $a^m \cdot a^n = a^{m+n}$
> 2. $a^0 = 1$
> 3. $a^{-n} = \dfrac{1}{a^n}$
> 4. $\dfrac{a^m}{a^n} = a^{m-n}$

In each of these properties, the base of each exponent was one number or variable. Now we will consider expressions such as $(5x)^3$, $(3xy)^4$, $\left(\dfrac{-2}{x}\right)^3$, and $(x^2)^4$, in which the base is a product or a quotient.

For example,

$$\begin{aligned}(5x)^3 &= (5x)\cdot(5x)\cdot(5x)\\ &= 5\cdot 5\cdot 5\cdot x\cdot x\cdot x \quad \text{using the associative and commutative properties of multiplication}\\ &= 5^3\cdot x^3\\ &= 125x^3\end{aligned}$$

$$\begin{aligned}(-2y)^5 &= (-2y)\cdot(-2y)\cdot(-2y)\cdot(-2y)\cdot(-2y)\\ &= (-2)(-2)(-2)(-2)(-2)\cdot y\cdot y\cdot y\cdot y\cdot y\\ &= (-2)^5\cdot y^5\\ &= -32y^5\end{aligned}$$

[Note carefully that in the expression $(-2y)^5$, the negative sign goes with 2, and -2 is treated as a factor of $-2y$.]

$$\begin{aligned}(3xy)^4 &= (3xy)(3xy)(3xy)(3xy)\\ &= 3\cdot 3\cdot 3\cdot 3\cdot x\cdot x\cdot x\cdot x\cdot y\cdot y\cdot y\cdot y\\ &= 3^4x^4y^4\\ &= 81x^4y^4\end{aligned}$$

These examples lead to property 5.

Property 5 of Exponents

If a and b are nonzero numbers and n is an integer, then

$$(ab)^n = a^nb^n$$

EXAMPLES

1. $(5x)^2 = 5^2\cdot x^2 = 25x^2$
2. $(xy)^3 = x^3\cdot y^3 = x^3y^3$
3. $(3xy)^4 = 3^4\cdot x^4\cdot y^4 = 81x^4y^4$
4. $(-7ab)^3 = (-7)^3a^3b^3 = -343a^3b^3$
5. $(-x)^3 = (-1\cdot x)^3 = (-1)^3\cdot x^3 = -1x^3 = -x^3$
6. $(ab)^{-5} = a^{-5}\cdot b^{-5} = \dfrac{1}{a^5}\cdot\dfrac{1}{b^5} = \dfrac{1}{a^5b^5}$

or, using property 3 first and then property 5,

$$(ab)^{-5} = \frac{1}{(ab)^5} = \frac{1}{a^5b^5}$$

Fractions raised to a power can be handled in a similar manner. For example,

$$\left(\frac{2}{x}\right)^3 = \frac{2}{x} \cdot \frac{2}{x} \cdot \frac{2}{x} = \frac{2 \cdot 2 \cdot 2}{x \cdot x \cdot x} = \frac{2^3}{x^3} = \frac{8}{x^3}$$

and

$$\left(\frac{3x}{y}\right)^4 = \frac{3x}{y} \cdot \frac{3x}{y} \cdot \frac{3x}{y} \cdot \frac{3x}{y} = \frac{3x \cdot 3x \cdot 3x \cdot 3x}{y \cdot y \cdot y \cdot y}$$

$$= \frac{(3x)^4}{y^4} = \frac{3^4x^4}{y^4} = \frac{81x^4}{y^4}$$

Thus, we now have property 6.

Property 6 of Exponents

If a and b are nonzero numbers and n is an integer, then

$$\left(\frac{a}{b}\right)^n = \frac{a^n}{b^n}$$

EXAMPLES

7. $\left(\frac{y}{x}\right)^5 = \frac{y^5}{x^5}$

8. $\left(\frac{2a}{b}\right)^4 = \frac{(2a)^4}{b^4} = \frac{2^4a^4}{b^4} = \frac{16a^4}{b^4}$

9. $\left(\frac{2x}{-3y}\right)^2 = \frac{(2x)^2}{(-3y)^2} = \frac{2^2x^2}{(-3)^2y^2} = \frac{4x^2}{9y^2}$

In an expression such as $-x^2$, you know that -1 is the understood coefficient of x^2 and the exponent 2 refers only to x as its base. Thus, $-x^2 = -1 \cdot x^2$.

The same is true for an expression such as -7^2 or -2^0. We have

$$-7^2 = -1 \cdot 7^2 = -1 \cdot 49 = -49$$

and

$$-2^0 = -1 \cdot 2^0 = -1 \cdot 1 = -1$$

But if parentheses are used, as in $(-7)^2$ and $(-2)^0$, then the base is the negative number:

$$(-7)^2 = (-7)(-7) = +49$$

and

$$(-2)^0 = 1$$

What happens when a power is raised to a power? For example, if you want to simplify the expressions $(x^2)^3$ and $(2^5)^2$, you might try writing

$$(x^2)^3 = x^2 \cdot x^2 \cdot x^2 = x^{2+2+2} = x^6$$

and

$$(2^5)^2 = 2^5 \cdot 2^5 = 2^{5+5} = 2^{10} = 1024$$

However, this technique can be quite time-consuming when the exponent is large, such as in $(3x^2y^3)^{17}$. Property 7 gives a convenient way to handle powers raised to powers.

Property 7 of Exponents

If a is a nonzero number and m and n are integers, then

$$(a^m)^n = a^{mn}$$

EXAMPLES

10. $(x^2)^4 = x^{2 \cdot 4} = x^8$

11. $(3^2)^3 = 3^{2 \cdot 3} = 3^6 = 729$

12. $(x^5)^{-2} = x^{5(-2)} = x^{-10} = \dfrac{1}{x^{10}}$

or

$$(x^5)^{-2} = \frac{1}{(x^5)^2} = \frac{1}{x^{5 \cdot 2}} = \frac{1}{x^{10}}$$

13. $(y^{-7})^2 = y^{-7 \cdot 2} = y^{-14} = \dfrac{1}{y^{14}}$

or

$$(y^{-7})^2 = \left(\frac{1}{y^7}\right)^2 = \frac{1^2}{(y^7)^2} = \frac{1}{y^{14}}$$

Various combinations of all seven properties of exponents are needed to simplify the following examples. There may be more than one correct sequence of steps to follow. **You should apply whichever property you "see" first.** The simplified form will be the same in any case.

EXAMPLES

14. $\left(\frac{-2x}{y^2}\right)^3 = \frac{(-2x)^3}{(y^2)^3} = \frac{(-2)^3x^3}{y^6} = \frac{-8x^3}{y^6}$

15. $\left(\frac{x^3}{y}\right)^{-4} = \frac{(x^3)^{-4}}{y^{-4}} = \frac{x^{-12}}{y^{-4}} = \frac{\frac{1}{x^{12}}}{\frac{1}{y^4}} = \frac{1}{x^{12}} \cdot \frac{y^4}{1} = \frac{y^4}{x^{12}}$

(Note that, as was mentioned in Section 4.1, we could have written $\frac{x^{-12}}{y^{-4}} = \frac{y^4}{x^{12}}$ directly since switching between numerator and denominator changes the sign of the exponent.) A general approach is to note that

$$\left(\frac{a}{b}\right)^{-n} = \frac{a^{-n}}{b^{-n}} = \frac{b^n}{a^n} = \left(\frac{b}{a}\right)^n$$

With this formula, we can write

$$\left(\frac{x^3}{y}\right)^{-4} = \left(\frac{y}{x^3}\right)^4 = \frac{y^4}{(x^3)^4} = \frac{y^4}{x^{12}}$$

16. $\left(\frac{3a^5b}{a^0b^2}\right)^2 = \left(\frac{3a^5}{b^{2-1}}\right)^2 = \frac{3^2(a^5)^2}{(b^1)^2} = \frac{9a^{10}}{b^2}$

One of the most fundamental uses of exponents on an elementary level occurs in the **scientific notation** used in physics, chemistry, and other scientific and technological fields. This notation is particularly convenient because it simplifies calculations with very large and very small numbers.

For example, you might want to multiply the distance to the sun from the earth by the distance to the moon from the earth. The calculations in scientific notation would look like this:

$$\begin{aligned} 93{,}000{,}000 \times 250{,}000 &= 9.3 \times 10^7 \times 2.5 \times 10^5 \\ &= 9.3 \times 2.5 \times 10^7 \times 10^5 \\ &= 23.25 \times 10^{12} \\ &= 2.325 \times 10^{13} \end{aligned}$$

(In scientific notation, a cross, $\times$, is used to indicate multiplication.)

In scientific notation, all decimal numbers are written as the product of a number between 1 and 10 and a power of 10. This notation is quite common in the hand-held calculators available to most students.

$$56{,}000 = 5.6 \times 10^4 \qquad (\textbf{Note: } 1 < 5.6 < 10)$$

$$0.00056 = 5.6 \times 10^{-4}$$

The exponent tells how many places the decimal point is to be moved and in what direction. In 5.6×10^4, the decimal point should go four places to the right (positive direction): $5.6 \times 10^4 = 5.6000. = 56{,}000$. In 5.6×10^{-4}, the decimal point should go four places to the left (negative direction): $5.6 \times 10^{-4} = .0005.6 = .00056$.

EXAMPLES

Write the following decimals in scientific notation.

17. $8{,}720{,}000 = 8.72 \times 10^6$

8.72 is between 1 and 10. If we move the decimal point 6 places to the right, we get the original number:

$$8.72 \times 10^6 = 8.\underset{1}{7}\,\underset{2}{2}\,\underset{3}{0}\,\underset{4}{0}\,\underset{5}{0}\,\underset{6}{0}\,. = 8{,}720{,}000.$$

18. $0.00381 = 3.81 \times 10^{-3}$

3.81 is between 1 and 10. If we move the decimal point 3 places to the left, we get the original number:

$$3.81 \times 10^{-3} = 0.\,\underset{3}{0}\,\underset{2}{0}\,\underset{1}{3}\,.81 = 0.00381.$$

19. $$\begin{aligned} 300{,}000 \times 700 &= 3 \times 10^5 \times 7 \times 10^2 \\ &= 21 \times 10^7 \\ &= 2.1 \times 10^1 \times 10^7 \\ &= 2.1 \times 10^8 \end{aligned}$$

20. $$\begin{aligned} \frac{0.0042 \times 0.003}{0.21} &= \frac{4.2 \times 10^{-3} \times 3 \times 10^{-3}}{2.1 \times 10^{-1}} \\ &= \frac{\overset{2}{\cancel{4.2}} \times 3}{\cancel{2.1}} \times \frac{10^{-6}}{10^{-1}} \\ &= 6 \times 10^{-6+1} \\ &= 6 \times 10^{-5} \end{aligned}$$

21. A light-year is the distance traveled by a particle of light in one year. Find the length of a light-year in scientific notation if light travels 186,000 miles per second.

Solution:

$$\begin{aligned} 60 \text{ sec} &= 1 \text{ min} \\ 60 \text{ min} &= 1 \text{ hr} \\ 24 \text{ hr} &= 1 \text{ day} \\ 365 \text{ days} &= 1 \text{ year} \\ 1 \text{ light-year} &= 186{,}000 \times 60 \times 60 \times 24 \times 365 \\ &= 5{,}865{,}696{,}000{,}000 \\ &= 5.865696 \times 10^{12} \text{ mi} \end{aligned}$$

PRACTICE QUIZ

Questions	**Answers**
Simplify each expression.	
1. $\dfrac{x^{-2}x^{5}}{x^{-7}}$	1. x^{10}
2. $\left(\dfrac{a^2b^3}{4}\right)^0$	2. 1
3. $\dfrac{-3^2 \cdot 5}{2 \cdot (-3)^2}$	3. $\dfrac{-5}{2}$ or $-\dfrac{5}{2}$
4. $\left(\dfrac{5x}{3b}\right)^{-2}$	4. $\dfrac{9b^2}{25x^2}$
Write the number in scientific notation.	
5. 186,000 (speed of light in miles per second)	5. 1.86×10^5

EXERCISES 4.2

Simplify each expression. Assume that all variables represent nonzero numbers.

1. $(6x^3)^2$ **2.** $(-3x^4)^2$ **3.** $(-3x^2)^3$ **4.** $(x^2y)^5$

5. $(4^0xy^2)^3$ **6.** $-(2x^3y^0)^2$ **7.** $(8a^4b)^2$ **8.** $(-2^2x^5)^3$

9. $-(7xy^2)^0$ **10.** $(6a^2b^{-3})^{-1}$ **11.** $(4m^2n^{-3})^{-2}$ **12.** $5(x^2y^{-1})^{-2}$

13. $-2(3x^5y^{-2})^{-3}$ **14.** $\left(\frac{3x}{y}\right)^3$ **15.** $\left(\frac{-4x}{y^2}\right)^2$ **16.** $\left(\frac{6m^3}{n^5}\right)^0$

17. $\left(\frac{3x^2}{y^3}\right)^2$ **18.** $\left(\frac{-2x^2}{y^{-2}}\right)^2$ **19.** $\left(\frac{x}{y}\right)^{-2}$ **20.** $\left(\frac{2a}{b}\right)^{-1}$

21. $\left(\frac{2x}{y^5}\right)^{-2}$ **22.** $\left(\frac{3x}{y^{-2}}\right)^{-1}$ **23.** $\left(\frac{4a^2}{b^{-3}}\right)^{-3}$ **24.** $\left(\frac{-3}{xy^2}\right)^{-3}$

25. $\left(\frac{5xy^3}{y}\right)^2$ **26.** $\left(\frac{m^2n^3}{mn}\right)^2$ **27.** $\left(\frac{2ab^3}{b^2}\right)^4$ **28.** $\left(\frac{-7^2x^2y}{y^3}\right)^{-1}$

29. $\left(\frac{2ab^4}{b^2}\right)^{-3}$ **30.** $\left(\frac{5x^3y}{y^2}\right)^{-2}$ **31.** $\left(\frac{2x^2y}{y^3}\right)^{-4}$ **32.** $\left(\frac{x^3y^{-1}}{y^2}\right)^2$

33. $\left(\frac{2a^2b^{-1}}{b^2}\right)^3$ **34.** $\left(\frac{6y^5}{x^2y^{-2}}\right)^2$ **35.** $\left(\frac{7x^{-2}y}{xy^{-1}}\right)^2$ **36.** $\left(\frac{3xy^{-1}}{7x^2y}\right)^{-2}$

37. $\left(\frac{2x^2y^{-3}}{3x^{-1}y}\right)^{-1}$ **38.** $\left(\frac{4x^3y^{-1}}{xy^{-2}}\right)^{-1}$ **39.** $\left(\frac{3x^3y}{2x^{-2}y}\right)^{-2}$ **40.** $\left(\frac{2x^{-3}y}{x^2y^{-2}}\right)^{-3}$

Write the decimal numbers in Exercises 41–68 in scientific notation.

41. 86,000 **42.** 927,000 **43.** 0.0362

44. 0.0061 **45.** 18,300,000 **46.** 376,000,000

47. 0.000217 **48.** 0.00000143 **49.** 500×9000

50. $35{,}000 \times 2000$ **51.** $410{,}000 \times 30{,}000$ **52.** $14{,}000 \times 200{,}000$

53. $300 \times .00015$ **54.** $0.000024 \times 40{,}000$ **55.** 0.0003×0.000025

56. $0.00005 \times .00013$ **57.** $\frac{3900}{0.003}$ **58.** $\frac{4800}{12{,}000}$

59. $\frac{125}{50{,}000}$ **60.** $\frac{0.0046}{230}$ **61.** $\frac{0.02 \times 3900}{0.013}$

62. $\frac{0.0084 \times 0.003}{0.21 \times 60}$ **63.** $\frac{0.005 \times 650 \times 3.3}{0.0011 \times 2500}$ **64.** $\frac{5.4 \times 0.003 \times 50}{15 \times 0.0027 \times 200}$

65. $\dfrac{160 \times 0.09 \times 4600}{0.00012 \times 0.0023}$

66. $\dfrac{5.2 \times 68{,}000 \times 0.042}{0.017 \times 2100 \times 0.013}$

67. $\dfrac{880 \times 3200 \times 0.0006}{640 \times 1200 \times 0.044}$

68. $\dfrac{0.008 \times 1600 \times 4200}{0.28 \times 0.0012 \times 0.02}$

69. Light travels approximately 3×10^{10} centimeters per second. How many centimeters would this be per minute? per hour?

70. An atom of gold weighs approximately 3.25×10^{-22} grams. What would be the weight of 2000 atoms of gold?

4.3 ADDING AND SUBTRACTING POLYNOMIALS

In Section 3.1 we defined a **term** as an expression that involves only multiplication and/or division with constants and/or variables. Examples of terms are $3x$, $-5y^2$, 17, and $\dfrac{x}{y}$. **Like terms** (or **similar terms**) are terms that contain variables that are of the same power or are constants. We combined like terms using the distributive property. For example,

$$3x + 5x = (3 + 5)x = 8x$$

and

$$14x^3 - 17x^3 = (14 - 17)x^3 = -3x^3$$

A **monomial** is a single term with only whole-number exponents for its variables and no variable in a denominator. The general form of a **monomial in x** is

$$\boldsymbol{kx^n} \qquad \textbf{where } \boldsymbol{n} \textbf{ is a whole number and } \boldsymbol{k} \textbf{ is any number}$$

$\boldsymbol{n}$ is called the **degree** of the monomial, and $\boldsymbol{k}$ is the **coefficient.**

A monomial may have more than one variable, but only monomials in one variable will be discussed in the remainder of this chapter. More variables will be used in Chapter 5.

In the case of a constant monomial, such as 6, we can write $6 = 6 \cdot 1 = 6 \cdot x^0$. So, we say a nonzero constant is a **monomial of 0 degree.** In the case of 0, we can write $0 = 0x = 0x^2 = 0x^{17}$. Because of all the possible ways of writing 0, we say that 0 is a **monomial of no degree.**

Terms That ARE Monomials

$13x^7$	seventh degree
$\frac{1}{2}x^2$	second degree
$-5y^6$	sixth degree

Monomials do not have fractional exponents or negative exponents and do not have any variables with positive exponents in a denominator.

Terms That ARE NOT Monomials

$2x^{-1}$	negative exponent
$x^{1/2}$	fractional exponent
$\frac{3}{x^2}$	variable in the denominator

Any monomial or algebraic sum of monomials is a **polynomial.** For example, $3x$, $x + 5$, and $a^3 + 5a - 7$ are all polynomials. (**Note:** Since we are dealing only with monomials in one variable, we will also consider only polynomials in one variable. Polynomials such as $5xy^2 + z$ will be discussed in later courses in mathematics.)

The **degree of a polynomial** is the largest of the degrees of its terms after like terms have been combined. Generally, for convenience, a polynomial will be written so that the degrees of its terms decrease from left to right. We say the powers are **descending.** For example,

$$7x^3 - 8 + x^2 - x^4 \qquad \text{fourth degree}$$

is the same as

$$-x^4 + 7x^3 + x^2 - 8 \qquad \text{written in descending order}$$

Some polynomial forms are used so frequently in algebra that they have been given special names, as follows:

Classification of Polynomials	***Examples***
• Monomial: polynomial with one term	$3x$
• Binomial: polynomial with two terms	$x + 5$
• Trinomial: polynomial with three terms	$a^3 + 5a - 7$

EXAMPLES

Simplify the following polynomials.

1. $5x^3 + 7x^3 = (5 + 7)x^3 = 12x^3$ third-degree monomial

2. $5x^3 + 7x^3 - 2x = 12x^3 - 2x$ third-degree binomial

3. $\frac{1}{2}y + 3y - \frac{2}{3}y^2 - 7 = -\frac{2}{3}y^2 + \frac{7}{2}y - 7$ second-degree trinomial

4. $x^2 + 8x - 15 - x^2 = 8x - 15$ first-degree binomial

The **sum** of two or more polynomials is found by combining like terms. For example,

$$\begin{aligned}
&(x^2 - 5x + 3) + (2x^2 - 8x - 4) + (3x^3 + x^2 - 5) \\
&= 3x^3 + (x^2 + 2x^2 + x^2) + (-5x - 8x) + (3 - 4 - 5) \\
&= 3x^3 + 4x^2 - 13x - 6
\end{aligned}$$

We can also write like terms one beneath the other and add the like terms in each column.

$$\begin{array}{rrrr}
 & x^2 & -\ 5x & +\ 3 \\
 & 2x^2 & -\ 8x & -\ 4 \\
3x^3 & +\ x^2 & & -\ 5 \\
\hline
3x^3 & +\ 4x^2 & -\ 13x & -\ 6
\end{array}$$

EXAMPLES

5. Add as indicated.

$$\begin{aligned}
&(5x^3 - 8x^2 + 12x + 13) + (-2x^2 - 8) + (4x^3 - 5x + 14) \\
&= (5x^3 + 4x^3) + (-8x^2 - 2x^2) + (12x - 5x) + (13 - 8 + 14) \\
&= 9x^3 - 10x^2 + 7x + 19
\end{aligned}$$

6. Find the sum.

$$\begin{array}{rrrr}
x^3 & -\ x^2 & +\ 5x & \\
4x^3 & +\ 5x^2 & -\ 8x & +\ 9 \\
\hline
5x^3 & +\ 4x^2 & -\ 3x & +\ 9
\end{array}$$

If a negative sign is written in front of a polynomial in parentheses, the meaning is the opposite of the entire polynomial. The opposite can be found by changing the sign of every term in the polynomial.

$$-(2x^2 + 3x - 7) = -2x^2 - 3x + 7$$

We can also think of the opposite of a polynomial as -1 times the polynomial.

$$\begin{aligned} -(2x^2 + 3x - 7) &= -1(2x^2 + 3x - 7) \\ &= -1(2x^2) - 1(3x) - 1(-7) \\ &= -2x^2 - 3x + 7 \end{aligned}$$

The result is the same with either approach. So the **difference** between two polynomials can be found by changing the sign of each term of the second polynomial and then combining like terms.

$$\begin{aligned} (5x^2 - 3x - 7) - (2x^2 + 5x - 8) &= 5x^2 - 3x - 7 - 2x^2 - 5x + 8 \\ &= 3x^2 - 8x + 1 \end{aligned}$$

If the polynomials are written one beneath the other, we change the signs of the terms of the polynomial being subtracted and then combine like terms. Subtract:

$$\begin{array}{r} 5x^2 - 3x - 7 \\ \underline{-(2x^2 + 5x - 8)} \end{array} \qquad \begin{array}{r} 5x^2 - 3x - 7 \\ \underline{-2x^2 - 5x + 8} \\ 3x^2 - 8x + 1 \end{array}$$

EXAMPLES

7. Subtract as indicated.

$$\begin{aligned} &(9x^4 - 22x^3 + 3x^2 + 10) - (5x^4 + 2x^3 + 5x^2 - x) \\ &= 9x^4 - 22x^3 + 3x^2 + 10 - 5x^4 - 2x^3 - 5x^2 + x \\ &= 4x^4 - 24x^3 - 2x^2 + x + 10 \end{aligned}$$

8. Find the difference.

$$\begin{array}{r} 8x^3 + 5x^2 - 14 \\ \underline{-(-2x^3 + x^2 + 6x)} \end{array} \qquad \begin{array}{r} 8x^3 + 5x^2 + 0x - 14 \\ \underline{2x^3 - x^2 - 6x + 0} \\ 10x^3 + 4x^2 - 6x - 14 \end{array}$$

Write in 0's for help in alignment in order to add like terms.

If an expression contains more than one pair of grouping (or inclusion) symbols, such as parentheses ( ), brackets [ ], or braces

{ }, simplify by removing the innermost pair of symbols first. As examples,

a. $5x - [2x + 3(4 - x) + 1] - 9$

$= 5x - [2x + 12 - 3x + 1] - 9$	parentheses first
$= 5x - [-x + 13] - 9$	
$= 5x + x - 13 - 9$	brackets second
$= 6x - 22$	

b. $10 - x + 2[x + 3(x - 5) + 7]$

$= 10 - x + 2[x + 3x - 15 + 7]$	parentheses first
$= 10 - x + 2[4x - 8]$	
$= 10 - x + 8x - 16$	brackets second
$= 7x - 6$	

PRACTICE QUIZ

Questions	**Answers**
1. Combine like terms and state the degree of the polynomial. $8x^3 - 3x^2 - x^3 + 5 + 3x^2$	1. $7x^3 + 5$; third degree
2. Add. $(15x + 4) + (3x^2 - 9x - 5)$	2. $3x^2 + 6x - 1$
3. Subtract. $(-5x^3 - 3x + 4) - (3x^3 - x^2 + 4x - 7)$	3. $-8x^3 + x^2 - 7x + 11$
4. Simplify. $2 - [3a - (4 - 7a) + 2a]$	4. $-12a + 6$

EXERCISES 4.3

Simplify the polynomials in Exercises 1–15 and state the degree of the result.

1. $x + 3x$

2. $4x^2 - x + x^2$

3. $x^3 + 3x^2 - 2x$

4. $3x^2 - 8x + 8x$

5. $x^4 - 4x^2 + 2x^2 - x^4$

6. $2 - 6x + 5x - 2$

7. $-x^3 + 6x + x^3 - 6x$

8. $11x^2 - 3x + 2 - 7x^2$

9. $6x^5 + 2x^2 - 7x^5 - 3x^2$

10. $2x^2 - 3x^2 + 2 - 4x^2 - 2 + 5x^2$

11. $4x - 8x^2 + 2x^3 + 8x^2$

12. $2x + 9 - x + 1 - 2x$

13. $5x^2 + 3 - 2x^2 + 1 - 3x^2$

14. $13x^2 - 6x - 9x^2 - 4x$

15. $7x^3 + 3x^2 - 2x^3 + x - 5x^3 + 1$

Add in Exercises 16–35.

16. $(2x^2 + 5x - 1) + (x^2 + 2x + 3)$

17. $(x^2 + 2x - 3) + (x^2 + 5)$

18. $(x^2 + 7x - 7) + (x^2 + 4x)$

19. $(x^2 + 3x - 8) + (3x^2 - 2x + 4)$

20. $(2x^2 - x - 1) + (x^2 + x + 1)$

21. $(3x^2 + 5x - 4) + (2x^2 + x - 6)$

22. $(-2x^2 - 3x + 9) + (3x^2 - 2x + 8)$

23. $(x^2 + 6x - 7) + (3x^2 + x - 1)$

24. $(-4x^2 + 2x - 1) + (3x^2 - x + 2) + (x - 8)$

25. $(8x^2 + 5x + 2) + (-3x^2 + 9x - 4)$

26. $(x^2 + 2x - 1) + (3x^2 - x + 2) + (x - 8)$

27. $(x^3 + 2x - 9) + (x^2 - 5x + 2) + (x^3 - 4x^2 + 1)$

28.
$$\begin{array}{r} x^2 + 4x - 6 \\ -2x^2 + 3x + 1 \\ \hline \end{array}$$

29.
$$\begin{array}{r} 5x^2 - 3x + 11 \\ -2x^2 + x - 6 \\ \hline \end{array}$$

30.
$$\begin{array}{l} 2x^2 + 4x - 3 \\ 3x^2 - 9x + 2 \\ \hline \end{array}$$

31.
$$\begin{array}{rrrr} x^3 & + 3x^2 & + x & \\ 2x^3 & - x^2 & + 2x & - 4 \\ \hline \end{array}$$

32.
$$\begin{array}{rrrr} x^3 & + 3x^2 & & - 4 \\ & 7x^2 & + 2x & + 1 \\ x^3 & + x^2 & - 6x & \\ \hline \end{array}$$

33.
$$\begin{array}{rrrr} x^3 & + 5x^2 & + 7x & - 3 \\ & 4x^2 & + 3x & - 9 \\ 4x^3 & + 2x^2 & & - 2 \\ \hline \end{array}$$

34.
$$\begin{array}{rrrr} 7x^3 & + 5x^2 & + x & - 6 \\ & - 3x^2 & + 4x & + 11 \\ -3x^3 & - 2x^2 & - 5x & + 2 \\ \hline \end{array}$$

35.
$$\begin{array}{rrrr} x^3 & + 2x^2 & & - 5 \\ -2x^3 & & + x & - 9 \\ x^3 & - 2x^2 & & + 14 \\ \hline \end{array}$$

Subtract in Exercises 36–55.

36. $(2x^2 + 4x + 8) - (x^2 + 3x + 2)$

37. $(3x^2 + 7x - 6) - (x^2 + 2x + 5)$

38. $(x^2 + 8x - 3) - (x^2 + 5x - 7)$

39. $(3x^2 - 5x - 11) - (2x^2 - 4x + 1)$

40. $(x^2 - 9x + 2) - (4x^2 - 3x + 4)$

41. $(2x^2 - x - 10) - (-x^2 + 3x - 2)$

42. $(7x^2 + 4x - 9) - (-2x^2 + x - 9)$

43. $(6x^2 + 11x + 2) - (4x^2 - 2x - 7)$

44. $(10x^2 - 2x + 9) - (-3x^2 + 5x - 2)$

45. $(x^2 - 12x + 3) - (8x^2 - 11x - 6)$

46. $(9x^2 + 6x - 5) - (13x^2 + 6)$

47. $(8x^2 + 9) - (4x^2 - 3x - 2)$

48. $(x^3 + 4x^2 - 7) - (3x^3 + x^2 + 2x + 1)$

49. $\begin{array}{r} 14x^2 - 6x + 9 \\ \underline{8x^2 + x - 9} \end{array}$

50. $\begin{array}{r} 9x^2 - 3x + 2 \\ \underline{4x^2 - 5x - 1} \end{array}$

51. $\begin{array}{r} 11x^2 + 5x - 13 \\ \underline{-3x^2 + 5x + 2} \end{array}$

52. $\begin{array}{r} -3x^2 + 7x - 6 \\ \underline{2x^2 - x + 4} \end{array}$

53. $\begin{array}{r} x^3 + 6x^2 - 3 \\ \underline{-x^3 + 2x^2 - 3x + 7} \end{array}$

54. $\begin{array}{r} 5x^2 + 8x + 11 \\ \underline{-3x^2 + 2x - 4} \end{array}$

55. $\begin{array}{r} 3x^3 + 9x - 17 \\ \underline{x^3 + 5x^2 - 2x - 6} \end{array}$

Remove the symbols of inclusion and combine like terms in Exercises 56–70.

56. $5x + 2(x - 3) - (3x + 7)$

57. $-4(x - 6) - (8x + 2)$

58. $11 + [3x - 2(1 + 5x)]$

59. $2x + [9x - 4(3x + 2) - 7]$

60. $8x - [2x + 4(x - 3) - 5]$

61. $17 - [-3x + 6(2x - 3) + 9]$

62. $3x - [5 - 7(x + 2) - 6x]$

63. $10x - [8 - 5(3 - 2x) - 7x]$

64. $(2x + 4) - [-8 + 2(7 - 3x) + x]$

65. $-[6x - 3(4 + 2x) + 9] - (x + 5)$

66. $2[3x + (x - 8) - (2x + 5)] - (x - 7)$

67. $-3[-x + (10 - 3x) - (8 - 3x)] + (2x - 1)$

68. $(x^2 - 1) + x[4 + (3 - x)]$

69. $x(x - 5) + [6x - x(4 - x)]$

70. $x(2x + 1) - [5x - x(2x + 3)]$

4.4 MULTIPLYING POLYNOMIALS

We have multiplied terms such as $5x^2 \cdot 3x^4 = 15x^6$. Also, we have applied the distributive property to expressions such as $5(2x + 3) = 10x + 15$. Now we will use both of these procedures to multiply polynomials. We will discuss first the product of a monomial with a polynomial of two or more terms; second, the product of two binomials; and third, the product of a binomial with a polynomial of more than two terms.

Using the distributive property $a(b + c) = ab + ac$, we can find the product of a monomial with a polynomial of two or more terms as follows:

$$5x(2x + 3) = 5x \cdot 2x + 5x \cdot 3 = 10x^2 + 15x$$

$$3x^2(4x - 1) = 3x^2 \cdot 4x + 3x^2(-1) = 12x^3 - 3x^2$$

$$\begin{aligned} -4a^5(a^2 - 8a + 5) &= -4a^5 \cdot a^2 - 4a^5(-8a) - 4a^5(5) \\ &= -4a^7 + 32a^6 - 20a^5 \end{aligned}$$

Now suppose that we want to multiply two binomials, say, $(x + 3)(x + 7)$. We will apply the distributive law in a very subtle way.

Compare $\qquad (x + 3)(x + 7)$

to $\qquad a(b + c)$

Think of $(x + 3)$ as taking the place of a. Thus,

$$a(b + c) \quad = \quad ab + ac$$

takes the form $\quad (x + 3)(x + 7) = (x + 3)x + (x + 3)7$

Completing the products on the right, using the distributive property twice again, gives

$$\begin{aligned} (x + 3)(x + 7) &= (x + 3)x + (x + 3)7 \\ &= x^2 + 3x + x \cdot 7 + 3 \cdot 7 \\ &= x^2 + 3x + 7x + 21 \\ &= x^2 + 10x + 21 \end{aligned}$$

In the same manner

$$\begin{aligned} (x + 2)(3x + 4) &= (x + 2)3x + (x + 2)4 \\ &= x \cdot 3x + 2 \cdot 3x + x \cdot 4 + 2 \cdot 4 \\ &= 3x^2 + 6x + 4x + 8 \\ &= 3x^2 + 10x + 8 \end{aligned}$$

Similarly,

$$\begin{aligned} (2x - 1)(x^2 + x - 5) &= (2x - 1)x^2 + (2x - 1)x + (2x - 1)(-5) \\ &= 2x \cdot x^2 - 1 \cdot x^2 + 2x \cdot x - 1 \cdot x + 2x(-5) - 1(-5) \\ &= 2x^3 - x^2 + 2x^2 - x - 10x + 5 \\ &= 2x^3 + x^2 - 11x + 5 \end{aligned}$$

One quick way to check if your products are correct is to substitute some convenient number for x into the original two factors and into the product. Choose any nonzero number that you like. The values of both expressions should be the same. For example, let $x = 1$. Then

$$(x + 2)(3x + 4) = (1 + 2)(3 \cdot 1 + 4) = (3)(7) = 21$$

$$3x^2 + 10x + 8 = 3 \cdot 1^2 + 10 \cdot 1 + 8 = 3 + 10 + 8 = 21$$

The product $(x + 2)(3x + 4) = 3x^2 + 10x + 8$ seems to be correct. We could double check by letting $x = 5$.

$$(x + 2)(3x + 4) = (5 + 2)(3 \cdot 5 + 4) = (7)(19) = 133$$

$$3x^2 + 10x + 8 = 3 \cdot 5^2 + 10 \cdot 5 + 8 = 75 + 50 + 8 = 133$$

Convinced? This is just a quick check, however, and is not foolproof unless you try more numbers than the degree of the product.

The product of two polynomials can also be found by writing one polynomial under the other. The distributive law is applied by multiplying each term of one polynomial by each term of the other. Writing in the usual manner, we have

$$\begin{aligned}(2x^2 + 3x - 4)(3x + 7) &= (2x^2 + 3x - 4)3x + (2x^2 + 3x - 4)7 \\ &= 2x^2 \cdot 3x + 3x \cdot 3x - 4 \cdot 3x + 2x^2 \cdot 7 + 3x \cdot 7 - 4 \cdot 7 \\ &= 6x^3 + 9x^2 - 12x + 14x^2 + 21x - 28 \\ &= 6x^3 + 23x^2 + 9x - 28\end{aligned}$$

Now, writing one under the other, we obtain

$$\begin{array}{r} 2x^2 + 3x - 4 \\ 3x + 7 \\ \hline 14x^2 + 21x - 28 \end{array}$$

Multiply by +7.

$$\begin{array}{r} 2x^2 + 3x - 4 \\ 3x + 7 \\ \hline 14x^2 + 21x - 28 \\ 6x^3 + 9x^2 - 12x \phantom{{}-28} \\ \hline \end{array}$$

Multiply by $3x$.

Align the like terms so that they can be easily combined.

$$
\begin{array}{rrrrl}
 & 2x^2 + & 3x - & 4 & \\
 & & 3x + & 7 & \\
\hline
 & 14x^2 + & 21x - & 28 & \text{Combine like terms.} \\
6x^3 + & 9x^2 - & 12x & & \\
\hline
6x^3 + & 23x^2 + & 9x - & 28 &
\end{array}
$$

EXAMPLES

Find each product.

1. $-4x(x^2 - 3x + 12) = -4x \cdot x^2 - 4x(-3x) - 4x \cdot 12$
   $= -4x^3 + 12x^2 - 48x$

2. $(3a + 4)(2a^2 + a + 5) = (3a + 4)2a^2 + (3a + 4)a + (3a + 4)5$
   $= 6a^3 + 8a^2 + 3a^2 + 4a + 15a + 20$
   $= 6a^3 + 11a^2 + 19a + 20$

3.
$$
\begin{array}{rrrrl}
 & 7y^2 - & 3y + & 2 & \\
 & & 2y + & 3 & \\
\hline
 & +\ 21y^2 - & 9y + & 6 & \text{Multiply by 3.} \\
14y^3 - & 6y^2 + & 4y & & \text{Multiply by } 2y. \\
\hline
14y^3 + & 15y^2 - & 5y + & 6 & \text{Combine like terms.}
\end{array}
$$

EXERCISES 4.4

Find the product in Exercises 1–45 and simplify if possible.

1. $-3x^2(-2x^3)$

2. $4x^5(x^2)$

3. $5x(-4x^2)$

4. $9x^3(2x)$

5. $-1(3x^2 + 2x)$

6. $-7(2x^2 + 3x)$

7. $-4x(x^2 + 1)$

8. $x^2(x^2 + 4x)$

9. $7x^2(x^2 + 2x - 1)$

10. $5x^3(5x^2 - x + 2)$

11. $-x(x^3 + 5x - 4)$

12. $-2x^4(x^3 - x^2 + 2x)$

13. $3x(2x + 1) - 2(2x + 1)$

14. $x(3x + 4) + 7(3x + 4)$

15. $3x(3x - 5) + 5(3x - 5)$

16. $6x(x - 1) + 5(x - 1)$

17. $5x(-2x + 7) - 2(-2x + 7)$

18. $x(x^2 + 1) - 1(x^2 + 1)$

19. $x(x^2 + 3x + 2) + 2(x^2 + 3x + 2)$

20. $4x(x^2 - x + 1) + 3(x^2 - x + 1)$

21. $(x + 4)(x - 3)$

22. $(x + 7)(x - 5)$

23. $(x + 6)(x - 8)$

24. $(x + 2)(x - 4)$

25. $(x - 2)(x - 1)$

26. $(x - 7)(x - 8)$

27. $3(x + 4)(x - 5)$

28. $-4(x + 6)(x - 7)$

29. $x(x + 3)(x + 8)$

30. $x(x - 4)(x - 7)$

31. $(2x + 1)(x - 4)$

32. $(3x - 1)(x + 4)$

33. $(6x - 1)(x + 3)$

34. $(3x + 5)(3x - 5)$

35. $(2x + 3)(2x - 3)$

36. $(8x + 15)(x + 1)$

37. $(4x + 1)(4x + 1)$

38. $(5x - 2)(5x - 2)$

39. $(x + 3)(x^2 - x + 4)$

40. $(2x + 1)(x^2 - 7x + 2)$

41. $$\begin{array}{r} 3x + 7 \\ x - 5 \\ \hline \end{array}$$

42. $$\begin{array}{r} x^2 + 3x + 1 \\ 5x - 9 \\ \hline \end{array}$$

43. $$\begin{array}{r} 8x^2 + 3x - 2 \\ -2x + 7 \\ \hline \end{array}$$

44. $$\begin{array}{r} 2x^2 + 3x + 5 \\ x^2 + 2x - 3 \\ \hline \end{array}$$

45. $$\begin{array}{r} 6x^2 - x + 8 \\ 2x^2 + 5x + 6 \\ \hline \end{array}$$

Find the product in Exercises 46–65 and simplify if possible. Check by letting $x = 2$.

46. $(3x - 4)(x + 2)$

47. $(x + 6)(4x - 7)$

48. $(2x + 5)(x - 1)$

49. $(5x - 3)(x + 4)$

50. $(2x + 1)(3x - 8)$

51. $(x - 2)(3x + 8)$

52. $(7x + 1)(x - 2)$

53. $(3x + 7)(2x - 5)$

54. $(2x + 3)(2x + 3)$

55. $(5x + 2)(5x + 2)$

56. $(x + 3)(x^2 - 4)$

57. $(x^2 + 2)(x - 4)$

58. $(2x + 7)(2x - 7)$

59. $(3x - 4)(3x + 4)$

60. $(x + 1)(x^2 - x + 1)$

61. $(x - 2)(x^2 + 2x + 4)$

62. $(7x - 2)(7x - 2)$

63. $(5x - 6)(5x - 6)$

64. $(2x + 3)(x^2 - x - 1)$

65. $(3x + 1)(x^2 - x + 9)$

4.5 SPECIAL PRODUCTS OF BINOMIALS

Certain types of products of binomials occur so frequently in algebra that they deserve special mention. Their basic forms (or formulas) should be memorized. The forms are these:

I. $(x + a)(x + b) = x^2 + (a + b)x + ab$

II. $(x + a)(x - a) = x^2 - a^2$

III. $(x + a)^2 = x^2 + 2ax + a^2$

IV. $(x - a)^2 = x^2 - 2ax + a^2$

I. $(x + a)(x + b) = x^2 + (a + b)x + ab$

Consider the product

$$\begin{aligned}(x + 4)(x + 6) &= (x + 4)x + (x + 4)6 \\ &= x^2 + 4x + 6x + 24 \\ &= x^2 + (4 + 6)x + 24 \\ &= x^2 + 10x + 24\end{aligned}$$

In this example, $a = 4$ and $b = 6$. Note that $a + b = 10$ and $a \cdot b = 24$. With practice, you should be able to go directly to the answer by working mentally.

$$\begin{aligned}(x + 7)(x + 8) &= x^2 + (7 + 8)x + 7 \cdot 8 \qquad \text{Here, } a = 7 \text{ and } b = 8. \\ &= x^2 + 15x + 56\end{aligned}$$

a and b can be negative as well as positive. The basic form still applies.

$$\begin{aligned}(x - 7)(x + 8) &= x^2 + (-7 + 8)x + (-7)(8) \qquad \text{Here, } a = -7 \text{ and } b = 8. \\ &= x^2 + x - 56\end{aligned}$$

$$\begin{aligned}(x + 7)(x - 8) &= x^2 + (7 - 8)x + 7(-8) \qquad \text{Here, } a = 7 \text{ and } b = -8. \\ &= x^2 - x - 56\end{aligned}$$

II. $(x + a)(x - a) = x^2 - a^2$

The result, $x^2 - a^2$, is the **difference of two squares.**

Consider the product as a special case of $(x + a)(x + b)$ where $b = -a$.

$$(x + 5)(x - 5) = x^2 + (5 - 5)x + 5(-5)$$
$$= x^2 + 0x - 25$$
$$= x^2 - 25$$
$$(x + 8)(x - 8) = x^2 - 8^2 = x^2 - 64$$
$$(x - 9)(x + 9) = x^2 - 9^2 = x^2 - 81$$

III. $\mathbf{(x + a)^2 = x^2 + 2ax + a^2}$

The result, $x^2 + 2ax + a^2$, is called a **perfect square trinomial** because it is the square of a binomial.

Consider the product as a special case of $(x + a)(x + b)$ where $b = a$.

$$(x + 5)(x + 5) = x^2 + (5 + 5)x + 5 \cdot 5$$
$$= x^2 + 2 \cdot 5 \cdot x + 5^2$$
$$= x^2 + 10x + 25$$
$$(x + 7)(x + 7) = x^2 + 14x + 49$$
$$(x + 10)(x + 10) = x^2 + 20x + 100$$

One interesting device that might help in remembering $(x + a)^2 = x^2 + 2ax + a^2$ is the square shown in Figure 4.1. The area of the square is $(x + a)^2$ because each side has length $x + a$. But the area is also equal to the sum of the areas of the smaller rectangles and squares

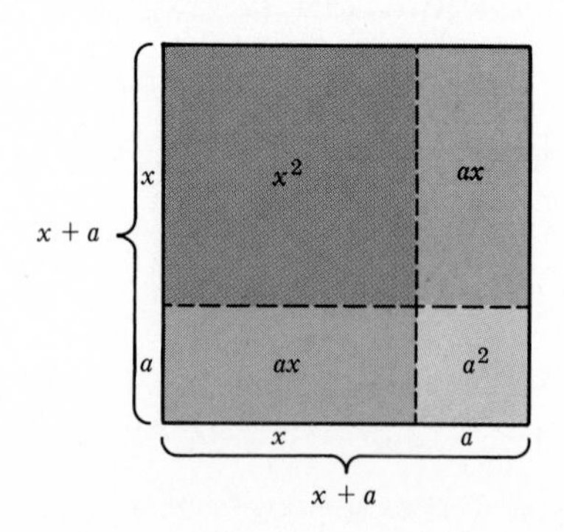

FIGURE 4.1

inside the large square. That is,

$$x^2 + ax + ax + a^2 \qquad \text{or} \qquad (x + a)^2 = x^2 + 2ax + a^2$$

IV. $(\mathbf{x} - \mathbf{a})^2 = \mathbf{x}^2 - 2\mathbf{ax} + \mathbf{a}^2$

The result, $x^2 - 2ax + a^2$, is also called a **perfect square trinomial.**

$$\begin{aligned}(x - 5)(x - 5) &= x^2 + (-5 - 5)x + (-5)(-5)\\ &= x^2 - 2(5)x + (-5)^2\\ &= x^2 - 10x + 25\end{aligned}$$

$$(x - 7)(x - 7) = x^2 - 14x + 49$$

$$(x - 10)(x - 10) = x^2 - 20x + 100$$

In the special cases noted by formulas I, II, III, and IV, the coefficient of x in each binomial was 1. If either coefficient is not 1, the following technique for multiplying binomials is useful:

$$\begin{aligned}(2x + 3)(5x + 2) &= (2x + 3)5x + (2x + 3)^2\\ &= 2x \cdot 5x + 3 \cdot 5x + 2x \cdot 2 + 3 \cdot 2\\ &= \underbrace{2x \cdot 5x}_{\substack{\textbf{F}\text{irst}\\ \text{terms}\\ \textbf{F}}} + \underbrace{2x \cdot 2}_{\substack{\textbf{O}\text{utside}\\ \text{terms}\\ \textbf{O}}} + \underbrace{3 \cdot 5x}_{\substack{\textbf{I}\text{nside}\\ \text{terms}\\ \textbf{I}}} + \underbrace{3 \cdot 2}_{\substack{\textbf{L}\text{ast}\\ \text{terms}\\ \textbf{L}}}\end{aligned}$$

This technique, illustrated by the following diagramed equations, is called the **FOIL method** of mentally multiplying two binomials.

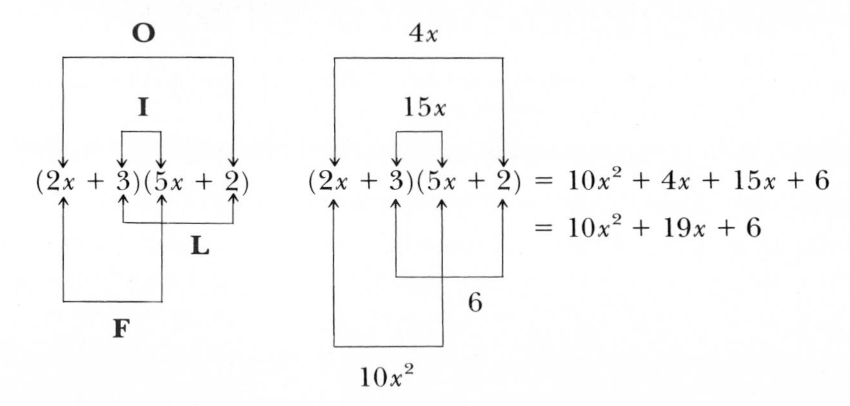

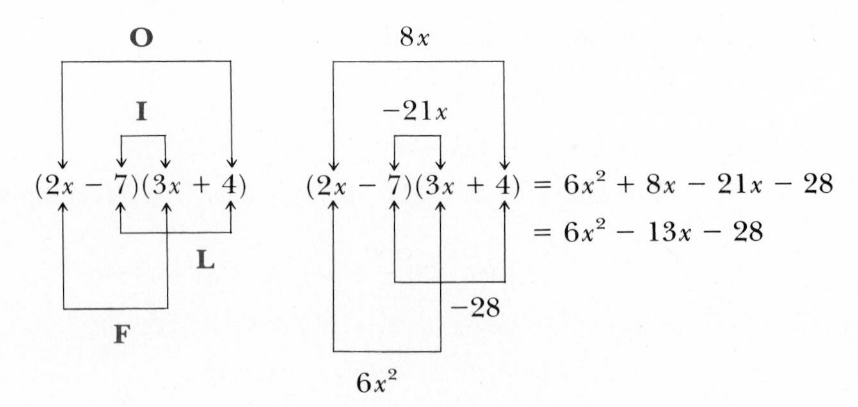

You must understand that in all the formulas, I, II, III, and IV, the letters are placeholders and x or a or b may be replaced by other terms. For example, formula II applies to the product

$$(3y + 5)(3y - 5) = (3y)^2 - 5^2 = 9y^2 - 25$$

Similarly, formula III applies to

$$\begin{aligned}(3y + 5)^2 &= (3y)^2 + 2 \cdot 5 \cdot 3y + 5^2 \\ &= 9y^2 + 30y + 25\end{aligned}$$

and

$$\begin{aligned}(y^3 + 1)^2 &= (y^3)^2 + 2 \cdot 1 \cdot y^3 + 1^2 \\ &= y^6 + 2y^3 + 1\end{aligned}$$

Although all four formulas can be considered as special cases of the FOIL method, they still should be memorized.

EXAMPLES

Find each product and identify those that are the difference of two squares or perfect square trinomials.

1. $$\begin{aligned}(x + 5)(x - 1) &= (x + 5)x + (x + 5)(-1) \\ &= x^2 + 5x - x - 5 \\ &= x^2 + 4x - 5\end{aligned}$$

   or, using the special form I directly,

   $$\begin{aligned}(x + 5)(x - 1) &= x^2 + (5 - 1)x + 5(-1) \\ &= x^2 + 4x - 5\end{aligned}$$

2. $\left(x + \frac{2}{5}\right)\left(x - \frac{2}{5}\right) = x^2 - \left(\frac{2}{5}\right)^2$

$= x^2 - \frac{4}{25}$ difference of two squares

3. $\left(x + \frac{2}{3}\right)^2 = x^2 + 2 \cdot \frac{2}{3} x + \left(\frac{2}{3}\right)^2$

$= x^2 + \frac{4}{3} x + \frac{4}{9}$ perfect square trinomial

4. $(2x + 1)(4x + 3) = (2x + 1)4x + (2x + 1)3$

$= 8x^2 + 4x + 6x + 3$

$= 8x^2 + 10x + 3$

or, using the FOIL setup,

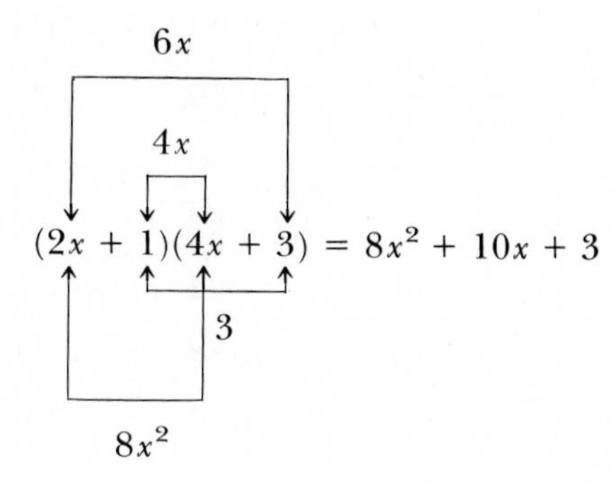

EXERCISES 4.5

Write the products of Exercises 1–30 and identify those that are the difference of two squares or perfect square trinomials.

1. $(x + 3)(x - 3)$

2. $(x - 7)^2$

3. $(x - 5)^2$

4. $(x + 4)(x + 4)$

5. $(x - 6)(x + 6)$

6. $(x + 9)(x - 9)$

7. $(x + 8)(x + 8)$

8. $(x + 12)(x - 12)$

9. $(2x + 3)(x - 1)$

10. $(3x + 1)(2x + 5)$

11. $(x - c)(x + c)$

12. $(x + c)^2$

13. $(2x + 1)(2x - 1)$

14. $(3x + 1)^2$

15. $(3x - 2)(3x - 2)$

16. $(4x + 5)(4x - 5)$

17. $(3 + x)^2$

18. $(8 - x)(8 - x)$

19. $(5 - x)(5 - x)$

20. $(11 - x)(11 + x)$

21. $(5x - 9)(5x + 9)$

22. $(4 - x)^2$

23. $(2x + 7)(2x + 7)$

24. $(3x + 2)^2$

25. $(9x + 2)(9x - 2)$

26. $(6x + 5)(6x - 5)$

27. $(5x + 2)(2x - 3)$

28. $(4x + 7)(2x + 1)$

29. $(1 + 7x)^2$

30. $(2 - 5x)^2$

Write the special products of Exercises 31–65.

31. $(x + 2)(x + 1)$

32. $(x - 2)(x - 3)$

33. $(x - 3)(x + 4)$

34. $(x + 8)(x - 11)$

35. $(x - 7)(x - 6)$

36. $(x + 7)(x + 9)$

37. $(5 + x)(5 + x)$

38. $(3 - x)(6 - x)$

39. $(x^2 + 1)(x^2 - 1)$

40. $(x^2 + 5)(x^2 - 5)$

41. $(x^3 + 2)(x^3 + 2)$

42. $(x^2 - 6)(x^2 + 9)$

43. $(x^3 - 6)^2$

44. $(x^3 - 4)(x^3 - 4)$

45. $(x^2 + 3)(x^2 - 5)$

46. $(x^3 + 8)(x^3 - 8)$

47. $(x^3 - 4)(x^3 - 7)$

48. $(x^3 + 5)(x^3 + 9)$

49. $(x^4 + 1)(x^4 - 1)$

50. $(x^4 - 3)(x^4 + 3)$

51. $\left(x + \frac{2}{3}\right)\left(x - \frac{2}{3}\right)$

52. $\left(x - \frac{1}{2}\right)\left(x + \frac{1}{2}\right)$

53. $\left(x + \frac{3}{4}\right)\left(x - \frac{3}{4}\right)$

54. $\left(x + \frac{3}{8}\right)\left(x - \frac{3}{8}\right)$

55. $\left(x + \frac{3}{5}\right)\left(x + \frac{3}{5}\right)$

56. $\left(x + \frac{4}{3}\right)\left(x + \frac{4}{3}\right)$

57. $\left(x - \frac{5}{6}\right)^2$

58. $\left(x - \frac{2}{7}\right)^2$

59. $\left(x + \frac{3}{2}\right)^2$

60. $\left(x + \frac{5}{9}\right)^2$

61. $\left(x + \frac{1}{4}\right)\left(x - \frac{1}{2}\right)$

62. $\left(x - \frac{1}{5}\right)\left(x + \frac{2}{3}\right)$

63. $\left(x + \frac{1}{3}\right)\left(x + \frac{1}{2}\right)$

64. $\left(x + \frac{3}{8}\right)\left(x + \frac{1}{2}\right)$

65. $\left(x - \frac{4}{5}\right)\left(x - \frac{3}{10}\right)$

CHAPTER 4 SUMMARY

Properties of Exponents

For nonzero real numbers a and b and integers m and n,

1. $a^m \cdot a^n = a^{m+n}$
2. $a^0 = 1$
3. $a^{-n} = \dfrac{1}{a^n}$
4. $\dfrac{a^m}{a^n} = a^{m-n}$
5. $(ab)^n = a^n b^n$
6. $\left(\dfrac{a}{b}\right)^n = \dfrac{a^n}{b^n}$
7. $(a^m)^n = a^{mn}$

The expression 0^0 is undefined.

In **scientific notation,** decimal numbers are written as the product of a number between 1 and 10 and a power of 10.

The general form of a **monomial in x** is

$$kx^n \quad \text{where } n \text{ is a whole number and } k \text{ is any number}$$

n is called the **degree** of the monomial and ***k*** is the **coefficient.**

A nonzero constant is a monomial of **0 degree.**

Zero is a monomial of **no degree.**

A monomial or algebraic sum of monomials is a **polynomial.**

The **degree of a polynomial** is the largest of the degrees of its terms after like terms have been combined.

Classification of Polynomials

- Monomial: polynomial with one term
- Binomial: polynomial with two terms
- Trinomial: polynomial with three terms

The **sum** of two or more polynomials is found by combining like terms. The **difference** of two polynomials is found by changing the sign of each term of the second polynomial and then combining like terms.

The **product** of two polynomials can be found by applying the distributive property $a(b + c) = ab + ac$.

The following basic forms of the products of binomials should be memorized:

I. $(x + a)(x + b) = x^2 + (a + b)x + ab$

II. $(x + a)(x - a) = x^2 - a^2$ **Difference of two squares**

III. $(x + a)^2 = x^2 + 2ax + a^2$ **Perfect square trinomial**

IV. $(x - a)^2 = x^2 - 2ax + a^2$ **Perfect square trinomial**

In the special cases I, II, III, and IV, the coefficient of x in each binomial is 1. If either coefficient is not 1, the **FOIL method** can be used to find the product of two binomials.

CHAPTER 4 REVIEW

Simplify each expression in Exercises 1–20 so that it has only positive exponents.

1. $6^4 \cdot 6^3$

2. $(-3)^3 \cdot (-3)^2$

3. $\frac{7^4}{7^5}$

4. $\frac{5^3 \cdot 5^0}{5^4}$

5. $y^3 \cdot y^2$

6. $y^{-3} \cdot y^4$

7. $\frac{2x^3}{x^{-1}}$

8. $\frac{2y^{-5}}{3y^{-7}}$

9. $\frac{x^0 \cdot x^{-4}}{x^3}$

10. $\frac{4x^3}{2x^{-2}x^4}$

11. $(4x^2y)^3$

12. $(7x^5y^{-2})^2$

13. $\left(\dfrac{6x^2}{y^5}\right)^2$ **14.** $\dfrac{x^{-3}}{x^3y^2}$ **15.** $(a^{-3}b^2)^{-2}$

16. $\left(\dfrac{3xy^4}{x^2}\right)^{-1}$ **17.** $\left(\dfrac{2x^2y^0}{x^{-1}y}\right)^{-3}$ **18.** $\left(\dfrac{8x^{-3}y^2}{xy^{-1}}\right)^0$

19. $\left(\dfrac{x^3y^{-1}}{xy^{-2}}\right)^2$ **20.** $\left(\dfrac{3^{-1}x^3y^{-1}}{x^{-1}y^2}\right)^2$

Write each of the Exercises 21–25 in scientific notation.

21. 4,270,000 **22.** 0.00023 **23.** 0.0015×4200

24. $\dfrac{840}{0.00021}$ **25.** $\dfrac{0.005 \times 66}{0.011 \times 250}$

Simplify the polynomials in Exercises 26–31 and state the degree of the result.

26. $7x^2 - 2x^2 + 1$ **27.** $3x + 4x^2 + 6x$

28. $-x^2 + 5x + 2x^2 - x$ **29.** $9x - x^3 + 3x^2 - x + x^3$

30. $-x^3 + 4x^2 - 3 + x - x^2 + 1$ **31.** $8x - 7x^2 + x^2 - x^3 - 6x - 4$

Add or subtract as indicated in Exercises 32–39.

32. $(7x^2 + 2x - 1) + (8x - 4)$

33. $(6x^2 - 5x + 3) + (-4x^2 - 3x + 9)$

34. $(-2x^2 - 11x + 1) + (x^3 + 3x - 7) + (2x - 1)$

35. $(4x^2 + 2x - 7) - (5x^2 + x - 2)$

36. $(x^3 + 4x^2 - x) - (-2x^3 + 6x + 3)$

37. $(6x^2 + x - 10) - (x^3 - x^2 + x - 4)$

38. $(x^2 + 2x + 6) + (5x^2 - x - 2) - (8x + 3)$

39. $(2x^2 - 5x - 7) - (3x^2 - 4x + 1) + (x^2 - 9)$

Remove the symbols of inclusion in Exercises 40–45 and combine like terms.

40. $3x(x + 4) - 5(x^2 + 3x)$

41. $-2[7x - (2x + 5) + 3]$

42. $6x - [9 - 2(3x - 1) + 7x]$

43. $2x^2 + x[4x - (8x - 3) + x(2x + 7)]$

44. $5 - 3[4x - (7 - 2x)]$

45. $3x + 2[x - 3(x + 5) + 4(x - 1)]$

Find the products in Exercises 46–55 and identify those products that are the difference of squares or perfect square trinomials.

46. $-3x(x^2 - 4)$

47. $x^2(x - x^3)$

48. $(x + 6)(x - 6)$

49. $(x + 4)(x - 3)$

50. $(3x + 7)(3x + 7)$

51. $(2x - 1)(2x + 1)$

52. $(x^2 + 5)(x^2 - 5)$

53. $(x^2 - 2)(x^2 - 2)$

54. $\left(x + \frac{2}{5}\right)\left(x - \frac{2}{5}\right)$

55. $\left(x + \frac{5}{8}\right)\left(x + \frac{5}{8}\right)$

Find the products in Exercises 56–65 and check by letting $x = 3$.

56. $-1(x^2 - 5x + 2)$

57. $3x(x^2 + 2x - 1)$

58. $(3x + 2)(4x - 7)$

59. $(2x - 9)(x + 4)$

60. $(x - 6)(5x + 3)$

61. $(3x - 4)(2x + 3)$

62. $(x^2 + 4)(x^2 - 4)$

63. $(x - 6)(x^2 + 8)$

64. $(4x + 1)(x^2 - x)$

65. $(5x - 2)(x^2 + x - 2)$

CHAPTER 4 TEST

Simplify each expression in Exercises 1–4 so that it has only positive exponents.

1. $(4x^4)(-3x^2)$

2. $\dfrac{8x^5 \cdot x^2}{4x^{-2}}$

3. $\left(\dfrac{4xy^2}{x^3}\right)^{-1}$

4. $\left(\dfrac{2x^0y^3}{x^{-1}y}\right)^2$

Write Exercises 5 and 6 in scientific notation.

5. 230×0.005

6. $\dfrac{65 \times 0.012}{150}$

Simplify the polynomials in Exercises 7 and 8, and state the degree of the result.

7. $3x + 4x^2 - x^3 + 4x^2 + x^3$

8. $2x^2 + 3x - x^3 + x^2 - 1$

9. Add: $(x^2 + 3x - 8) + (2x^2 - x + 1)$.

10. Add: $(-x^2 + 9x - 6) + (3x^2 - 2)$.

11. Add: $(5x^3 - 2x + 7) + (-x^2 + 8x - 2)$.

12. Subtract: $(4x^2 + 3x - 1) - (6x^2 + 2x + 5)$.

13. Subtract: $(x^2 + 3x + 9) - (-6x^2 - 11x + 5)$.

14. Subtract: $(3x^3 - 2x) - (4x^2 + 3x - 8)$.

Remove the symbols of inclusion in Exercises 15 and 16, and combine like terms.

15. $7x + [2x - 3(4x + 1) + 5]$

16. $12x - 2[5 - (7x + 1) + 3x]$

Find the product in Exercises 17–20.

17. $5x^2(-3x^2 + 9x)$

18. $(5x + 4)(5x - 4)$

19. $(4 - 3x)^2$

20. $3x(x - 7)(2x + 5)$

USING THE COMPUTER

```
10  REM
20  REM This program will determine whether or not an integer
30  REM is prime.
40  '*******************************************************
50  '* NUMBER : a number which is input                    *
60  '*******************************************************
70  INPUT "Type in a positive integer. ( 0 to quit ) ",NUMBER
80  IF NUMBER < 0 THEN GOTO 70
90  IF NUMBER = 0 THEN GOTO 190
100 IF NUMBER = 1 THEN GOTO 170
110 FOR COUNT = 2 TO NUMBER-1
120 REMAINDER = NUMBER MOD COUNT
130 IF REMAINDER = 0 THEN GOTO 170
140 NEXT COUNT
150 PRINT "The number "NUMBER" is a prime number."
160 PRINT : GOTO 70
170 PRINT "The number "NUMBER" is not a prime number."
180 PRINT : GOTO 70
190 PRINT "Bye............."
200 END
```

```
Type in a positive integer. ( 0 to quit )  1
The number  1  is not a prime number.

Type in a positive integer. ( 0 to quit )  2
The number  2  is a prime number.

Type in a positive integer. ( 0 to quit )  3
The number  3  is a prime number.

Type in a positive integer. ( 0 to quit )  4
The number  4  is not a prime number.

Type in a positive integer. ( 0 to quit )  5
The number  5  is a prime number.

Type in a positive integer. ( 0 to quit )  6
The number  6  is not a prime number.

Type in a positive integer. ( 0 to quit )  7
The number  7  is a prime number.

Type in a positive integer. ( 0 to quit )  8
The number  8  is not a prime number.

Type in a positive integer. ( 0 to quit )  9
The number  9  is not a prime number.
```

```
Type in a positive integer. ( 0 to quit )   10
The number  10  is not a prime number.

Type in a positive integer. ( 0 to quit )   11
The number  11  is a prime number.

Type in a positive integer. ( 0 to quit )  0
Bye.............
```

Algebra is the intellectual instrument which has been created for rendering clear the quantitative aspect of the world.

ALFRED NORTH WHITEHEAD 1861–1947

DID YOU KNOW?

You have noticed by now that almost every algebraic skill somehow relates to equation solving and applied problems. The emphasis on equation solving has always been a part of classical algebra.

In Italy, during the Renaissance, it was the custom for one mathematician to challenge another mathematician to an equation-solving contest. A large amount of money, often in gold, was applied by patrons or sponsoring cities as the prize. At that time, it was important not to publish equation-solving methods, since mathematicians could earn large amounts of money if they could solve problems that their competitors could not. Equation-solving techniques were passed down from a mathematician to an apprentice, but they were never shared.

A Venetian mathematician, Niccolò Fontana (1500?–57), known as Tartaglia, "the stammerer," discovered how to solve third-degree or cubic equations. At that time, everyone could solve first- and second-degree equations and special kinds of equations of higher degree. Tartaglia easily won equation-solving contests simply by giving his opponents third-degree equations to solve.

Tartaglia planned to keep his method secret, but, after receiving a pledge of secrecy, he gave his method to Girolamo Cardano (1501–76). Cardano broke his promise by publishing one of the first successful Latin algebra texts, *Ars Magna*, "The Great Art." In it, he included not only Tartaglia's solution to the third-degree equations but also a pupil's (Ferrari) discovery of the general solution to fourth-degree equations. Until recently, Cardano received credit for discovering both methods.

It was not until 300 years later that it was shown that there are no general algebraic methods for solving fifth- or higher-degree equations. Thus, a great deal of time and energy has gone into developing the methods of equation solving that you are learning.

5 Factoring Polynomials

5.1 COMMON MONOMIAL FACTORS

In Section 4.1, we discussed the property of exponents $\frac{a^m}{a^n} = a^{m-n}$. This property, along with reducing fractions and/or dividing numbers, can be used when dividing two monomials. For example,

$$\frac{35x^8}{5x^2} = 7x^6 \qquad \text{and} \qquad \frac{16a^5}{-8a} = -2a^4$$

To divide a polynomial by a monomial, a procedure that will be discussed in more detail in Chapter 6, we divide each term in the polynomial by the monomial. For example,

$$\frac{8x^3 - 14x^2 + 10x}{2x} = \frac{8x^3}{2x} - \frac{14x^2}{2x} + \frac{10x}{2x}$$

$$= 4x^2 - 7x + 5$$

This concept of dividing each term by a monomial is part of finding a monomial factor of a polynomial. Finding the greatest common monomial factor in a polynomial means to **choose the monomial with the highest degree and the largest integer coefficient that will divide into each term of the polynomial.**

This monomial will be one factor, and the sum of the various quotients will be the other factor. For example, factor

$$8x^3 - 14x^2 + 10x$$

We have already seen that $2x$ will divide into each term. Note carefully that this division yields **integer** coefficients. The factors of $8x^3 - 4x^2 + 10x$ are $2x$ and $4x^2 - 7x + 5$. We write

$$8x^3 - 14x^2 + 10x = 2x(4x^2 - 7x + 5)$$

Now factor $24x^6 - 12x^4 - 18x^3$. Look closely at each term. Each term is divisible by several different monomials, such as x, x^2, $3x^2$, $2x^3$, and $6x^3$. We want the one with the largest coefficient and the largest exponent, namely, $6x^3$. Thus,

$$24x^6 - 12x^4 - 18x^3 = 6x^3 \cdot 4x^3 + 6x^3(-2x) + 6x^3(-3)$$
$$= 6x^3(4x^3 - 2x - 3)$$

$$\left(\textbf{Note: } \frac{24x^6}{6x^3} = 4x^3, \quad \frac{-12x^4}{6x^3} = -2x, \quad \frac{-18x^3}{6x^3} = -3\right)$$

With practice, all this division can be done mentally.

If all the terms are negative or if the leading term (the term of highest degree) is negative, we will generally factor a negative common monomial. This will leave a positive coefficient for the first term in parentheses.

EXAMPLES

Factor each polynomial by finding the greatest common monomial factor.

1. $x^3 - 7x = x \cdot x^2 + x(-7) = x(x^2 - 7)$
2. $5x^3 - 15x^2 = 5x^2 \cdot x + 5x^2(-3) = 5x^2(x - 3)$
3. $2x^4 - 3x^2 + 1$ no common monomial factor
4. $-4a^5 + 2a^3 - 6a^2 = -2a^2(2a^3 - a + 3)$

   (Here a negative term was factored because the leading term was negative. Although this method is optional for you, it is recommended.)

A polynomial may be **in more than one variable.** For example, $5x^2y + 10xy^2$ is in the two variables **x** and **y**. Thus, a common monomial factor may have more than one variable.

$$5x^2y + 10xy^2 = 5xy \cdot x + 5xy \cdot 2y$$
$$= 5xy(x + 2y)$$

Similarly,

$$4xy^3 - 2x^2y^2 + 8xy^2 = 2xy^2 \cdot 2y + 2xy^2(-x) + 2xy^2 \cdot 4$$
$$= 2xy^2(2y - x + 4)$$

$$\left(\textbf{Note: } \frac{4xy^3}{2xy^2} = 2y, \quad \frac{-2x^2y^2}{2xy^2} = -x, \quad \frac{8xy^2}{2xy^2} = 4\right)$$

EXAMPLES

5. $4ax^3 + 4ax = 4ax(x^2 + 1)$

6. $3x^2y^2 - 6xy^2 = 3xy^2(x - 2)$

7. $14by^3 + 7b^2y - 21by^2 = 7by(2y^2 + b - 3y)$

8. $-2x^2y^3 + 4xy^4 = -2xy^3(x - 2y)$

PRACTICE QUIZ

Questions	Answers
Factor each polynomial by finding the greatest common monomial factor.	
1. $2x^2 - 4$	1. $2(x^2 - 2)$
2. $-5x^2 - 5x$	2. $-5x(x + 1)$
3. $7ax^2 - 7ax$	3. $7ax(x - 1)$
4. $9x^2y^2 + 12x^2y - 6x^2$	4. $3x^2(3y^2 + 4y - 2)$

EXERCISES 5.1

Simplify the expressions in Exercises 1–10. Assume that no denominator is equal to 0.

1. $\frac{x^3}{x}$ **2.** $\frac{y^6}{y^4}$ **3.** $\frac{x^7}{x^3}$ **4.** $\frac{x^8}{x^3}$ **5.** $\frac{-8y^3}{2y^2}$

6. $\frac{12x^2}{2x}$ **7.** $\frac{9x^5}{3x^2}$ **8.** $\frac{-10x^5}{2x}$ **9.** $\frac{4x^3y^2}{2xy}$ **10.** $\frac{21x^4y^3}{-3xy^2}$

Factor each of the polynomials in Exercises 11–70 by finding the greatest common monomial factor.

11. $6x - 21$
12. $-14x - 21$
13. $-9x^2 + 36$
14. $11x^2 - 121$
15. $6x + 12y$
16. $-8a - 16b$
17. $x^3 - 9x$
18. $16y^3 - 9y$
19. $a^2b - a^2c$
20. $4ax - 8ay$

21. $-3x^2 + 6x$

22. $-6ax - 9ay$

23. $7xy - 14yz$

24. $10x^2y - 25xy$

25. $24y^2z - 12yz$

26. $-42by^2 + 21by$

27. $24ax^3 - 54ax$

28. $16x^4y - 14x^2y$

29. $44xy^5 - 121xy^3$

30. $18y^2z^2 - 2yz$

31. $80m^2y^3 - 5y^3$

32. $-14x^2y^3 + 56x^2y$

33. $11x^2 - 22x + 11$

34. $8y^2 - 32y + 32$

35. $-3x^2 + 6x - 9$

36. $10x^2 - 15x - 20$

37. $a^2 - a + 2ab$

38. $-cx - cy + cz$

39. $x^3 - 4x^2 + 6x$

40. $ad^2 + 10ad + 25a$

41. $14ab + 6ac - 2ad$

42. $12cm + 3cm + 18dm$

43. $2bc^2 + 6bc + 8b$

44. $8mx - 12my + 4mz$

45. $-8y^3 - 16y^2 + 24y$

46. $36tx - 45ty + 24tz$

47. $14x^4 + 27x^3 + 9x^2$

48. $34x^5 - 51x^4 + 17x^3$

49. $110x^3 - 121x^2 + 11x$

50. $-56x^4 - 98x^3 - 35x^2$

51. $15x^7 + 24x^6 - 32x^4$

52. $108x^6 - 72x^5 - 135x^4$

53. $50axb^2 - 2axc^2$

54. $a^2x^2z - y^2z$

55. $-8x^3y + 32xy^3$

56. $7x^4y^6 - 28x^2z^4$

57. $9axy^3 - 9axy$

58. $-4cx^4 - 12cx^3 + 16cx^2$

59. $4c^2x^2 - 8c^2xy - 12c^2y^2$

60. $36x^2y^2 + 12x^2y - 15x^2$

61. $-3x^2y^2 - 6x^3y^3 - 9xy$

62. $15xy^2 - 20x^2y^3 - 25x^5y^7$

63. $abx^2 - ab^2x^2 + a^2bx^2$

64. $-16x^5y - 15x^4y - 3x^2y$

65. $12xy^5 - 18xy^4 + 24y^3$

66. $22x^3z + 11x^2z - 33xz$

67. $x^5 - 3x^4 + 7x^2 - 21x$

68. $16y^6 - 56y^5 - 120y^4 + 64y^3$

69. $4x^4 - 6x^3 + 14x^2 - 2x$

70. $10x^5 - 25x^4 + 60x^3 - 45x^2$

5.2 FACTORING SPECIAL PRODUCTS

In Section 4.5, we discussed the following special products of binomials:

I. $(x + a)(x + b) = x^2 + (a + b)x + ab$

II. $(x + a)(x - a) = x^2 - a^2$ difference of two squares

III. $(x + a)^2 = x^2 + 2ax + a^2$ perfect square trinomial

IV. $(x - a)^2 = x^2 - 2ax + a^2$ perfect square trinomial

Suppose the product polynomial $x^2 + 9x + 20$ is known. We can find the factors by reversing the multiplication procedure. **If all four forms have been memorized,** then we might recognize $x^2 + 9x + 20$ as being in form I. We need to know the factors of 20 that add to be 9. They are 5 and 4, so, using form I,

$$\begin{aligned} x^2 + 9x + 20 &= x^2 + (4 + 5)x + 4 \cdot 5 \\ &= (x + 4)(x + 5) \end{aligned}$$

Other examples are

a. $x^2 - 12x + 20 = (x - 2)(x - 10)$

since $(-2)(-10) = 20$ and $(-2) + (-10) = -12$

b. $x^2 - x - 20 = (x - 5)(x + 4)$

since $(-5)(+4) = -20$ and $-5 + 4 = -1$

(Note: Since we are interested in learning the skills related to factoring, the polynomials discussed here are all factorable in some way. However, you should be aware of the fact that not all polynomials are factorable. Such polynomials are said to be **irreducible.** There are a few of these in the exercises.)

If the polynomial is the difference of two squares, we know from form II that the factors are the sum and difference of the terms that were squared.

$$\begin{aligned} x^2 - a^2 &= (x + a)(x - a) \\ x^2 - 9 &= (x + 3)(x - 3) \\ x^2 - y^2 &= (x + y)(x - y) \\ 25y^2 - 4 &= (5y + 2)(5y - 2) \end{aligned}$$

If the polynomial is a perfect square trinomial, then the last term must be a perfect square and the middle coefficient must be twice the term that was squared. (**Note:** We are assuming here that the coefficient of x^2 is 1. The case where the coefficient is not 1 will be covered in Section 5.3.) Using form III and form IV,

$$x^2 + 6x + 9 = (x + 3)^2; \qquad 9 = 3^2 \text{ and } 6 = 2 \cdot 3$$

$$x^2 - 14x + 49 = (x - 7)^2; \qquad 49 = (-7)^2 \text{ and } -14 = 2(-7)$$

Recognizing the form of the polynomial is the key to factoring. Sometimes the form may be disguised by a common monomial factor or by a rearrangement of the terms. **Always look for a common monomial factor first.** For example,

$$5x^2y - 20y = 5y(x^2 - 4) \qquad \text{factoring the common monomial } 5y$$

$$= 5y(x + 2)(x - 2) \qquad \text{difference of two squares}$$

EXAMPLES

Factor each of the following polynomials completely.

1. $x^2 - x - 12$

Solution:

$$x^2 - x - 12 = (x - 4)(x + 3) \qquad -4(3) = -12 \text{ and } -4 + 3 = -1$$

2. $y^2 - 10y + 25$

Solution:

$$y^2 - 10y + 25 = (y - 5)^2 \qquad \text{perfect square trinomial}$$

3. $6a^2b - 6b$

Solution:

$$6a^2b - 6b = 6b(a^2 - 1) \qquad \text{common monomial factor}$$

$$= 6b(a + 1)(a - 1) \qquad \text{difference of two squares}$$

4. $3x^2 - 15 + 12x$

Solution:

$$3x^2 - 15 + 12x = 3(x^2 - 5 + 4x) \qquad \text{common monomial factor}$$

$$= 3(x^2 + 4x - 5) \qquad \text{rearranging terms}$$

$$= 3(x + 5)(x - 1) \qquad -1(15) = -5 \text{ and } -1 + 5 = 4$$

5. $a^6 - 100$ $\qquad a^6 = (a^3)^2$

Solution:

$a^6 - 100 = (a^3 + 10)(a^3 - 10)$ difference of two squares

Procedures to Factor Special Products

1. Look for a common monomial factor.
2. Check the number of terms:
   a. Two terms—Is it the difference of two squares?
   b. Three terms—Is it a perfect square trinomial?
3. Check the possibility of factoring any of the factors.

Closely related to factoring special products is the procedure of **completing the square.** This procedure involves adding a square term to a binomial so that the resulting trinomial is a perfect square trinomial, thus "completing the square." For example,

$$x^2 + 10x + \underline{\qquad} = (\qquad)^2$$

The middle coefficient, 10, is twice the number that is to be squared. So, by taking half this coefficient and squaring the result, we will have the missing constant.

$$x^2 + 10x + \underline{\qquad} = (\qquad)^2$$

$$x^2 + 10x + \underline{\ 25\ } = (x + 5)^2; \quad \tfrac{1}{2}(10) = 5 \text{ and } 5^2 = 25$$

For $x^2 + 18x$, we get

$$x^2 + 18x + \underline{\qquad} = (\qquad)^2$$

$$x^2 + 18x + \underline{\ 81\ } = (x + 9)^2; \quad \tfrac{1}{2}(18) = 9 \text{ and } 9^2 = 81$$

EXAMPLES

Complete the square as indicated.

6. $y^2 + 20y + \underline{\qquad} = (\qquad)^2$

Solution: Since $\frac{1}{2}(20) = 10$ and $10^2 = 100$, then

$$y^2 + 20y + \underline{\ 100\ } = (y + 10)^2$$

7. $x^2 + 3x + \_\_\_\_\_ = (\qquad)^2$

Solution: Since $\frac{1}{2}(3) = \frac{3}{2}$ and $\left(\frac{3}{2}\right)^2 = \frac{9}{4}$, then

$$x^2 + 3x + \underline{\frac{9}{4}} = \left(x + \frac{3}{2}\right)^2$$

(Note: We introduce fractions in this situation because they seem to cause considerable difficulty later on for many students.)

8. $a^2 - 5a + \_\_\_\_\_ = (\qquad)^2$

Solution: Since $\frac{1}{2}(-5) = -\frac{5}{2}$ and $\left(\frac{-5}{2}\right)^2 = \frac{25}{4}$, then

$$a^2 - 5a + \underline{\frac{25}{4}} = \left(a - \frac{5}{2}\right)^2$$

PRACTICE QUIZ

Questions	Answers
Factor completely.	
1. $x^2 - 16 =$	1. $(x + 4)(x - 4)$
2. $25 - x^2 =$	2. $(5 + x)(5 - x)$
3. $3y^2 + 6y + 3 =$	3. $3(y + 1)^2$
4. $y^{10} - 81 =$	4. $(y^5 + 9)(y^5 - 9)$
Complete the square.	
5. $x^2 + 6x + \_\_\_\_ = (\qquad)^2$	5. $x^2 + 6x + \underline{9} = (x + 3)^2$
6. $x^2 + 7x + \_\_\_\_ = (\qquad)^2$	6. $x^2 + 7x + \underline{\frac{49}{4}} = \left(x + \frac{7}{2}\right)^2$

EXERCISES 5.2

Factor each of the polynomials in Exercises 1–60 completely.

1. $x^2 - 3x + 2$

2. $x^2 + 5x + 6$

3. $x^2 - 1$

4. $x^2 - 4$

5. $x^2 + 7x + 12$

6. $y^2 + y - 20$

7. $y^2 - 3y - 10$

8. $t^2 - 2t - 8$

9. $t^2 - t - 2$

10. $y^2 - 4y - 12$

11. $a^2 + 16$

12. $x^2 - 10x + 21$

13. $y^2 - 5y - 24$

14. $x^2 - 4x + 4$

15. $x^2 + 10x + 25$

16. $t^2 + 11t + 10$

17. $c^2 - d^2$

18. $y^2 + 8y + 64$

19. $9a^2 - 1$

20. $z^2 - 12z + 36$

21. $x^2 - 12x + 32$

22. $y - 9x^2y$

23. $a^2 - 2a + 2$

24. $x^2 + 11x + 24$

25. $x^2 + 13x + 22$

26. $x^2 - 4y^2$

27. $y^2 - 20y + 100$

28. $c^2 + 3c - 18$

29. $25 - 9x^2$

30. $2ax + 4ay + 10az$

31. $z^2 - 13z - 14$

32. $4 - 4c + c^2$

33. $3x^2 - 27y^2$

34. $x^2 + 2x - 12$

35. $cx^2 + 3cx - 4c$

36. $ax^2 + ax + 2a$

37. $12ax^2 + 75ay^2$

38. $16x^2 - y^2z^2$

39. $ay^2 + 2ay + a$

40. $4y^2 + 12y - 40$

41. $m^3 - mn^2$

42. $nz^2 - 4nz - 5n$

43. $2ay^2 - 14ay + 24a$

44. $3by^2 + 18by + 24b$

45. $5x^2y - 10xy + 20y$

46. $100x^2 - 81y^2$

47. $27m^2 - 48n^2$

48. $ax^2 - 11ax + 28a$

49. $6 - 5x + x^2$

50. $4x^2y^2 - 24xy^2 + 36y^2$

51. $2ay^2 - 44ay + 242a$

52. $3x^3 + 12x^2 + 18x$

53. $a^4 - 1$

54. $24 - 5x - x^2$

55. $-3x^2 + 36x - 105$

56. $x^4 - y^4$

57. $3x^4 - 24x^2 + 48$

58. $6ay^4 + 42ay^2 + 36a$

59. $a^6 - b^4$

60. $x^4 - 5x^2 + 4$

Complete the square by adding the correct constant in Exercises 61–75 so that the trinomial will factor as indicated.

61. $x^2 - 6x +$ _____ $= (\quad)^2$

62. $x^2 - 12x +$ _____ $= (\quad)^2$

63. $x^2 + 4x +$ _____ $= (\quad)^2$

64. $x^2 + 20x +$ _____ $= (\quad)^2$

65. $x^2 - 8x +$ _____ $= (\quad)^2$

66. $x^2 + 14x +$ _____ $= (\quad)^2$

67. $x^2 - 18x +$ _____ $= (\quad)^2$

68. $x^2 + 22x +$ _____ $= (\quad)^2$

69. $x^2 - 16x +$ _____ $= (\quad)^2$

70. $x^2 - 24x +$ _____ $= (\quad)^2$

71. $x^2 + x +$ _____ $= (\quad)^2$

72. $y^2 - 7y +$ _____ $= (\quad)^2$

73. $y^2 - 9y +$ _____ $= (\quad)^2$

74. $y^2 - 3y +$ _____ $= (\quad)^2$

75. $x^2 + 11x +$ _____ $= (\quad)^2$

5.3 MORE ON FACTORING POLYNOMIALS

Using the FOIL method of multiplication discussed in Section 4.5, we can find the product

$$(2x + 5)(3x + 1) = 6x^2 + 17x + 5$$

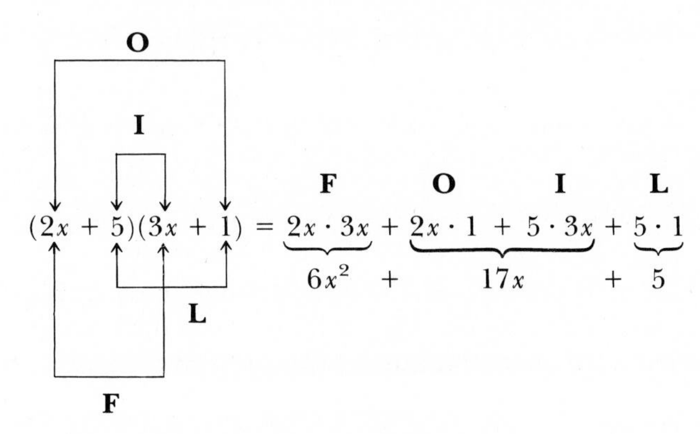

- **F:** the product of the first two terms is $6x^2$.
- **O:** }
- **I:** } the sum of the inner and outer products is $17x$.
- **L:** the product of the last two terms is 5.

To factor the trinomial $6x^2 + 31x + 5$ as a product of two binomials, we know that the product of the first two terms must be $6x^2$. By **trial and error** we try all combinations of factors of $6x^2$, namely, $6x$ and x or $3x$ and $2x$, along with the factors of 5. This will guarantee that the first product, F, and the last product, L, are correct.

a. $(3x + 1)(2x + 5)$
b. $(3x + 5)(2x + 1)$
c. $(6x + 1)(x + 5)$
d. $(6x + 5)(x + 1)$

Now, for these possibilities, we need to check the sums of the inner and outer products to find $31x$.

15x
2x

a. $(3x + 1)(2x + 5)$; $15x + 2x = 17x$

3x
10x

b. $(3x + 5)(2x + 1)$; $3x + 10x = 13x$

30x
x

c. $(6x + 1)(x + 5)$; $30x + x = 31x$

We have found the correct combination of factors, so we need not try $(6x + 5)(x + 1)$. So,

$$6x^2 + 31x + 5 = (6x + 1)(x + 5)$$

With practice the inner and outer sums can be found mentally and much time can be saved; but the method is still basically trial and error. To help limit the trial-and-error search, look at the constant term:

1. If the sign of the constant term is positive (+), the signs in both factors will be the same (either both + or both −).

2. If the sign of the constant term is negative (−), the signs in the factors will differ.

EXAMPLES

1. Factor $6x^2 - 31x + 5$.

Solution: Since the middle term is $-31x$ and the constant is $+5$, we know that the two factors of 5 must both be negative, -5 and -1.

$$6x^2 - 31x + 5 = (6x - 1)(x - 5); \qquad -30x - x = -31x$$

2. Factor $2x^2 + 12x + 10$ completely.

Solution:

$$2x^2 + 12x + 10 = 2(x^2 + 6x + 5)$$ First find any common monomial factor.

$$= 2(x + 5)(x + 1); \qquad x + 5x = 6x$$

Special Note: To **factor completely** means to find factors of the polynomial none of which are themselves factorable. Thus, $2x^2 + 12x + 10 = (2x + 10)(x + 1)$ is **not** factored completely since $2x + 10 = 2(x + 5)$. We could write

$$\begin{aligned} 2x^2 + 12x + 10 &= (2x + 10)(x + 1) \\ &= 2(x + 5)(x + 1) \end{aligned}$$

Finding the greatest common monomial factor first generally makes the problem easier. The trial-and-error method may seem difficult at first, but with practice you will learn to "guess" better and to eliminate certain combinations quickly. For example, to factor $10x^2 + x - 2$, do we use $10x$ and x or $5x$ and $2x$; and for -2, do we use -2 and $+1$ or $+2$ and -1? The terms $5x$ and $2x$ are more likely candidates since they are closer together than $10x$ and x and the middle term is small, $1x$. So,

$(5x + 1)(2x - 2)$	$-10x + 2x = -8x$	**Reject.**
$(5x - 1)(2x + 2)$	$+10x - 2x = 8x$	**Reject.**
$(5x + 2)(2x - 1)$	$-5x + 4x = -x$	**Reject.**
$(5x - 2)(2x + 1)$	$5x - 4x = x$	**Accept!**

$$10x^2 + x - 2 = (5x - 2)(2x + 1)$$

Not all polynomials are factorable. For example, no matter what combinations we try, $3x^2 - 3x + 4$ will not have two binomial factors with integer coefficients. This polynomial is **irreducible; it cannot be factored as a product of polynomials with integer coefficients.**

An important irreducible polynomial is the sum of two squares, $a^2 + b^2$. For example, $x^2 + 4$ is irreducible. There are no factors with integer coefficients whose product is $x^2 + 4$.

EXAMPLES

Factor completely. Look first for the greatest common monomial factor.

3. $4x^2 - 100 = 4(x^2 - 25) = 4(x + 5)(x - 5)$

4. $6x^3 - 8x^2 + 2x = 2x(3x^2 - 4x + 1) = 2x(3x - 1)(x - 1)$

5. $2x^2 + x - 6 = (2x - 3)(x + 2)$

6. $x^2 + x + 1 = x^2 + x + 1$ irreducible

The product

$$(x + 3)(y + 5) = (x + 3)y + (x + 3)5$$
$$= xy + 3y + 5x + 15$$

has four terms and none of the terms are like terms; therefore, none can be combined. Factoring polynomials with four or more terms can **sometimes** be accomplished by grouping the terms and using the distributive property, as in the following examples.

EXAMPLES

7. $xy + 5x + 3y + 15 = (xy + 5x) + (3y + 15)$ Group terms that have a common factor.

$= x(y + 5) + 3(y + 5)$ Use the distributive property.

$= (y + 5)(x + 3)$ Note that $y + 5$ is a common factor.

8. $ax + ay + bx + by = (ax + ay) + (bx + by)$ Group.

$= a(x + y) + b(x + y)$ Use the distributive property.

$= (x + y)(a + b)$ Use the distributive property again.

Some other grouping may yield the same results. For example,

$$\begin{aligned} ax + ay + bx + by &= (ax + bx) + (ay + by) \\ &= x(a + b) + y(a + b) \\ &= (a + b)(x + y) \end{aligned}$$

9. $$\begin{aligned} x^2 - xy - 5x + 5y &= (x^2 - xy) + (-5x + 5y) \\ &= x(x - y) + 5(-x + y) \end{aligned}$$

This does not work because $x - y \neq -x + y$. Try factoring -5 instead of $+5$ from the last two terms.

$$\begin{aligned} x^2 - xy - 5x + 5y &= (x^2 - xy) + (-5x + 5y) \\ &= x(x - y) - 5(x - y) \\ &= (x - y)(x - 5) \qquad \text{Success!} \end{aligned}$$

Procedures to Follow in Factoring

1. Look for a common monomial factor.
2. Check the number of terms:
   a. Two terms—Is it the difference of two squares?
   b. Three terms—Is it a perfect square trinomial? Do you have to use the trial-and-error method?
   c. Four terms—Group terms that have a common factor.
3. Check the possibility of factoring any of the factors.

PRACTICE QUIZ

Questions	**Answers**
Factor completely.	
1. $3x^2 + 7x - 6$	1. $(3x - 2)(x + 3)$
2. $x^2 + 25$	2. irreducible
3. $3x^2 + 15x + 18$	3. $3(x + 2)(x + 3)$
4. $10x^2 - 41x - 18$	4. $(5x + 2)(2x - 9)$
5. $5x + 35 - xy - 7y$	5. $(x + 7)(5 - y)$

EXERCISES 5.3

Factor as completely as possible.

1. $2x^2 - x - 1$

2. $2y^2 + 3y + 1$

3. $4t^2 - 3t - 1$

4. $2x^2 - 3x - 2$

5. $2a^2 + 7a + 3$

6. $2y^2 - 9y + 9$

7. $5a^2 - a - 6$

8. $3a^2 + 4a + 1$

9. $7x^2 + 5x - 2$

10. $8x^2 - 10x - 3$

11. $4x^2 + 23x + 15$

12. $6x^2 + 23x + 21$

13. $x^2 + 6x - 16$

14. $9x^2 - 3x - 20$

15. $6x^2 - 19x + 10$

16. $4b^2 - 4b + 1$

17. $10y^2 - y - 24$

18. $5y^2 - 7y - 6$

19. $3x^2 - 7x + 2$

20. $7x^2 - 11x - 6$

21. $9x^2 - 6x + 1$

22. $4x^2 - 25$

23. $4x^2 + 25$

24. $x^2 + 81$

25. $12y^2 - 7y - 12$

26. $x^2 - 46x + 45$

27. $5x^2 + 45$

28. $4x^2 + 64$

29. $4a^2 - 11a + 6$

30. $6m^2 - 73m + 12$

31. $x^2 + x + 1$

32. $x^2 + 2x + 2$

33. $16x^2 - y^2$

34. $3x^2 - 11x - 4$

35. $12x^2 - 25x + 7$

36. $16x^2 - 16x - 5$

37. $64x^2 - 48x + 9$

38. $35x^2 + 74x + 35$

39. $15m^2 - 19m - 10$

40. $9x^2 - 12x + 4$

41. $6x^2 + 2x - 20$

42. $12y^2 - 15y + 3$

43. $10x^2 + 35x + 30$

44. $24y^2 + 4y - 4$

45. $18x^2 - 8y^2$

46. $15y^2 - 10y - 40$

47. $7x^4 - 5x^3 + 3x^2$

48. $12x^2 - 60x + 75$

49. $-12m^2 + 22m + 4$

50. $32y^2 + 50z^2$

51. $6x^3 + 9x^2 - 6x$

52. $-5y^2 + 40y - 60$

53. $9x^3y^3 + 9xy$

54. $30a^3 + 51a^2 + 9a$

55. $12x^3 - 108x^2 + 243x$

56. $48x^2y - 354xy + 126y$

57. $48xy^3 - 100xy^2 + 48xy$

58. $24a^2x^2 + 72a^2x + 48a^2$

59. $21y^4 - 98y^3 + 56y^2$

60. $72a^3 - 306a^2 + 189a$

61. $bx + b + cx + c$

62. $3x + 3y + ax + ay$

63. $x^3 + 3x^2 + 6x + 18$

64. $2z^3 - 14z^2 + 3z - 21$

65. $x^2 - 4x + 6xy - 24y$

66. $3x + 3y - bx - by$

67. $5xy + yz - 20x - 4z$

68. $x - 3xy + 2z - 6zy$

69. $24y + 2xz - 3yz - 16x$

70. $10xy - 2y^2 + 7yz - 35xz$

5.4 SOLVING QUADRATIC EQUATIONS BY FACTORING

How would you solve the equation $5(x - 2) = 0$? Would you proceed in either of the following ways?

$$5(x - 2) = 0 \qquad\qquad 5(x - 2) = 0$$

$$\frac{5(x - 2)}{5} = \frac{0}{5} \qquad\qquad 5x - 10 = 0$$

$$x - 2 = 0 \qquad\qquad 5x = 10$$

$$x = 2 \qquad\qquad \frac{5x}{5} = \frac{10}{5}$$

$$x = 2$$

Both ways are correct and yield the solution $x = 2$. But did you think that $x - 2$ had to be 0? This is true because $5 \cdot 0 = 0$, and 0 is the only number multiplied by 5 that will give a product of 0. You could have written

$$5(x - 2) = 0$$

$$x - 2 = 0$$

$$x = 2$$

Now consider an equation involving a product of two polynomials such as

$$(x - 3)(x - 2) = 0$$

If you multiply $(x - 3)(x - 2)$, you get

$$(x - 3)(x - 2) = 0$$

$$x^2 - 5x + 6 = 0$$

This procedure does not help because $x^2 - 5x + 6 = 0$ is not easier to solve than the original equation.

Since we have a product that equals 0, we allow one of the factors to be 0. Which one? Choose $x - 2$.

$$(x - 3)(x - 2) = 0$$
$$x - 2 = 0$$
$$x = 2$$

Checking, we have

$$(2 - 3)(2 - 2) \stackrel{?}{=} 0$$
$$(-1)(0) = 0$$

So, 2 is *one* solution.

Suppose we choose to let $x - 3 = 0$.

$$(x - 3)(x - 2) = 0$$
$$x - 3 = 0$$
$$x = 3$$

Checking, we have

$$(3 - 3)(3 - 2) \stackrel{?}{=} 0$$
$$0(1) = 0$$

So, 3 is also a solution.

Thus, to solve an equation involving a product of polynomials equal to 0, we can let each factor in turn equal 0 to find all possible solutions. The reason is that a product is 0 only if at least one of the factors is 0.

> If $a \cdot b = 0$, then $a = 0$ or $b = 0$.

EXAMPLES

Solve the following equations.

1. $(x - 5)(x + 4) = 0$

$x - 5 = 0$ or $x + 4 = 0$

$x = 5$ $\qquad$ $x = -4$

Check: $(5 - 5)(5 + 4) = 0$ or $(4 - 5)(4 - 4) = 0$

$0(9) = 0$ $(-1)(0) = 0$

$0 = 0$ $0 = 0$

2. $x(x - 3) = 0$

$x = 0$ or $x - 3 = 0$

$x = 3$

Check: $0(0 - 3) = 0$ or $3(3 - 3) = 0$

$0(-3) = 0$ $3(0) = 0$

$0 = 0$ $0 = 0$

Polynomials of second degree are called quadratics.
Equations of the form

$$ax^2 + bx + c = 0 \qquad (a \neq 0)$$

are called **quadratic equations.**

Factoring the quadratic expression, **when possible,** gives two factors of first degree. Putting each factor equal to 0 gives two first-degree equations that can easily be solved. Each of these solutions is a solution of the quadratic equation.

Not all quadratics can be factored using integer coefficients. Using techniques other than factoring to solve quadratic equations is discussed in Chapter 10.

EXAMPLES

Solve the following quadratic equations by factoring.

3. $x^2 - 7x + 12 = 0$

$(x - 4)(x - 3) = 0$

$x - 4 = 0$ or $x - 3 = 0$

$x = 4$ $x = 3$

4. $$x^2 + 9x = 22$$

$$x^2 + 9x - 22 = 0$$ One side of the equation must be 0.

$$(x + 11)(x - 2) = 0$$

$$x + 11 = 0 \quad \text{or} \quad x - 2 = 0$$

$$x = -11 \qquad\qquad x = 2$$

5. $$x^2 - 16x + 64 = 0$$

$$(x - 8)^2 = 0$$

$$x - 8 = 0$$

$$x = 8$$ Since both factors are the same, there is only one solution.

6. $$x^2 = 11x - 28$$

$$x^2 - 11x + 28 = 0$$

$$(x - 4)(x - 7) = 0$$

$$x - 4 = 0 \quad \text{or} \quad x - 7 = 0$$

$$x = 4 \qquad\qquad x = 7$$

7. $$4x^2 - 4x - 24 = 0$$

$$4(x^2 - x - 6) = 0$$

$$4(x - 3)(x + 2) = 0$$ The constant factor 4 can never be 0 and does not affect the solution.

$$x - 3 = 0 \quad \text{or} \quad x + 2 = 0$$

$$x = 3 \qquad\qquad x = -2$$

8. $$\frac{x^2}{3} + 5x - 18 = 0$$

$$\mathbf{3} \cdot \frac{x^2}{3} + \mathbf{3} \cdot 5x - \mathbf{3} \cdot 18 = \mathbf{3} \cdot 0$$ Multiply each term on both sides of the equation by 3. If there is more than one denominator, multiply by the LCM of the denominators.

$$x^2 + 15x - 54 = 0$$

$$(x + 18)(x - 3) = 0$$

$$x + 18 = 0 \quad \text{or} \quad x - 3 = 0$$

$$x = -18 \qquad\qquad x = 3$$

9.
$$x^2 = -8x$$
$$x^2 + 8x = 0$$
$$x(x + 8) = 0$$
$$x = 0 \quad \text{or} \quad x + 8 = 0$$
$$x = -8$$

Solving by factoring applies to equations with more than two factors. Their product must equal 0, just as with quadratics.

10.
$$2x^3 - 4x^2 - 6x = 0$$
$$2x(x^2 - 2x - 3) = 0$$
$$2x(x - 3)(x + 1) = 0$$
$$2x = 0 \quad \text{or} \quad x - 3 = 0 \quad \text{or} \quad x + 1 = 0$$
$$x = 0 \qquad x = 3 \qquad x = -1$$

To Solve a Quadratic Equation by Factoring

1. Add or subtract terms as necessary so that 0 is on one side of the equation and the equation is in the form
$$ax^2 + bx + c = 0 \qquad (a \neq 0)$$
2. If there are any fractional coefficients, multiply each term by the least common denominator so that all coefficients will be integers.
3. Factor the quadratic, if possible.
4. Set each nonconstant factor equal to 0 and solve for the unknown.
5. Remember, there will never be more than two solutions.

PRACTICE QUIZ

Questions	**Answers**
Solve the following quadratic equations.	
1. $x^2 - 6x = 0$	1. $x = 0, x = 6$
2. $x^2 - 6x = -5$	2. $x = 1, x = 5$
3. $x^2 - 8x = -16$	3. $x = 4$

EXERCISES 5.4

Solve the equations in Exercises 1–60.

1. $(x - 3)(x + 4) = 0$

2. $(x + 6)(x + 7) = 0$

3. $(x - 9)(x + 11) = 0$

4. $x(x - 4) = 0$

5. $7(x + 6)(x - 2) = 0$

6. $8(x + 3)(x + 1) = 0$

7. $(x + 3)(x - 1) = 0$

8. $(x - 1)(x + 7) = 0$

9. $3x(x + 9) = 0$

10. $(x - 2)(x + 3) = 0$

11. $(x - 7)(x + 3) = 0$

12. $4x(x + 6) = 0$

13. $6x(x - 5) = 0$

14. $4(x + 2)(x + 1) = 0$

15. $\frac{1}{2}(x - 8)(x + 3) = 0$

16. $8(x + 4)(x - 9) = 0$

17. $7x(x + 5)(x + 3) = 0$

18. $\frac{3}{4}x(x + 5)(x - 1) = 0$

19. $\frac{4}{5}(x + 11)(x - 5) = 0$

20. $\frac{2}{3}(x + 9)(x - 6) = 0$

21. $x^2 - 5x + 6 = 0$

22. $x^2 + 3x - 10 = 0$

23. $x^2 - 5x - 14 = 0$

24. $x^2 + 3x - 28 = 0$

25. $x^2 - 6x + 9 = 0$

26. $x^2 + 8x + 12 = 0$

27. $x^2 - 25 = 0$

28. $x^2 + 10x + 25 = 0$

29. $2x^2 - 4x = 0$

30. $3x^2 + 9x = 0$

31. $x^3 - 3x^2 = 4x$

32. $x^3 + 7x^2 = -12x$

33. $5x^2 + 15x = 0$

34. $x^2 - 11x = -18$

35. $2x^2 - 24 = 2x$

36. $4x^2 - 12x = 0$

37. $x^2 + 8 = 6x$

38. $x^2 = x + 30$

39. $2x^2 + 2x - 24 = 0$

40. $9x^2 + 63x + 90 = 0$

41. $4x^2 - 8 = 4x$

42. $20x^2 = 60x$

43. $9x^2 - 36 = 0$

44. $2x^2 + 18x + 36 = 0$

45. $3x^2 + 15x = 42$

46. $4x^2 - 16 = 0$

47. $5x^2 = 30x$

48. $5x^2 + 10x + 5 = 0$

49. $2x^2 + 4x = 6$

50. $3x^2 + 9x = 30$

51. $8x^2 + 32 = 32x$

52. $-6x^2 + 18x + 24 = 0$

53. $9x^2 = 81$

54. $3x^2 = 15x$

55. $64 - 24x - 4x^2 = 0$

56. $63 - 12x - 3x^2 = 0$

57. $\frac{x^2}{3} - 2x + 3 = 0$

58. $\frac{x^2}{9} = 1$

59. $\frac{x^2}{5} - x - 10 = 0$

60. $\frac{2}{3}x^2 + 2x - \frac{20}{3} = 0$

5.5 APPLICATIONS

Whether or not word problems cause you difficulty depends a great deal on your personal experiences and general reasoning abilities. These abilities are developed over a long period of time. A problem that is easy for you, possibly because you have had experience in a particular situation, might be quite difficult for a friend, and vice versa.

Most problems do not say specifically to add, subtract, multiply, or divide. You are to know from the nature of the problem what to do. You are to ask yourself, "What information is given? What am I trying to find? What tools, skills, and abilities do I need to use?"

Word problems should be approached in an orderly manner. Have an "attack plan."

Attack Plan for Word Problems

1. Read the problem carefully at least twice.
2. Decide what is asked for and assign a variable to the unknown quantity.
3. Organize a chart, or a table, or a diagram relating all the information provided.
4. Form an equation. (Possibly a formula of some type is necessary.)
5. Solve the equation.
6. Check your solution with the wording of the problem to be sure it makes sense.

Several types of problems lead to quadratic equations. The problems in this section are set up so that the equations can be solved by factoring. More general problems and approaches to solving quadratic equations are discussed in Chapter 10.

EXAMPLES

1. One number is four more than another, and the sum of their squares is 296. What are the numbers?

Solution: Let x = smaller number
$x + 4$ = larger number

$$x^2 + (x + 4)^2 = 296 \qquad \text{Add the squares.}$$

$$x^2 + x^2 + 8x + 16 = 296$$

$$2x^2 + 8x - 280 = 0$$

$$x^2 + 4x - 140 = 0$$

The constant factor 2 does not affect the solution.

$$(x + 14)(x - 10) = 0$$

$$x + 14 = 0 \quad \text{or} \quad x - 10 = 0$$

$$x = -14 \qquad\qquad x = 10$$

$$x + 4 = -10 \qquad\qquad x + 4 = 14$$

There are two sets of answers to the problem, 10 and 14 or -14 and -10.

Check:

$$10^2 + 14^2 = 100 + 196 = 296$$

and

$$(-14)^2 + (-10)^2 = 196 + 100 = 296$$

2. In an orange grove, there are 10 more trees in each row than there are rows. How many rows are there if there are 96 trees in the grove?

Solution: Let r = number of rows
$r + 10$ = number of trees per row

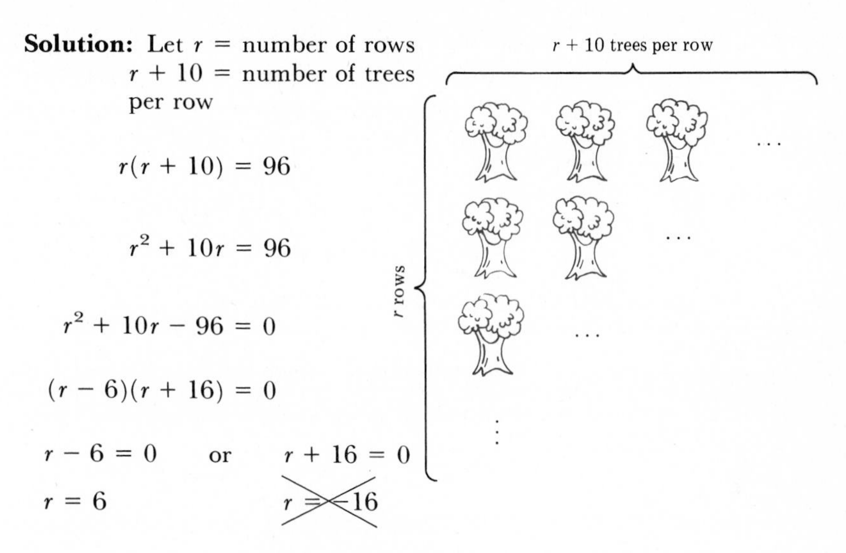

$$r(r + 10) = 96$$

$$r^2 + 10r = 96$$

$$r^2 + 10r - 96 = 0$$

$$(r - 6)(r + 16) = 0$$

$$r - 6 = 0 \quad \text{or} \quad r + 16 = 0$$

$$r = 6 \qquad\qquad \cancel{r = -16}$$

($r = -16$ is an extraneous solution because -16 does not fit the conditions of the problem even though -16 is a solution to the equation.)

There are 6 rows in the grove ($6 \cdot 16 = 96$ trees).

3. A rectangle has an area of 135 square meters and a perimeter of 48 meters. What are the dimensions of the rectangle?

Solution: The area of a rectangle is the product of its length and width ($A = lw$). The perimeter of a rectangle is given by $P = 2l + 2w$. Since the perimeter is 48 meters, then the length plus the width must be 24 meters.

Let w = width

$24 - w$ = length

$$w(24 - w) = 135 \qquad \text{Area = width times length}$$

$$24w - w^2 = 135$$

$$0 = w^2 - 24w + 135$$

$$0 = (w - 9)(w - 15)$$

$$w - 9 = 0 \quad \text{or} \quad w - 15 = 0$$

$$w = 9 \qquad\qquad w = 15$$

$$24 - w = 15 \qquad\qquad 24 - w = 9$$

w

$24 - w$

The width is 9 meters and the length is 15 meters ($9 \cdot 15 = 135$).

4. A man wants to build a block wall shaped like a rectangle along three sides of his property. If 180 feet of fencing are needed and the area of the lot is 4000 square feet, what are the dimensions of the lot?

Solution: Let

$$x = \text{one of two equal sides}$$

$$180 - 2x = \text{third side}$$

The product of width times length gives the area.

$$x(180 - 2x) = 4000$$

$$180x - 2x^2 = 4000$$

$$0 = 2x^2 - 180x + 4000$$

$$0 = x^2 - 90x + 2000$$

$$0 = (x - 50)(x - 40)$$

$$x - 50 = 0 \quad \text{or} \quad x - 40 = 0$$

$$x = 50 \qquad x = 40$$

$$180 - 2x = 80 \qquad 180 - 2x = 100$$

From this information, there are two possible answers: the lot is 50 feet by 80 feet, or the lot is 40 feet by 100 feet.

5. The sum of the squares of two positive consecutive odd integers is 202. Find the integers.

Solution: Let

$$n = \text{first odd integer}$$

$$n + 2 = \text{next consecutive odd integer}$$

$$n^2 + (n + 2)^2 = 202$$

$$n^2 + n^2 + 4n + 4 = 202$$

$$2n^2 + 4n - 198 = 0$$

$$2(n^2 + 2n - 99) = 0$$

$$2(n - 9)(n + 11) = 0$$

$n = 9$ ~~$n = -11$~~ The problem asked for positive integers.

$$n + 2 = 11$$

The first integer is 9 and the next consecutive odd integer is 11 $(9^2 + 11^2 = 81 + 121 = 202)$.

Special Note: Many of the problems in Exercises 5.5 relate to geometric figures. These figures and related formulas are discussed in detail in Section 1.8. You should review Section 1.8 briefly before working on these problems.

EXERCISES 5.5

Determine a quadratic equation for each of the following problems. Then solve the equation.

1. The square of an integer is equal to seven times the integer. Find the integer.

2. The square of an integer is equal to twice the integer. Find the integer.

3. The square of a positive integer is equal to the sum of the integer and twelve. Find the integer.

4. If the square of a positive integer is added to three times the integer, the result is 28. Find the integer.

5. One number is 7 more than another. Their product is 78. Find the numbers.

6. One positive number is three more than twice another. If the product is 27, find the numbers.

7. If the square of a positive integer is added to three times the number, the result is 54. Find the number.

8. One number is six more than another. The difference between their squares is 132. What are the numbers?

9. The difference between two positive integers is 8. If the smaller is added to the square of the larger, the sum is 124. Find the numbers.

10. The sum of two integers is 17, and their product is -110. Find the integers.

11. One positive number is four more than twice a second positive number. The product of the two numbers is 160. Find the two positive numbers.

12. Find a positive integer such that the product of the integer with a number three less than the integer is equal to the integer increased by thirty-two.

13. The product of two consecutive positive integers is 72. Find the integers.

14. Find two consecutive integers whose product is 110.

15. Find two consecutive positive integers such that the sum of their squares is 85.

16. Find two consecutive positive integers such that the square of the second integer added to four times the first is equal to 41.

17. Find two consecutive positive integers such that the difference between their squares is 17.

18. The product of two consecutive odd integers is 63. Find the integers.

19. The product of two consecutive even integers is 120. Find the integers.

20. The product of two consecutive even integers is 168. Find the integers.

21. The length of a rectangle is twice the width. The area is 72 square inches. Find the length and width of the rectangle.

22. The length of a rectangle is three times the width. If the area is 147 square centimeters, find the length and width.

23. The length of a rectangular yard is 12 meters greater than the width. If the area of the yard is 85 square meters, find the length and width.

24. The length of a rectangle is three centimeters greater than the width. The area is 108 square centimeters. Find the length and width.

25. The width of a rectangle is 4 feet less than the length. The area is 117 square feet. Find the length and width.

26. The altitude of a triangle is 4 feet less than the base. The area of the triangle is 16 square feet. Find the length of the base and altitude.

27. The base of a triangle exceeds the altitude by 5 meters. If the area is 42 square meters, find the length of the base and altitude.

28. The base of a triangle is 15 inches greater than the altitude. If the area is 63 square inches, find the length of the base.

29. The base of a triangle is 6 feet less than the altitude. The area is 56 square feet. Find the length of the altitude.

30. The perimeter of a rectangle is 32 inches. The area of the rectangle is 48 square inches. Find the dimensions.

31. The area of a rectangle is 104 square centimeters. If the perimeter is 42 centimeters, find the length and width.

32. The area of a rectangle is 24 square centimeters. If the perimeter is 20 centimeters, find the length and width.

33. The perimeter of a rectangle is 40 meters and the area is 96 square meters. Find the dimensions.

34. An orchard has 140 orange trees. The number of rows exceeds the number of trees per row by 13. How many trees are there in each row?

35. One formation for a drill team is rectangular. The number of members in each row exceeds the number of rows by 3. If there is a total of 108 members in the formation, how many rows are there?

36. A theater can seat 144 people. The number of rows is 7 less than the number of seats in each row. How many rows of seats are there?

37. The length of a rectangle is 7 centimeters greater than the width. If 4 centimeters are added to both the length and width, the new area would be 98 square centimeters. Find the dimensions of the original rectangle.

38. The width of a rectangle is 5 meters less than the length. If 6 meters were added to both the length and width, the new area would be 300 square meters. Find the dimensions of the original rectangle.

39. Suzie is going to fence a rectangular flower garden in her back yard. She has 50 feet of fencing, and she plans to use the house as one side of the garden. If the area is 300 square feet, what are the dimensions of the flower garden?

40. A rancher is going to build a corral with 52 yards of fencing. He is planning to use the barn as one side of the corral. If the area is 320 square yards, what are the dimensions?

5.6 ADDITIONAL APPLICATIONS (OPTIONAL)

The volume of a cylinder is given by the formula

$$V = \pi r^2 h$$

where V is the volume
$\pi = 3.14$ (3.14 is an approximation for π)
r is the radius
h is the height

1. Find the volume of a cylinder with a radius of 6 in. and a height of 20 in.

2. Find the height of a cylinder if the volume is 282.6 in. and the radius is 3 in.

3. A cylinder has a height of 14 in. and a volume of 1099 cu in. Find the radius.

4. Find the radius of a cylinder whose volume is 2512 cu cm and whose height is 8 cm.

The safe load L of a horizontal wooden beam supported at both ends is expressed by the formula

$$L = \frac{kbd^2}{l}$$

where L is expressed in pounds
b is the breadth or width in inches
d is the depth in inches
l is the length in inches
k, a constant, depends on the grade of the beam and is expressed in pounds per square inch

5. What is the safe maximum load of a white pine beam 180 in. long, 3 in. wide, and 4 in. deep if $k = 3000$ lb per sq in.?

6. Find the constant k if a beam 144 in. long, 2 in. wide, and 6 in. deep supports a maximum load of 1100 lb.

7. A solid oak beam is required to support a load of 12,000 lb. It can be no more than 8 in. deep and is 192 in. long. For this grade of oak, $k = 6000$ lb per sq in. How wide should the beam be?

8. The safe maximum load of a white pine beam 4 in. wide and 150 in. long is 2880 lb. For white pine, $k = 3000$ lb per sq in. Find the depth of the beam.

9. A Douglas fir beam is required to support a load of 20,000 lb. For Douglas fir, $k = 4800$ lb per sq in. If the beam is 6 in. wide and 144 in. long, what is the minimum depth for the beam to support the required load?

The equation

$$h = -16t^2 + v_0t$$

gives the height h, in feet, that a body will be above the earth at time t, in seconds, if it is projected upward with an initial velocity v_0, in feet per second.

10. Find the height of an object 3 seconds after it has been projected upward at a rate of 56 feet per second.

11. Find the height of an object 5 seconds after it has been projected upward at a rate of 120 feet per second.

12. A ball is thrown upward with a velocity of 144 feet per second. When will it strike the ground?

13. An object is projected upward at a rate of 160 feet per second. Find the time when it is 384 feet above the ground.

14. An object is projected upward at a rate of 96 feet per second. Find the time when it is 144 feet above the ground.

The power output of a generator armature is given by the equation

$$P_o = E_gI - r_gI^2$$

where P_o is measured in kilowatts
E_g is measured in volts
r_g is measured in ohms
I is measured in amperes

15. Find I if $P_o = 120$ kilowatts, $E_g = 16$ volts, and $r_g = \frac{1}{2}$ ohm.

16. Find I if $P_o = 180$ kilowatts, $E_g = 22$ volts, and $r_g = \frac{2}{3}$ ohm.

The demand for a product is the number of units of the product that consumers are willing to buy when the market price is p dollars. The consumers' total expenditure for the product is found by multiplying the price times the demand.

17. When fishing reels are priced at p dollars, local consumers will buy $36 - p$ fishing reels. What is the price if total sales were \$320?

18. A manufacturer can sell $100 - 2p$ lamps at p dollars each. If the receipts from the lamps total \$1200, what is the price of the lamps?

The demand for a certain commodity is given by

$$D = -20p^2 + ap + 1200 \text{ units per month}$$

where p is the selling price
a is a constant

19. Find the selling price if 1120 units are sold and $a = 60$.

20. Find the selling price if 1860 units are sold and $a = 232$.

5.7 ADDITIONAL FACTORING REVIEW (OPTIONAL)

Factor completely.

1. $m^2 + 7m + 6$

2. $a^2 - 4a + 3$

3. $x^2 + 11x + 18$

4. $y^2 + 8y + 15$

5. $x^2 - 100$

6. $n^2 - 8n + 12$

7. $m^2 - m - 6$

8. $y^2 - 49$

9. $a^2 + 2a + 24$

10. $x^2 + 12x + 35$

11. $64x^2 - 1$

12. $x^2 - 4x - 21$

13. $x^2 + 10x + 25$

14. $x^2 + 3x - 10$

15. $x^2 + 9x - 36$

16. $x^2 + 17x + 72$

17. $x^2 + 13x + 36$

18. $2y^2 - 24y + 70$

19. $5x^2 - 70x + 240$

20. $7x^2 + 14x - 168$

21. $200 + 20x - 4x^2$

22. $64 + 49x^2$

23. $3x^2 - 147$

24. $x^3 - 4x^2 - 12x$

25. $3x^3 + 15x^2 + 18x$

26. $112x - 2x^2 - 2x^3$

27. $16x^3 - 100x$

28. $2x^2 - 3x + 1$

29. $3x^2 - 17x + 10$

30. $2x^2 + 7x + 3$

31. $4x^2 - 14x + 6$

32. $6x^2 - 11x + 4$

33. $12x^2 - 32x + 5$

34. $12x^2 + x - 6$

35. $6x^2 + x - 35$

36. $4x^2 - 18x + 20$

37. $8x^2 + 6x - 35$

38. $12x^2 + 5x - 3$

39. $20x^2 - 21x - 54$

40. $150x^2 - 96$

41. $12x^2 - 60x - 75$

42. $14 + 11x - 15x^2$

43. $24 + x - 3x^2$

44. $21x^2 - x - 10$

45. $8x^2 - 22x + 15$

46. $63x^2 - 40x - 12$

47. $20x^2 + 9x - 20$

48. $35x^2 - x - 6$

49. $18x^2 - 15x + 2$

50. $12x^2 - 47x + 11$

51. $252x - 175x^3$

52. $12x^3 + 2x^2 + 70x$

53. $21x^3 - 13x^2 + 2x$

54. $36x^3 + 120x^2 + 100x$

55. $36x^3 + 21x^2 - 30x$

56. $63x - 3x^2 - 30x^3$

57. $16x^3 - 52x^2 + 22x$

58. $24x^3 - 4x^2 - 160x$

59. $75 + 10x - 120x^2$

60. $144x^3 - 10x^2 - 50x$

61. $xy + 3y - 4x - 12$

62. $2xz + 10x + z + 5$

63. $x^2 + 2xy - 6x - 12y$

64. $2y^2 + 6yz + 5y + 15z$

65. $x^3 - 8x^2 - 5x + 40$

66. $2x^3 - 14x^2 - 3x + 21$

CHAPTER 5 SUMMARY

Finding the greatest common monomial factor in a polynomial means to **choose the monomial with the highest degree and the largest integer coefficient that will divide into each term of the polynomial.**

As mentioned in Chapter 4, the following special forms should be memorized:

I. $x^2 + (a + b)x + ab = (x + a)(x + b)$
II. $x^2 - a^2 = (x + a)(x - a)$
III. $x^2 + 2ax + a^2 = (x + a)^2$
IV. $x^2 - 2ax + a^2 = (x - a)^2$

Adding a square term to a binomial so that the resulting trinomial is a perfect square trinomial is called **completing the square.**

Polynomials that are not of the forms I, II, III, or IV can be factored using the **trial-and-error** method. To **factor completely** means to find factors of the polynomial, none of which are themselves factorable.

A polynomial is **irreducible** if it cannot be factored as a product of polynomials with integer coefficients. For example, $x^2 + a^2$ is irreducible.

Procedures to follow in factoring

1. Look for a common monomial factor.
2. Check the number of terms:
   a. Two terms—Is it the difference of two squares?
   b. Three terms—Is it a perfect square trinomial?
   c. Four terms—Group terms that have a common factor.
3. Check the possibility of factoring any of the factors.

If $a \cdot b = 0$, then $a = 0$ or $b = 0$.

Equations of the form

$$ax^2 + bx + c = 0 \qquad (a \neq 0)$$

are called **quadratic equations.**

To Solve a Quadratic Equation by Factoring

1. Add or subtract terms as necessary so that 0 is on one side of the equation and the equation is in the form

$$ax^2 + bx + c = 0 \qquad (a \neq 0)$$

2. If there are any fractional coefficients, multiply each term by the least common denominator so that all coefficients will be integers.
3. Factor the quadratic, if possible.
4. Set each nonconstant factor equal to 0 and solve for the unknown.
5. Remember, there will never be more than two solutions.

Attack Plan for Word Problems

1. Read the problem carefully at least twice.
2. Decide what is asked for and assign a variable to the unknown quantity.
3. Organize a chart, or a table, or a diagram relating all the information provided.
4. Form an equation. (Possibly a formula of some type is necessary.)
5. Solve the equation.
6. Check your solution with the wording of the problem to be sure it makes sense.

CHAPTER 5 REVIEW

Simplify the expressions in Exercises 1–5. Assume that no denominator is equal to 0.

1. $\dfrac{x^5}{x^2}$ **2.** $\dfrac{x^4}{x}$ **3.** $\dfrac{4x^3}{2x^2}$

4. $\dfrac{-24x^4y^2}{3x^2y}$ **5.** $\dfrac{36x^3y^5}{9xy^4}$

Factor completely Exercises 6–30.

6. $5x - 10$ **7.** $-12x^2 - 16x$ **8.** $16x^2y - 24xy$

9. $10x^4 - 25x^3 + 5x^2$ **10.** $4x^2 - 1$ **11.** $y^2 - 20y + 100$

12. $81x^2 - 4y^2$ **13.** $4x^2 - 12x + 9$ **14.** $ac^2 + a^2b^2 + ad$

15. $3x^2 - 48y^2$ **16.** $x^2 - 7x - 18$ **17.** $5x^2 + 40x + 80$

18. $5x^2 + 17x + 6$ **19.** $x^2 + x - 30$ **20.** $x^2 + x + 3$

21. $25x^2 + 20x + 4$ **22.** $3x^2 + 5x + 2$

23. $2x^3 - 20x^2 + 50x$ **24.** $4x^3 + 4xz^2$

25. $6x^2 - x - 2$ **26.** $8x^3 - 10x^2 - 12x$

27. $x^4 + 3x^2 - 28$ **28.** $x^4 - 2x^2 - 8$

29. $xy + 3x + 2y + 6$ **30.** $ax - 2a - 2b + bx$

Complete the square by adding the correct constant in Exercises 31–35 so that the trinomials in each will factor as indicated.

31. $x^2 - 4x +$ ______ $= (\quad)^2$

32. $x^2 + 18x +$ ______ $= (\quad)^2$

33. $x^2 - 8x +$ ______ $= (\quad)^2$

34. $x^2 - 10x +$ ______ $= (\quad)^2$

35. $x^2 + 5x +$ ______ $= (\quad)^2$

Solve the equations in Exercises 36–50.

36. $(x - 7)(x + 1) = 0$ **37.** $x(3x + 5) = 0$

38. $2x(x + 5)(x - 2) = 0$ **39.** $21x - 3x^2 = 0$

40. $x^2 + 8x + 12 = 0$ **41.** $x^2 = 3x + 28$

42. $\dfrac{x^2}{3} - 3 = 0$ **43.** $x^3 + 5x^2 - 6x = 0$

44. $x^2 - 55 = 6x$

45. $\frac{1}{4}x^2 + x - 15 = 0$

46. $15 - 12x - 3x^2 = 0$

47. $x^3 + 14x^2 + 49x = 0$

48. $8x = 12x + 2x^2$

49. $6x^2 = 24x$

50. $\frac{1}{2}x^2 + x - 24 = 0$

51. The difference between two positive numbers is 9. If the smaller is added to the square of the larger, the result is 147. Find the numbers.

52. The length of a rectangle exceeds the width by 5 cm. If both dimensions were increased by 3 cm, the area would be increased by 96 square centimeters. Find the dimensions of the original rectangle.

53. Find two consecutive whole numbers such that the sum of their squares is equal to 145.

54. The length of a rectangle is 1 foot more than twice the width. If the area is 78 square feet, find the length and width.

55. The base of a triangle is 5 inches longer than the altitude. If the area is 187 square inches, find the length of the base and altitude.

Optional

56. A ball is thrown upward with an initial velocity of 88 feet per second. When will it be 120 feet above the ground? ($h = -16t^2 + v_0t$.)

CHAPTER 5 TEST

Simplify the expressions in Exercises 1 and 2. Assume that no denominator is 0.

1. $\dfrac{16x^4}{8x}$

2. $\dfrac{42x^3y^3}{-6x^3y}$

Factor each of the polynomials in Exercises 3–10 completely.

3. $20x^3y^2 + 18x^3y - 15x^4y$

4. $x^2 - 9x + 20$

5. $x^2 + 14x + 49$

6. $36x^2 - 1$

7. $6x^2 + x - 5$

8. $3x^2 + x - 24$

9. $16x^2 - 25y^2$

10. $2x^3 - x^2 - 3x$

Add the correct constants in Exercises 11 and 12 so that the trinomials will factor as indicated.

11. $x^2 - 10x +$ _____ $= (\qquad)^2$

12. $x^2 + 16x +$ _____ $= (\qquad)^2$

Solve the equations in Exercises 13–17.

13. $(x + 2)(3x - 5) = 0$

14. $x^2 - 7x - 8 = 0$

15. $-3x^2 = 18x$

16. $\dfrac{2x^2}{5} - 6 = \dfrac{4x}{5}$

17. $4x^2 + 17x - 15 = 0$

18. One number is 10 less than five times another number. Their product is 120. Find the numbers.

19. The product of two consecutive positive integers is 342. Find the two integers.

20. The length of a rectangle is 7 centimeters less than twice the width. If the area of the rectangle is 165 square centimeters, find the length and width.

USING THE COMPUTER

```
10  REM
20  REM   This program will find the factors
30  REM   of a given positive integer NUM.
40  REM
50  '**************************************************
60  '*   FACTOR : factor of number N                  *
70  '**************************************************
80  INPUT "Type in a positive integer (0 to quit):",NUM
90  IF NUM = 0 THEN GOTO 190
100 IF NUM < 0 THEN GOTO 80
110 PRINT "The factors of number ";NUM;" are"
120 FOR COUNT = 1 TO NUM/2
130 LET REMAINDER = NUM MOD COUNT
140 IF REMAINDER = 0 THEN PRINT COUNT;
150 NEXT COUNT
160 PRINT NUM
170 PRINT
180 GOTO 80
190 PRINT "Thank you...."
200 END
```

```
Type in a positive integer (0 to quit): 1
The factors of number  1  are
 1

Type in a positive integer (0 to quit): 2
The factors of number  2  are
 1  2

Type in a positive integer (0 to quit): 3
The factors of number  3  are
 1  3

Type in a positive integer (0 to quit): 4
The factors of number  4  are
 1  2  4

Type in a positive integer (0 to quit): 8
The factors of number  8  are
 1  2  4  8

Type in a positive integer (0 to quit): 9
The factors of number  9  are
 1  3  9

Type in a positive integer (0 to quit): 1024
The factors of number  1024  are
 1  2  4  8  16  32  64  128  256  512  1024

Type in a positive integer (0 to quit): 0
Thank you....
```

Number rules the Universe.

THE PYTHAGOREANS

DID YOU KNOW?

Pythagoras (c. 550 B.C.) was a Greek mathematician who founded a secret brotherhood whose objective was to investigate music, astronomy, and mathematics, especially geometry. The Pythagoreans believed that all physical phenomena were explainable in terms of "arithmos," the basic properties of whole numbers and their ratios.

In Chapter 6, you will be studying rational numbers (fractions), numbers written as the ratio of two integers. The Pythagorean Society attached mystical significance to these rational numbers. Their investigations of musical harmony revealed that a vibrating string must be divided exactly into halves, thirds, fourths, fifths, etc., to produce tones in harmony with the string vibrating as a whole.

The recognition that the chords which are pleasing to the ear correspond to exact divisions of the vibrating string by whole numbers stimulated Pythagoras to propose that all physical properties could be described using rational numbers. The Society attempted to compute the orbits of the planets by relating them to the musical intervals. The Pythagoreans thought that as the planets moved through space they produced music, the music of the spheres.

Unfortunately, a scandalous idea soon developed within the Pythagorean Society. It became apparent that there were some numbers that could not be represented as the ratio of two whole numbers. This shocking idea came

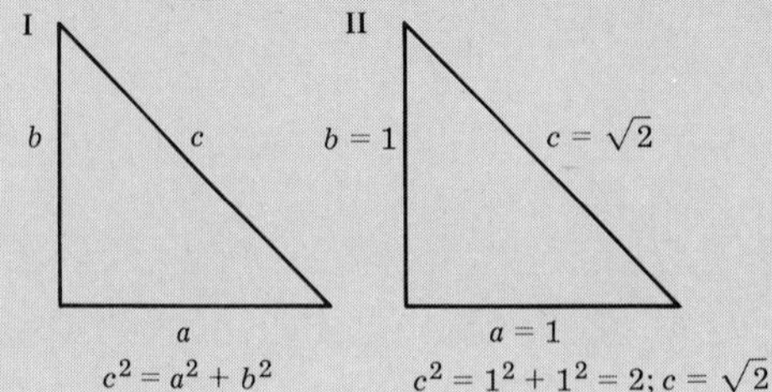

6 Rational Expressions

about when the Pythagoreans investigated their master's favorite theorem, the Pythagorean Theorem:

Given a right triangle, the length of the hypotenuse squared is equal to the sum of the lengths squared of the other two sides (as illustrated in the diagrams on the opposite page). The hypotenuse of the second triangle is some number which, when squared, gives two.

It can be proven (and it was by the Pythagoreans) that no such rational number exists. Imagine the amazement within the Brotherhood. Legend has it that the Pythagoreans attempted to keep the "irrational" numbers a secret. Hippaes, the brother, who first told an outsider about the new numbers, was supposedly expelled from the Society and punished for his unfaithfulness by death.

6.1 DIVISION WITH POLYNOMIALS

In Chapter 1 we reviewed the rules of arithmetic for working with fractions.

Summary of Arithmetic Rules for Fractions

A **fraction** is a number that can be written in the form $\frac{a}{b}$ and means $a \div b$ ($b \neq 0$; no denominator can be 0).

Multiplication: $$\frac{a}{b} \cdot \frac{c}{d} = \frac{a \cdot c}{b \cdot d}$$

The Fundamental Principle:

$$\frac{a}{b} = \frac{a \cdot k}{b \cdot k} \quad \text{where } k \neq 0$$

The **reciprocal** of $\frac{a}{b}$ is $\frac{b}{a}$.

Division: $$\frac{a}{b} \div \frac{c}{d} = \frac{a}{b} \cdot \frac{d}{c}$$

Addition: $$\frac{a}{b} + \frac{c}{b} = \frac{a + c}{b}$$

Subtraction: $$\frac{a}{b} - \frac{c}{b} = \frac{a - c}{b}$$

Fractions in which the numerator and denominator are integers are called **rational numbers.** Fractions in which the numerator and denominator are polynomials are called **rational expressions.** For example, all of the following are rational expressions:

$$\frac{x}{x^2 + 1}, \quad \frac{x^2 + 5x + 6}{x^2 + 7x + 12}, \quad \frac{1}{5x}, \quad \text{and} \quad \frac{x^2 + 2x + 1}{x}$$

In this chapter, we will develop the rules for arithmetic (adding, subtracting, multiplying, and dividing) with rational expressions. You should find the work quite similar to what you did with fractions in Chapter 1. Obviously, all your new knowledge of exponents and factoring will also be very helpful.

The sum of fractions with common denominators can be written as a single fraction by adding the numerators and using the common denominator. For example,

$$\frac{3}{a} + \frac{2b}{a} + \frac{c}{a} = \frac{3 + 2b + c}{a}$$

Reversing the process, we can write

$$\frac{3x^3 + 6x^2 + 9x}{3x} = \frac{3x^3}{3x} + \frac{6x^2}{3x} + \frac{9x}{3x} = x^2 + 2x + 3$$

Thus, we can express a single fraction with a monomial denominator as the sum of fractions with that denominator. This expression is sometimes preferred if some of the fractions can be simplified as we just did.

Similarly,

$$\frac{x^2 + 2x + 1}{x} = \frac{x^2}{x} + \frac{2x}{x} + \frac{1}{x} = x + 2 + \frac{1}{x}$$

and

$$\frac{3xy + 6xy^2 - 9}{3y} = \frac{3xy}{3y} + \frac{6xy^2}{3y} - \frac{9}{3y} = x + 2xy - \frac{3}{y}$$

EXAMPLES

Divide each polynomial by the monomial denominator by writing a sum of fractions. Reduce each fraction, if possible.

1. $$\frac{8x^2 - 14x + 1}{2} = \frac{8x^2}{2} - \frac{14x}{2} + \frac{1}{2} = 4x^2 - 7x + \frac{1}{2}$$

2. $$\frac{12x^3 + 9x^2 - 9x}{3x^2} = \frac{12x^3}{3x^2} + \frac{9x^2}{3x^2} - \frac{9x}{3x^2}$$
$$= 4x + 3 - \frac{3}{x}$$

3. $$\frac{10x^2y + 25xy + 3y^2}{5xy^2} = \frac{10x^2y}{5xy^2} + \frac{25xy}{5xy^2} + \frac{3y^2}{5xy^2}$$
$$= \frac{2x}{y} + \frac{5}{y} + \frac{3}{5x}$$

The procedure we follow when dividing one number by another is called the **division algorithm.** By this division algorithm, we can find $135 \div 8$ as follows:

$$\begin{array}{r} 16 \\ 8\overline{)135} \\ \underline{8} \\ 55 \\ \underline{48} \\ 7 \end{array}$$

The dividend is 135.
The divisor is 8.
The quotient is 16.
The remainder is 7.

We can also write the answer as $16\frac{7}{8}$.
(**Note:** The remainder is smaller than the divisor.)

Check: $135 = 8 \cdot 16 + 7$

In algebra, the division algorithm with polynomials is quite similar. If we divide one polynomial by another, the quotient will be another polynomial with a remainder. Symbolically, if we have $P \div D$, then

$$P = Q \cdot D + R$$

or

$$\frac{P}{D} = Q + \frac{R}{D}$$

where P is the dividend, D is the divisor, Q is the quotient, and R is the remainder. **The remainder must be of smaller degree than the divisor. If the remainder is 0, then the divisor and quotient are factors of the dividend.**

The division algorithm is illustrated in a step-by-step form in the following two examples. In this chapter, we will divide only by first-degree binomials. You should understand, however, that the basic procedure is the same as long as the degree of the dividend is greater than or equal to the degree of the divisor.

EXAMPLES

4. $\dfrac{3x^2 - 5x + 4}{x + 2}$ or $(3x^2 - 5x + 4) \div (x + 2)$

Steps	***Explanation***
Step 1 $x + 2\overline{)3x^2 - 5x + 4}$	Write both polynomials in order of descending powers. If any powers are missing, fill in with 0's.
Step 2 $\begin{array}{r} 3x \\ x + 2\overline{)3x^2 - 5x + 4} \end{array}$	Mentally divide $3x^2$ by x. $\dfrac{3x^2}{x} = 3x$. Write $3x$ above $3x^2$.

Steps		***Explanation***
Step 3	$\begin{array}{r} 3x \\ x + 2 \overline{\big) 3x^2 - 5x + 4} \\ \underline{3x^2 + 6x} \end{array}$	Multiply $3x$ times $(x + 2)$, and write the terms under the like terms in the dividend.
Step 4	$\begin{array}{r} 3x \\ x + 2 \overline{\big) 3x^2 - 5x + 4} \\ \underline{\mp 3x^2 \mp 6x} \\ - 11x \end{array}$	Subtract $3x^2 + 6x$ by changing signs and adding.
Step 5	$\begin{array}{r} 3x \\ x + 2 \overline{\big) 3x^2 - 5x + 4} \\ \underline{\mp 3x^2 \mp 6x} \\ - 11x + 4 \end{array}$	Bring down the 4.
Step 6	$\begin{array}{r} 3x - 11 \\ x + 2 \overline{\big) 3x^2 - 5x + 4} \\ \underline{\mp 3x^2 \mp 6x} \\ - 11x + 4 \end{array}$	Mentally divide $-11x$ by x. $\dfrac{-11x}{x} = -11$. Write -11 in the quotient directly above $-11x$.
Step 7	$\begin{array}{r} 3x - 11 \\ x + 2 \overline{\big) 3x^2 - 5x + 4} \\ \underline{\mp 3x^2 \mp 6x} \\ - 11x + 4 \\ \underline{- 11x - 22} \end{array}$	Multiply -11 times $(x + 2)$, and write the terms under like terms in the expression $-11x + 4$.
Step 8	$\begin{array}{r} 3x - 11 \\ x + 2 \overline{\big) 3x^2 - 5x + 4} \\ \underline{\mp 3x^2 \mp 6x} \\ - 11x + 4 \\ \underline{\pm 11x \pm 22} \\ 26 \end{array}$	Subtract $-11x - 22$ by changing signs and adding. The remainder is 26.

Step 9 **Check:** Multiply the divisor and quotient, and add the remainder; the result should be the dividend.

$$\begin{aligned} (3x - 11)(x + 2) + 26 &= 3x^2 + 6x - 11x - 22 + 26 \\ &= 3x^2 - 5x + 4 \end{aligned}$$

The answer can also be written in the form $Q + \dfrac{R}{D}$ as

$$3x - 11 + \frac{26}{x + 2}$$

5. $\dfrac{4x^3 - 7x + 5}{2x + 1}$ or $(4x^3 - 7x + 5) \div (2x + 1)$

This example will be done in fewer steps. Note that $0x^2$ is inserted so that like terms will be aligned.

	Steps	***Explanation***
Step 1	$2x + 1\overline{)4x^3 + 0x^2 - 7x + 5}$	x^2 is a missing power, so $0x^2$ is supplied.
Step 2	$\begin{array}{r} 2x^2 \qquad\qquad\quad \\ 2x + 1\overline{)4x^3 + 0x^2 - 7x + 5} \\ \underline{4x^3 + 2x^2} \qquad\quad \end{array}$	$\dfrac{4x^3}{2x} = 2x^2$ $2x^2(2x + 1) = 4x^3 + 2x^2$
Step 3	$\begin{array}{r} 2x^2 \qquad\qquad\quad \\ 2x + 1\overline{)4x^3 + 0x^2 - 7x + 5} \\ \underline{\mp 4x^3 \mp 2x^2} \qquad\quad \\ -2x^2 - 7x \quad \end{array}$	Subtract $4x^3 + 2x^2$ and bring down $-7x$.
Step 4	$\begin{array}{r} 2x^2 - x \qquad\quad \\ 2x + 1\overline{)4x^3 + 0x^2 - 7x + 5} \\ \underline{\mp 4x^3 \mp 2x^2} \qquad\quad \\ -2x^2 - 7x \quad \\ \underline{-2x^2 - x} \quad \end{array}$	$\dfrac{-2x^2}{2x} = -x$ Multiply $-x$ times $(2x + 1)$.
Step 5	$\begin{array}{r} 2x^2 - x - 3 \qquad \\ 2x + 1\overline{)4x^3 + 0x^2 - 7x + 5} \\ \underline{\mp 4x^3 \mp 2x^2} \qquad\quad \\ -2x^2 - 7x \quad \\ \underline{\pm 2x^2 \pm x} \quad \\ -6x + 5 \\ \underline{\pm 6x \pm 3} \\ 8 \end{array}$	Continue dividing, using the same procedure. The remainder is 8.

Step 6 **Check:**

$(2x + 1)(2x^2 - x - 3) + 8$

$= 4x^3 - 2x^2 - 6x + 2x^2 - x - 3 + 8$

$= 4x^3 - 7x + 5$

The answer can also be written in the form

$$2x^2 - x - 3 + \frac{8}{2x + 1}$$

EXERCISES 6.1

Express each quotient in Exercises 1–19 as a sum of fractions and simplify if possible.

1. $\dfrac{4x^2 + 8x + 3}{4}$ **2.** $\dfrac{6x^2 - 10x + 1}{2}$ **3.** $\dfrac{10x^2 - 15x - 3}{5}$

4. $\dfrac{9x^2 - 12x + 5}{3}$ **5.** $\dfrac{2x^2 + 5x}{x}$ **6.** $\dfrac{8x^2 - 7x}{x}$

7. $\dfrac{x^2 + 6x - 3}{x}$ **8.** $\dfrac{-2x^2 - 3x + 8}{x}$ **9.** $\dfrac{4x^2 + 6x - 3}{2x}$

10. $\dfrac{3x^3 - 2x^2 + x}{x^2}$ **11.** $\dfrac{6x^3 - 9x^2 - 3x}{3x^2}$ **12.** $\dfrac{5x^2y - 10xy^2 - 3y}{5xy}$

13. $\dfrac{7x^2y^2 + 21xy^3 - 11y^3}{7xy^2}$ **14.** $\dfrac{12x^3y + 6x^2y^2 - 3xy}{6x^2y}$

15. $\dfrac{3x^2y - 8xy^2 - 4y^3}{4xy}$ **16.** $\dfrac{2x^3y^2 - 6x^2y^3 + 15xy}{3xy^2}$

17. $\dfrac{5x^2y^2 - 8xy^2 + 16xy}{8xy^2}$ **18.** $\dfrac{3x^3y - 14x^2y - 7xy}{7x^2y}$

19. $\dfrac{8x^3y^2 - 9x^2y^3 + 5xy}{9xy^2}$

Divide in Exercises 20–45 using the long division procedure, and check your answer.

20. $119 \div 17$ **21.** $278 \div 23$

22. $326 \div 64$ **23.** $437 \div 59$

24. $(x^2 + 3x + 2) \div (x + 2)$ **25.** $(x^2 - 5x - 6) \div (x + 3)$

26. $(y^2 + 8y + 15) \div (y + 4)$ **27.** $(a^2 - 2a - 15) \div (a - 2)$

28. $(x^2 - 7x - 18) \div (x + 5)$ **29.** $(y^2 - y - 42) \div (y - 4)$

30. $(4a^2 - 21a + 2) \div (a - 6)$ **31.** $(5y^2 + 14y - 7) \div (y + 5)$

32. $(8x^2 + 10x - 4) \div (2x + 3)$ **33.** $(8c^2 + 2c - 14) \div (2c + 3)$

34. $(6x^2 + x - 4) \div (2x - 1)$ **35.** $(10m^2 - m - 6) \div (5m - 3)$

36. $(x^2 - 6) \div (x + 2)$ **37.** $(x^2 + 3x) \div (x + 5)$

38. $(2x^3 + 4x^2 - x + 1) \div (x - 3)$ **39.** $(y^3 - 9y^2 + 26y - 24) \div (y - 2)$

40. $(3t^3 + 10t^2 + 6t + 3) \div (3t + 1)$ **41.** $(12a^3 - 3a^2 + 4a + 1) \div (4a - 1)$

42. $(2x^3 + x^2 - 6) \div (x + 4)$

43. $(x^3 + 2x^2 - 5) \div (x - 5)$

44. $(x^3 - 8) \div (x - 2)$

45. $(x^3 + 27) \div (x + 3)$

6.2 MULTIPLICATION AND DIVISION

To reduce a fraction such as $\frac{6}{8}$ or $\frac{18}{15}$, we can factor both the numerator and denominator and "divide out" common factors. For example,

$$\frac{6}{8} = \frac{\cancel{2} \cdot 3}{\cancel{2} \cdot 4} = \frac{3}{4}$$

$$\frac{18}{15} = \frac{2 \cdot \cancel{3} \cdot 3}{\cancel{3} \cdot 5} = \frac{6}{5}$$

Not all fractions can be reduced. For example, $\frac{9}{10}$ cannot be reduced because 9 and 10 do not have any common factors other than $+1$ and -1.

To reduce a rational expression, proceed just as with rational numbers. Factor both the numerator and denominator, and "divide out" any common factors. For example,

$$\frac{x^2 + 5x + 6}{x^2 + 7x + 12} = \frac{\cancel{(x + 3)}(x + 2)}{\cancel{(x + 3)}(x + 4)} = \frac{x + 2}{x + 4}$$

(**Note:** x can have any value here except $x \neq -3$ and $x \neq -4$ because no denominator can have a value of 0.)

Be careful. **"Divide out" only common factors.** Do not divide out a term unless it is a **factor** of both the numerator and denominator.

RIGHT

$$\frac{5x + 15}{5x + 20} = \boxed{\frac{\cancel{5}(x + 3)}{\cancel{5}(x + 4)}} = \frac{x + 3}{x + 4}$$ 5 is a common **factor**

WRONG

$$\frac{5x}{x + 5} = \boxed{\frac{\cancel{5}x}{x + \cancel{5}}} = \frac{x}{x + 1}$$ 5 is **not** a **factor** of the denominator.

WRONG

$$\frac{5x}{x + 5} = \boxed{\frac{5\cancel{x}}{\cancel{x} + 5}} = \frac{5}{1 + 5}$$ x is **not** a **factor** of the denominator.

EXAMPLES

Reduce each rational expression to lowest terms. **No denominator can be 0.**

1. $\dfrac{2x+4}{3x+6} = \dfrac{2(\cancel{x+2})}{3(\cancel{x+2})} = \dfrac{2}{3} \qquad (x \neq -2)$

2. $\dfrac{x^2-16}{x+4} = \dfrac{(\cancel{x+4})(x-4)}{\cancel{x+4}} = x - 4 \qquad (x \neq -4)$

3. $\dfrac{a}{a^2-5a} = \dfrac{\overset{1}{\cancel{a}}}{\cancel{a}(a-5)} = \dfrac{1}{a-5} \qquad (a \neq 0, 5)$

   This example illustrates the importance of writing 1 in the numerator if all the factors divide out. 1 is an understood factor.

4. $\dfrac{3-y}{y-3} = \dfrac{-1(-3+y)}{y-3} = \dfrac{-1(\cancel{y-3})}{\cancel{y-3}} = -1 \qquad (y \neq 3)$

Example 4 illustrates a relationship that is very useful in working with rational expressions. That is, **when opposites are divided, the quotient is −1.**

$$\frac{-8}{+8} = -1$$

$$\frac{14}{-14} = -1$$

$$\frac{x-5}{5-x} = \frac{\overset{1}{(\cancel{x-5})}}{-(\cancel{x-5})} = \frac{1}{-1} = -1$$

$$\frac{4-y}{y-4} = \frac{-(\cancel{y-4})}{(\cancel{y-4})} = \frac{-1}{1} = -1$$

You should observe that $x - 5$ and $5 - x$ are opposites:

$$-1(x-5) = -x + 5 = 5 - x$$

Also, $4 - y$ and $y - 4$ are opposites:

$$-1(y-4) = -y + 4 = 4 - y$$

In general, for $b \neq a$,

$$\frac{a - b}{b - a} = -1$$

Next, consider the three fractions

$$\frac{-6}{2}, \quad \frac{6}{-2}, \quad \text{and} \quad -\frac{6}{2}$$

Are these equal? Yes! The negative sign can be with the numerator, with the denominator, or in front of the fraction, and the meaning or result will be the same.

$$\frac{-6}{2} = -3, \quad \frac{6}{-2} = -3, \quad \text{and} \quad -\frac{6}{2} = -3$$

In fact, we can make the following general statements:

For integers a and b, $b \neq 0$,

$$-\frac{a}{b} = \frac{-a}{b} = \frac{a}{-b}$$

For polynomials P and Q, $Q \neq 0$,

$$-\frac{P}{Q} = \frac{-P}{Q} = \frac{P}{-Q}$$

At this stage of the discussion, we will assume that no denominator is 0. You should keep in mind that there are certain restrictions on the variable, but these restrictions will not be stated.

EXAMPLES

Simplify the following expressions.

5. $$-\frac{-2x + 6}{x^2 - 3x} = \frac{-1(-2x + 6)}{x^2 - 3x} = \frac{2x - 6}{x^2 - 3x}$$

$$= \frac{2\cancel{(x - 3)}}{x\cancel{(x - 3)}} = \frac{2}{x}$$

6. $$-\frac{5-10x}{6x-3}=\frac{-1(5-10x)}{6x-3}=\frac{-1(5)\overset{-1}{(\cancel{1-2x})}}{3(\cancel{2x-1})}=\frac{5}{3}$$

or

$$-\frac{5-10x}{6x-3}=\frac{5-10x}{-1(6x-3)}=\frac{5\overset{-1}{(\cancel{1-2x})}}{-1(3)(\cancel{2x-1})}=\frac{-5}{-3}=\frac{5}{3}$$

To **multiply** any two rational expressions, multiply the numerators and multiply the denominators, keeping the expressions in factored form. Then "divide out" any common factors.

> If P, Q, R, and S are polynomials with $Q, S \neq 0$, then
>
> $$\frac{P}{Q}\cdot\frac{R}{S}=\frac{P\cdot R}{Q\cdot S}$$

EXAMPLES

Multiply and reduce if possible.

7. $$\frac{x+3}{x}\cdot\frac{x-3}{x+5}=\frac{(x+3)(x-3)}{x(x+5)}=\frac{x^2-9}{x(x+5)}$$

(There are no common factors in the numerator and denominator.)

8. $$\frac{x+5}{7x}\cdot\frac{49x^2}{x^2-25}=\frac{\overset{7x}{\cancel{49x^2}}(\cancel{x+5})}{\cancel{7x}(\cancel{x+5})(x-5)}=\frac{7x}{x-5}$$

9. $$\frac{x^2+5x+6}{3x+6}\cdot\frac{x^2-4}{x^2-2x-8}=\frac{(\cancel{x+2})(x+3)(\cancel{x+2})(x-2)}{3(\cancel{x+2})(x-4)(\cancel{x+2})}$$
$$=\frac{(x+3)(x-2)}{3(x-4)}$$
$$=\frac{x^2+x-6}{3(x-4)}$$

As shown in Examples 7 and 9, we will multiply the factors in the numerator, yet leave the denominator in factored form. This form is not necessary, but it is useful for adding and subtracting, as we will see in Section 6.3.

To divide any two rational expressions, multiply by the **reciprocal** of the divisor.

> If P, Q, R, and S are polynomials with $Q, R, S \neq 0$, then
>
> $$\frac{P}{Q} \div \frac{R}{S} = \frac{P}{Q} \cdot \frac{S}{R}$$

EXAMPLES

Divide and reduce if possible.

10. $$\frac{a^2 - 49}{12a^2} \div \frac{a^2 + 8a + 7}{18a} = \frac{a^2 - 49}{12a^2} \cdot \frac{18a}{a^2 + 8a + 7}$$

$$= \frac{\cancel{(a+7)}(a - 7) \cdot \cancel{6} \cdot 3 \cdot \cancel{a}}{\cancel{6} \cdot 2 \cdot \cancel{a}^2 \cancel{(a+7)}(a + 1)}$$

$$= \frac{3(a - 7)}{2a(a + 1)}$$

$$= \frac{3a - 21}{2a(a + 1)}$$

11. $$\frac{2x^2 + 3x - 2}{x^2 + 3x + 2} \div \frac{1 - 2x}{x - 2} = \frac{2x^2 + 3x - 2}{x^2 + 3x + 2} \cdot \frac{x - 2}{1 - 2x}$$

$$= \frac{\overset{-1}{\cancel{(2x - 1)}}\cancel{(x + 2)}(x - 2)}{(x + 1)\cancel{(x + 2)}\cancel{(1 - 2x)}}$$

$$= \frac{-1(x - 2)}{x + 1}$$

$$= \frac{-x + 2}{x + 1}$$

PRACTICE QUIZ

Questions	**Answers**
Perform the indicated operations and simplify. Assume that no denominator is 0.	
1. $\dfrac{2y^2 - 16y}{6y^2 + 7y - 3} \cdot \dfrac{2y^2 + 11y + 12}{y^2 - 9y + 8}$	1. $\dfrac{2y^2 + 8y}{(3y - 1)(y - 1)}$

2. $\dfrac{a-b}{b-a} \div \dfrac{a^2+2ab+b^2}{a^2+ab}$ 2. $\dfrac{-a}{a+b}$

3. $\dfrac{x^2+3x-4}{x^2-1} \div \dfrac{x^2+6x+8}{x+1}$ 3. $\dfrac{1}{x+2}$

EXERCISES 6.2

Reduce Exercises 1–20. State any restrictions on the variable.

1. $\dfrac{3x-6}{6x+3}$ **2.** $\dfrac{5x+20}{6x+24}$ **3.** $\dfrac{x}{x^2-4x}$

4. $\dfrac{4-2x}{2x-4}$ **5.** $\dfrac{7x+14}{x+2}$ **6.** $\dfrac{3x^2-3x}{2x-2}$

7. $\dfrac{5+3x}{3x+5}$ **8.** $\dfrac{4xy+y^2}{3y^2+2y}$ **9.** $\dfrac{4-4x^2}{4x^2-4}$

10. $\dfrac{4x-8}{(x-2)^2}$ **11.** $\dfrac{2x+6}{(x+3)^2}$ **12.** $\dfrac{x^2-4}{2x+4}$

13. $\dfrac{x^2+7x+10}{x^2-25}$ **14.** $\dfrac{x^2-3x-10}{x^2-7x+10}$ **15.** $\dfrac{x^2-3x-18}{x^2+6x+9}$

16. $\dfrac{8x^2+6x-9}{16x^2-9}$ **17.** $\dfrac{x^2-5x+6}{8x-2x^3}$ **18.** $\dfrac{6x^2-11x+3}{4x^2-12x+9}$

19. $\dfrac{16x^2+40x+25}{4x^2+13x+10}$ **20.** $\dfrac{6x^2-11x+4}{3x^2-7x+4}$

Perform the indicated operations in Exercises 21–60. Assume that no denominator is zero.

21. $\dfrac{2x-4}{3x+6} \cdot \dfrac{3x}{x-2}$ **22.** $\dfrac{5x+20}{2x} \cdot \dfrac{4x}{2x+4}$

23. $\dfrac{4x^2}{x^2+3x} \cdot \dfrac{x^2-9}{2x-2}$ **24.** $\dfrac{x^2-x}{x-1} \cdot \dfrac{x+1}{x}$

25. $\dfrac{x^2-4}{x} \cdot \dfrac{3x}{x+2}$ **26.** $\dfrac{x-3}{15x} \div \dfrac{3x-9}{30x^2}$

27. $\dfrac{x^2-1}{5} \div \dfrac{x^2+2x+1}{10}$ **28.** $\dfrac{x-5}{3} \div \dfrac{x^2-25}{6x+30}$

29. $\frac{10x^2 - 5x}{6x^2 + 12x} \div \frac{2x - 1}{x^2 + 2x}$

30. $\frac{x^2 + 2x - 8}{4x} \div \frac{2x^2 + 5x + 2}{3x^2}$

31. $\frac{x^2 + x}{x^2 + 2x + 1} \cdot \frac{x^2 - x - 2}{x^2 - 1}$

32. $\frac{x^2 - x - 6}{x^2 - 4} \cdot \frac{x^2 - 25}{x^2 + 2x - 15}$

33. $\frac{x + 3}{x^2 + 3x - 4} \cdot \frac{x^2 + x - 2}{x + 2}$

34. $\frac{x^2 + 5x + 6}{2x + 4} \cdot \frac{5x}{x + 3}$

35. $\frac{6x^2 - 7x - 3}{x^2 - 1} \cdot \frac{x - 1}{2x - 3}$

36. $\frac{x^2 - 9}{2x^2 + 7x + 3} \cdot \frac{2x^2 + 11x + 5}{x^2 - 3x}$

37. $\frac{x^2 - 8x + 15}{x^2 - 9x + 14} \cdot \frac{7 - x}{x^2 + 4x - 21}$

38. $\frac{x^2 - 16x + 39}{6 + x - x^2} \cdot \frac{4x + 8}{x + 1}$

39. $\frac{3x^2 - 7x + 2}{1 - 9x^2} \cdot \frac{3x + 1}{x - 2}$

40. $\frac{16 - x^2}{x^2 + 2x - 8} \cdot \frac{4 - x^2}{x^2 - 2x - 8}$

41. $\frac{4x^2 - 1}{x^2 - 16} \div \frac{2x + 1}{x^2 - 4x}$

42. $\frac{4x^2 - 13x + 3}{16x^2 - 4x} \div \frac{x^2 - 6x + 9}{8x^2}$

43. $\frac{x^2 + x - 6}{2x^2 + 6x} \div \frac{x^2 - 5x + 6}{8x^2}$

44. $\frac{x^2 - 4}{x^2 - 5x + 6} \div \frac{x^2 + 3x + 2}{x^2 - 2x - 3}$

45. $\frac{2x^2 - 5x - 12}{x^2 - 10x + 24} \div \frac{4x^2 - 9}{x^2 - 9x + 18}$

46. $\frac{2x^2 - 7x + 3}{x^2 - 3x + 2} \div \frac{x^2 - 6x + 9}{x^2 - 4x + 3}$

47. $\frac{6x^2 - x - 2}{12x^2 + 5x - 2} \div \frac{4x^2 - 1}{8x^2 - 6x + 1}$

48. $\frac{8x^2 + 6x - 9}{8x^2 - 26x + 15} \div \frac{4x^2 + 12x + 9}{16x^2 + 18x - 9}$

49. $\frac{3x^2 + 11x + 6}{4x^2 + 16x + 7} \div \frac{3x^2 - x - 2}{2x^2 - x - 28}$

50. $\dfrac{2x^2 + 5x - 3}{5x^2 + 17x - 12} \div \dfrac{2x^2 + 3x - 2}{5x^2 + 7x - 6}$

51. $\dfrac{x^2 - 16}{x^2 - 4x} \cdot \dfrac{x^2}{x + 4} \cdot \dfrac{2x^2 - 2x}{x - 1}$

52. $\dfrac{10x^2 + 3x - 1}{6x^2 + x - 2} \cdot \dfrac{2x^2 - x}{2x^2 - x - 1} \cdot \dfrac{3x + 2}{5x^2 - x}$

53. $\dfrac{x^2 - 3x}{x^2 + 2x + 1} \cdot \dfrac{x - 3}{x^2 - 9} \cdot \dfrac{x^2 - 3x - 18}{x - 6}$

54. $\dfrac{x^2 + 2x - 3}{2x^2 + 3x - 2} \cdot \dfrac{3x + 5}{2x - 3} \div \dfrac{3x^2 + 2x - 5}{2x^2 + x - 6}$

55. $\dfrac{2x^2 + 5x - 3}{x^2 + 2x - 3} \cdot \dfrac{x^2 + 3x - 4}{4x^2 - 1} \div \dfrac{x^2 + 4x}{x^3 + 5x^2}$

56. $\dfrac{6x^2 + 7x - 3}{2x^2 + 5x + 3} \div \dfrac{x^2 + 5x + 6}{4x^2 + 3x - 1} \cdot \dfrac{x^2 + 2x - 3}{4x^2 + 7x - 2}$

57. $\dfrac{x^2 + 4x + 3}{x^2 + 8x + 7} \div \dfrac{x^2 - 7x - 8}{35 + 12x + x^2} \div \dfrac{x^2 + 8x + 15}{x^2 - 9x + 8}$

58. $\dfrac{12x^2}{x^2 - 1} \div \dfrac{4x^2 + 4x - 3}{2x^2 - 5x + 3} \div \dfrac{6x^2 - 9x}{1 - 4x^2}$

59. $\dfrac{2x^2 - 7x + 3}{36x^2 - 1} \div \dfrac{6x^2 + 5x + 1}{2x + 1} \div \dfrac{2x^2 - 5x - 3}{18x^2 + 3x - 1}$

60. $\dfrac{12x^2 - 8x - 15}{4x^2 - 8x + 3} \cdot \dfrac{6x^2 - 13x + 6}{4x^2 + 5x + 1} \div \dfrac{18x^2 + 3x - 10}{8x^2 - 2x - 1}$

6.3 ADDITION AND SUBTRACTION

To add rational expressions with a common denominator, proceed just as with fractions: add the numerator and use the common denominator. For example,

$$\frac{5}{x + 1} + \frac{6}{x + 1} = \frac{5 + 6}{x + 1} = \frac{11}{x + 1}$$

Sometimes the sum can be reduced.

$$\frac{x^2}{x + 1} + \frac{2x + 1}{x + 1} = \frac{x^2 + 2x + 1}{x + 1} = \frac{(x + 1)^2}{x + 1} = x + 1$$

> For polynomials P, Q, and R, with $Q \neq 0$,
>
> $$\frac{P}{Q} + \frac{R}{Q} = \frac{P + R}{Q}$$

To add $\frac{5}{x^2} + \frac{3}{x^2 - 4x}$, find a common denominator and change each fraction to an equal fraction with that denominator. The new denominator should be the least common multiple (LCM) of the original denominators. The LCM was discussed in Sections 1.2 and 1.4. To find the LCM for a set of polynomials,

1. Completely factor each polynomial (including prime factors for numerical factors).
2. Form the product of all factors that appear, using each factor the most number of times it appears in any one polynomial.

Thus, for the denominators x^2 and $x^2 - 4x$, we have

$$\left.\begin{aligned} x^2 &= x^2 \\ x^2 - 4x &= x(x - 4) \end{aligned}\right\} \quad \text{LCM} = x^2(x - 4)$$

So $x^2(x - 4)$ is to be the common denominator:

$$\begin{aligned} \frac{5}{x^2} + \frac{3}{x^2 - 4x} &= \frac{5\mathbf{(x - 4)}}{x^2\mathbf{(x - 4)}} + \frac{3 \cdot \mathbf{x}}{x(x - 4) \cdot \mathbf{x}} \\ &= \frac{5x - 20}{x^2(x - 4)} + \frac{3x}{x^2(x - 4)} \\ &= \frac{5x - 20 + 3x}{x^2(x - 4)} = \frac{8x - 20}{x^2(x - 4)} \end{aligned}$$

To subtract one rational expression from another with the same denominator, simply subtract the numerators and use the common denominator. Since the numerators will be polynomials, **a good idea is to put both numerators in parentheses so that all changes in signs will be done correctly.**

For example, many beginning students will make a **mistake** like the following:

WRONG

$$\frac{7}{x + 5} - \frac{2 - x}{x + 5} = \frac{\boxed{7 - 2 - x}}{x + 5}$$

By using parentheses, you can avoid such mistakes:

RIGHT

$$\frac{7}{x+5} - \frac{2-x}{x+5} = \frac{\boxed{7-(2-x)}}{x+5} = \frac{7-2+x}{x+5} = \frac{5+x}{x+5}$$

$$= \frac{x+5}{x+5} = 1$$

> For polynomials P, Q, and R with $Q \neq 0$,
>
> $$\frac{P}{Q} - \frac{R}{Q} = \frac{P-R}{Q}$$

If the rational expressions do not have the same denominator, find the LCM of the denominators, and change each fraction to an equal fraction with the LCM as denominator. For example,

$$\frac{x}{x-1} - \frac{2}{x+1} \qquad \left.\begin{matrix} x-1 \\ x+1 \end{matrix}\right\} \text{ LCM} = (x-1)(x+1)$$

$$\frac{x}{x-1} - \frac{2}{x+1} = \frac{x\mathbf{(x+1)}}{(x-1)\mathbf{(x+1)}} - \frac{2\mathbf{(x-1)}}{(x+1)\mathbf{(x-1)}}$$

$$= \frac{x^2+x}{(x-1)(x+1)} - \frac{2x-2}{(x+1)(x-1)}$$

$$= \frac{(x^2+x)-(2x-2)}{(x-1)(x+1)} = \frac{x^2+x-2x+2}{(x-1)(x+1)}$$

$$= \frac{x^2-x+2}{(x-1)(x+1)}$$

Another way to do the same problem is to place the negative sign with the numerator and add. The results are the same.

$$\frac{x}{x-1} - \frac{2}{x+1} = \frac{x}{x-1} + \frac{-2}{x+1}$$

$$= \frac{x\mathbf{(x+1)}}{(x-1)\mathbf{(x+1)}} + \frac{-2\mathbf{(x-1)}}{(x+1)\mathbf{(x-1)}}$$

$$= \frac{x^2+x}{(x-1)(x+1)} + \frac{-2x+2}{(x+1)(x-1)}$$

$$= \frac{x^2+x-2x+2}{(x+1)(x-1)} = \frac{x^2-x+2}{(x+1)(x-1)}$$

EXAMPLES

Perform the indicated operations and reduce if possible.

1. $\dfrac{x^2+5}{x+5}+\dfrac{6x}{x+5}$

Solution:

$$\frac{x^2+5}{x+5}+\frac{6x}{x+5}=\frac{x^2+5+6x}{x+5}=\frac{x^2+6x+5}{x+5}$$

$$=\frac{\cancel{(x+5)}(x+1)}{\cancel{x+5}}=x+1$$

2. $\dfrac{5}{x^2-1}+\dfrac{4}{(x-1)^2}$

Solution: $x^2-1=(x+1)(x-1)$, $(x-1)^2=(x-1)^2$ } LCM $=(x+1)(x-1)^2$

$$\frac{5}{x^2-1}+\frac{4}{(x-1)^2}=\frac{5\mathbf{(x-1)}}{(x+1)(x-1)\mathbf{(x-1)}}+\frac{4\mathbf{(x+1)}}{(x-1)^2\mathbf{(x+1)}}$$

$$=\frac{5x-5+4x+4}{(x-1)^2(x+1)}=\frac{9x-1}{(x-1)^2(x+1)}$$

3. $\dfrac{x+4}{x+3}-\dfrac{5x+1}{x+3}$

Solution:

$$\frac{x+4}{x+3}-\frac{5x+1}{x+3}=\frac{(x+4)-(5x+1)}{x+3}$$

$$=\frac{x+4-5x-1}{x+3}=\frac{-4x+3}{x+3}$$

4. $\dfrac{3}{x-2}+\dfrac{5}{4-x^2}$

Solution: Factoring, we have $4-x^2=(2+x)(2-x)$ and $x-2=-1(2-x)$. Change the second fraction to avoid a factor of -1 in the denominator.

$$\frac{3}{x-2}+\frac{5}{4-x^2}=\frac{3}{x-2}+\frac{-5}{x^2-4}$$

$$=\frac{3\mathbf{(x+2)}}{(x-2)\mathbf{(x+2)}}+\frac{-5}{(x+2)(x-2)}$$

$$=\frac{3x+6-5}{(x+2)(x-2)}=\frac{3x+1}{(x+2)(x-2)}$$

5. $\dfrac{2x}{x^2 - 9} - \dfrac{1}{x^2 + 7x + 12}$

Solution: $\left.\begin{aligned} x^2 - 9 &= (x + 3)(x - 3) \\ x^2 + 7x + 12 &= (x + 4)(x + 3) \end{aligned}\right\}$ LCM $= (x + 3)(x - 3)(x + 4)$

$$\frac{2x}{x^2 - 9} - \frac{1}{x^2 + 7x + 12}$$

$$= \frac{2x\mathbf{(x + 4)}}{(x + 3)(x - 3)\mathbf{(x + 4)}} + \frac{-1\mathbf{(x - 3)}}{(x + 4)(x + 3)\mathbf{(x - 3)}}$$

$$= \frac{2x^2 + 8x - x + 3}{(x + 3)(x - 3)(x + 4)} = \frac{2x^2 + 7x + 3}{(x + 3)(x - 3)(x + 4)}$$

$$= \frac{(2x + 1)\cancel{(x + 3)}}{\cancel{(x + 3)}(x - 3)(x + 4)} = \frac{2x + 1}{(x - 3)(x + 4)}$$

PRACTICE QUIZ

Questions	Answers
Perform the indicated operations and reduce if possible.	
1. $\dfrac{5}{x - 1} - \dfrac{4 + x}{x - 1}$	1. -1
2. $\dfrac{5}{x + 1} + \dfrac{10x}{x^2 + 4x + 3}$	2. $\dfrac{15}{x + 3}$
3. $\dfrac{1}{x^2 + x} + \dfrac{4}{x^2} - \dfrac{2}{x^3 - x}$	3. $\dfrac{5x^2 - 3x - 4}{x^2(x + 1)(x - 1)}$

EXERCISES 6.3

Perform the indicated operations and reduce if possible.

1. $\dfrac{4x}{x + 2} + \dfrac{8}{x + 2}$

2. $\dfrac{x - 1}{x + 4} + \dfrac{x + 9}{x + 4}$

3. $\dfrac{3x - 1}{2x - 6} + \dfrac{x - 11}{2x - 6}$

4. $\dfrac{2x + 5}{4(x + 1)} + \dfrac{-x + 3}{4x + 4}$

5. $\dfrac{x^2}{x^2 + 2x + 1} + \dfrac{-2x - 3}{x^2 + 2x + 1}$

6. $\dfrac{2x - 1}{x^2 - x - 6} + \dfrac{1 - x}{x^2 - x - 6}$

7. $\dfrac{-(3x + 2)}{x^2 - 7x + 6} + \dfrac{4x - 4}{x^2 - 7x + 6}$

8. $\dfrac{x + 1}{x^2 - 2x - 3} + \dfrac{-(3x - 5)}{x^2 - 2x - 3}$

9. $\dfrac{x^2 + 2}{x^2 - 4} + \dfrac{-(4x - 2)}{x^2 - 4}$

10. $\dfrac{2x + 5}{2x^2 - x - 1} + \dfrac{-(4x + 2)}{2x^2 - x - 1}$

11. $\dfrac{3x + 2}{4x - 2} - \dfrac{x + 2}{4x - 2}$

12. $\dfrac{2x + 1}{x^2 - x - 6} - \dfrac{x - 1}{x^2 - x - 6}$

13. $\dfrac{2x^2 - 3x + 1}{x^2 - 3x - 4} - \dfrac{x^2 + x + 6}{x^2 - 3x - 4}$

14. $\dfrac{x^2 - x - 2}{x^2 - 4} - \dfrac{x^2 + x - 2}{x^2 - 4}$

15. $\dfrac{3}{x - 3} - \dfrac{2}{2x - 6}$

16. $\dfrac{2x}{3x + 6} + \dfrac{5}{2x + 4}$

17. $\dfrac{8}{5x - 10} - \dfrac{6}{3x - 6}$

18. $\dfrac{7}{4x - 20} - \dfrac{1}{3x - 15}$

19. $\dfrac{2}{x} + \dfrac{1}{x + 4}$

20. $\dfrac{8x + 3}{x + 1} + \dfrac{x - 1}{x}$

21. $\dfrac{x}{x + 4} + \dfrac{x}{x - 4}$

22. $\dfrac{5}{x - 2} + \dfrac{x}{x + 3}$

23. $\dfrac{3}{2 - x} + \dfrac{6}{x - 2}$

24. $\dfrac{5}{2x - 3} + \dfrac{2}{3 - 2x}$

25. $\dfrac{x}{5 - x} + \dfrac{2x + 3}{x - 5}$

26. $\dfrac{x - 1}{3x - 1} + \dfrac{4}{x + 2}$

27. $\dfrac{x}{x - 1} - \dfrac{1}{x + 2}$

28. $\dfrac{x + 2}{x + 3} + \dfrac{4}{3 - x}$

29. $\dfrac{x + 1}{x + 4} + \dfrac{2x}{4 - x}$

30. $\dfrac{8}{x} - \dfrac{x + 1}{x - 6}$

31. $\dfrac{8}{x - 2} - \dfrac{4}{x + 2}$

32. $\dfrac{8}{2x + 2} + \dfrac{3}{3x + 3}$

33. $\dfrac{x}{4x - 8} - \dfrac{3x + 1}{3x - 6}$

34. $\dfrac{7x - 1}{x^2 - 25} + \dfrac{4}{x + 5}$

35. $\dfrac{2x + 3}{x^2 + 4x - 5} - \dfrac{4}{x - 1}$

36. $\dfrac{2x - 3}{x + 1} - \dfrac{x + 1}{2x - 2}$

37. $\dfrac{3x}{6 + x} - \dfrac{2x}{x^2 - 36}$

38. $\dfrac{x}{7 - x} - \dfrac{3x + 2}{x^2 - 49}$

39. $\frac{2x+1}{x-7}+\frac{3x}{x^2-8x+7}$

40. $\frac{x-4}{x-2}-\frac{x-7}{2x-10}$

41. $\frac{x+1}{x+2}-\frac{x+2}{2x+6}$

42. $\frac{3x-4}{x^2-x-20}-\frac{2}{x-5}$

43. $\frac{x-3}{x^2-16}-\frac{x+2}{8-2x}$

44. $\frac{7}{x-9}-\frac{x-1}{3x+6}$

45. $\frac{1}{x^2+x-2}-\frac{1}{x^2-1}$

46. $\frac{4}{x^2+x-6}+\frac{4}{x^2+5x+6}$

47. $\frac{2x}{x^2+x-12}+\frac{3x}{x^2-9}$

48. $\frac{x}{x^2+4x-21}+\frac{1-x}{x^2+8x+7}$

49. $\frac{x-3}{x^2+4x+4}-\frac{3x}{x^2+3x+2}$

50. $\frac{x-1}{2x^2+3x-2}+\frac{x}{2x^2-3x+1}$

51. $\frac{x}{x^2-16}-\frac{3x}{x^2+5x+4}$

52. $\frac{2x}{x^2-2x-15}-\frac{5}{x^2-6x+5}$

53. $\frac{x-2}{2x^2+5x-3}+\frac{2x-5}{2x^2-9x+4}$

54. $\frac{7}{12+4x-x^2}+\frac{x+1}{x^2-3x-18}$

55. $\frac{x}{x^2+3x-10}+\frac{3x}{4-x^2}$

56. $\frac{2x+3}{x^2-6x-7}+\frac{x-1}{x^2-5x-14}$

57. $\frac{3x}{x-4}+\frac{7x}{x+4}-\frac{x+3}{x^2-16}$

58. $\frac{x}{x+3}-\frac{x+1}{x-3}+\frac{x^2+4}{x^2-9}$

59. $\frac{1}{x-2}-\frac{1}{x-1}+\frac{x}{x^2-3x+2}$

60. $\frac{2}{x^2-9}-\frac{3}{x^2-4x+3}+\frac{x-1}{x^2+2x-3}$

6.4 COMPLEX ALGEBRAIC EXPRESSIONS

An expression that contains various combinations of addition, subtraction, multiplication, and division with rational expressions is called a **complex algebraic expression.** Examples of complex algebraic expressions are

$$\frac{\frac{1}{x}+\frac{1}{y}}{x+y}, \quad \frac{\frac{1}{x+2}-\frac{1}{x}}{1+\frac{2}{x}}, \quad \text{and} \quad \frac{4-x}{x+3}+\frac{x}{x+3}\div\frac{x}{x-3}$$

In an expression such as $\dfrac{\frac{1}{x} + \frac{1}{y}}{x + y}$, the large fraction bar is a symbol of inclusion. The expression could be written as follows:

$$\frac{\frac{1}{x} + \frac{1}{y}}{x + y} = \left(\frac{1}{x} + \frac{1}{y}\right) \div (x + y)$$

Similarly,

$$\frac{\frac{1}{x + 2} - \frac{1}{x}}{1 + \frac{2}{x}} = \left(\frac{1}{x + 2} - \frac{1}{x}\right) \div \left(1 + \frac{2}{x}\right)$$

Simplify by working with the numerator and denominator separately as if they were in parentheses. Make the numerator a single fraction and the denominator a single fraction. Then divide.

a. $$\frac{\frac{1}{x} + \frac{1}{y}}{x + y} = \frac{\frac{1 \cdot \mathbf{y}}{x \cdot \mathbf{y}} + \frac{1 \cdot \mathbf{x}}{y \cdot \mathbf{x}}}{\frac{x + y}{1}}$$

$$= \frac{\frac{y + x}{xy}}{\frac{x + y}{1}} = \frac{\cancel{y + x}}{xy} \cdot \frac{1}{\cancel{x + y}} = \frac{1}{xy}$$

b. $$\frac{\frac{1}{x + 2} - \frac{1}{x}}{1 + \frac{2}{x}} = \frac{\frac{1 \cdot \mathbf{x}}{(x + 2)\mathbf{x}} - \frac{1(\mathbf{x + 2})}{x(\mathbf{x + 2})}}{\frac{x}{x} + \frac{2}{x}}$$

$$= \frac{\frac{x - (x + 2)}{x(x + 2)}}{\frac{x + 2}{x}} = \frac{\frac{x - x - 2}{x(x + 2)}}{\frac{x + 2}{x}}$$

$$= \frac{-2}{\cancel{x}(x + 2)} \cdot \frac{\cancel{x}}{x + 2} = \frac{-2}{(x + 2)^2}$$

A second method is to find the LCM of the denominators in the fractions in both the original numerator and the original denominator, and then multiply both the numerator and the denominator by this LCM.

a′. $$\frac{\frac{1}{x}+\frac{1}{y}}{x+y} = \frac{\frac{1}{x}+\frac{1}{y}}{\frac{x+y}{1}} \qquad \left.\begin{matrix} x \\ y \\ 1 \end{matrix}\right\} \text{LCM} = xy$$

$$= \frac{\left(\frac{1}{x}+\frac{1}{y}\right)\mathbf{xy}}{\left(\frac{x+y}{1}\right)\mathbf{xy}} = \frac{\frac{1}{x}\cdot xy + \frac{1}{y}\cdot xy}{(x+y)xy}$$

$$= \frac{\cancel{y+x}}{(\cancel{x+y})xy} = \frac{1}{xy}$$

b′. $$\frac{\frac{1}{x+2}-\frac{1}{x}}{1+\frac{2}{x}} = \frac{\left(\frac{1}{x+2}-\frac{1}{x}\right)\cdot \mathbf{x(x+2)}}{\left(1+\frac{2}{x}\right)\cdot \mathbf{x(x+2)}}$$

$$\left.\begin{matrix} x \\ x+2 \end{matrix}\right\} \text{LCM} = x(x+2)$$

$$= \frac{\frac{1}{\cancel{x+2}}\cdot x(\cancel{x+2}) - \frac{1}{\cancel{x}}\cdot \cancel{x}(x+2)}{1\cdot x(x+2) + \frac{2}{\cancel{x}}\cdot \cancel{x}(x+2)}$$

$$= \frac{x-(x+2)}{x(x+2)+2(x+2)} = \frac{x-x-2}{(x+2)(x+2)} = \frac{-2}{(x+2)^2}$$

Each of the techniques just described is valid. Sometimes one is easier to use than the other, but the choice is up to you.

EXAMPLES

1. Simplify the following complex algebraic expression:

$$\frac{\frac{1}{x+y}-\frac{1}{x-y}}{\frac{2y}{x^2-y^2}} = \frac{\frac{1(\mathbf{x}-\mathbf{y})}{(x+y)(\mathbf{x}-\mathbf{y})}-\frac{1(\mathbf{x}+\mathbf{y})}{(x-y)(\mathbf{x}+\mathbf{y})}}{\frac{2y}{x^2-y^2}}$$

$$= \frac{\frac{(x-y)-(x+y)}{(x-y)(x+y)}}{\frac{2y}{x^2-y^2}} = \frac{x-y-x-y}{(x-y)(x+y)}\cdot\frac{x^2-y^2}{2y}$$

$$= \frac{\overset{-1}{\cancel{-2y}}}{\cancel{(x-y)}\cancel{(x+y)}}\cdot\frac{\cancel{(x-y)}\cancel{(x+y)}}{\cancel{2y}} = -1$$

Or use the technique of multiplying the numerator and the denominator by the LCM of the denominators of the various fractions.

$$\frac{\frac{1}{x+y}-\frac{1}{x-y}}{\frac{2y}{x^2-y^2}} = \frac{\left(\frac{1}{x+y}-\frac{1}{x-y}\right)(\mathbf{x}+\mathbf{y})(\mathbf{x}-\mathbf{y})}{\left(\frac{2y}{x^2-y^2}\right)(\mathbf{x}+\mathbf{y})(\mathbf{x}-\mathbf{y})}$$

$$= \frac{\frac{1}{\cancel{x+y}}\cancel{(x+y)}(x-y)-\frac{1}{\cancel{x-y}}(x+y)\cancel{(x-y)}}{\frac{2y\cancel{(x+y)}\cancel{(x-y)}}{\cancel{(x+y)}\cancel{(x-y)}}}$$

$$= \frac{(x-y)-(x+y)}{2y}$$

$$= \frac{x-y-x-y}{2y} = \frac{-2y}{2y} = -1$$

2. In an expression such as

$$\frac{4-x}{x+3}+\frac{x}{x+3}\div\frac{x}{x-3}$$

the rules for order of operations indicate that the division is to be done first.

$$\frac{4-x}{x+3}+\frac{x}{x+3}\div\frac{x}{x-3}=\frac{4-x}{x+3}+\frac{\cancel{x}}{x+3}\cdot\frac{x-3}{\cancel{x}}$$

$$=\frac{4-x}{x+3}+\frac{x-3}{x+3}$$

$$=\frac{4-x+x-3}{x+3}$$

$$=\frac{1}{x+3}$$

PRACTICE QUIZ

Questions	**Answers**
Simplify the following expressions.	
1. $\dfrac{\frac{1}{x}}{1+\frac{1}{x}}$	1. $\dfrac{1}{x+1}$
2. $\dfrac{1+\frac{3}{x-3}}{x+\frac{x^2}{3-x}}$	2. $-\dfrac{1}{3}$

EXERCISES 6.4

Simplify the complex algebraic expressions in Exercises 1–32.

1. $\dfrac{\frac{4}{5}}{\frac{7}{10}}$ **2.** $\dfrac{\frac{5}{6}}{\frac{2}{3}}$ **3.** $\dfrac{\frac{1}{2}+\frac{1}{3}}{\frac{5}{6}-\frac{1}{4}}$ **4.** $\dfrac{\frac{3}{4}-\frac{7}{8}}{\frac{1}{3}-\frac{1}{6}}$

5. $\dfrac{1+\frac{1}{3}}{2+\frac{2}{3}}$ **6.** $\dfrac{2+\frac{5}{7}}{3-\frac{2}{7}}$ **7.** $\dfrac{\frac{3x}{5}}{\frac{x}{5}}$ **8.** $\dfrac{\frac{x}{2y}}{\frac{3x}{y}}$

9. $\dfrac{\dfrac{4}{xy^2}}{\dfrac{6}{x^2y}}$ 10. $\dfrac{\dfrac{x}{2y^2}}{\dfrac{5x^2}{6y}}$ 11. $\dfrac{\dfrac{8x^2}{5y}}{\dfrac{x}{10y^2}}$ 12. $\dfrac{\dfrac{15y^2}{2x^3}}{8y}$

13. $\dfrac{\dfrac{12x^3}{7y^4}}{\dfrac{3x^5}{2y}}$ 14. $\dfrac{\dfrac{9x^2}{5y^3}}{\dfrac{3xy}{10}}$ 15. $\dfrac{\dfrac{3x}{x+1}}{\dfrac{x-2}{x+1}}$ 16. $\dfrac{\dfrac{x-3}{x+4}}{\dfrac{x-2}{x+4}}$

17. $\dfrac{\dfrac{x+3}{2x}}{\dfrac{2x-1}{4x^2}}$ 18. $\dfrac{\dfrac{x-2}{6x}}{\dfrac{x+3}{3x^2}}$ 19. $\dfrac{\dfrac{x^2-9}{x}}{x-3}$ 20. $\dfrac{\dfrac{x^2-3x+2}{x-4}}{x-2}$

21. $\dfrac{\dfrac{x+2}{x-2}}{\dfrac{4x^2+8x}{x^2-4}}$ 22. $\dfrac{\dfrac{x-1}{x+3}}{\dfrac{x+2}{x^2+2x-3}}$ 23. $\dfrac{\dfrac{1}{x}-\dfrac{1}{3x}}{\dfrac{x+6}{x^2}}$ 24. $\dfrac{\dfrac{1}{3}+\dfrac{1}{x}}{\dfrac{1}{2}-\dfrac{1}{x}}$

25. $\dfrac{1+\dfrac{1}{x}}{1-\dfrac{1}{x^2}}$ 26. $\dfrac{\dfrac{3}{x}-\dfrac{6}{x^2}}{\dfrac{1}{x}-\dfrac{2}{x^2}}$ 27. $\dfrac{\dfrac{1}{x}-\dfrac{1}{y}}{\dfrac{y}{x^2}-\dfrac{1}{y}}$ 28. $\dfrac{\dfrac{x}{y}-\dfrac{1}{3}}{\dfrac{6}{y}-\dfrac{2}{x}}$

29. $\dfrac{2-\dfrac{4}{x}}{\dfrac{x^2-4}{x^2+x}}$ 30. $\dfrac{\dfrac{1}{x}}{1-\dfrac{1}{x-2}}$ 31. $\dfrac{x-\dfrac{2}{x+1}}{x+\dfrac{x-3}{x+1}}$ 32. $\dfrac{x-\dfrac{2x-3}{x-2}}{2x-\dfrac{x+3}{x-2}}$

Write each of the following as a single fraction reduced to lowest terms.

33. $\dfrac{1}{x+1}-\dfrac{3}{2x}\cdot\dfrac{4x}{x+1}$ 34. $\dfrac{4}{x}-\dfrac{2}{x^2-2x}\cdot\dfrac{x-2}{5}$

35. $\left(\dfrac{8}{x}-\dfrac{3}{4x}\right)\div\dfrac{4x+5}{x}$ 36. $\left(\dfrac{2}{x}+\dfrac{5}{x-3}\right)\cdot\dfrac{2x-6}{x}$

37. $\dfrac{x}{x-1}-\dfrac{3}{x-1}\cdot\dfrac{x+2}{x}$ 38. $\dfrac{x+3}{x+2}+\dfrac{x}{x+2}\div\dfrac{x^2}{x-3}$

39. $\dfrac{x-1}{x+4}+\dfrac{x-6}{x^2+3x-4}\div\dfrac{x-4}{x-1}$ 40. $\dfrac{x}{x+3}-\dfrac{3}{x-5}\cdot\dfrac{x^2-3x-10}{x-2}$

6.5 PROPORTIONS WITH APPLICATIONS

Consider the statement, "At that university, there are two male students for every three female students." Does this mean there are only 5 students, 2 of whom are men and 3 of whom are women? Not very likely. Suppose the odds on a horse race are 10 to 1 on Old Blue. If Old Blue wins and you bet \$2, what do you win?

These are examples of **ratios.** A **ratio** is a comparison of two numbers, usually written as $a:b$ or $\frac{a}{b}$. The ratio of men to women students can be written 2:3 or $\frac{2}{3}$. The ratio of dollars paid to dollars bet on Old Blue would be 10:1 or $\frac{10}{1}$.

We are interested in **proportions.** A **proportion** is an equation that states two ratios are equal. Examples are

$$\frac{3}{4} = \frac{6}{8}, \quad \frac{10}{1} = \frac{x}{2}, \quad \frac{3n}{5n} = \frac{3}{5}, \quad \text{and} \quad \frac{x+3}{2} = \frac{x-5}{4}$$

Is the proportion $\frac{6}{10} = \frac{9}{15}$ true? One way to answer this question is to reduce both fractions.

$$\frac{6}{10} = \frac{\cancel{2} \cdot 3}{\cancel{2} \cdot 5} = \frac{3}{5} \quad \text{and} \quad \frac{9}{15} = \frac{\cancel{3} \cdot 3}{\cancel{3} \cdot 5} = \frac{3}{5}$$

So, $\frac{6}{10} = \frac{9}{15}$ is true.

Another technique is to note $6 \cdot 15 = 10 \cdot 9$ or $6 \cdot 15 = 90$ and $10 \cdot 9 = 90$. We say **the product of the extremes is equal to the product of the means.** The following format can be used in identifying the extremes and means of a proportion.

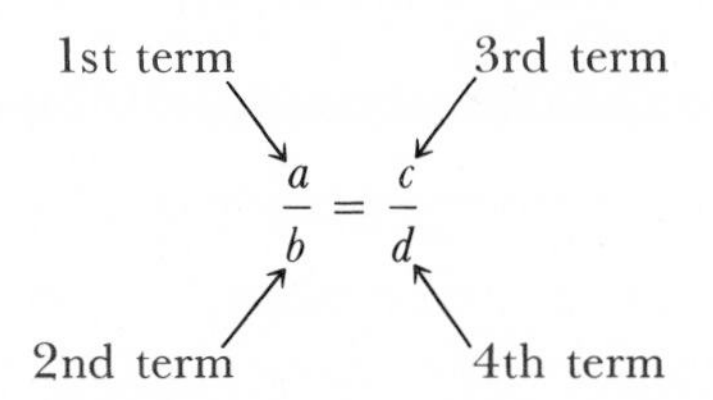

a and d (the first and fourth terms) are called the **extremes;** b and c (the second and third terms) are called the **means.**

$$\text{If } \frac{a}{b} = \frac{c}{d}, \text{ then } a \cdot d = b \cdot c$$

This relationship can be shown in the following way.

$$\frac{a}{b} = \frac{c}{d}$$

$$\frac{a}{b} \cdot \frac{bd}{1} = \frac{c}{d} \cdot \frac{bd}{1} \quad \text{Multiply both sides by } bd.$$

$$\frac{a}{\cancel{b}} \cdot \frac{\cancel{b}d}{1} = \frac{c}{\cancel{d}} \cdot \frac{b\cancel{d}}{1} \quad \text{Simplify.}$$

$$a \cdot d = b \cdot c$$

If one or more of the terms are not known, the equation should be solved to find a value for the unknown represented in the terms. The first step should be to find the product of the means and the product of the extremes and then set these equal to each other.

EXAMPLES

Solve the following proportions for the unknown number.

1. $\frac{12}{15} = \frac{8}{x}$

$$12x = 15 \cdot 8$$

$$\frac{12x}{12} = \frac{15 \cdot 8}{12}$$

$$x = \frac{\cancel{3} \cdot 5 \cdot \cancel{4} \cdot 2}{\cancel{3} \cdot \cancel{4}} = 10$$

Or, reducing $\frac{12}{15}$ first to $\frac{4}{5}$:

$$\frac{12}{15} = \frac{8}{x}$$

$$\frac{4}{5} = \frac{8}{x}$$

$$4x = 5 \cdot 8$$

$$\frac{4x}{4} = \frac{40}{4}$$

$$x = 10$$

2. $\frac{x+3}{2} = \frac{x-5}{4}$

$$4(x+3) = 2(x-5)$$

$$4x + 12 = 2x - 10$$

$$4x + 12 - 2x - 12 = 2x - 10 - 2x - 12$$

$$2x = -22$$

$$\frac{2x}{2} = \frac{-22}{2}$$

$$x = -11$$

Check: $\frac{-11+3}{2} \stackrel{?}{=} \frac{-11-5}{4}$

$$\frac{-8}{2} \stackrel{?}{=} \frac{-16}{4}$$

$$-4 = -4$$

Proportions arise naturally in everyday problems. For example if tires are on sale at 2 for \$75 but you need 5 new tires, what would you pay for the 5 tires? Setting up a proportion gives:

$$\underbrace{\frac{2 \text{ tires}}{75 \text{ dollars}} = \frac{5 \text{ tires}}{x \text{ dollars}}}_{\substack{\text{Numerators agree in type} \\ \text{and} \\ \text{denominators agree in type.}}} \quad \text{or} \quad \underbrace{\frac{75 \text{ dollars}}{x \text{ dollars}} = \frac{2 \text{ tires}}{5 \text{ tires}}}_{\substack{\text{Numerators correspond} \\ \text{and} \\ \text{denominators correspond.}}}$$

Solving, we have

$$\frac{2}{75} = \frac{5}{x}$$

$$2x = 5 \cdot 75$$

$$\frac{2x}{2} = \frac{375}{2}$$

$$x = \$187.50$$

The cost of 5 tires would be \$187.50.

Be very careful with the units in the setup of a proportion. One of the following conditions must be true:

1. The numerators agree in type and the denominators correspond.

2. The numerators correspond and the denominators correspond.

Any other arrangement will give a wrong answer. For example,

$$\frac{75 \text{ dollars}}{x \text{ dollars}} = \frac{5 \text{ tires}}{2 \text{ tires}} \text{ is } \textbf{WRONG}$$

because 75 dollars does not correspond to 5 tires.

EXAMPLES

3. On an architect's scale drawing of a building, $\frac{1}{2}$ inch represents 12 feet. What does 3 inches represent?

Solution: $$\frac{\frac{1}{2}\text{ inch}}{12 \text{ feet}} = \frac{3 \text{ inches}}{x \text{ feet}}$$

$$\frac{1}{2}x = 3 \cdot 12$$

$$2 \cdot \frac{1}{2}x = 36 \cdot 2$$

$$x = 72$$

Three inches represents 72 feet.

4. Making a statistical analysis, Mike finds three defective computer disks in a sample of 20 disks. If this ratio is consistent, how many bad disks does he expect to find in an order of 2400?

Solution: $$\frac{3 \text{ defective disks}}{20 \text{ disks}} = \frac{x \text{ defective disks}}{2400 \text{ disks}}$$

$$3 \cdot 2400 = 20x$$

$$\frac{7200}{20} = \frac{20x}{20}$$

$$360 = x$$

He expects to find 360 defective disks. (**Note:** He probably will return the order to the manufacturer.)

PRACTICE QUIZ

Questions	Answers
Solve the following proportions.	
1. $\frac{10}{x} = \frac{15}{48}$	1. $x = 32$
2. $\frac{x-4}{7} = \frac{3}{5}$	2. $x = \frac{41}{5}$
3. $\frac{3}{5x+2} = \frac{2}{x-1}$	3. $x = -1$

EXERCISES 6.5

Solve the proportions in Exercises 1–30 for the missing number.

1. $\frac{7}{8} = \frac{x}{24}$

2. $\frac{20}{x} = \frac{5}{7}$

3. $\frac{y}{17} = \frac{6}{-51}$

4. $\frac{9}{5} = \frac{36}{x}$

5. $\frac{12}{16} = \frac{36}{x}$

6. $\frac{x}{72} = \frac{100}{144}$

7. $\frac{5}{12} = \frac{9}{x}$

8. $\frac{6}{x} = \frac{9}{20}$

9. $\frac{7}{8} = \frac{x}{7}$

10. $\frac{7}{3} = \frac{15}{x}$

11. $\frac{x}{4} = \frac{20}{3}$

12. $\frac{8}{6} = \frac{x}{13}$

13. $\frac{4}{5x} = \frac{2}{25}$

14. $\frac{15}{2} = \frac{3x}{8}$

15. $\frac{21}{x+4} = \frac{7}{8}$

16. $\frac{8}{3} = \frac{x-1}{24}$

17. $\frac{x+3}{5} = \frac{11}{6}$

18. $\frac{9}{17} = \frac{3}{x-4}$

19. $\frac{10}{x} = \frac{5}{x-2}$

20. $\frac{3x}{8} = \frac{x+3}{2}$

21. $\frac{4x}{7} = \frac{x+5}{3}$

22. $\frac{x-6}{5x} = \frac{2}{7}$

23. $\frac{2x}{x-1} = \frac{3}{2}$

24. $\frac{x+7}{x-4} = \frac{5}{6}$

25. $\frac{x-4}{2x+1} = \frac{6}{7}$

26. $\frac{x-3}{8} = \frac{2x-3}{12}$

27. $\frac{5x+2}{x-6} = \frac{11}{4}$

28. $\dfrac{x+9}{3x+2} = \dfrac{5}{8}$ **29.** $\dfrac{2x+3}{3x-1} = \dfrac{6}{7}$ **30.** $\dfrac{4x-5}{2x+3} = \dfrac{9}{8}$

Set up proportions and solve the word problems in Exercises 31–45.

31. The local sales tax is 6 cents for each dollar. What is the price of an item if the sales tax is 51 cents?

32. The property tax rate is $10.87 on every $100 of assessed valuation. Find the property tax on a house assessed at $38,000.

33. On a map, each inch represents $7\frac{1}{2}$ miles. What is the distance represented by 6 inches?

34. An elementary school has a ratio of 1 teacher for each 24 children. If the school presently has 21 teachers, how many students are enrolled?

35. At the Bright-As-Day light bulb plant, three out of each 100 bulbs produced are defective. If the daily production is 4800 bulbs, how many are defective?

36. A floor plan is drawn to scale where 1 inch represents 4 feet. What size will the drawing be for a room that is 26 feet by 32 feet?

37. Twelve pounds of oranges cost $1.20. How much will 30 pounds cost?

38. Pete is averaging 17 hits for each 50 times at bat. If he maintains this average, how many "at bats" will he need in order to get 204 hits?

39. Every 50 pounds of a certain alloy contain 6 pounds of copper. To have 27 pounds of copper, how many pounds of alloy will you need?

40. Instructions for Never-Ice Antifreeze state that 4 quarts of antifreeze are needed for each 10 quarts of radiator capacity. If Sal's car has a 22-quart radiator, how many quarts of antifreeze will it need?

41. The negative of a picture is $1\frac{1}{2}$ in. by 2 in. If the largest paper available is 11 in. long, what would be the maximum width of the enlargement?

42. A flagpole casts a 30-foot shadow at the same time as a 6-foot man casts a $4\frac{1}{2}$-foot shadow. How tall is the flagpole?

43. In a group of 30 people, there are 3 left-handed people. How many left-handed people would you expect to find in a group of 180 people?

44. In a class of 40 students, 21 are under 20 years of age, 14 are 20–29 years of age, 4 are 30–39 years of age, and 1 is over 40 years of age. If this represents a "typical"class, and if there is a total of 9840 students in the entire school, how many are in each age group?

45. The ratio of two numbers is 7 to 9. If the sum of the numbers is 48, find the numbers.

46. The ratio of two numbers is 3 to 7. If the sum of the numbers is 60, find the numbers.

47. In a math class, the ratio of women to men is 6 to 5. If there are 44 students in the class, how many men are there?

48. In a certain class, the ratio of success to failure is 7 to 2. How many can be expected to successfully complete the class if 36 people are enrolled?

49. An office has two copiers. One can make 8 copies while the other makes 5. If both are used to make 650 copies, how many are made by each machine?

50. In an iodine-alcohol mixture there are 3 parts of iodine for each 8 parts of alcohol. If you want a total of 220 milliliters of mixture, how much iodine and alcohol are needed?

6.6 ADDITIONAL APPLICATIONS

If a gear with T_1 teeth is meshed with a gear having T_2 teeth, the numbers of revolutions of the two gears are related by the proportion

$$\frac{T_1}{T_2} = \frac{R_2}{R_1} \qquad \text{or} \qquad T_1R_1 = T_2R_2$$

where R_1 and R_2 are the numbers of revolutions.

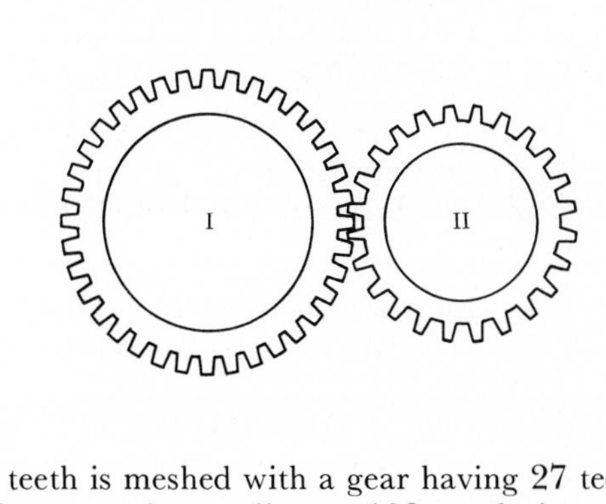

1. A gear having 96 teeth is meshed with a gear having 27 teeth. Find the speed of the small gear if the large one is traveling at 108 revolutions per minute.

2. What size gear (how many teeth) is needed to turn a pulley on a feed grinder at 24 revolutions per minute if it is driven by a 15-tooth gear turning at 56 revolutions per minute?

3. What size gear (how many teeth) is needed to turn the reel on a harvester at 20 revolutions per minute if it is driven by a 12-tooth gear turning at 60 revolutions per minute?

The lifting force P in pounds exerted by the atmosphere on the wings of an airplane is related to the area A of the wings in square feet and the speed of the plane V in miles per hour by the formula

$$P = kAV^2$$

where k is the constant of proportionality.

(**Hint for Problems 4–9:** First substitute the given information to find the value for k, the constant of proportionality. Then answer the question using the given change in conditions.)

4. If the lift is 9600 lb for a wing area of 120 sq ft and a speed of 80 mph, find the lift of the same airplane at a speed of 100 mph.

5. If a lift is 12,000 lb for a wing area of 110 sq ft and a speed of 90 mph, what would be the necessary wing area to attain the same lift at 80 mph?

6. The lift for a wing of area 280 sq ft is 34,300 lb when the plane is going 210 mph. What is the lift if the speed is decreased to 180 mph?

Boyle's Law states that if the temperature of a gas sample remains the same, the pressure of the gas is related to the volume by the formula

$$P = \frac{k}{V}$$

where k is the constant of proportionality.

7. The temperature of a gas remains the same. The pressure of the gas is 16 lb/sq ft when the volume is 300 cu ft. What will be the pressure if the gas is compressed into 25 cu ft?

8. A pressure of 1600 lb/sq ft is exerted by 2 cu ft of air in a cylinder. If a piston is pushed into the cylinder until the pressure is 1800 lb/sq ft, what will be the volume of the air?

9. If 1000 cu ft of gas exerting a pressure of 140 lb/sq ft must be placed in a container which has a capacity of 200 cu ft, what is the new pressure exerted by the gas?

In order to administer the medication prescribed by a doctor, a nurse must often figure the dosage in relation to the drugs available. The following formula is sometimes used:

$$\underset{\textbf{\textit{Labeling Information}}}{\frac{\text{stated dosage}}{\text{stated amount of solution}}} = \underset{\textbf{\textit{Prescribed Dosage}}}{\frac{\text{desired dosage}}{\text{amount of solution needed}}}$$

10. If a doctor orders 1.5 mg of reserpine, how many milliliters of a solution labeled 2.5 mg in 1 ml should be given?

11. The doctor orders 450,000 units of penicillin. The dose on hand is in a solution labeled 3,000,000 units per 10 ml. How many milliliters should be administered?

12. A doctor orders 25 units of insulin. How many milliliters of a solution labeled 40 units per milliliter should be given?

The ingredients in concrete are cement, sand, and gravel mixed in water. The proportions of each ingredient, except water, are expressed as

cement : sand : gravel

13. Find the number of sacks of cement (one cubic foot per sack), cubic feet of sand, and cubic feet of gravel needed to make 81 cubic feet of 1:3:5 concrete.

14. At the cost of \$1.30 per sack, determine the cost of the cement required to make 60 cubic feet of concrete using a 1:3:4 mix (1 sack = 1 cubic foot).

If a lever is balanced, with weight on opposite sides of the balance points, then the following proportion exists:

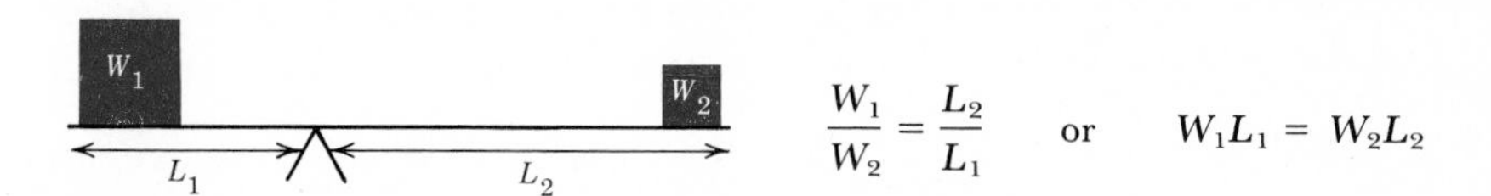

(Ignore the weight of the lever.)

15. How much weight can be raised at one end of an 8-ft bar by the downward force of 60 lb when the balance point is $\frac{1}{2}$ ft from the unknown weight?

16. A force of 40 lb at one end of a 5-ft bar is to balance 160 lb at the other end of the bar. Ignoring the weight of the bar, how far from the 40-lb weight should a hole be drilled in the bar for a bolt to serve as a balance point?

17. Where should the balance point of a 12-ft bar be located if a 120-lb force is to raise a load weighing 960 lb?

The resistance, R, in a wire is given by the formula

$$R = \frac{kL}{d^2}$$

where L is the length of the wire and d is the diameter.

18. The resistance of a wire 500 ft long and with a diameter of 0.01 in. is 20 ohms. What is the resistance of a wire 1500 ft long and with a diameter of 0.02 in.?

19. The resistance of a wire 100 ft long and 0.01 in. in diameter is 8 ohms. What is the resistance of a piece of the same type of wire with a length of 150 ft and a diameter of 0.015 in.?

20. The resistance is 2.6 ohms when the diameter of a wire is 0.02 in. and the wire is 10 ft long. Find the resistance of the same type of wire with a diameter of 0.01 in. and a length of 5 ft.

CHAPTER 6 SUMMARY

Rational expressions are fractions in which the numerator and denominator are polynomials.

The **division algorithm** with polynomials is the procedure followed when dividing a polynomial by a polynomial: For $P \div D$,

$$P = Q \cdot D + R \qquad \text{or} \qquad \frac{P}{D} = Q + \frac{R}{D}$$

where P is the dividend, D is the divisor, Q is the quotient, and R is the remainder. **The remainder must be of smaller degree than the divisor.**

In general, for $b \neq a$,

$$\frac{a-b}{b-a} = -1$$

For integers a and b, $b \neq 0$,

$$-\frac{a}{b} = \frac{-a}{b} = \frac{a}{-b}$$

For polynomials P and Q, $Q \neq 0$,

$$-\frac{P}{Q} = \frac{-P}{Q} = \frac{P}{-Q}$$

If P, Q, R, and S are polynomials with $Q, S \neq 0$, then

1. $\dfrac{P}{Q} \cdot \dfrac{R}{S} = \dfrac{P \cdot R}{Q \cdot S}$

2. $\dfrac{P}{Q} \div \dfrac{R}{S} = \dfrac{P}{Q} \cdot \dfrac{S}{R}$

3. $\dfrac{P}{Q} + \dfrac{R}{Q} = \dfrac{P + R}{Q}$

4. $\dfrac{P}{Q} - \dfrac{R}{Q} = \dfrac{P - R}{Q}$

An expression that contains various combinations of addition, subtraction, multiplication, and division with rational expressions is called a **complex algebraic expression.**

A **ratio** is a comparison of two numbers, usually written as $a:b$ or $\dfrac{a}{b}$.

A **proportion** is an equation that states two ratios are equal.

In a proportion, the first and fourth terms are called the **extremes** and the second and third terms are called the **means.** In a true proportion, the product of the extremes is equal to the product of the means.

$$\text{If } \frac{a}{b} = \frac{c}{d}, \text{ then } a \cdot d = b \cdot c.$$

CHAPTER 6 REVIEW

Express each quotient in Exercises 1–5 as a sum of fractions and simplify if possible.

1. $\dfrac{6x^2 + 3x - 5}{3}$

2. $\dfrac{8x^2 - 14x + 6}{2x}$

3. $\dfrac{3x^3y + 10x^2y^2 + 5xy^3}{5x^2y}$

4. $\dfrac{7x^2y^2 - 21x^2y^3 + 8x^2y^4}{7x^2y^2}$

5. $\dfrac{20x^3y^2 - 15x^2y^3 + 10xy^4}{10xy^2}$

Divide Exercises 6–15 using the long division procedure and check your answers.

6. $378 \div 37$

7. $792 \div 61$

8. $(x^2 + 7x - 18) \div (x + 9)$

9. $(x^2 - 14x - 16) \div (x + 1)$

10. $(4x^2 + 5x - 2) \div (4x - 3)$

11. $(2x^3 + 5x^2 + 7) \div (x + 3)$

12. $\dfrac{4x^2 + 5x + 9}{x - 2}$

13. $\dfrac{x^2 + 9}{x + 3}$

14. $(6x^3 - 5x^2 + 1) \div (3x + 2)$

15. $(x^3 + 64) \div (x + 4)$

Reduce each fraction in Exercises 16–23 to its lowest terms. State any restrictions on the variable.

16. $\dfrac{x}{x^2 + x}$

17. $\dfrac{4 - x}{3x - 12}$

18. $\dfrac{2x + 6}{3x + 9}$

19. $\dfrac{x^2 + 3x}{x^2 + 7x + 12}$

20. $\dfrac{x^2 + 2x - 15}{2x^2 - 12x + 18}$

21. $\dfrac{x^2 - 9x + 20}{16 - x^2}$

22. $\dfrac{8x^2 - 2x - 3}{4x^2 + x - 3}$

23. $\dfrac{2x^2 - 5x^3}{10x^2 - 4x}$

Perform the indicated operations in Exercises 24–42. Assume that no denominator is 0.

24. $\dfrac{x^2}{x + y} - \dfrac{y^2}{x + y}$

25. $\dfrac{x^2 + x - 7}{x^2 + 3x - 10} + \dfrac{x^2 + 5x - 13}{x^2 + 3x - 10}$

26. $\dfrac{3x}{x + 2} \cdot \dfrac{x^2 + 2x}{x^2}$

27. $\dfrac{3x}{x + 2} \div \dfrac{4x}{x + 2}$

28. $\dfrac{2y + 4}{5} \cdot \dfrac{35}{6y + 12}$

29. $\dfrac{7}{x} + \dfrac{9}{x - 4}$

30. $\dfrac{4}{y + 2} - \dfrac{4}{y + 3}$

31. $\dfrac{4x}{3x + 6} - \dfrac{x}{x + 1}$

32. $\dfrac{4x}{x - 4} \div \dfrac{12x^2}{x^2 - 16}$

33. $\dfrac{x^2 + 3x + 2}{x + 3} \div \dfrac{x + 1}{3x^2 + 6x}$

34. $\dfrac{2x+1}{x^2+5x-6} \cdot \dfrac{x^2+6x}{x}$

35. $\dfrac{5}{x-5} + \dfrac{x}{25-x^2}$

36. $\dfrac{4}{x+6} + \dfrac{8x}{x^2+x-6}$

37. $\dfrac{8}{x^2+x-6} + \dfrac{2x}{x^2-3x+2}$

38. $\dfrac{x}{x^2+3x-4} - \dfrac{x+1}{x^2-1}$

39. $\dfrac{x^2+3x-10}{x^2-9} \cdot \dfrac{x^2-x-6}{x^2+7x+10}$

40. $\dfrac{x+1}{x^2+4x+4} \div \dfrac{x^2-x-2}{x^2-2x-8}$

41. $\dfrac{7-x}{7x^3} \div \dfrac{7x^2+48x-7}{x^3+7x^2} \cdot \dfrac{7x^2-50x+7}{x^2-7x}$

42. $\dfrac{3}{x+2} + \dfrac{2x}{x^2-3x-10} - \dfrac{1}{x^2-2x-15}$

Simplify the complex algebraic expressions in Exercises 43–47.

43. $\dfrac{1-\dfrac{2}{3}}{\dfrac{1}{4}+\dfrac{5}{6}}$

44. $\dfrac{\dfrac{3}{x}+\dfrac{1}{6x}}{\dfrac{7}{3x}}$

45. $\dfrac{\dfrac{2x}{x-4}}{\dfrac{x}{x^2-3x-4}}$

46. $\dfrac{\dfrac{1}{x}-\dfrac{1}{x^2}}{\dfrac{1}{x}+\dfrac{1}{x^2}}$

47. $\dfrac{\dfrac{4}{3x}+\dfrac{1}{6x}}{\dfrac{1}{x^2}-\dfrac{1}{2x}}$

Solve each proportion in Exercises 48–51 for the missing number.

48. $\dfrac{4}{7} = \dfrac{x}{91}$

49. $\dfrac{5}{8} = \dfrac{9}{x}$

50. $\dfrac{3}{x} = \dfrac{5}{x+3}$

51. $\dfrac{x-2}{x+4} = \dfrac{3}{7}$

Set up proportions for the word problems in Exercises 52–55 and then solve them.

52. One number is 12 more than twice another. Their ratio is 2 to 5. Find the numbers.

53. A set of house plans is drawn to scale, where $\frac{1}{4}$ inch represents 1 foot. What will be the size of the drawing of a room that is 18 feet by 24 feet?

54. During basketball season, Rod averaged 5 successful free throws for each 7 attempted. If he made 115 free throws, how many did he attempt?

55. Anna saved 30% on a blouse she bought. If she saved \$5.40, what was the original price of the blouse?

Optional

56. If the lift is 16,000 lb for a wing area of 180 sq ft and a speed of 120 mph, find the lift of the same plane at a speed of 150 mph ($P = kAv^2$).

57. The resistance of a wire 250 ft long and with a diameter of 0.01 in. is 10 ohms. What is the resistance of a piece of the same type of wire with a length of 300 ft and a diameter of 0.02 in.? $\left(R = \frac{kL}{d^2}.\right)$

CHAPTER 6 TEST

Express each quotient in Exercises 1 and 2 as a sum of fractions and simplify if possible.

1. $\dfrac{4x^3 + 3x^2 - 6x}{2x^2}$

2. $\dfrac{5x^2y + 6x^2y^2 + 3xy^3}{3x^2y}$

Divide Exercises 3–5 using long division.

3. $(x^2 + 5x - 36) \div (x - 4)$

4. $(2x^2 - 9x - 20) \div (2x + 3)$

5. $(x^3 - 27) \div (x - 3)$

6. Reduce and state any restrictions on the variable:

$$\frac{2x^2 + 5x - 3}{2x^2 - x}$$

Reduce the fractions in Exercises 7 and 8. Assume that no denominator is 0.

7. $\dfrac{8x^2 - 2x - 3}{4x^2 + x - 3}$

8. $\dfrac{15x^4 - 18x^3}{36x - 30x^2}$

Perform the indicated operations in Exercises 9–14 and reduce your answer. Assume that no denominator is 0.

9. $\dfrac{16x^2 + 24x + 9}{3x^2 + 14x + 8} \cdot \dfrac{21x + 14}{4x^2 + 3x}$

10. $\dfrac{3x - 3}{3x^2 - x - 2} \div \dfrac{24x^2 - 6}{6x^2 + x - 2}$

11. $\dfrac{3}{x^2 - 1} + \dfrac{7}{x^2 - x - 2}$

12. $\dfrac{8x}{x^2 - 25} - \dfrac{3}{x^2 + 3x - 10}$

13. $\dfrac{1}{x - 1} + \dfrac{2x}{x^2 - x - 12} - \dfrac{5}{x^2 - 5x + 4}$

14. $\dfrac{x^2 + 8x + 15}{x^2 + 6x + 5} \cdot \dfrac{x - 3}{2x} \div \dfrac{x + 3}{x^2}$

Simplify the complex fractions in Exercises 15 and 16.

15. $\dfrac{\dfrac{1}{x} - \dfrac{1}{x^2}}{1 - \dfrac{1}{x^2}}$

16. $\dfrac{\dfrac{6}{x - 5}}{\dfrac{x + 18}{x^2 - 2x - 15}}$

Solve the proportions in Exercises 17 and 18.

17. $\dfrac{9}{3 - x} = \dfrac{8}{2x + 1}$

18. $\dfrac{3}{4} = \dfrac{2x + 5}{5 - x}$

Set up proportions for Exercises 19 and 20 and then solve them.

19. A 20-foot flagpole casts a shadow 12 feet long. If a man 6 feet tall is standing beside the flagpole, how long is his shadow?

20. Eighty percent of a math class received a passing grade on the last exam. If 32 people passed, how many are in the class?

USING THE COMPUTER

```
10  REM
20  REM  This program will make a multiplication table
30  REM  for the integers from 1 to 11 times the integers
40  REM  from -5 to 9.
50  REM
60  PRINT TAB(23);"Multiplication table" ' Print the title...
70  PRINT                                ' Print a blank line
80  REM The number will be fit into 5 spaces
90  WRITING.FORMAT$ = "#####"
100 PRINT "     :";
110 FOR COLUMN = 1 TO 11               ' Loop for column number
120 PRINT USING WRITING.FORMAT$; COLUMN;
130 NEXT COLUMN                          ' End loop
140 PRINT                                ' End line
150 PRINT
"   -------------------------------------------------------"
160 FOR ROW = -5 TO 9
170 PRINT USING WRITING.FORMAT$; ROW;    ' Print row number
180 PRINT ":";
189 REM Loop for calculating the products
190 FOR EACH.COLUMN = 1 TO 11
200 PRODUCT = ROW * EACH.COLUMN
210 PRINT USING WRITING.FORMAT$; PRODUCT;   ' Print product
220 NEXT EACH.COLUMN                     ' End EACH.COLUMN loop
230 PRINT                                ' End line
240 NEXT ROW                             ' End ROW loop
250 END                                  ' End program
```

```
                       Multiplication table

     :    1    2    3    4    5    6    7    8    9   10   11
---------------------------------------------------------------
   -5:   -5  -10  -15  -20  -25  -30  -35  -40  -45  -50  -55
   -4:   -4   -8  -12  -16  -20  -24  -28  -32  -36  -40  -44
   -3:   -3   -6   -9  -12  -15  -18  -21  -24  -27  -30  -33
   -2:   -2   -4   -6   -8  -10  -12  -14  -16  -18  -20  -22
   -1:   -1   -2   -3   -4   -5   -6   -7   -8   -9  -10  -11
    0:    0    0    0    0    0    0    0    0    0    0    0
    1:    1    2    3    4    5    6    7    8    9   10   11
    2:    2    4    6    8   10   12   14   16   18   20   22
    3:    3    6    9   12   15   18   21   24   27   30   33
    4:    4    8   12   16   20   24   28   32   36   40   44
    5:    5   10   15   20   25   30   35   40   45   50   55
    6:    6   12   18   24   30   36   42   48   54   60   66
    7:    7   14   21   28   35   42   49   56   63   70   77
    8:    8   16   24   32   40   48   56   64   72   80   88
    9:    9   18   27   36   45   54   63   72   81   90   99
```

The stages of invention: 1. preparation,
2. incubation, 3. illumination, 4. verification.

JACQUES HADAMARD 1865–1963

DID YOU KNOW?

The way in which individuals approach problem-solving situations has itself been a subject of research. In 1944, the French mathematician Jacques Hadamard wrote *An Essay on the Psychology of Invention in the Mathematical Field.* One of Hadamard's more interesting ideas was that the subconscious mind continues to work on a difficult unsolved problem even while the conscious mind is involved in other activities. Hadamard cites instances of scientists experiencing "flashes," in which the complete solution to a problem came to them when they were not involved actively in their research. Some scientists even report instances of thinking carefully about a problem before going to sleep and then waking with knowledge of the solution to the problem.

Hadamard points out, however, that it is impossible for the unconscious mind to assist in problem solving unless the conscious mind has done the necessary work of preparation. When Isaac Newton was asked how he discovered gravity, he is reported to have said, "by constantly thinking it over." Hadamard claims that sudden inspirations never happen except after voluntary efforts at consciously solving the problem. This means that solutions to problems are never chance happenings, but rather are the result of hard work, both conscious and unconscious.

As you use the attack plan for word problems, you will see that there is a definite method for the preparatory work in solving word problems. Your conscious mind can be made to work along certain lines of attack that are most likely to succeed. The important idea is that the problems must be actively attacked before any inspiration can occur.

7 Applications with First-Degree Equations and Inequalities

7.1 REVIEW OF SOLVING EQUATIONS (INCLUDING SOME QUADRATICS)

In Chapter 3, we developed the techniques for solving first-degree equations. First-degree equations can be written in the form $ax + b = c$ where a, b, and c are constants and $a \neq 0$. The basic procedures for solving first-degree equations are listed in Section 3.4 and again here for convenient reference.

To Solve a First-Degree Equation

1. Simplify each side of the equation by removing any grouping symbols and combining like terms. (In some cases, you may want to multiply each term by an expression to clear fractions or decimal coefficients.)
2. Use the addition property of equality to add the opposites of constants and/or variables so that variables are on one side and constants on the other.
3. Use the multiplication property of equality to multiply both sides by the reciprocal of the coefficient of the variable (or divide both sides by the coefficient).
4. Check your answer by substituting it into the original equation.

Remember, the objective is to get the variable by itself on one side of the equation.

EXAMPLES

Solve the following equations.

1. $3x + 14 = x - 2(x + 1)$ Write the equation.

 $3x + 14 = x - 2x - 2$ Use the distributive property to remove parentheses.

 $3x + 14 = -x - 2$ Simplify.

$4x + 14 = -2$ Add x to both sides.

$4x = -16$ Add -14 to both sides.

$x = -4$ Divide both sides by 4.

2. $1 + 2x + 3 - 3x = 20 - x + 6x$ Write the equation.

$4 - x = 20 + 5x$ Simplify.

$4 = 20 + 6x$ Add x to both sides.

$-16 = 6x$ Add -20 to both sides.

$-\frac{8}{3} = x$ Divide both sides by 6 and reduce.

3. $\frac{3x}{4} - 7 = -1$ Write the equation.

$\frac{3x}{4} = 6$ Add $+7$ to both sides.

$3x = 24$ Multiply both sides by 4.

$x = 8$ Divide both sides by 3 and reduce.

Since $\frac{3x}{4} = \frac{3}{4} \cdot \frac{x}{1} = \frac{3}{4}x$, we could solve an equation such as $\frac{3x}{4} = 6$ in one step by multiplying both sides by $\frac{4}{3}$, the reciprocal of $\frac{3}{4}$, as follows:

$$\frac{3x}{4} = 6$$

$$\left(\frac{4}{3} \cdot \frac{3}{4}\right)x = \frac{4}{3} \cdot 6$$

$$x = 8$$

Example 3 can also be solved by first multiplying by 4 instead of adding +7 first. In this procedure, however, we must be sure to **multiply each term by 4 on both sides of the equation.**

$$\frac{3x}{4} - 7 = -1 \qquad \text{Write the equation.}$$

$$\frac{3x}{4}\cdot\mathbf{4} - 7\cdot\mathbf{4} = -1\cdot\mathbf{4} \qquad \text{Multiply each term by 4.}$$

$$3x - 28 = -4 \qquad \text{Simplify.}$$

$$3x = 24 \qquad \text{Add +28 to both sides.}$$

$$x = 8 \qquad \text{Divide both sides by 3 and reduce.}$$

If an equation contains fractions, you should multiply each term by the LCM (least common multiple) of the denominators. Generally, this makes the equation easier to solve. **No denominator can have a value of 0.** So, if you multiply by a variable, be sure to check your answers.

EXAMPLES

Solve the following equations.

4. $$\frac{3}{2x} + \frac{1}{x} = 1 \qquad \text{Here the LCM of the denominators is } 2x\ (x \neq 0).$$

$$2x\cdot\frac{3}{2x} + 2x\cdot\frac{1}{x} = 2x\cdot 1 \qquad \text{Multiply each term by } 2x.$$

$$3 + 2 = 2x \qquad \text{The result is a linear equation.}$$

$$5 = 2x$$

$$\frac{5}{2} = \frac{\cancel{2}x}{\cancel{2}}$$

$$\frac{5}{2} = x$$

5. $$0.3x - 0.5x + 0.4 = 0.2(x + 1) \qquad \text{Here the equation is linear with decimal coefficients.}$$

$$10(0.3x) - 10(0.5x) + 10(0.4) = 10(0.2)(x + 1) \qquad \text{Multiply each term by 10 so that the coefficients will be integers.}$$

$$3x - 5x + 4 = 2(x + 1)$$

$$-2x + 4 = 2x + 2$$

$$-2x - 2x = 2 - 4$$

$$-4x = -2$$

$$\frac{-4x}{-4} = \frac{-2}{-4}$$

$$x = \frac{1}{2}$$

or $$x = 0.5$$

6. After you have multiplied by the LCM, the resulting equation may be quadratic. We know how to solve by factoring (see Section 5.4). We will do more with quadratics in Chapter 11.

$$\frac{2}{x} + \frac{2}{x+3} = 1$$ LCM $= x(x + 3)$. Also, $x \neq 0$ and $x \neq -3$.

$$\frac{2}{\cancel{x}} \cdot \cancel{x}(x + 3) + \frac{2}{\cancel{x+3}} \cdot x\cancel{(x+3)} = 1 \cdot x(x + 3)$$ Multiply each term by $x(x + 3)$.

$$2(x + 3) + 2x = x(x + 3)$$ Simplify.

$$2x + 6 + 2x = x^2 + 3x$$

$$0 = x^2 - x + 6$$ The result is a quadratic.

$$0 = (x + 2)(x - 3)$$ Factor.

$$x + 2 = 0 \quad \text{or} \quad x - 3 = 0$$

$$x = -2 \qquad\qquad x = 3$$

There are two solutions: -2 and 3. Both are valid solutions since neither gives a denominator value of 0. The two values for x that could not be solutions are 0 and -3.

7. $$\frac{x}{x-2} + \frac{x-6}{x(x-2)} = \frac{5x}{x-2} - \frac{10}{x-2}$$ LCM $= x(x - 2)$. Also, $x \neq 0$ and $x \neq 2$.

$$\frac{x}{\cancel{x-2}} \cdot x(x-2) + \frac{x-6}{\cancel{x(x-2)}} \cdot \cancel{x(x-2)} =$$

$$\frac{5x}{\cancel{x-2}} \cdot x(\cancel{x-2}) - \frac{10}{\cancel{x-2}} \cdot x(\cancel{x-2})$$ Multiply each term by $x(x-2)$.

$$x^2 + x - 6 = 5x^2 - 10x$$ Simplify.

$$0 = 4x^2 - 11x + 6$$

$$0 = (4x-3)(x-2)$$

$$4x - 3 = 0 \quad \text{or} \quad x - 2 = 0$$

$$4x = 3 \qquad \cancel{x = 2}$$

$$x = \frac{3}{4}$$

The only solution is $x = \frac{3}{4}$; 2 is **not** a solution since no denominator can be 0.

PRACTICE QUIZ

Questions	Answers
Solve the following equations.	
1. $9x = 11x + 11$	1. $x = -\frac{11}{2}$
2. $x + 17 = 2x + 7 - 3x$	2. $x = -5$
3. $2x^2 = x + 1$	3. $x = 1$ or $x = -\frac{1}{2}$
4. $0.3a + 0.6a = a + 0.007$	4. $a = -0.07$
5. $\frac{5}{8}x + 1 = x - \frac{1}{6}$	5. $x = \frac{28}{9}$
6. $\frac{x-5}{x} + \frac{7}{x} = x$	6. $x = 2$ or $x = -1$

EXERCISES 7.1

Solve the following equations.

1. $3x + 6 = x - 4$

2. $x + 9 = 4x - 3$

3. $7 - x = 2x - 8$

4. $11 - 2x = -3 + 5x$

5. $-3x - 5 = 7 - 5x$

6. $7x - 2 = -2x + 7$

7. $4x + 3 = 2x + 8$

8. $3x + 8 = -3x + 14$

9. $18x - 4 = 3x + 26$

10. $7x + 9 = 4x + 36$

11. $3x + 14 = 29 - 2x$

12. $8x + 5 = 5x - 6$

13. $11x - 3 = 6 + 5x$

14. $10x + 13 = 4x - 8$

15. $2 - 5x = 7x - 4$

16. $4x + 9 = 5 - 2x$

17. $\frac{5x}{6} - 1 = \frac{3}{2}$

18. $\frac{3x}{4} + \frac{1}{2} = -3$

19. $\frac{x}{3} - \frac{6}{7} = \frac{x}{21}$

20. $\frac{2x}{5} + \frac{2}{3} = \frac{x}{3}$

21. $\frac{3x}{14} - \frac{x}{7} = -\frac{1}{2}$

22. $\frac{4x}{3} - \frac{3}{4} = \frac{x}{6}$

23. $\frac{x}{4} - \frac{x}{3} = \frac{1}{12}$

24. $\frac{3x}{2} + \frac{x}{6} = -2$

25. $\frac{x}{3} - \frac{x}{7} = 4$

26. $2x - \frac{6x}{5} = \frac{9}{10}$

27. $\frac{x}{2} - \frac{2x}{3} = \frac{3}{4} + \frac{x}{3}$

28. $\frac{x}{4} - \frac{1}{2} = \frac{x}{6} - \frac{x}{3}$

29. $-0.15x = 9 - 0.6x$

30. $7 - 0.23x = 2.6 + 0.32x$

31. $0.4x - 2.2 = 0.48x + 5$

32. $0.2x - 0.04 = 0.15x + 0.02$

33. $3 - 0.2x = 0.05x - 1.25$

34. $15.6 + 0.4x = 1.15x - 8.4$

35. $3(2 - x) - 2x = -7$

36. $2x - 3(x + 2) = 10$

37. $-(x + 5) = 2x + 4$

38. $3(2x - 1) = 5 - x$

39. $3x + 8 = -3(2x - 3)$

40. $6(3x + 1) = 5(1 - 2x)$

41. $4(6 - x) = -2(3x + 1)$

42. $x - (2x + 5) = 7 - (4 - x) + 10$

43. $\frac{2}{3x} = \frac{1}{4} - \frac{1}{6x}$

44. $\frac{x - 4}{x} + \frac{3}{x} = 0$

45. $\frac{3}{8x} - \frac{7}{10} = \frac{1}{5x}$

46. $\frac{1}{x} - \frac{8}{21} = \frac{3}{7x}$

47. $\frac{5}{9x} + \frac{3}{2} = \frac{1}{2x}$

48. $\frac{3}{4x} - \frac{1}{2} = \frac{7}{8x} + \frac{1}{6}$

49. $\frac{2}{3x} + \frac{3}{4} = \frac{1}{6x} + \frac{3}{2}$

50. $\frac{5}{10x} - \frac{2}{5} = \frac{4}{5x} + \frac{1}{2}$

51. $3x^2 - 15x = 0$

52. $8x^2 + 24x = 0$

53. $5x^2 + 9x = 0$

54. $6x^2 - 14x = 0$

55. $x^2 - 6x - 7 = 0$

56. $x^2 + x - 56 = 0$

57. $x^2 + 3x = 10$

58. $2x^2 - 9x + 9 = 0$

59. $5x^2 - x - 6 = 0$

60. $4x^2 + 23x + 15 = 0$

61. $5x^2 - 7x - 6 = 0$

62. $8x^2 - 10x = 3$

63. $3x^2 = 7x - 2$

64. $3x^2 = 11x + 4$

65. $6x^2 = 20 - 2x$

66. $12x^2 = 10x + 12$

67. $\frac{3}{x - 6} = \frac{5}{x}$

68. $\frac{3}{x - 5} = \frac{-2}{x}$

69. $\frac{1}{x - 1} = \frac{2}{x - 2}$

70. $\frac{7}{x - 3} = \frac{6}{x - 4}$

71. $\frac{x}{x + 3} + \frac{1}{x + 2} = 1$

72. $\frac{1}{3} + \frac{1}{x + 4} = \frac{1}{x}$

73. $\frac{5}{x + 3} - \frac{2}{x + 1} = \frac{1}{4}$

74. $\frac{4}{x} - \frac{2}{x + 1} = \frac{4}{3}$

75. $\frac{2}{x + 2} + \frac{5}{x - 2} = -3$

76. $\frac{6}{x + 4} + \frac{2}{x - 3} = -1$

Calculator Exercises

77. $0.035x - 0.04 = 0.02x + 0.32$

78. $0.0031 - 0.012x = 0.024x - 0.0185$

79. $0.361x - 1.036 = 0.127x + 1.868$

80. $0.147x + 1.651 = 0.099x - 0.269$

7.2 FIRST-DEGREE INEQUALITIES

In Section 2.1, we discussed inequalities (using the symbols $<$, $>$, $\leq$, and $\geq$) and number lines. We stated that certain inequalities, such as $-2 < +5$, are true while others, such as $-3.1 \geq 0$, are false. We also graphed integers and rational numbers (fractions) on number lines. (Figure 7.1 is a graph of the set $-\frac{3}{2}$, -1, 0, 2.)

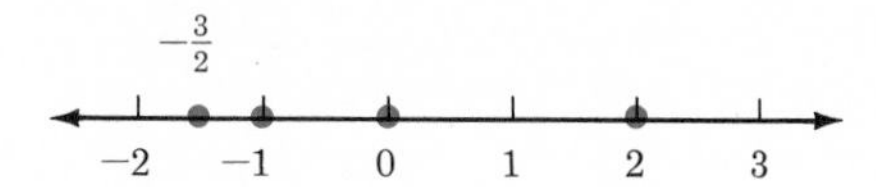

FIGURE 7.1

In this section we will encounter inequalities with variables and learn to solve these inequalities and graph the solutions. There is only one solution to a first-degree equation such as $x + 2 = 7$. That is, $x = 5$ is the one and only solution. However, an inequality, such as $x + 2 < 7$, may have an infinite number of solutions. In fact, any number less than ($<$) 5 is a solution.

Before we discuss the mechanics of solving first-degree inequalities, a comment concerning number lines is appropriate. The numbers that correspond to the points on a line are called **real numbers,** and number lines are called **real number lines.** In Chapter 2, we mentioned **rational numbers** (which include integers and fractions with integers as numerator and denominator) and **irrational numbers.** Together, these two types of numbers make up the **real number system.**

Irrational numbers, such as $\sqrt{2}$, $\sqrt{3}$, $-\sqrt{7}$, and π will be discussed in some detail in Chapter 10. The important idea here is that such numbers do indeed correspond to points on a number line, and when we consider inequalities, such as $x < 8$, all real numbers less than 8 are included (Figure 7.2).

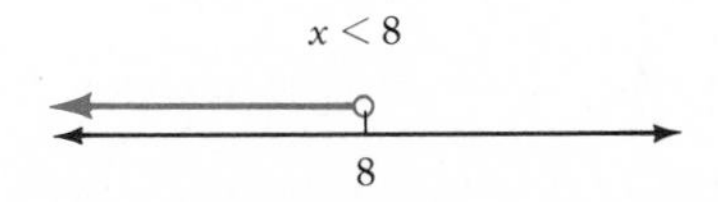

FIGURE 7.2

Remember, for every real number there is one corresponding point on a line, and for every point on a line there is one corresponding real number.

The graphs of the solutions to inequalities involve the following:

1. Putting a small circle on one point.
   a. An open circle indicates that the point is **not** included.
   b. A shaded circle indicates that the point is included.
2. Shading the portion of the number line corresponding to the solutions.

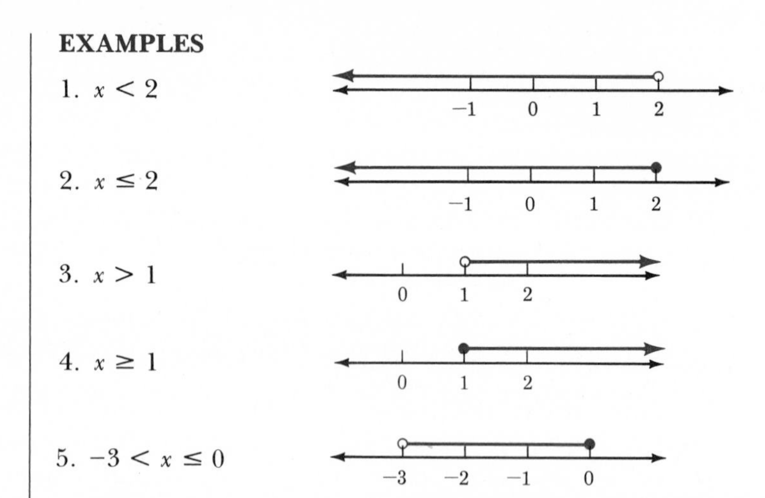

This case represents the numbers satisfying two inequalities at the same time:

$$-3 < x \quad \textbf{and} \quad x \leq 0$$

6. $x < -1$ or $x > 1$

−1 0 1

This case represents a combination (or union) of two intervals. The word "or" is necessary, and there is no way to write the two intervals together as there is in Example 5.

The graphs (or sets of numbers) shown in Examples 1–5 are all **intervals** of real numbers. Example 6 represents a combination of two intervals called a **union.**

An inequality that can be written in the form

$$ax + b < c \qquad (\text{or } ax + b > c)$$

and

$$ax + b \leq c \qquad (\text{or } ax + b \geq c)$$

where a, b, and c are constants and $a \neq 0$ is called a **first-degree inequality.** We are now interested in developing the techniques for solving first-degree inequalities or combinations of first-degree inequalities and graphing their solutions on number lines.

The following basic properties of inequalities are needed. Similar properties for equations were discussed in Chapter 3.

Addition Property of Inequalities

If A, B, and C are algebraic expressions, then the inequalities

$$A < B$$

and

$$A + C < B + C$$

have the same solutions.

Multiplication Property of Inequalities

If A, B, and C are algebraic expressions, then

1. $A < B$ and $AC < BC$ have the same solutions if $C > 0$.
2. $A < B$ and $AC > BC$ have the same solutions if $C < 0$.

Solving an inequality, such as $2x + 1 < 7$, is similar to solving a first-degree equation. The object is to get the variable by itself on one side of the inequality. The important difference involves multiplying or dividing by negative numbers. **Multiplying or dividing both sides of an inequality by a negative number reverses the sense of the inequality.** That is, "less than" becomes "greater than," and vice versa. For example,

a. $5 < 7$ 5 is **less than** 7.

$-2(5) \downarrow -2(7)$ Multiply both numbers by -2.

$-10 > -14$ Now -10 is **greater than** -14.

b. $6 \geq -12$ 6 is **greater than or equal to** -12.

$\dfrac{6}{-6} \downarrow \dfrac{-12}{-6}$ Divide both numbers by -6.

$-1 \leq 2$ -1 is **less than or equal** to 2.

c. $-2x > 6$ — $-2x$ is greater than 6.

$$\frac{-2x}{-2} < \frac{6}{-2}$$ Divide both sides by -2 and reverse the inequality.

$x < -3$ — The solution is "x is less than -3."

To Solve a First-Degree Inequality

1. Simplify each side of the inequality by removing any parentheses and combining like terms.
2. Add constants and/or variables to or subtract them from both sides of the inequality so that variables are on one side and constants on the other.
3. Divide both sides by the coefficient (or multiply them by the reciprocal) of the variable, and **reverse the sense of the inequality if this coefficient is negative.**
4. A quick (and generally satisfactory) check is to select any one number in your solution and substitute it into the original inequality.

EXAMPLES

Solve the following inequalities and graph the solutions.

7. $5x + 4 \leq -1$ — Write the inequality.

$5x + 4 \mathbf{- 4} \leq -1 \mathbf{- 4}$ — Add -4 to both sides.

$5x \leq -5$ — Simplify.

$$\frac{5x}{\mathbf{5}} \leq \frac{-5}{\mathbf{5}}$$ Divide both sides by 5.

$x \leq -1$ — Simplify.

−2 −1 0

(**Note:** The solid dot means that -1 is included.)

Check: As a quick check, pick any number less than -1, say, -4, and substitute it into the original inequality. If a false statement results, then you have made a mistake.

$$5(-4) + 4 \leq -1$$

$$-16 \leq -1$$ A true statement.

8.

$x + 1 > 3 - 2x$	Write the inequality.
$x + 1 + \mathbf{2x} > 3 - 2x + \mathbf{2x}$	Add $2x$ to both sides.
$3x + 1 > 3$	Simplify.
$3x + 1 \mathbf{- 1} > 3 \mathbf{- 1}$	Add -1 to both sides.
$3x > 2$	Simplify.
$\dfrac{3x}{\mathbf{3}} > \dfrac{2}{\mathbf{3}}$	Divide both sides by 3.
$x > \dfrac{2}{3}$	Simplify.

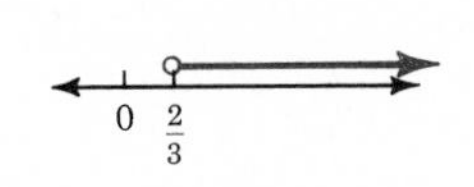

(**Note:** The open circle means that $\frac{2}{3}$ is not included in the solution.)

9. $-1 < -2(x + 1) \leq 3$

(**Note:** Here we are actually solving two inequalities at the same time: $-1 < -2(x + 1)$ **and** $-2(x + 1) \leq 3$. We want to find those numbers that satisfy both inequalities.

$-1 < -2(x + 1) \leq 3$	Write the expression.
$-1 < -2x - 2 \leq 3$	Use the distributive property.
$-1 \mathbf{+ 2} < -2x - 2 \mathbf{+ 2} \leq 3 \mathbf{+ 2}$	Add 2 to all three expressions.
$1 < -2x \leq 5$	Simplify.
$\dfrac{1}{\mathbf{-2}} > \dfrac{-2x}{\mathbf{-2}} \geq \dfrac{5}{\mathbf{-2}}$	Divide each expression by -2 and reverse the sense of each inequality.
$-\dfrac{1}{2} > x \geq -\dfrac{5}{2}$	Simplify.

Graph the solution.

Check: For a quick check, substitute $x = -2$.

$$-1 < -2(-2 + 1) \leq 3$$

$$-1 < -2(-1) \leq 3$$

$$-1 < 2 \leq 3 \quad \text{True statement.}$$

While this check does **not** guarantee that there are no errors, it is a good indicator.

EXERCISES 7.2

Graph the numbers that satisfy the inequalities in Exercises 1–10.

1. $-2 \le x < 3$

2. $-4 \le x \le -1$

3. $x > 2$ or $x \le -4$

4. $x \le 3$ or $x > 6$

5. $x \le 1$ and $x > -2$

6. $x > 2$ or $x \le -5$

7. $x < \frac{10}{3}$ and $x \ge 1$

8. $x \ge 6$ or $x < \frac{13}{4}$

9. $x < -1$ or $x > \frac{1}{3}$

10. $x \ge 3$ and $x \le 10$

Solve the inequalities in Exercises 11–50 and graph the solutions.

11. $8 - x < 4$

12. $3 - 2a < 21$

13. $3x + 14 < 0$

14. $3x + 8 < 4$

15. $4x + 5 \ge -6$

16. $2x + 3 > -8$

17. $3x + 2 > 2x - 1$

18. $3y + 2 \le y + 8$

19. $y - 6 \le 4 - y$

20. $4x - 2 < 6x + 6$

21. $3x - 5 > 3 - x$

22. $3y - 1 \ge 11 - 3y$

23. $5y + 6 < 2y - 2$

24. $4 - 2x < 5 + x$

25. $4 + x > 1 - x$

26. $x - 6 > 3x + 5$

27. $\frac{x}{4} + 1 \le 5 - \frac{x}{4}$

28. $\frac{x}{2} - 1 \le \frac{5x}{2} - 3$

29. $\frac{x}{3} - 2 > 1 - \frac{x}{3}$

30. $\frac{5x}{3} + 2 > \frac{x}{3} - 1$

31. $-(x + 5) \le 2x + 4$

32. $3(2x - 1) > 5 - x$

33. $3x + 8 \le -3(2x - 3)$

34. $6(3x + 1) < 5(1 - 2x)$

35. $4(6 - x) > -2(3x + 1)$

36. $-3(2x - 5) \le 3(x - 1)$

37. $x - (2x + 5) \ge 7 - (4 - x) + 10$

38. $x - 3(4 - x) + 5 \ge -2(3 - 2x) - x$

39. $-5(3 - y) \le 5 - (y + 10)$

40. $1 - (2x + 8) < (9 + x) - 4x$

41. $-3 \le 4x \le 8$

42. $3 \le 3x < 8$

43. $-4 < -2x < 2$

44. $-5 < 1 - x < 3$

45. $2 < 2x - 4 \le 8$

46. $1 < 3x - 2 < 4$

47. $-5 < 4x + 1 < 7$

48. $-7 \le 5x + 3 < 9$

49. $-1 < 3 - 2x < 6$

50. $0 \le 4 - 7x \le 5$

7.3 APPLICATIONS: DISTANCE AND GEOMETRY

As was discussed in Section 5.5, the difficulty or ease with which you solve a particular problem depends on many factors, including your personal experiences and general reasoning abilities. For example, suppose you were given the following problem:

"A car travels 170 miles in 3 hours.
What was the average speed?"

The problem does not say directly to MULTIPLY, DIVIDE, ADD, or SUBTRACT. You must know that rate multiplied by time equals distance or, $r \cdot t = d$. You are given the distance (170 miles) and the time (3 hours). You are to find the average speed. The tool you need is the formula $r \cdot t = d$.

Let r = average speed. Then,

$$3 \cdot r = 170$$
$$r = 56\tfrac{2}{3} \text{ mph}$$

EXAMPLE 1: Distance

A man leaves on a business trip, and at the same time his wife takes their children to visit their grandparents. The cars, traveling in opposite directions, are 360 miles apart at the end of 3 hours. If the man's average speed is 10 mph more than his wife's, what is her average speed?

Solution: Let x = average speed of wife. Then

	rate	· time	= distance
wife	x	3	$3x$
man	$(x + 10)$	3	$3(x + 10)$

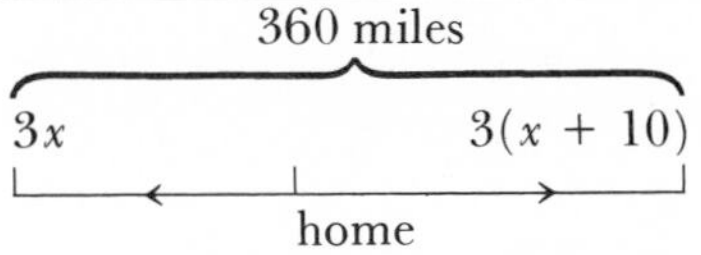

360 miles

$3x$ $3(x + 10)$

home

$$\underbrace{\text{distance for wife}}_{3x} + \underbrace{\text{distance for man}}_{3(x+10)} = \underbrace{\text{distance apart}}_{360}$$

$$3x + 3x + 30 = 360$$
$$6x = 330$$
$$x = 55 \text{ mph}$$

The wife's average speed is 55 mph.

EXAMPLE 2: Distance

Two trains, A and B, are 540 kilometers apart and travel toward each other on parallel tracks. Train A travels at 40 kilometers per hour, and train B travels at 50 kilometers per hour. In how many hours will they meet?

Solution: Let x = time. Then

	rate	· time	= distance
train A	40	x	$40x$
train B	50	x	$50x$

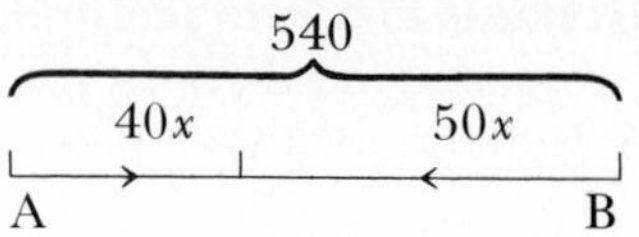

540

$40x$ $50x$

A B

$$40x + 50x = 540$$

$$90x = 540$$

$$x = 6 \text{ hr}$$

The trains will meet in 6 hours.

EXAMPLE 3: Geometry

A rectangle with a perimeter of 140 meters has a length that is 20 meters less than twice the width. Find the dimensions of the rectangle.

Solution: Draw a diagram and use the formula $P = 2l + 2w$.

Let w = width

$2w - 20$ = length

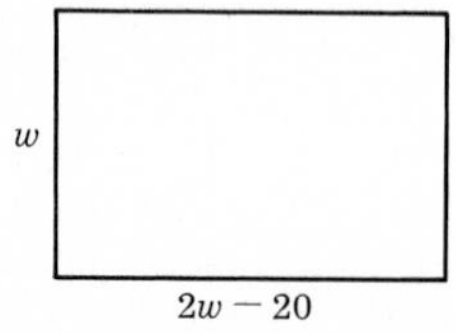

w

$2w - 20$

$$2(w) + 2(2w - 20) = 140$$

$$2w + 4w - 40 = 140$$

$$6w = 180$$

$$w = 30 \text{ meters}$$

$$2w - 20 = 40 \text{ meters}$$

The width is 30 meters and the length is 40 meters.

EXAMPLE 4: Distance

Arno can paddle his canoe 4 mph in still water. After paddling downstream for 3.5 hours, Arno takes 4.5 hours to return to the same place he started by paddling upstream. What is the rate of the current?

Solution: Let c = rate of the current. Then

rate · time = distance

	rate	time	distance
downstream	$4 + c$	3.5	$3.5(4 + c)$
upstream	$4 - c$	4.5	$4.5(4 - c)$

$$3.5(4 + c) = 4.5(4 - c) \quad \text{The two distances are equal.}$$
$$14 + 3.5c = 18 - 4.5c$$
$$4.5c + 3.5c = 18 - 14$$
$$8c = 4$$
$$c = 0.5$$

The rate of the current is 0.5 mph.

EXERCISES 7.3

1. The length of a rectangle is 5 times the width. The perimeter of the rectangle is 96 cm. Find the dimensions of the rectangle.

2. The length of a rectangle is 11 ft more than the width. If the perimeter is 90 ft, what are the dimensions?

3. The perimeter of a rectangular parcel of land is 720 ft. The length is 60 ft less than twice the width. What are the dimensions of the parcel?

4. The length of a rectangle is twice the width. If the perimeter is 96 cm, what are the dimensions of the rectangle?

5. The length of a rectangle is 18 meters more than the width. Find the dimensions of the rectangle if the perimeter is 104 meters.

6. The perimeter of a triangle is 51 meters. If the second side is twice the first side and the third side is 1 meter longer than the second side, how long is each side? (**Hint:** $P = a + b + c$ where a, b, and c are the sides of a triangle.)

7. The length of side b of a triangle is 4 cm more than the length of a side a. The length of side c is 1 cm less than twice the length of side a. If the perimeter is 43 cm find the length of each side.

8. The second side of a triangle is three meters more than twice the first. The third side is 11 m less than the sum of the lengths of the other two sides. If the perimeter is 49 m, what is the length of each side?

9. An isosceles triangle is a triangle with two sides equal. The perimeter of an isosceles

triangle is 51 cm. If the third side is 5 cm less than twice the length of one of the equal sides, how long is each side?

10. The ratio of the width of a rectangle to the length of a rectangle is 5 to 9. If the perimeter of the rectangle is 84 cm, find the length and width of the rectangle. (**Hint:** $5a$ and $9a$ are in the ratio of 5 to 9.)

11. Two trains leave Kansas City at the same time. One train travels east and the other, west. The speed of the west-bound train is 5 mph greater than the speed of the east-bound train. After 6 hours, they are 510 miles apart. Find the rate of each train. Assume the trains travel in a straight line in directly opposite directions. **Hint:** Complete the following chart.

	rate	$\cdot$ time	= distance
west-bound	$x + 5$		
east-bound	x		

12. Steve travels 4 times as fast as Fred. Traveling in opposite directions, they are 105 miles apart after 3 hours. Find their rates of travel.

13. Sue travels 5 mph less than twice as fast as June. Starting at the same point and traveling in the same direction, they are 80 miles apart after 4 hours. Find their speeds.

14. Mary and Linda live 324 miles apart. They start at the same time and travel toward each other. Mary's speed is 8 mph greater than Linda's. If they meet in 3 hours, find their speeds.

15. Two planes leave from points 1860 miles apart at the same time and travel toward each other—at slightly different altitudes, of course. If the rates are 220 mph and 400 mph, how soon will they meet? **Hint:** Complete the following chart.

	rate	$\cdot$ time	= distance
plane 1	220	x	
plane 2	400	x	

16. Two buses leave Ocarche at the same time traveling in opposite directions. One bus travels at 55 mph, and the other at 59 mph. How soon will they be 285 miles apart?

17. A motor boat crossed a lake traveling 8 mph and returned along the same route at

12 mph. If it took $\frac{3}{4}$ of an hour less for the return trip, how far was the distance across the lake? **Hint:** Complete the following chart.

	rate ·	time =	distance
across	8	x	
back again	12	$x - \frac{3}{4}$	

18. A jogger runs out into the country at a rate of 10 mph. He returns along the same route at 6 mph. If the total trip took 1 hour, 36 minutes, how far did he jog?

19. Mr. Green leaves El Paso on a business trip to Los Angeles, traveling at 56 mph. Fifteen minutes later, Mrs. Green discovers that he forgot his briefcase, and, jumping into their second car, she begins to pursue him. What was her average speed if she caught him 84 miles outside El Paso? $\left(\textbf{Hint: } \text{His time was } \frac{\text{distance}}{\text{rate}} = \frac{84}{56} \text{ hours.}\right)$

20. An airliner's average speed is $3\frac{1}{2}$ times the average speed of a private plane. Two hours after they leave the same airport at the same time, traveling in the same direction, they are 580 miles apart. What is the average speed of each plane? (**Hint:** Since they are traveling in the same direction, the distance between them will be the difference of their distances.)

21. A plane left Denver for Hawaii flying at 480 mph. Thirty minutes later, a second plane followed, traveling at 520 mph. How long will it take the second plane to overtake the first?

22. A train left Strong City traveling 48 mph. Forty-five minutes later, a second train left traveling in the opposite direction at 60 mph. How long will it take for the trains to be 279 miles apart?

23. The length of a rectangle is 10 meters more than one-half the width. If the perimeter is 44 meters, what is the length and the width?

24. The length of a rectangle is 1 meter less than twice the width. If each side is increased by 4 meters, the perimeter will be 116 meters. Find the length and the width of the original rectangle.

25. The River Queen tour boat can travel 12 mph in still water. After traveling for 3 hr downstream, it takes 5 hr to return. What is the rate of the current? (**Hint:** If c is the rate of the current, then $12 + c$ is the rate going downstream and $12 - c$ is the rate going upstream.)

26. An airplane can travel 320 mph in still air. If it travels 690 miles with the wind in the same length of time it travels 510 miles against the wind, what is the speed of the wind? $\left(\textbf{Hint: } \text{time} = \dfrac{\text{distance}}{\text{rate}}.\right)$

27. The length of a rectangle is three times the width. If each side is increased by 4 cm, the perimeter will be doubled. Find the length and the width of the original rectangle.

28. A farmer has 160 meters of chain link fencing to build a rectangular corral. If he uses the barn for one of the longer sides, what will be the dimensions of the corral if the length is 5 meters less than three times the width?

29. Mr. Snyder had a sales meeting scheduled 80 miles from his home. After traveling the first 30 miles at a rather slow rate due to freeway traffic, he found that, in order to make it on time, he had to increase his speed by 25 mph. If he traveled the same length of time at each rate, find the two rates. $\left(\textbf{Hint: } \text{time} = \dfrac{\text{distance}}{\text{rate}}.\right)$

30. Mike rides the ski lift to the top of Snowy Peak, a distance of $1\frac{3}{4}$ miles. He then skis directly down the hill. If he skis five times as fast as the lift runs and the round trip takes 45 minutes, find the rate at which he skis. $\left(\textbf{Hint: } \text{time} = \dfrac{\text{distance}}{\text{rate}}.\right)$

7.4 APPLICATIONS: INTEREST AND WORK

People in business know several formulas involving the principal (amount of money invested), rate (percent or rate of interest), and interest (the actual profit or interest earned). These formulas can depend on such related topics as the way the interest is paid on a loan (monthly, daily, yearly, etc.), whether or not there are penalties for early payment of a loan, or escalation clauses if an investment is particularly profitable.

In this section, we will use only the basic formula that calculates interest on an annual basis: $P \cdot R = I$, or principal times rate equals interest.

EXAMPLE 1: Interest

A man invests in a certain bond yielding 9% interest and then invests \$500 in a high-risk stock yielding 12%. After one year, his total interest from the two investments is \$240. What amount did he invest in the bond?

Solution: Let P = principal invested at 9%. Then

	principal	· rate	= interest
bond	P	0.09	$0.09P$
high-risk stock	500	0.12	0.12(500)

$$\underbrace{\text{interest on bond}}_{0.09P} + \underbrace{\text{interest on stock}}_{0.12(500)} = \underbrace{\text{total income}}_{240}$$

$$0.09P + 60 = 240$$

$$0.09P = 180$$

$$\frac{0.09P}{\mathbf{0.09}} = \frac{180}{\mathbf{0.09}}$$

$$P = \$2000$$

He invested $2000 in the bond yielding 9% interest.

EXAMPLE 2: Interest

A woman has $7000. She decides to separate her funds into two investments. One yields an interest of 6%, and the other, 10%. If she wants an annual income from the investments to be $580, how should she split the money?

Solution: Since we know that the total to be invested is $7000, if one investment is $\$x$, the other must be $\$7000 - \$x$.

Let x = amount invested at 10%

$7000 - x$ = amount invested at 6%

	principal	· rate	= interest
10% investment	x	0.10	$0.10x$
6% investment	$(7000 - x)$	0.06	$0.06(7000 - x)$

$$\underbrace{\text{interest on 10\% investment}}_{0.10x} + \underbrace{\text{interest on 6\% investment}}_{0.06(7000 - x)} = \underbrace{\text{total income}}_{580}$$

$$10x + 6(7000 - x) = 58{,}000$$ Multiply each term by 100 to eliminate decimals.

$$10x + 42{,}000 - 6x = 58{,}000$$
$$4x = 16{,}000$$
$$x = \$4000 \text{ @ } 10\%$$
$$7000 - x = \$3000 \text{ @ } 6\%$$

She should invest $4000 at 10% and $3000 at 6%.

Problems involving "work" can be very sophisticated and require calculus and physics. The problems we will be concerned with relate to the time involved to complete the work on a particular job. These problems involve only the idea of what fraction of the job is done in one unit of time (hours, minutes, days, weeks, and so on). For example, if a man can dig a ditch in 4 hours, what part (of the ditch-digging job) did he do in one hour? The answer is $\frac{1}{4}$. If the job took 5 hours, he would do $\frac{1}{5}$ in one hour. If the job took x hours, he would do $\frac{1}{x}$ in one hour.

EXAMPLE 3: Work

Mike can clean his family's pool in 2 hours. His younger sister, Stacey, can do it in 3 hours. If they work together, how long will it take them to clean the pool?

Solution: Let x = number of hours working together. Then

	hours	part in 1 hour
Mike	2	$\frac{1}{2}$
Stacey	3	$\frac{1}{3}$
together	x	$\frac{1}{x}$

$$\underbrace{\text{part done in 1 hr by Mike}}_{\frac{1}{2}} + \underbrace{\text{part done in 1 hr by Stacey}}_{\frac{1}{3}} = \underbrace{\text{part done in 1 hr together}}_{\frac{1}{x}}$$

$$\frac{1}{2}(\mathbf{6x}) + \frac{1}{3}(\mathbf{6x}) = \frac{1}{x}(\mathbf{6x})$$

Multiply each term on both sides of the equation by $6x$, the LCM of the denominators.

$$3x + 2x = 6$$

$$5x = 6$$

$$x = \frac{6}{5}\text{ hr}$$

Together they can clean the pool in $\frac{6}{5}$ hours, or 1 hour, 12 minutes.

EXAMPLE 4: Work

A man was told that his new Jacuzzi® pool would fill through an inlet valve in 3 hours. He knew something was wrong when the pool took 8 hours to fill. He found he had left the drain valve open. How long will it take to drain the pool?

Solution: Let t = time to drain pool. (**Note:** In this case, the inlet and outlet valves work against each other.)

	hours	part in 1 hour
inlet	3	$\frac{1}{3}$
outlet	t	$\frac{1}{t}$
together	8	$\frac{1}{8}$

$$\underbrace{\text{part filled by inlet}} - \underbrace{\text{part emptied by outlet}} = \underbrace{\text{part filled together}}$$

$$\frac{1}{3} - \frac{1}{t} = \frac{1}{8}$$

$$\frac{1}{3}(\mathbf{24t}) - \frac{1}{t}(\mathbf{24t}) = \frac{1}{8}(\mathbf{24t})$$

$$8t - 24 = 3t$$

$$5t = 24$$

$$t = \frac{24}{5}$$

The pool will drain in $\frac{24}{5}$ hours, or 4 hours, 48 minutes.

EXERCISES 7.4

1. Amy receives $273 annually from two investments. If she has $1500 invested at 11%, how much does she have invested at 9%?

2. A company has $12,000 invested in a project yielding a 40% return. How much should it invest in a project having a 30% return to earn $10,200 annually from both investments?

3. Mr. Jackson invested $900 at 11% interest. How much did he invest at 9% if his yearly interest from both investments is $202.50?

4. Mildred has money in two savings accounts. One rate is 8% and the other is 10%. If she has $200 more in the 10% account, how much is invested at 8% if the total interest is $101?

5. Money is invested at two rates. One rate is 9% and the other is 13%. If there is $700 more invested at 9%, find the amount invested at each rate if the annual interest is $239.

6. A total of $6000 is invested, part at 8% and the remainder at 12%. How much is invested at each rate if the annual interest is $620?

7. Frank has half of his investments in stock paying an 11% dividend and the other half in a debentured stock paying 13% interest. If his total annual interest is $840, how much does he have invested?

8. Betty invested some of her money at 12% interest. She invested $300 more than twice that amount at 10%. How much is invested at each rate if her income is $318 annually?

9. Mrs. Brown has $12,000 invested. Part is invested at 9% and the remainder at 11%. If the interest from the 9% investment exceeds the interest from the 11% investment by $380, how much is invested at each rate?

10. GFA invested some money in a development yielding 24% and $9000 less in a development yielding 18%. If the first investment produces $2820 more per year than the second, how much is invested in each development?

11. Eight thousand dollars is to be invested, part at 15% and the remainder at 12%. If the annual income from the 15% investment exceeds the income from the 12% investment by $66, how much is invested at each rate?

12. Tom can fix a car in 4 hours. Mike can do it in 6 hours. How long will it take them working together?

13. A carpenter can build a wall in 2 hours. His apprentice can build it in 6 hours. How long will it take them working together?

14. A brick mason can build a retaining wall in 8 hours. If his partner helps him, they can complete it in $2\frac{2}{3}$ hours. How long would it take the partner working alone?

15. One pipe can fill a tank in 15 minutes. The tank can be drained by another pipe in 45 minutes. If both pipes are opened, how long will it take to fill the tank?

16. One company can grade a parcel of land in 25 hours. Another company requires 30 hours to do the job. If both are hired, how long will it take them to grade the parcel working together?

17. Sonny needs 4 hours to complete the yardwork. His wife, Toni, needs 5 hours to do the work. How long will it take if they work together?

18. Working together, Rick and Rod can clean the snow from the driveway in 20 minutes. It would have taken Rick, working alone, 36 minutes. How long would it have taken Rod alone?

19. A carpenter and his partner can put up a patio cover in $3\frac{3}{7}$ hours. If the partner needs 8 hours to complete the patio alone, how long will it take the carpenter working alone?

20. A man can wax his car three times as fast as his daughter can. Together they can do the job in 4 hours. How long does it take each of them working alone?

21. A contractor hires two bulldozers to clear a 10-acre tract of land. One works twice as fast as the other. It takes them 3 days to clear the land working together. How long would it take each of them working alone?

22. It takes Marie twice as long to complete a research project as it takes Ellen to do the same project. If they both work on it, it takes them 1 hour. How long would it take each of them working alone?

23. One pipe can fill a tank three times as fast as another. When both pipes are used, it takes $1\frac{1}{2}$ hours to fill the tank. How long does it take each pipe alone?

24. Richard had \$6400 in a savings account drawing 9% interest. He withdrew enough to buy a car. If the annual interest was still \$261, how much did the car cost?

25. On an investment of \$9500, Bill lost 3% on one part and earned 6% on the remainder. If his net annual receipts were \$282, how much was each investment?

26. Mr. Matthews invested some money at 10%. He invested \$200 less than twice that amount at 13%. If he had reversed the two investments, he would have received \$150 less annually. How much is invested at each rate?

27. Chris invested \$20,000, part at 9% and the remainder at 7%. If he had invested twice as much at 9% and the rest at 7%, he would have received \$152 more interest. How much was invested at each rate? [**Hint:** The interest on the original investment is represented as $0.09x + 0.07\ (20{,}000 - x)$.]

28. It takes Bob four hours longer to repair a car than it takes Ken. Working together, they can complete the job in $1\frac{1}{2}$ hours. How long would it take each of them working alone?

29. Working together, Alice and Sharon can type the company's sales report in $2\frac{2}{5}$ hours.

Working alone takes Sharon two hours longer to type the report than it would take Alice. How long would it take each of them working alone?

30. A power plant has two coal-burning boilers. If both boilers are in operation, a load of coal is burned in 6 days. If only one boiler is used, a load of coal will last the new "fuel-efficient" boiler 5 days longer than the older boiler. How long will it take each boiler to burn a load of coal?

7.5 APPLICATIONS: MIXTURE AND INEQUALITIES

Problems involving mixtures occur in physics and chemistry and in such places as a candy store or a tobacco shop. Two or more items of a different percentage of concentration of a chemical such as salt, chlorine, or antifreeze are to be mixed; or two or more types of tobacco are to be mixed to form a final mixture that satisfies certain conditions of percentage of concentration.

The basic plan is to write an equation that deals only with one part of the mixture. The following examples explain how this can be accomplished.

EXAMPLE 1: Mixture

A particular experiment in chemistry demands a 10% solution of acid. If the lab assistant has 9 ounces of a 5% solution on hand, how much acid should be added to get the 10% solution? (**Hint:** Write an equation that deals only with amounts of acid.)

Solution: Let x = amount of acid to be added. Then

	amount of solution $\cdot$	percent acid $=$	amount of acid
original solution	9	0.05	0.05(9)
added solution	x	1.00	$1.00(x)$
final solution	$(x + 9)$	0.10	$0.10(x + 9)$

$$\underbrace{\text{acid in 9 oz}} + \underbrace{\text{acid added}} = \underbrace{\text{acid in final solution}}$$

$$0.05(9) + 1.00(x) = 0.10(x + 9)$$

$$5(9) + 100(x) = 10(x + 9) \qquad \text{Multiply each term by 100.}$$

$$45 + 100x = 10x + 90$$
$$90x = 45$$
$$x = \frac{45}{90}$$
$$x = 0.5 \text{ oz of acid}$$

Check:

$$\underbrace{\text{acid in 9 oz}} + \underbrace{\text{acid added}} = \underbrace{\text{acid in final mix}}$$
$$0.05(9) + 0.5 \stackrel{?}{=} 0.10(0.5 + 9)$$
$$0.45 + 0.5 \stackrel{?}{=} 0.10(9.5)$$
$$0.95 = 0.95$$

The 10% solution can be had by adding 0.5 oz of acid.

EXAMPLE 2: Mixture

How many gallons of a 20% salt solution should be mixed with a 30% salt solution to produce 50 gallons of 23% solution? (**Hint:** Write an equation that deals only with amounts of salt.)

Solution:

Let x = amount of 20% solution

$50 - x$ = amount of 30% solution

Note: Since the total number of gallons is known, one amount is found by **subtracting** the other amount from the total.

amount of solution · percent salt = amount of salt

20% solution	x	0.20	$0.20x$
30% solution	$50 - x$	0.30	$0.30(50 - x)$
23% solution	50	0.23	$0.23(50)$

$$\underbrace{\text{salt in 20\% solution}}_{0.20x} + \underbrace{\text{salt in 30\% solution}}_{0.30(50 - x)} = \underbrace{\text{salt in 23\% solution}}_{0.23(50)}$$

$$20x + 30(50 - x) = 23(50)$$
$$20x + 1500 - 30x = 1150$$
$$-10x = 1150 - 1500$$
$$-10x = -350$$
$$x = 35 \text{ gal of } 20\% \text{ solution}$$

Check:

$$\underbrace{\text{salt in } 20\% \text{ solution}} + \underbrace{\text{salt in } 30\% \text{ solution}} = \underbrace{\text{salt in } 50 \text{ gal}}$$
$$0.20(35) + 0.30(50 - 35) \stackrel{?}{=} .23(50)$$
$$7.0 + 0.30(15) \stackrel{?}{=} 11.5$$
$$7.0 + 4.5 \stackrel{?}{=} 11.5$$
$$11.5 = 11.5$$

Thirty-five gallons of the 20% solution should be added to 15 gallons of the 30% solution.

The following example using inequalities is self-explanatory. Study it carefully.

EXAMPLE 3: Inequalities

A physics student has grades of 85, 98, 93, and 90 on four examinations. If he must average 90 or better to receive an A for the course, what scores can he receive on the final exam and get an A?

Solution: Let x = score on final exam. The average is found by adding the scores and dividing by 5.

$$\frac{85 + 98 + 93 + 90 + x}{5} \geq 90$$
$$\frac{366 + x}{5} \geq 90$$
$$5\left(\frac{366 + x}{5}\right) \geq 5 \cdot 90$$
$$366 + x \geq 450$$
$$x \geq 450 - 366$$
$$x \geq 84$$

If the student scores 84 or more on the final exam, he will average 90 or more and receive an A in physics.

EXERCISES 7.5

1. How many ounces of a 15% hydrochloric acid solution should be mixed with 24 ounces of a 10% solution to make a 12% solution?

2. Sixty liters of a 30% acid solution is to be reduced to a 20% acid solution by adding water. How much water should be added?

3. A pharmacist wishes to reduce 36 ounces of a 10% iodine in alcohol solution to a 3% solution of iodine in alcohol. How many ounces of alcohol will he need? (**Hint:** The amount of iodine is the same in both the original solution and the final solution.)

4. How many gallons of water must be added to 4 gallons of a 30% salt brine to produce a 20% brine?

5. How many quarts of pure antifreeze must be added to 16 quarts of a 30% antifreeze solution to produce a 50% solution? (**Hint:** Pure antifreeze is a 100% solution.)

6. In preparing for an experiment in chemistry, the lab assistant noticed that he needed a 10% acid solution. If he has 20 ounces of a 12% acid solution, how much water must he add to reduce it to a 10% solution? (**Hint:** The amount of acid is the same in both the original solution and the final solution.)

7. A manufacturer has received an order for 24 tons of a 60% copper alloy. His stock contains only 80% copper alloy and 50% copper alloy. How much of each will be needed to fill the order? (**Hint:** 24 tons is the total amount of alloy.)

8. A tobacco shop wants 50 ounces of tobacco that is 24% of a rare Turkish blend. If a 30% Turkish blend and a 20% Turkish blend are mixed, how much of each blend will be needed?

9. To receive a B grade, a student must average 80 or more but less than 90. If John received a B in the course and had five grades of 94, 78, 91, 86, and 87 before taking the final exam, what were the possible grades for his final?

10. If in Exercise 9 the final exam were counted as two tests, what would be the possible grades on the final exam for the student to receive a B?

11. The range for a C grade is 70 or more but less than 80. Before taking the final exam, Clyde had grades of 59, 68, 76, 84, and 69. If the final exam is counted as two tests, what is the minimum grade he could make on the final to receive a C? If there were 100 points possible, could he receive a B grade if an average of at least 80 were required?

12. The temperature of a mixture in a chemistry experiment varied from 15° Celsius to 65° Celsius. What is this temperature expressed in degrees Fahrenheit? $\left[C = \frac{5}{9}(F - 32).\right]$

13. The temperature at Braver Lake ranged from a low of 23°F to a high of 59°F. What is the equivalent range of temperatures in degrees Celsius? $\left[F = \frac{9}{5}C + 32.\right]$

14. The sum of the lengths of any two sides of a triangle must be greater than the third side. If a triangle has one side that is 17 cm and a second side that is 1 cm less than twice the third side, what are the smallest possible lengths for the second and third sides?

15. The sum of four times a number and 21 is greater than 45 and less than 73. What are the possible values for the number?

16. In order for Chuck to receive a B in his math class, he must have a total of at least 400 points. If he has scores of 72, 68, 85, and 89, what scores can he make on the final and receive a B?

17. In Exercise 16, if the final were counted twice, could he receive an A in the class if it takes at least 540 points for an A? Assume that the maximum possible score on the final is 100 points.

18. Twice a certain number increased by 26 is greater than four times the number increased by 2. Find those whole numbers that satisfy this condition.

19. How many pounds each of a 12% zinc alloy and a 30% zinc alloy must be used to produce 90 pounds of a 22% zinc alloy?

20. How many liters each of a 40% acid solution and a 55% acid solution must be used to produce 60 liters of a 45% acid solution?

21. Grade B milk is 3.2% butterfat. How many quarts each of Grade A milk, testing at 4.0%, and Grade C milk, testing at 3.0% must be used to obtain 40 quarts of Grade B milk?

22. To meet the government's specifications, an alloy must be 65% aluminum. How many pounds each of a 70% aluminum alloy and a 54% aluminum alloy will be needed to produce 640 pounds of the 65% aluminum alloy?

23. The length of a rectangle is 5 feet less than three times the width. If the perimeter must be less than 86 feet what is the largest possible length for the rectangle?

24. Ellen's secretary is going to buy 20 stamps, some 13¢ and some 10¢. If she has $2.35, what is the maximum number of 13¢ stamps she can get?

25. The Concert Hall has 400 seats. For a concert, the admission will be $6.00 for adults and $3.50 for students. If the expense of producing the concert is $1825, what is the least number of adult tickets that must be sold to realize a profit?

26. The Pep Club is selling candied apples to raise money. The price per apple is 25¢ until Friday, when they will sell for 20¢. If the Pep Club sells 200 apples, what is the minimum number they must sell at 25¢ each in order to raise at least $47.50?

27. How many gallons of gasoline priced at $1.30 a gallon should be mixed with 300 gallons of gasoline priced at $1.20 to obtain a mixture worth $1.24 a gallon?

28. The Coffee Grinder has two grades of coffee, one priced at $2.80 per pound and another priced at $3.20 per pound. How many pounds of each should be used to make 12 pounds of a blend worth $2.90 per pound?

29. A candy store has two kinds of chocolate. One type sells for $3.90 per pound and the other sells for $3.50 per pound. If $151.20 were received from the sale of 40 pounds of candy, how many pounds of each kind were sold?

30. Raggs & Associates sells one style of trousers for $35 per pair and another style for $28 per pair. How many pairs of each style were sold if a total of 16 pairs yielded a revenue of $483?

CHAPTER 7 SUMMARY

First-degree equations can be written in the form

$$ax + b = c$$

where a, b, and c are constants and $a \neq 0$.

For every real number there is one corresponding point on a line, and for every point on a line there is one corresponding real number.

A **first-degree inequality** can be written in the form

$$ax + b < c \qquad (\text{or } ax + b > c)$$

or

$$ax + b \leq c \qquad (\text{or } ax + b \geq c)$$

where a, b, and c are constants and $a \neq 0$.

Addition Property of Inequalities

If A, B, and C are algebraic expressions, then the inequalities

$$A < B$$

and

$$A + C < B + C$$

have the same solutions.

Multiplication Property of Inequalities

If A, B, and C are algebraic expressions, then

1. $A < B$ and $AC < BC$ have the same solutions if $C > 0$.
2. $A < B$ and $AC > BC$ have the same solutions if $C < 0$.

Multiplying or dividing both sides of an inequality by a negative number reverses the sense of the inequality.

To Solve a First-Degree Inequality

1. Simplify each side of the inequality by removing any parentheses and combining like terms.
2. Add constants and/or variables to or subtract them from both sides of the inequality so that variables are on one side and constants on the other.
3. Divide both sides by the coefficient (or multiply them by the reciprocal) of the variable, and **reverse the sense of the inequality if this coefficient is negative.**
4. A quick (and generally satisfactory) check is to select any one number in your solution and substitute it into the original inequality.

Some formulas related to word problems are

$r \cdot t = d$	Rate times time equals distance.
$P \cdot R = I$	Principal times rate of interest equals interest.
$P = 2l + 2w$	Perimeter of a rectangle equals twice the length plus twice the width.

Other formulas related to geometric figures are listed in Chapter 1.

Problems with work involve the idea of what fraction of the job is done in one unit of time (hours, minutes, days, weeks, and so on).

Problems with mixture involve writing an equation that deals with only one quantity in the mixture.

CHAPTER 7 REVIEW

Find the solution for the equations in Exercises 1–20.

1. $4x + 1 = 17$

2. $\frac{x}{2} + 9 = 6$

3. $6x + 2 = 3x - 1$

4. $\frac{5x}{3} + 2 = 12$

5. $\frac{2x}{5} + 3 = -3$

6. $2x + 7 = 5(x - 1)$

7. $37 - 4x = 20 + (x + 12)$

8. $2(x + 6) + 14 = 5(x + 2) - 18$

9. $\frac{1}{2}(x - 7) = \frac{1}{3}(2x + 5)$

10. $x - \frac{x}{5} = 10 - \frac{x}{2}$

11. $\frac{2}{5}(x - 4) = \frac{3}{4}x + 2$

12. $\frac{x}{3} + 1 = \frac{x + 4}{5}$

13. $\frac{3}{2x} = \frac{4}{5} + \frac{1}{2x}$

14. $\frac{2}{3x} = \frac{1}{2} + \frac{1}{6x}$

15. $1.3x + 11 = 0.5x + 3.8$

16. $2.4x - 4 = 0.8(x + 3)$

17. $3x^2 + 8x = 0$

18. $4x^2 - 5x - 6 = 0$

19. $\frac{7}{x + 3} = \frac{5}{x - 3}$

20. $\frac{4}{x + 1} + \frac{2}{x - 3} = \frac{-2}{3}$

Graph the numbers that satisfy the inequalities in Exercises 21–27.

21. $-5 < x < 7$

22. $-1 \leq x \leq 4$

23. $x < -2$ or $x \geq 5$

24. $x < -2$ or $x > \frac{4}{5}$

25. $x < \frac{7}{8}$ and $x > -3$

26. $2 \leq x < 14$

27. $-3 < x \leq 2.7$

Solve the inequalities and graph the solutions in Exercises 28–40.

28. $3x + 2 \leq 8$

29. $4x - 7 > 9$

30. $\frac{x}{3} + 1 > 2$

31. $3x + 5 \leq 6$

32. $2x + 3 \geq 4x + 5$

33. $5x + 1 \geq 2x + 6$

34. $5x + 3 \geq 2x + 15$

35. $2x - 5 < 3x + 2$

36. $2(x - 7) < 4(2x + 3)$

37. $7x + 4 \leq \frac{1}{2}(5x - 1)$

38. $x - (4 - 2x) \leq 2(x + 6) + 1$

39. $-2 \leq 4x + 7 \leq 3$

40. $-6 < 4 - 2x \leq 1$

41. A man has a rectangular parcel of land whose length is 40 meters less than 10 times its width. The perimeter is 250 meters. What are the dimensions of the parcel?

42. The sum of two numbers is 30. If $\frac{3}{4}$ of the smaller is equal to $\frac{1}{2}$ the larger one, find the numbers.

43. A man has three times as much money invested at 6% as he has invested at 4%. If his yearly income is \$1430, how much is invested at each rate?

44. Mark paddles his canoe downstream for $1\frac{1}{2}$ hours. The return trip takes 6 hours. If the speed of the current is 3 mph, how fast does Mark row in still water? (**Hint:** If r is his rate in still water, $r + 3$ is his rate downstream, and $r - 3$ is his rate upstream.)

45. Minh's father is 11 years more than three times as old as Minh. If the father's age is between 47 and 62 years inclusive, what is the range for Minh's age?

46. A meat market has ground beef that is 40% fat and extra lean ground beef that is only 15% fat. How many pounds of each will be needed to obtain 50 pounds of lean ground beef that is 25% fat?

47. Jimmy can clean the apartment in 6 hours. It takes Jerry 12 hours to clean it. If they would work together, how long would it take them?

48. The length of a rectangle is 7 centimeters more than twice the width. If both are increased by 3 centimeters the area will be increased by 84 square centimeters. What are the length and width of the rectangle?

49. Lisa can travel 228 miles in the same time that Soo travels 168 miles. If Lisa's speed is 15 mph faster than Soo's, find their rates.

50. Admission to the baseball game is \$2.00 for general admission and \$3.50 for reserved seats. The receipts were \$36,250 for 12,500 paid admissions. How many of each ticket, general and reserved, were sold?

CHAPTER 7 TEST

Solve the equations in Exercises 1–6.

1. $5x - 3 = 4 - 2x$

2. $\dfrac{5x}{6} + 4 = 9$

3. $-2(5 - 3x) = 6 - (4x - 3)$

4. $0.4x - 6 = 8 - 0.3x$

5. $6x^2 - 9x = 0$

6. $\dfrac{x}{x-2} + \dfrac{3}{x-1} = 1$

Graph the numbers that satisfy the inequalities in Exercises 7–9.

7. $x < -4$ or $x \geq 1.5$

8. $x < \dfrac{3}{2}$ and $x > -1$

9. $-\dfrac{3}{4} < x \leq 3$

Solve the inequalities in Exercises 10–13.

10. $4x - 5 > 12 - 5x$

11. $-2 < 3 - 5x < 6$

12. $\dfrac{2x}{5} - 3 \leq \dfrac{3}{10} - \dfrac{x}{3}$

13. $-2(5 - 3x) > 6 - (4x - 3)$

Set up equations for Exercises 14–20 and solve them.

14. The length of a rectangle is 9 feet more than twice its width. The perimeter of the rectangle is 66 feet. Find the dimensions.

15. Two automobiles leave Phoenix at the same time and travel in opposite directions. After two hours, they are 234 miles apart. If one automobile travels 13 mph faster than the other, find the rate of each.

16. Lynda invested some money yielding 18% and $1200 more at a yield of 12%. If the 18% investment produces $156 more each year than the second, how much was invested at each rate?

17. A plumber can complete a job in 2 hours. If an apprentice helps, it takes only $1\frac{1}{3}$ hours. How long would it take the apprentice working alone?

18. A metallurgist needs 2000 pounds of an alloy that is 80% copper. In stock he has only 83% copper and 68% copper. How many pounds of each must be used?

19. The range for a B grade is 75 or more but less than 85. Eric has grades of 73, 65, 77, and 74 before taking the final exam. What are the possible grades on the final that will earn Eric a B grade?

20. A men's clothing store sells two styles of sports jackets, one selling for $95 and one selling for $120. Last month, the store sold 40 jackets, with receipts totaling $4250. How many of each style did the store sell?

USING THE COMPUTER

```
10  REM
20  REM This program uses the formulas for conversion from
30  REM the Fahrenheit temperature scale to the Celsius scale
40  REM and vice versa.
50  '*********************************************************
60  '* FAHRENHEIT : temperature in the Fahrenheit scale      *
70  '* CELSIUS    : temperature in the Celsius scale         *
80  '*********************************************************
90  PRINT
100 PRINT "              Menu"
110 PRINT "1) Fahrenheit ———————> Celsius"
120 PRINT "2) Celsius ———————> Fahrenheit"
130 PRINT "3) Quit"
140 PRINT
150 INPUT "Type in your choice 1,2 or 3 :",CHOICE$
160 IF CHOICE$="3" THEN GOTO 220
170 INPUT "Type in the temperature      :",TEMPERATURE
180 LET FAHRENHEIT = 9*TEMPERATURE/5 + 32
190 LET CELSIUS = (5*(TEMPERATURE-32))/9
200 IF CHOICE$="1" THEN PRINT
"The temperature in Celsius scale is ";CELSIUS :GOTO 90
210 IF CHOICE$="2" THEN PRINT
"The temperature in Fahrenheit scale is ";FAHRENHEIT :GOTO 90
220 PRINT "Thank you..."
230 END
```

```
              Menu
1) Fahrenheit ———————> Celsius
2) Celsius ———————> Fahrenheit
3) Quit

Type in your choice 1,2 or 3 :1
Type in the temperature      : 32
The temperature in Celsius scale is  0

              Menu
1) Fahrenheit ———————> Celsius
2) Celsius ———————> Fahrenheit
3) Quit

Type in your choice 1,2 or 3 :2
Type in the temperature      : 64.4
The temperature in Fahrenheit scale is  147.92

              Menu
1) Fahrenheit ———————> Celsius
2) Celsius ———————> Fahrenheit
3) Quit

Type in your choice 1,2 or 3 :3
Thank you...
```

Divide each problem that you examine into as many parts as you can and as you need to solve them more easily.

RENÉ DESCARTES 1596–1650

DID YOU KNOW?

In Chapter 8 you will be introduced to the idea of a graph of an algebraic equation. A graph is simply a picture of an algebraic relationship. This topic is more formally called **analytic geometry.** It is a combination of algebra (the equation) and geometry (the picture).

The idea of combining algebra and geometry was not thought of until René Descartes wrote his famous *Discourse on the Method of Reasoning* in 1637. The third appendix in this book, "La Geometrie," made Descartes' system of analytic geometry known to the world. In fact, you will find that analytic geometry is sometimes called Cartesian geometry.

René Descartes is perhaps better known as a philosopher than as a mathematician; he is often referred to as the father of modern philosophy. His method of reasoning was to apply the same logical structure to philosophy that had been developed in mathematics and especially in geometry.

In the Middle Ages, the highest forms of knowledge were believed to be mathematics, philosophy, and theology. Many famous people in history who have reputations as poets, artists, philosophers, and theologians were also creative mathematicians. Almost every royal court had mathematicians whose work reflected glory on the royal sponsor who paid the mathematician for his research and his court presence.

Descartes, in fact, died in 1650 after accepting a position at the court of the young warrior-queen, Christina of Sweden. Apparently, the frail French philosopher-mathematician, who spent his mornings in bed doing mathematics, could not stand the climate of Sweden and the hardships imposed by Christina in her demand that Descartes tutor her in mathematics each morning at 5 o'clock in an unheated castle library.

8 Graphing Linear Equations

8.1 ORDERED PAIRS

Equations such as $d = 60t$, $I = 0.05P$, and $y = 2x + 3$ represent relationships between pairs of variables. For example, in the first equation, if $t = 3$, then $d = 60 \cdot 3 = 180$. With the understanding that t is first and d is second, we can write the pair $(3, 180)$ to represent $t = 3$ and $d = 180$. The pair $(3, 180)$ is called an **ordered pair.** Obviously, $(180, 3)$ is different from $(3, 180)$ if t is the first number and d is the second number.

We say that $(3, 180)$ satisfies the equation or is a solution of the equation $d = 60t$. Similarly, $(100, 5)$ satisfies $I = 0.05P$ where $P = 100$ and $I = 0.05(100) = 5$. Also, $(2, 7)$ satisfies $y = 2x + 3$ where $x = 2$ and $y = 2 \cdot 2 + 3 = 7$.

In an ordered pair such as (x, y) x is called the **first component** and y is called the **second component.** To find ordered pairs that satisfy an equation such as $y = 2x + 3$, we can choose **any** value for one variable and then find the corresponding value for the other variable by substituting into the equation. For example,

$y = 2x + 3$	*Ordered Pairs*
• Choose $x = 1$; then $y = 2 \cdot 1 + 3 = 5$.	$(1, 5)$
• Choose $x = -2$; then $y = 2(-2) + 3 = -1$.	$(-2, -1)$
• Choose $x = \frac{1}{2}$, then $y = 2(\frac{1}{2}) + 3 = 4$.	$(\frac{1}{2}, 4)$

The ordered pairs $(1, 5)$, $(-2, -1)$, and $(\frac{1}{2}, 4)$ all satisfy the equation $y = 2x + 3$.

The variable assigned to the first component is also called the **independent variable,** and the variable assigned to the second component is called the **dependent variable.** Thus, we associate (x, y) with $y = 2x + 3$, and y "depends" on the values assigned to x.

We can also write the pairings in the form of tables. The choices for the values of the independent variable are arbitrary.

$d = 60t$

t	d
5	$60 \cdot 5 = 300$
10	$60 \cdot 10 = 600$
12	$60 \cdot 12 = 720$
15	$60 \cdot 15 = 900$

$I = 0.05P$

P	I
100	$0.05(100) = 5$
200	$0.05(200) = 10$
500	$0.05(500) = 25$
1000	$0.05(1000) = 50$

$y = 2x + 3$

x	y
-2	$2(-2) + 3 = -1$
-1	$2(-1) + 3 = 1$
0	$2(0) + 3 = 3$
3	$2(3) + 3 = 9$

Another notation commonly used in mathematics to represent the dependent variable is **function notation.** The symbol $f(x)$, read "f of x", is used to represent the value of the dependent variable related to x. For example, if

$$f(x) = 2x - 5$$

then
$$f(1) = 2 \cdot 1 - 5 = -3$$

$$f(3) = 2 \cdot 3 - 5 = 1$$

$$f(6) = 2 \cdot 6 - 5 = 7$$

and
$$f(-1) = 2(-1) - 5 = -7$$

[**Note:** $f(x)$ does not mean to multiply f by x. $f(x)$ is a notation unto itself.]

Due to the work of René Descartes (1596–1650), we can represent ordered pairs as points on a plane by graphing, using the **Cartesian coordinate system.** In this system, two number lines intersect at right angles and separate the plane into four **quadrants.** The **origin,** designated by the ordered pair (0, 0), is the point of intersection. The horizontal number line is called the **horizontal axis** or **x-axis.** The vertical number line is called the **vertical axis** or **y-axis.** Points that lie on either axis are not in any quadrant. They are simply on an axis (Figure 8.1).

> **There is a one-to-one correspondence between points in a plane and ordered pairs of real numbers.** In other words, for each point there is one corresponding ordered pair of real numbers, and for each ordered pair of real numbers there is one corresponding point. This important relationship is the cornerstone of the Cartesian coordinate system.

The graphs of the points $A(2, 1)$, $B(-2, 3)$, $C(-3, -2)$, $D(1, -2)$, and $E(3, 0)$ are shown in Figure 8.2. [**Note:** An ordered pair of real

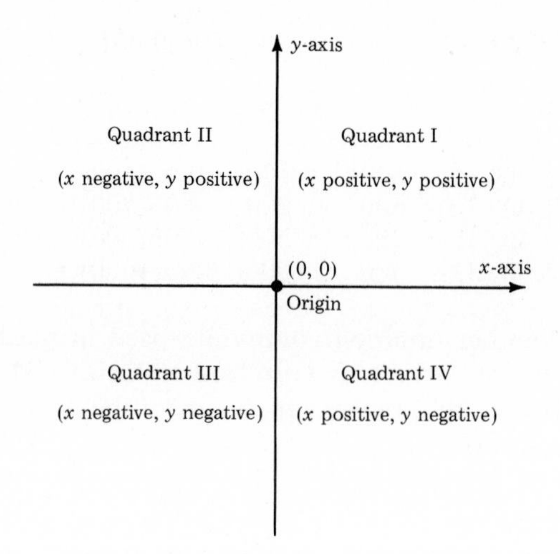

FIGURE 8.1

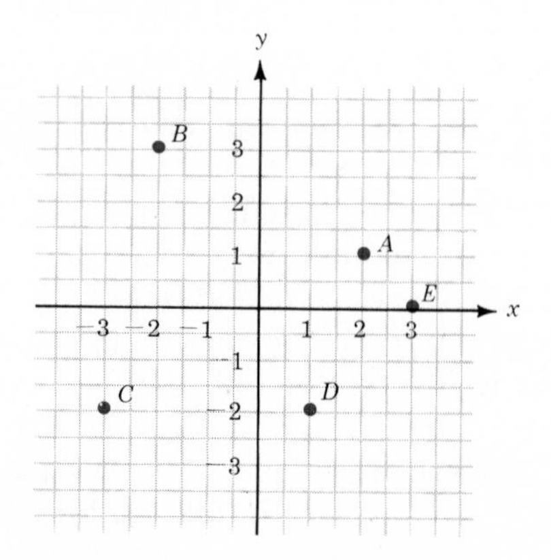

FIGURE 8.2

numbers and the corresponding point on the graph are frequently used to refer to each other. Thus, the ordered pair (2, 1) and the point (2, 1) are interchangeable ideas.]

EXAMPLES

Graph the sets of ordered pairs in Examples 1 and 2.

1. $\{(-2, 1), (0, 2), (1, 3), (2, -3)\}$

Solution:

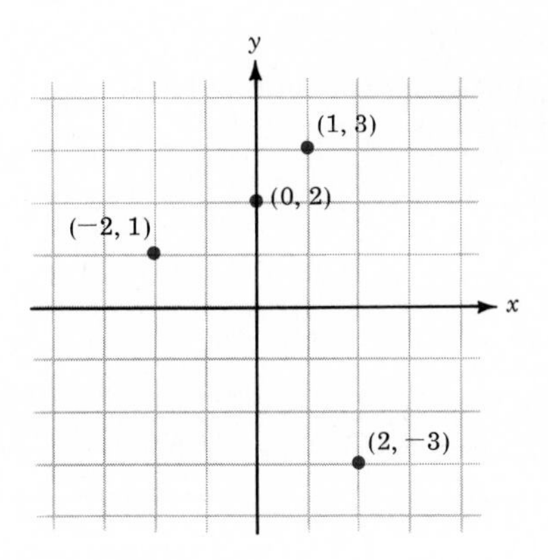

2. $\{(-1, 3), (0, 1), (1, -1), (2, -3), (3, -5)\}$

Solution:

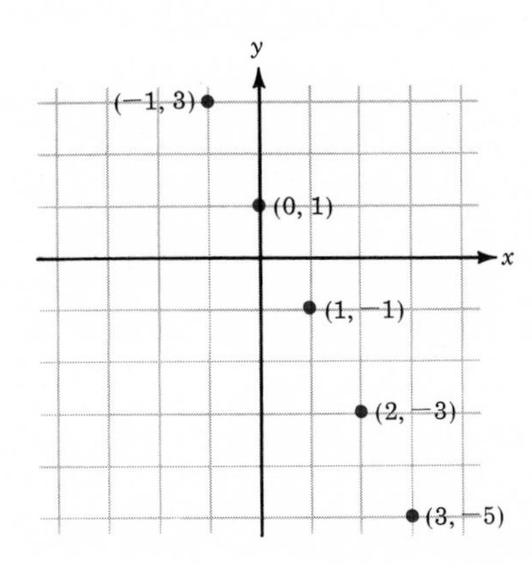

3. If $f(x) = -2x + 1$, find $f(-1)$, $f(0)$, $f(1)$, $f(2)$, and $f(3)$.

Solution:

$$f(-1) = -2(-1) + 1 = 3$$

$$f(0) = -2(0) + 1 = 1$$

$$f(1) = -2(1) + 1 = -1$$

$$f(2) = -2(2) + 1 = -3$$

$$f(3) = -2(3) + 1 = -5$$

EXERCISES 8.1

List the sets of ordered pairs corresponding to the graphs in Exercises 1–10. Assume that the graph lines are marked one unit apart.

1.

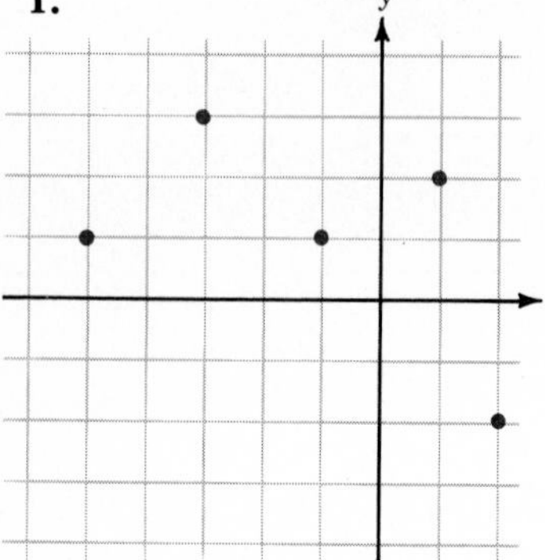

2.

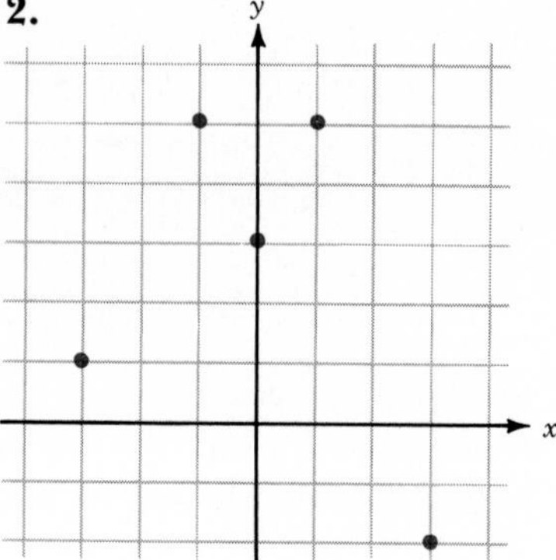

3.

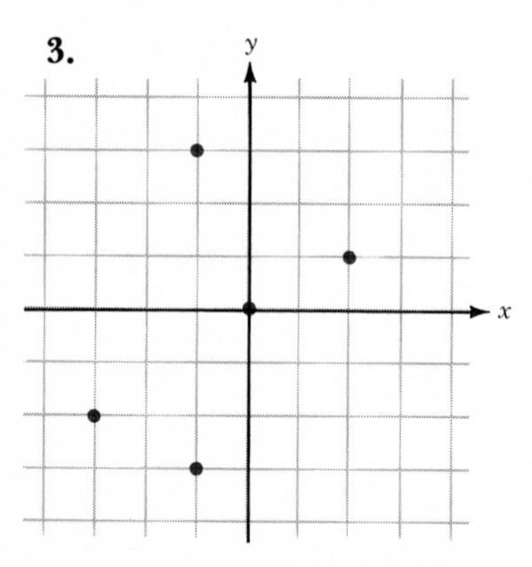

4.

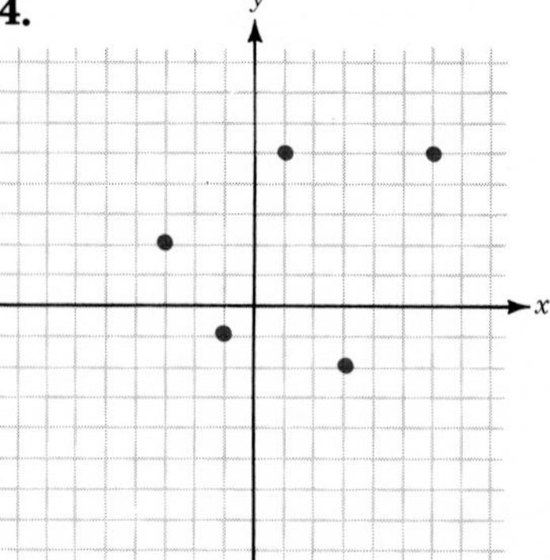

5.

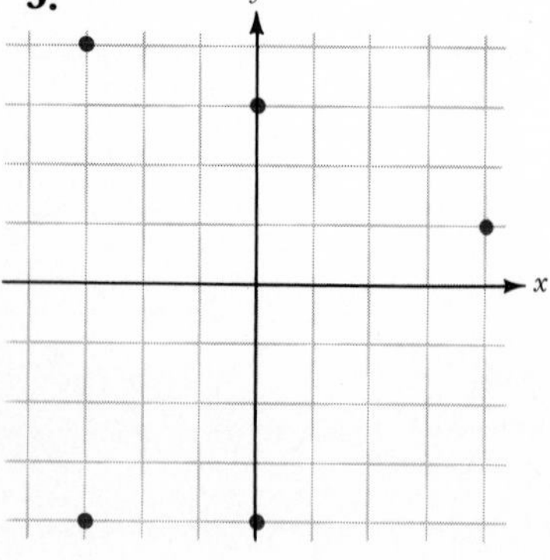

6.

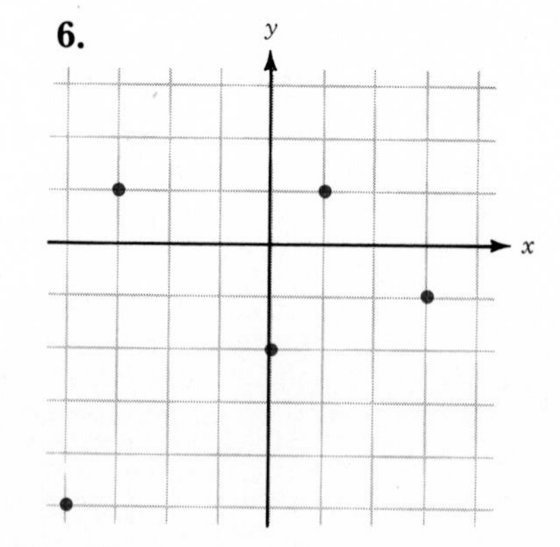

7.

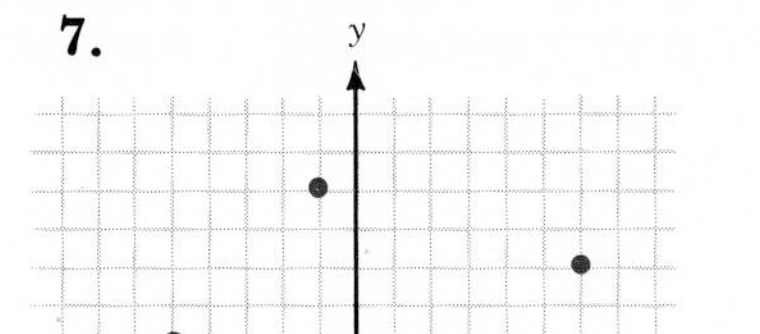

8.

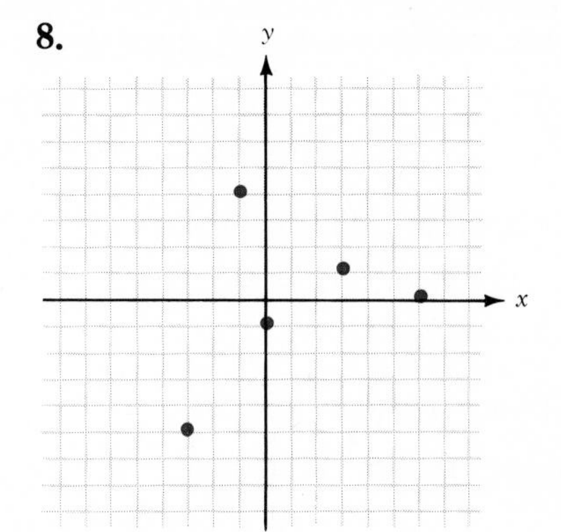

9.

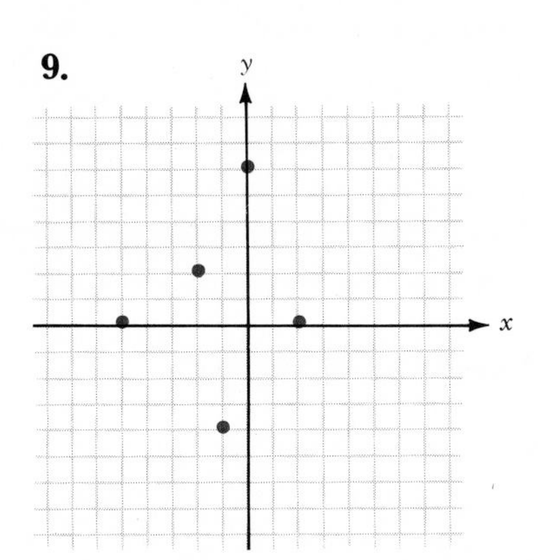

10.

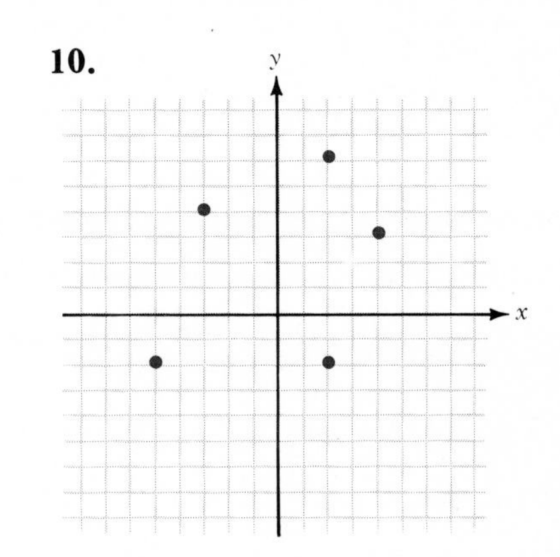

Graph the sets of ordered pairs in Exercises 11–35.

11. $\{(0, 0), (3, 1), (-2, 4), (1, 1), (2, 0)\}$

12. $\{(4, -1), (3, 2), (0, 5), (1, -1), (2, 4)\}$

13. $\{(1, 5), (-1, 3), (-2, 1), (2, 7), (0, 5)\}$

14. $\{(-1, -1), (-3, -2), (1, 3), (0, 0), (2, 5)\}$

15. $\{(0, 1), (-1, 3), (3, 4), (2, -5), (1, 2)\}$

16. $\{(1, 2), (0, 2), (-1, 2), (2, 2), (-3, 2)\}$

17. $\{(5, 3), (2, 1), (0, -3), (2, 4), (1, 0)\}$

18. $\{(1, 0), (3, 0), (-2, 1), (-1, 1), (0, 0)\}$

19. $\{(0, 2), (-1, 1), (2, 4), (3, 5)\}$

20. $\{(-1, -4), (0, -3), (2, -1), (4, 1)\}$

21. $\{(0, 0), (-1, -2), (1, 2), (3, 6)\}$

22. $\{(-1, 4), (0, 5), (2, 7), (-2, 3)\}$

23. $\{(0, -1), (1, 0), (-1, -2), (3, 2), (-2, -3)\}$

24. $\{(-4, -1), (-2, 1), (0, 3), (1, 4), (3, 6)\}$

25. $\{(-1, -3), (0, 0), (1, 3), (2, 6), (-2, -6)\}$

26. $\{(-1, -1), (0, 1), (1, 3), (2, 5), (3, 7)\}$

27. $\{(-1, 2), (-2, 5), (1, 2), (2, 5), (3, 10)\}$

28. $\{(4, 1), (0, -3), (1, -2), (2, -1)\}$

29. $\{(0, 1), (1, 0), (2, -1), (3, -2), (4, -3)\}$

30. $\{(1, 4), (-1, -2), (0, 1), (2, 7), (-2, -5)\}$

31. $\{(1, -3), (-4, \frac{3}{4}), (2, -2\frac{1}{2}), (\frac{1}{2}, 4)\}$

32. $\{(0, 0), (\frac{3}{2}, 2), (-1, \frac{7}{4}), (3, -\frac{1}{2})\}$

33. $\{(\frac{3}{4}, \frac{1}{2}), (2, -\frac{5}{4}), (\frac{1}{3}, -2), (-\frac{5}{3}, 2)\}$

34. $\{(1.6, -2), (3, 2.5), (-1, 1.5), (0, -2.3)\}$

35. $\{(-2, 2), (-3, 1.6), (3, 0.5), (1.4, 0)\}$

36. If $f(x) = x + 3$, find
a. $f(-2)$
b. $f(1)$
c. $f(2)$
d. $f(3)$
e. $f(4)$

37. If $f(x) = 3x - 2$, find
a. $f(-1)$
b. $f(0)$
c. $f(1)$
d. $f(2)$
e. $f(3)$

38. If $f(x) = \frac{x}{2} + 1$, find
a. $f(-2)$
b. $f(0)$
c. $f(2)$
d. $f(4)$
e. $f(-4)$

39. If $g(x) = 2 - 3x$, find
a. $g(-1)$
b. $g(0)$
c. $g(2)$
d. $g(-\frac{1}{3})$
e. $g(1)$

40. If $g(x) = 2x - 3$, find
a. $g(-1)$
b. $g(3)$
c. $g(2)$
d. $g(-2)$
e. $g(\frac{1}{2})$

41. If $h(x) = 1 - 2x$, find
a. $h(0)$
b. $h(\frac{1}{2})$
c. $h(3)$
d. $h(-1)$
e. $h(-3)$

42. If $f(x) = x^2 - 5$, find
a. $f(0)$
b. $f(-3)$
c. $f(2)$

43. If $f(x) = x^2 + 1$, find
a. $f(1)$
b. $f(0)$
c. $f(-1)$

d. $f(3)$
e. $f(-1)$

44. If $f(x) = x^2 + x$, find
a. $f(1)$
b. $f(-1)$
c. $f(2)$
d. $f(0)$
e. $f(-2)$

d. $f(2)$
e. $f(-2)$

45. If $f(x) = 2x^2 + 1$, find
a. $f(-2)$
b. $f(0)$
c. $f(-1)$
d. $f(3)$
e. $f(2)$

8.2 GRAPHING LINEAR EQUATIONS

Suppose we want to graph the points that satisfy the equation $y = 2x + 3$. There are an infinite number of such points. We will graph five to try to find a pattern (Figure 8.3).

The five points in Figure 8.3 appear to lie on a straight line. They in fact do lie on a straight line, and any ordered pair that satisfies the equation $y = 2x + 3$ will also lie on that same line.

What determines whether or not the points that satisfy an equation will lie on a straight line? The points that satisfy any equation of the form

$$Ax + By = C$$

will lie on a straight line. The equation is called a **linear equation** and is considered the **standard form** for the equation of a line. We can write the equation $y = 2x + 3$ in the standard form $-2x + y = 3$.

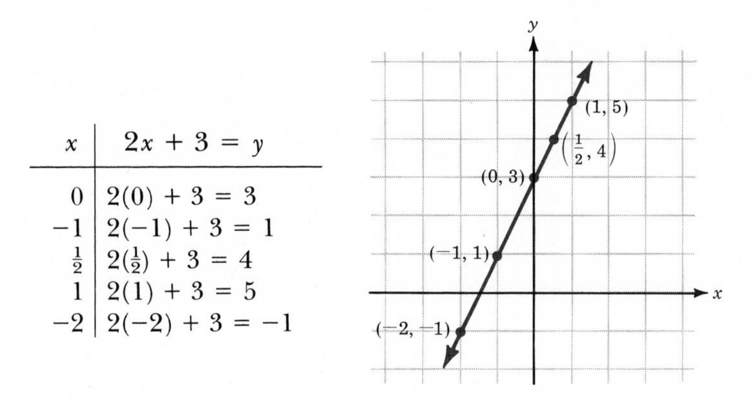

x	$2x + 3 = y$
0	$2(0) + 3 = 3$
-1	$2(-1) + 3 = 1$
$\frac{1}{2}$	$2(\frac{1}{2}) + 3 = 4$
1	$2(1) + 3 = 5$
-2	$2(-2) + 3 = -1$

FIGURE 8.3

Since we now know that the graph will be a straight line, only two points are necessary to determine the entire graph. (Two points determine a line.) The choice of the two points depends on the choice of any two values of x or any two values of y.

EXAMPLES

1. Draw the graph of the linear equation $x + 3y = 6$.

Solution:

$$x = 0 \qquad\qquad x = 3$$
$$0 + 3y = 6 \qquad\qquad 3 + 3y = 6$$
$$y = 2 \qquad\qquad 3y = 3$$
$$y = 1$$

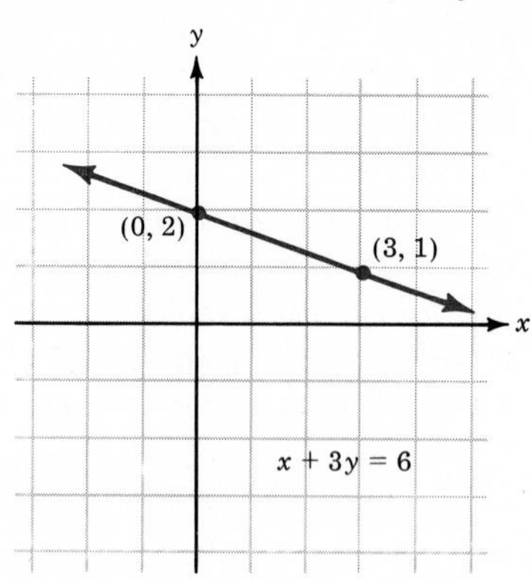

Two points on the graph are (0, 2) and (3, 1). You may have chosen two others. Avoid choosing two points close together.

2. Draw the graph of the linear equation $2x - 5y = 10$.

Solution:

$$x = 5 \qquad\qquad x = 0 \qquad\qquad x = -5$$
$$2 \cdot 5 - 5y = 10 \qquad\qquad 2 \cdot 0 - 5y = 10 \qquad\qquad 2(-5) - 5y = 10$$
$$10 - 5y = 10 \qquad\qquad 0 - 5y = 10 \qquad\qquad -10 - 5y = 10$$
$$-5y = 0 \qquad\qquad y = -2 \qquad\qquad -5y = 20$$
$$y = 0 \qquad\qquad\qquad\qquad y = -4$$

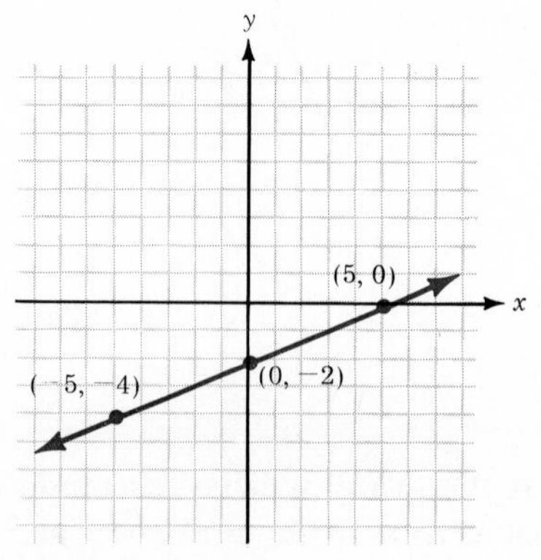

Graphing three points is a good idea, simply to be sure the graph is in the right position.

While the choice of the values for x or y can be arbitrary, letting the value of $x = 0$ will locate the point on the graph where the line crosses the y-axis. This point is called the **y-intercept**. The **x-intercept** is the point found by letting $y = 0$. These two points are generally easy to locate and are frequently used as the two points for drawing the graph of a linear equation.

EXAMPLES

Graph the following linear equations by locating the x-intercepts and the y-intercepts.

3. $3y + 2x = 6$

Solution:

$x = 0 \rightarrow y = 2$

$y = 0 \rightarrow x = 3$

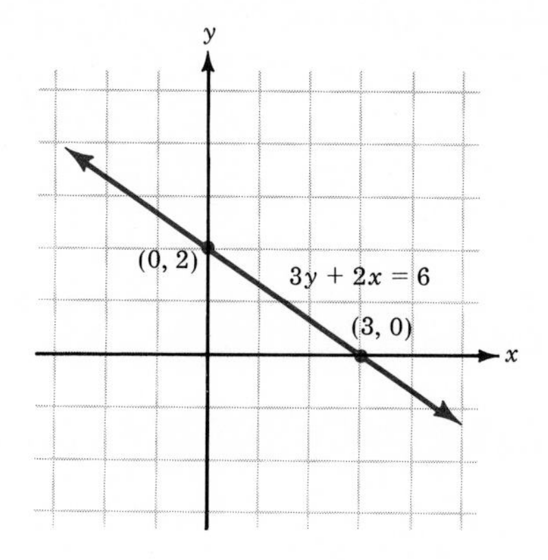

4. $3y - 2x = 6$

Solution:

$x = 0 \rightarrow y = 2$

$y = 0 \rightarrow x = -3$

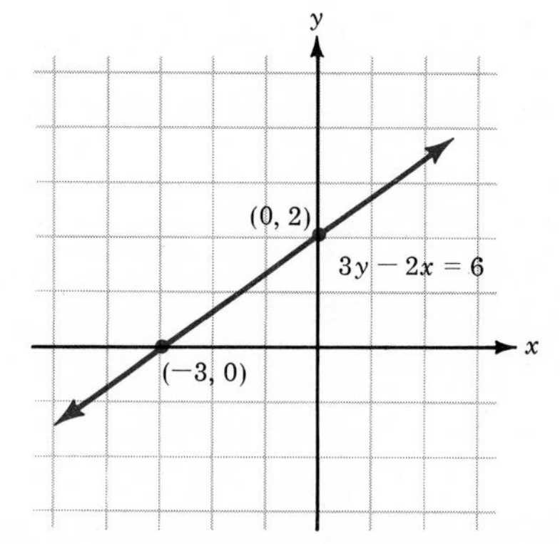

Since the x-coordinate of the y-intercept is always 0, we will agree to call the y-coordinate the y-intercept. In Examples 3 and 4 just given, we will say the y-intercept is 2 rather than giving the coordinates $(0, 2)$. Similarly, in Example 3 the x-intercept is 3, and in example 4 the x-intercept is -3.

If the line goes through the origin, both the x-intercept and the y-intercept will be 0. In this case, some other point must be used.

EXAMPLE

5. Graph the following linear equation: $y - 3x = 0$.

Solution:

$x = 0 \rightarrow y = 0$

$x = 2 \rightarrow y = 6$

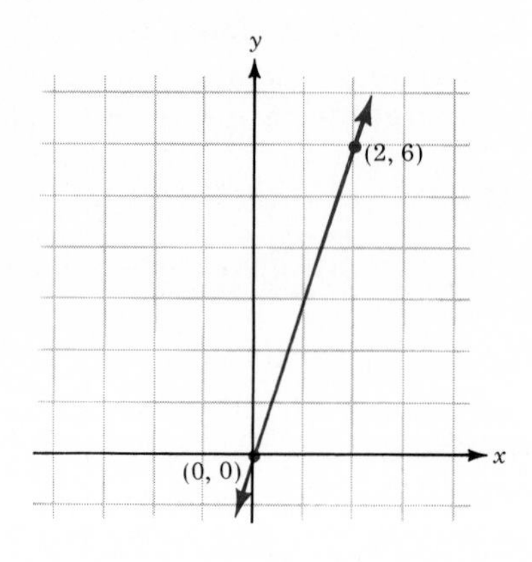

EXERCISES 8.2

Graph the linear equations in Exercises 1–30.

1. $y = 2x$ **2.** $y = 3x$ **3.** $y = -x$ **4.** $y = -5x$

5. $y = x - 4$ **6.** $y = x + 3$ **7.** $y = x + 2$ **8.** $y = x - 6$

9. $y = 4 - x$ **10.** $y = 8 - x$ **11.** $y = 2x - 1$ **12.** $y = 5 - 2x$

13. $x - 2y = 4$ **14.** $x + 3y = 5$ **15.** $x + y = 0$ **16.** $x - y = 0$

17. $x + y = 3$ **18.** $2x - y = 1$ **19.** $2y = x$ **20.** $2x + 3y = 7$

21. $4x + 3y = 11$ **22.** $3y = 2x - 4$ **23.** $3x - 2y = 6$

24. $5x + 2y = 9$ **25.** $2x - 7y = -14$ **26.** $4x + 2y = 10$

27. $y = \frac{1}{2}x + 1$ **28.** $y = \frac{1}{3}x - 3$ **29.** $\frac{2}{3}x + y = 4$

30. $\frac{1}{2}y = 2x - 1$

Graph the linear equations in Exercises 31–45 by locating the x-intercept and the y-intercept.

31. $x + y = 4$ **32.** $x - 2y = 6$ **33.** $3x - 2y = 6$

34. $3x - 4y = 12$ **35.** $5x + 2y = 10$ **36.** $3x + 7y = -21$

37. $2x - y = 9$ **38.** $4x + y = 7$ **39.** $x + 3y = 5$

40. $x - 6y = 3$ **41.** $\frac{1}{2}x - y = 4$ **42.** $\frac{2}{3}x - 3y = 4$

43. $\frac{1}{2}x - \frac{3}{4}y = 6$ **44.** $5x + 3y = 7$ **45.** $2x + 3y = 5$

8.3 THE SLOPE-INTERCEPT FORM: $y = mx + b$

A carpenter is given a set of house plans that call for a 5:12 roof. What does this mean to the carpenter? It means that he must construct the roof so that for every 5 inches of rise (vertical distance), there are 12 inches of run (horizontal distance). That is, the ratio of rise to run is $\frac{5}{12}$ (Figure 8.4).

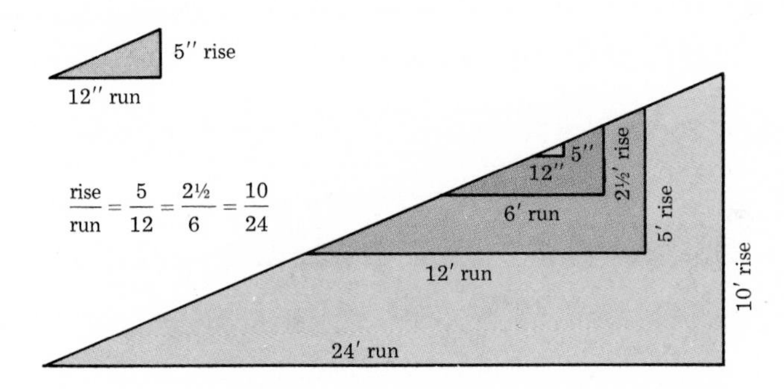

FIGURE 8.4

What if another roof is to be 7:12? Would the carpenter then construct the roof so that for every 7 inches of rise, there would be 12 inches of run? Of course. The ratio of rise to run would be $\frac{7}{12}$ (Figure 8.5).

Other examples involving the ratio of rise to run are the **slope** of a road and the **slope** of the side of a ditch.

For a straight line, the ratio of rise to run is called the **slope** of the line.

$$\text{slope} = \frac{\text{rise}}{\text{run}}$$

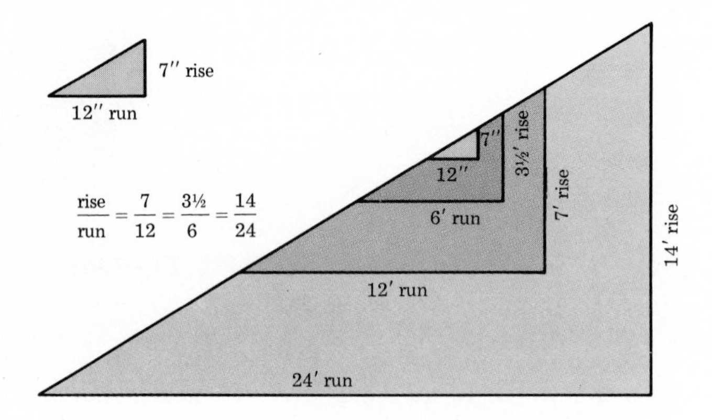

FIGURE 8.5

The graph of the linear equation $y = 3x - 1$ is given in Figure 8.6. What do you think is the slope of the line?

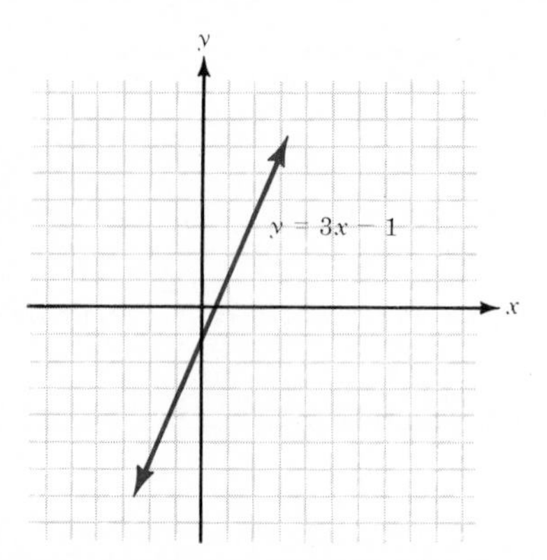

FIGURE 8.6

To calculate the slope, find **any two** points on the line; then find the rise and the run using those two points [Figure 8.7(a)]. (The ratio of rise to run is the same for any two points on a line.) For example,

$$x = 3 \qquad y = 3(3) - 1 = 9 - 1 = 8$$

$$x = -1 \qquad y = 3(-1) - 1 = -3 - 1 = -4$$

Now, using $P_1(-1, -4)$ and $P_2(3, 8)$, the coordinates of P_3 are $(3, -4)$, as shown in Figure 8.7(b). P_3 has the same x-coordinates as P_2 and the same y-coordinates as P_1.

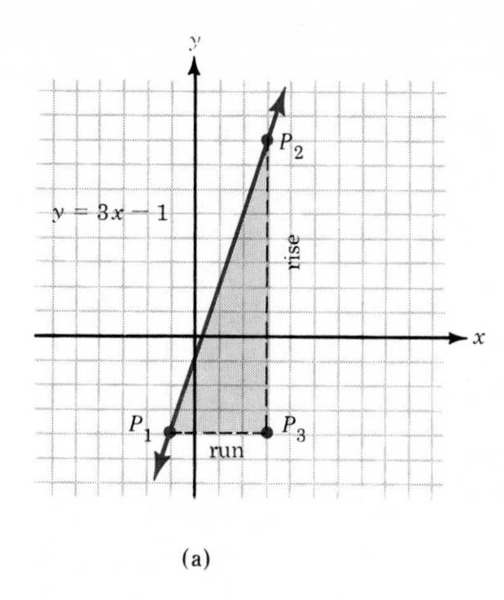

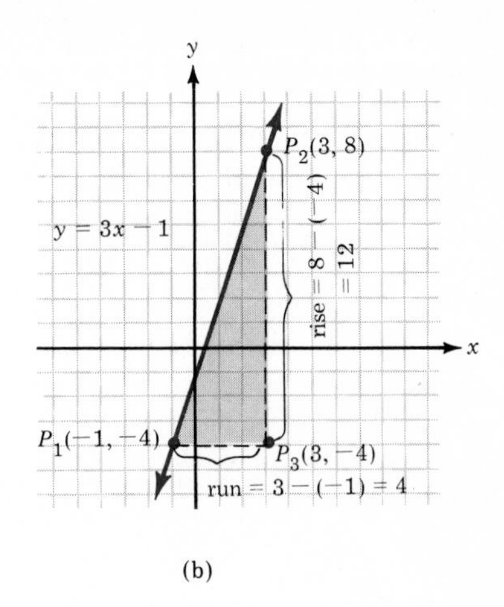

FIGURE 8.7

(**Note:** The notation P_1 is read "P sub 1," and the 1 is called a **subscript.** Similarly, P_2 is read "P sub 2," and P_3 is read "P sub 3.")

For the line $y = 3x - 1$,

$$\text{slope} = \frac{\text{rise}}{\text{run}} = \frac{8 - (-4)}{3 - (-1)} = \frac{12}{4} = 3$$

Now find the slope of the line $y = 3x + 2$. Locate two points on

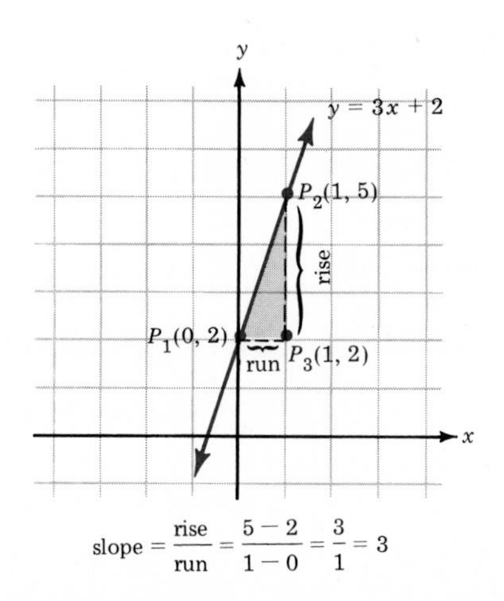

FIGURE 8.8

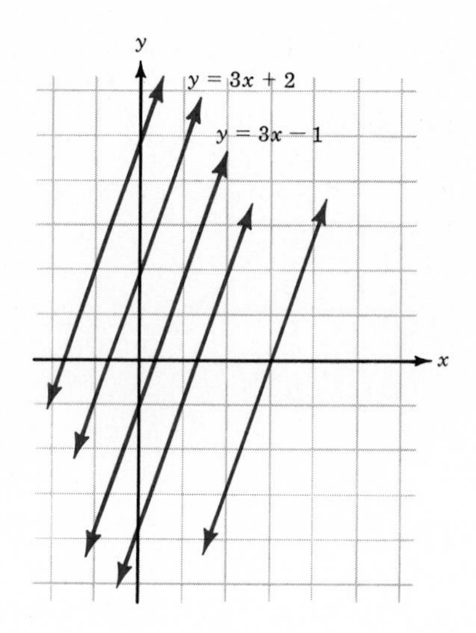

FIGURE 8.9

the line and calculate the ratio, $\frac{\text{rise}}{\text{run}}$. For example,

$$x = 0 \qquad y = 3 \cdot 0 + 2 = 0 + 2 = 2$$

$$x = 1 \qquad y = 3 \cdot 1 + 2 = 3 + 2 = 5$$

Thus, $P_1(0, 2)$, $P_2(1, 5)$, and $P_3(1, 2)$ are as shown in Figure 8.8 on page 333.

Notice that the two lines $y = 3x - 1$ and $y = 3x + 2$ have the same slope, 3 (Figure 8.9). This means that the lines are **parallel.** All

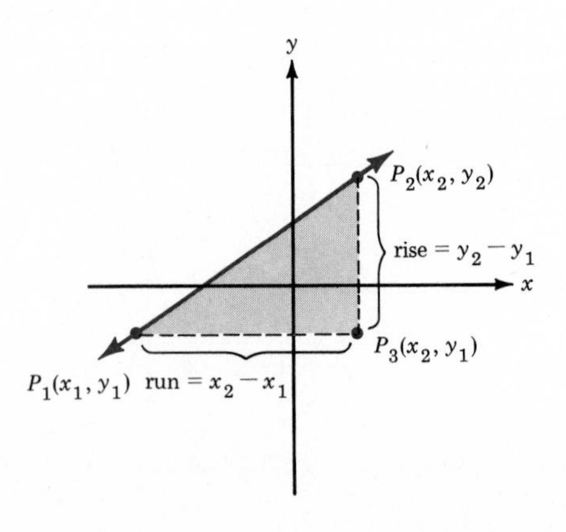

FIGURE 8.10

lines with the same slope are parallel or **coincide.** All the lines in Figure 8.9 have slope 3 and are parallel.

By using subscript notation, we can develop a formula for the slope of any line. Let $P_1(x_1, y_1)$ and $P_2(x_2, y_2)$ be two points on a line. Then $P_3(x_2, y_1)$ is at the right angle shown in Figure 8.10, and the slope can be calculated.

$$\text{slope} = \frac{\text{rise}}{\text{run}} = \frac{y_2 - y_1}{x_2 - x_1}$$

EXAMPLES

1. Using the formula for slope, find the slope of the line $2x + 3y = 6$.

Solution:

Let $x_1 = 0$

$$2 \cdot 0 + 3y_1 = 6$$
$$3y_1 = 6$$
$$y_1 = 2$$

Let $x_2 = -3$

$$2(-3) + 3y^2 = 6$$
$$-6 + 3y_2 = 6$$
$$3y_2 = 12$$
$$y_2 = 4$$

$(x_1, y_1) = (0, 2)$ and $(x_2, y_2) = (-3, 4)$

$$\text{slope} = \frac{y_2 - y_1}{x_2 - x_1} = \frac{4 - 2}{-3 - 0} = \frac{2}{-3} = -\frac{2}{3}$$

2. Suppose that the order of the points in Example 1 is changed. That is, $(x_1, y_1) = (-3, 4)$ and $(x_2, y_2) = (0, 2)$. Will this make a difference in the slope?

Solution:

$$\text{slope} = \frac{y_2 - y_1}{x_2 - x_1} = \frac{2 - 4}{0 - (-3)} = \frac{-2}{3} = -\frac{2}{3}$$

As demonstrated in Example 2, changing the order of the points does not make a difference in the calculation of the slope. Both the numerator and the denominator change sign, so the fraction has the same value. In Example 1, $\frac{2}{-3} = -\frac{2}{3}$, and in Example 2, $\frac{-2}{3} = -\frac{2}{3}$. The important procedure is that **the coordinates must be subtracted in the same order in both the numerator and the denominator.**

In general,

$$\text{slope} = \frac{y_2 - y_1}{x_2 - x_1} = \frac{y_1 - y_2}{x_1 - x_2}$$

The negative slope for the line $2x + 3y = 6$ means that the line slants (or slopes) downward to the right (Figure 8.11). **All lines with negative slopes slant downward to the right, and all lines with positive slopes slant upward to the right.**

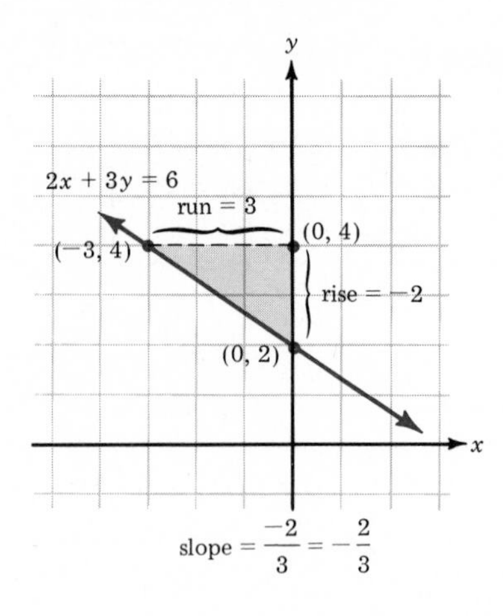

FIGURE 8.11

In the beginning of this section we graphed the line $y = 3x - 1$ and then found its slope to be 3. In Example 1, the line $2x + 3y = 6$ had a slope of $-\frac{2}{3}$. If we solve this equation for y, we get

$$2x + 3y = 6$$

$$3y = -2 + 6$$

$$y = \frac{-2x}{3} + \frac{6}{3}$$

$$y = -\frac{2}{3}x + 2$$

In both cases, when the equation is solved for y, the coefficient of x is the slope. This result is not an accident. The following discussion proves that this coefficient will always be the slope.

Given $y = mx + b$, then m is the **slope.**

Proof. Suppose that the equation is solved for y and $y = mx + b$. Let (x_1, y_1) and (x_2, y_2) be two points on the line where $x_1 \neq x_2$. (More will be said about this condition in Section 8.5.) Then,

$$y_1 = mx_1 + b \qquad \text{and} \qquad y_2 = mx_2 + b$$

$$\begin{aligned} \text{slope} &= \frac{y_2 - y_1}{x_2 - x_1} = \frac{(mx_2 + b) - (mx_1 + b)}{x_2 - x_1} \\ &= \frac{mx_2 + b - mx_1 - b}{x_2 - x_1} = \frac{mx_2 - mx_1}{x_2 - x_1} \\ &= \frac{m(\cancel{x_2 - x_1})}{\cancel{x_2 - x_1}} = m \end{aligned}$$

The slope is m, the coefficient of x, in the form $y = mx + b$.

For the line $y = mx + b$, the point where $x = 0$ is the point where the line will cross the y-axis. This point is called the **y-intercept.** By letting $x = 0$, we get

$$\begin{aligned} y &= mx + b \\ y &= m \cdot 0 + b \\ y &= b \end{aligned}$$

The point $(0, b)$ is the y-intercept. Generally, as discussed in Section 8.2, we say that b is the y-intercept. Thus, we have the following definition.

Definition: $y = mx + b$ is called the **slope-intercept form** for the equation of a line.

m is the **slope,** and b is the **y-intercept.**

EXAMPLES

Find the slope, m, and the y-intercept, b, of each of the following lines by rewriting the equation in the slope-intercept form. Then graph the line.

3. $2x + 3y = 3$

Solution:

$$3y = -2x + 3$$

$$y = \frac{-2x + 3}{3}$$

$$y = -\frac{2}{3}x + 1$$

$$m = -\frac{2}{3}$$

y-intercept $= 1$.

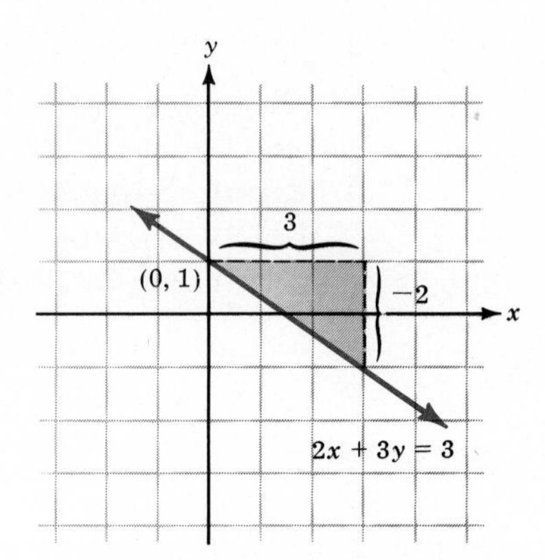

We have used the slope and y-intercept to help draw the graph. From the y-intercept $(0, 1)$, move 3 units to the right and 2 units down $\left(\text{since } m = -\frac{2}{3}\right)$ to locate another point on the line. Draw the line through this point and the y-intercept. We could also have moved 6 units right and 4 units down (or 3 units left and 2 units up) as long as the ratio remains $\frac{\text{rise}}{\text{run}} = -\frac{2}{3}$.

4. $x - 2y = 6$

Solution:

$$x = 6 + 2y$$

$$x - 6 = 2y$$

$$\frac{x - 6}{2} = y$$

$$\frac{1}{2}x - 3 = y$$

$$m = \frac{1}{2}$$

$$b = -3$$

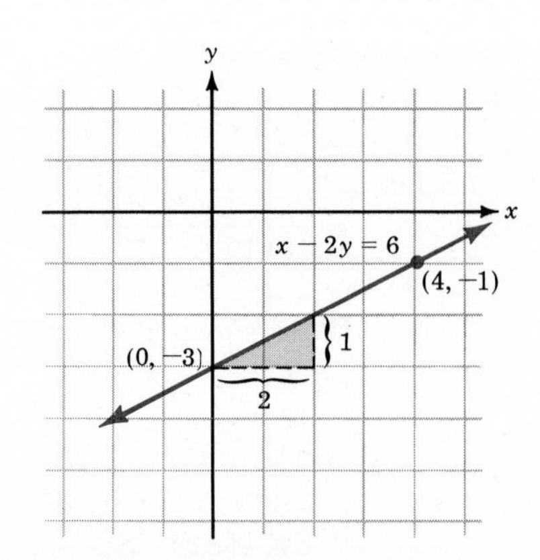

Since $m = \frac{1}{2}$, moving 2 units right and 1 unit up from the y-intercept locates another point on the graph. Of course, another point could be located by substituting some value for x in the equation. For example, if $x = 4$, then

$$x - 2y = 6$$
$$4 - 2y = 6$$
$$-2y = 6 - 4$$
$$-2y = 2$$
$$\frac{-2y}{-2} = \frac{2}{-2}$$

The point $(4, -1)$ is also on the graph.

5. $-4x + 2y = 7$

Solution:

$$2y = 4x + 7$$
$$y = \frac{4x + 7}{2}$$
$$y = 2x + \frac{7}{2}$$
$$m = 2$$
$$b = \frac{7}{2}$$

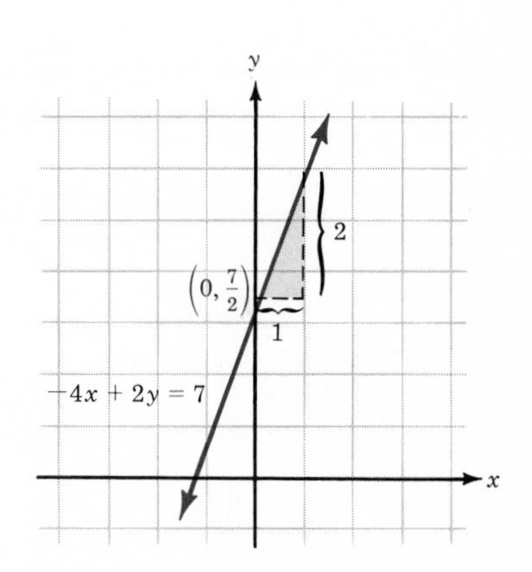

6. Find the equation of the line with y-intercept $(0, -2)$ and slope $m = \frac{3}{4}$.

Solution: In this case, we can substitute directly into the slope-intercept form $y = mx + b$ with $b = -2$ and $m = \frac{3}{4}$.

$$y = \frac{3}{4}x + (-2) \qquad \text{or} \qquad y = \frac{3}{4}x - 2$$

In standard form,

$$3x - 4y = 8$$

Both forms are correct, and either form is an acceptable answer.

EXERCISES 8.3

Find the slope of the line determined by each pair of points in Exercises 1–10.

1. $(1, 4), (2, -1)$ **2.** $(3, 1), (5, 0)$ **3.** $(4, 7), (-3, -1)$

4. $(-6, 2), (1, 3)$ **5.** $(-3, 8), (5, 9)$ **6.** $(0, 0), (-6, -4)$

7. $(4, 2), (-1, \frac{1}{2})$ **8.** $(\frac{3}{4}, 2), (1, \frac{3}{2})$ **9.** $(\frac{7}{2}, 3), (\frac{1}{2}, -\frac{3}{4})$

10. $(-2, \frac{4}{5}), (\frac{3}{2}, \frac{1}{10})$

Find the equation and draw the graph of the line passing through the given y-intercept with the given slope in Exercises 11–25.

11. $(0, 0), m = \frac{2}{3}$ **12.** $(0, 1), m = \frac{1}{5}$ **13.** $(0, -3), m = -\frac{3}{4}$

14. $(0, -2), m = \frac{4}{3}$ **15.** $(0, 3), m = -\frac{5}{3}$ **16.** $(0, 2), m = 4$

17. $(0, -1), m = 2$ **18.** $(0, 4), m = -\frac{3}{5}$ **19.** $(0, -5), m = -\frac{1}{4}$

20. $(0, 5), m = -3$ **21.** $(0, -4), m = \frac{3}{2}$ **22.** $(0, 0), m = -\frac{1}{3}$

23. $(0, 5), m = 1$ **24.** $(0, 6), m = -1$ **25.** $(0, -6), m = \frac{2}{5}$

Find the slope, m, and the y-intercept, b, for each of the equations in Exercises 26–45. Then graph the line.

26. $y = 2x - 3$ **27.** $y = 3x + 4$ **28.** $y = 2 - x$

29. $y = -\frac{2}{3}x + 1$ **30.** $x + y = 5$ **31.** $2x - y = 3$

32. $3x - y = 4$ **33.** $4x + y = 0$ **34.** $x - 3y = 9$

35. $x + 4y = -8$ **36.** $3x + 2y = 6$ **37.** $2x - 5y = 10$

38. $4x - 3y = -3$ **39.** $7x + 2y = 4$ **40.** $6x + 5y = -15$

41. $3x + 8y = -16$ **42.** $x + 4y = 6$ **43.** $2x + 3y = 8$

44. $-3x + 6y = 4$ **45.** $5x - 2y = 3$

8.4 THE POINT-SLOPE FORM: $y - y_1 = m(x - x_1)$

Lines represented by equations in the **standard form** $Ax + By = C$ and in the **slope-intercept form** $y = mx + b$ have been discussed in Sections 8.2 and 8.3.

Now, suppose you are given the slope of a line and a point on the line. Can you draw the graph of the line? Probably. The technique is the same as that discussed in the examples in Section 8.3, but the given point need not be the y-intercept. Consider the following example.

EXAMPLE

1. Graph the line with slope $m = \frac{3}{4}$ which passes through the point $(4, 5)$.

Solution: Start from the point $(4, 5)$ and locate another point on the line using the slope as $\frac{\text{rise}}{\text{run}} = \frac{3}{4}$.
Four units to the right (run) and three units up (rise) from $(4, 5)$ will give another point on the line.

From $(4, 5)$ you might have moved 4 units left and 3 units down, or 8 units right and then 6 units up. Just move so that the ratio of rise to run is 3 to 4, and you will locate a second point on the graph. (For a negative slope, move either to the right and then down or to the left and then up.)

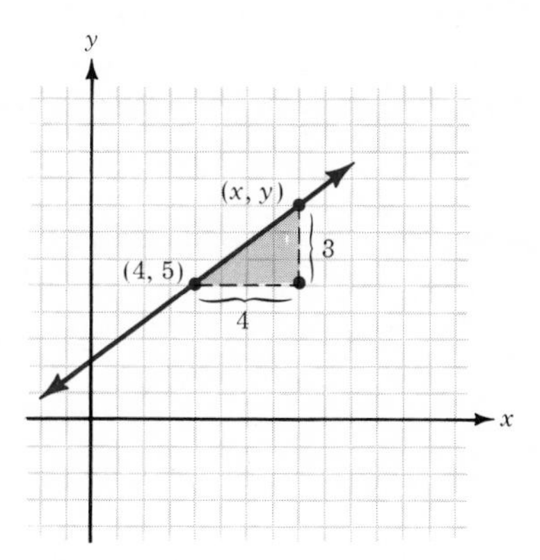

In Example 1, the graph is drawn given the slope $m = \frac{3}{4}$ and a point $(4, 5)$ on the line. How would you find the equation that corresponds to this line? If the slope is $\frac{3}{4}$ and (x, y) is to be a point on the line along with $(4, 5)$, then the following discussion will help.

$$\text{slope} = \frac{y - 5}{x - 4} \quad \text{and} \quad \text{slope} = \frac{3}{4}$$

Setting these equal to each other gives

$$\frac{y - 5}{x - 4} = \frac{3}{4}$$

This is the equation we want. However, to obtain a better form, multiply both sides by $x - 4$ and get

$$y - 5 = \frac{3}{4}(x - 4)$$

With this approach, we can say that a line with slope m that passes through a given fixed point (x_1, y_1) can be represented in the form

$$y - y_1 = m(x - x_1)$$

Definition: $y - y_1 = m(x - x_1)$ is called the **point-slope form** for the equation of a line.

m is the slope, and (x_1, y_1) is a given point on the line.

EXAMPLES

2. Find the equation of the line that has slope $\frac{2}{3}$ and passes through the point $(1, 5)$.

Solution: Substitute into the point-slope form $y - y_1 = m(x - x_1)$.

$$y - 5 = \frac{2}{3}(x - 1)$$

This same equation can be written in the slope-intercept form and in standard form.

$$y - 5 = \frac{2}{3}x - \frac{2}{3}$$

$$y = \frac{2}{3}x - \frac{2}{3} + 5$$

$$y = \frac{2}{3}x + \frac{13}{3} \qquad \text{slope-intercept form}$$

Or, multiplying each term by 3 gives

$$3y = 2x + 13$$

$$-2x + 3y = 13 \qquad \text{standard form}$$

3. Find three forms for the equation of the line with slope $-\frac{5}{2}$ that contains the point $(6, -1)$.

Solution: Use $y - y_1 = m(x - x_1)$.

$$y - (-1) = -\frac{5}{2}(x - 6)$$

$$y + 1 = -\frac{5}{2}(x - 6) \qquad \text{point-slope form}$$

$$y + 1 = -\frac{5}{2}x + 15$$

$$y = -\frac{5}{2}x + 14 \qquad \text{slope-intercept form}$$

$$2y = -5x + 28$$

$$5x + 2y = 28 \qquad \text{standard form}$$

You must understand that all three forms for the equation of a line are acceptable and correct. Your answer may be in a different **form** from that in the back of the text, but it should be algebraically equivalent.

If two points are given,

1. Use the formula $m = \dfrac{y_2 - y_1}{x_2 - x_1}$ to find the slope.

2. Use this slope m and either point in the point-slope formula $y - y_1 = m(x - x_1)$.

The following examples illustrate this common situation.

EXAMPLES

4. Find the equation of the line passing through the two points $(-1, -3)$ and $(5, -2)$.

Solution:

$$m = \frac{y_2 - y_1}{x_2 - x_1} = \frac{-2 - (-3)}{5 - (-1)} = \frac{-2 + 3}{5 + 1} = \frac{1}{6} \qquad \text{Find } m \text{ using the formula for slope.}$$

Using the point-slope form [either point $(-1, -3)$ or point $(5, -2)$ will give the same result] yields

$$y - (-2) = \frac{1}{6}(x - 5)$$

$$y + 2 = \frac{1}{6}(x - 5) \qquad \text{point-slope form}$$

or

$$y + 2 = \frac{1}{6}x - \frac{5}{6}$$

$$y = \frac{1}{6}x - \frac{17}{6} \qquad \text{slope-intercept form}$$

or

$$6y = x - 17$$

$$17 = x - 6y \qquad \text{standard form } Ax + By = C$$

5. Find the equation of the line that contains the points $(-2, 3)$ and $(5, 1)$.

Solution:

$$m = \frac{y_2 - y_1}{x_2 - x_1} = \frac{1 - 3}{5 - (-2)} = \frac{-2}{7} = -\frac{2}{7}$$

Using $(-2, 3)$ in the point-slope form gives

$$y - 3 = -\frac{2}{7}(x - (-2))$$

$$y - 3 = -\frac{2}{7}(x + 2)$$

You should be able to show that equivalent forms are

$$y = -\frac{2}{7}x + \frac{17}{7} \qquad \text{and} \qquad 2x + 7y = 17$$

We can also write linear equations based on information found in word problems. But we must be careful in our interpretation of the results. For example, projection beyond the given information can be very misleading in statistics, and points representing fractions of a sales item or parts of a person may not have a practical meaning.

EXAMPLE

6. After advertising shoes on sale at \$20 per pair, a store sold 100 pairs in a week. When the advertisement was changed to \$15 per pair, the store sold 300 pairs. Write a linear equation relating the price in dollars, d, and the number of pairs of shoes sold, s $(s = md + b)$.

Solution: The information gives us two ordered pairs in the form (d, s) where

$$d = \text{price in dollars}$$

$$s = \text{number of pairs of shoes sold}$$

If $d_1 = \$20$, then $s_1 = 100$ pairs of shoes $(20, 100)$. If $d_2 = \$15$, then $s_2 = 300$ pairs of shoes $(15, 300)$. Now use these two points to find the equation.

$$m = \frac{s_2 - s_1}{d_2 - d_1} = \frac{300 - 100}{15 - 20} = \frac{200}{-5} = -40$$

Using $s - s_1 = m(d - d_1)$, we get

$$s - 100 = -40(d - 20)$$
$$s - 100 = -40d + 800$$
$$s = -40d + 900$$

EXERCISES 8.4

In Exercises 1–12, write an equation (in slope-intercept form or standard form) for the line passing through the given point with the given slope.

1. $(0, 3)$, $m = 2$ **2.** $(1, 4)$, $m = -1$ **3.** $(-1, 3)$, $m = -\frac{2}{5}$

4. $(-2, -5)$, $m = \frac{7}{2}$ **5.** $(-3, 2)$, $m = \frac{3}{4}$ **6.** $(6, -1)$, $m = \frac{5}{3}$

7. $(1, -4)$, $m = -\frac{5}{6}$ **8.** $(4, 0)$, $m = -\frac{1}{5}$ **9.** $(1, \frac{2}{3})$, $m = \frac{3}{2}$

10. $(\frac{1}{2}, -2)$, $m = \frac{4}{3}$ **11.** $(\frac{1}{2}, -\frac{5}{4})$, $m = -\frac{1}{2}$ **12.** $(\frac{1}{3}, \frac{5}{6})$, $m = -\frac{4}{3}$

In Exercises 13–24, write an equation (in slope-intercept form or standard form) for the line passing through the given points.

13. $(-2, 3)$, $(1, 2)$ **14.** $(-5, 1)$, $(2, 0)$ **15.** $(3, 4)$, $(6, 2)$

16. $(-4, -4)$, $(3, 1)$ **17.** $(8, 2)$, $(3, -1)$ **18.** $(1, -2)$, $(4, -3)$

19. $(-5, 2)$, $(1, 4)$ **20.** $(-2, 6)$, $(3, 1)$ **21.** $(-4, 2)$, $(1, \frac{1}{2})$

22. $(0, 2)$, $(1, \frac{3}{4})$ **23.** $(\frac{3}{4}, 2)$, $(\frac{1}{2}, \frac{3}{2})$ **24.** $(\frac{5}{4}, \frac{2}{3})$, $(\frac{1}{4}, -\frac{1}{3})$

25. The rental rate schedule for a car rental is described as "\$30 per day plus 15¢ per mile driven." Write a linear equation for the daily cost of renting a car for x miles $(C = mx + b)$.

26. For each job, a print shop charges \$5.00 plus \$0.05 per page printed. Write a linear equation for the cost of printing x pages $(C = mx + b)$.

27. A salesperson's weekly salary depends on the amount of her sales. Her weekly salary is $200 plus 9% of her weekly sales. Write a linear equation for her weekly salary if her sales are x dollars ($S = mx + b$).

28. A manufacturer has determined that the cost, C, of producing x units is given by a linear equation ($C = mx + b$). If it costs $752 to produce 12 units and $800 to produce 15 units, find a linear equation for the cost of producing x units.

29. A local amusement park found that if the admission was $7, they had about 1000 customers per day. When the price was dropped to $6, they had about 1200 customers per day. Write a linear equation for the price in terms of attendence, x ($P = mx + b$).

30. A department store manager has determined that if the price for a necktie is $5, the store will sell 100 neckties per month. However, only 50 neckties per month are sold when the price is raised to $6. Write a linear equation for the price in terms of the number sold, x ($P = mx + b$).

8.5 HORIZONTAL AND VERTICAL LINES

Consider the equation $y = 3$. If $x = 5$, what is y? If $x = -2$, what is y? If $x = 0$, what is y? In each case, $y = 3$. The points $(5, 3)$, $(-2, 3)$, and $(0, 3)$ all satisfy the condition $y = 3$. x can have any value, but y must be 3. To emphasize this fact, the equation $y = 3$ could be written $0x + y = 3$ or $y = 0x + 3$. The graph of the equation $y = 3$ is a **horizontal line,** and each y-coordinate is 3 (Figure 8.12).

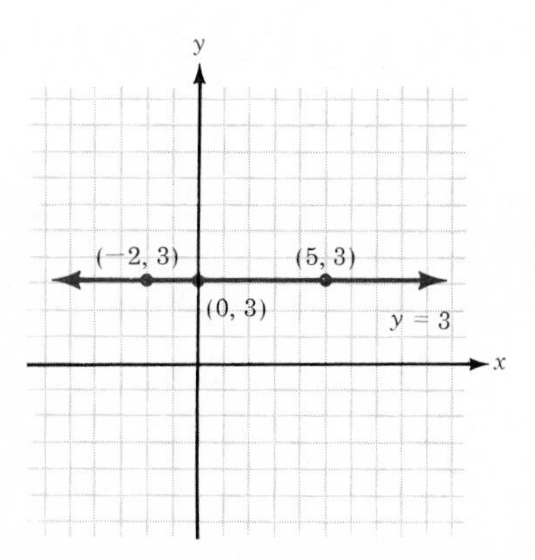

FIGURE 8.12

What is the slope of the line $y = 3$? By writing the equation in the form $y = 0x + 3$, we know that the slope is 0. Also, by the formula for slope, using $(-2, 3)$ and $(5, 3)$, the slope is indeed 0:

$$\text{slope} = m = \frac{y_2 - y_1}{x_2 - x_1} = \frac{3 - 3}{5 - (-2)} = \frac{0}{7} = 0$$

What are the graphs of the equations $y = -2$ and $y = \frac{3}{2}$? Both are horizontal lines with slope 0. In fact, **any equation of the form $y = b$ (or $y = 0x + b$) is a horizontal line with slope 0** (Figure 8.13).

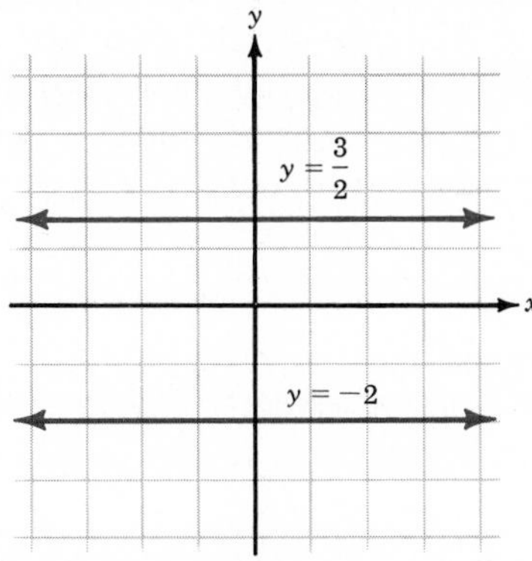

FIGURE 8.13

Now consider the equation $x = 2$. If $y = 4$, what is x? If $y = -3$, what is x? If $y = 0$, what is x? In each case, $x = 2$. The points $(2, 4)$, $(2, -3)$, and $(2, 0)$ all satisfy the condition $x = 2$. y can have any value, but x must be 2. We can write $x = 2$ in the form $0y + x = 2$ or $x = 0y + 2$. The graph of the equation $x = 2$ is a **vertical line,** and each x-coordinate is 2 (Figure 8.14).

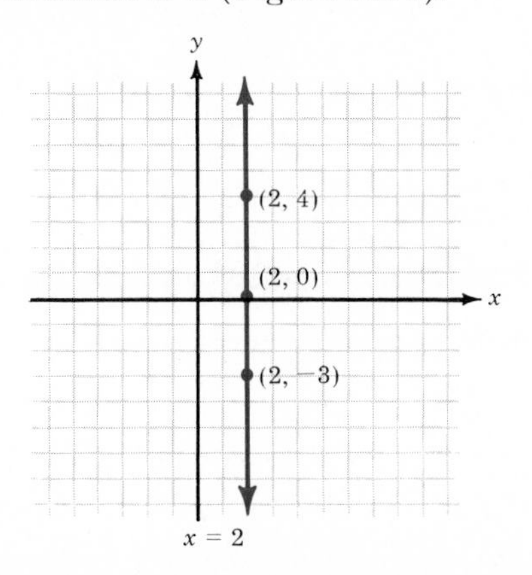

FIGURE 8.14

What is the slope of the line $x = 2$? To answer this question, let us use the slope formula and the two points (2, 4) and (2, −3):

$$\text{slope} = m = \frac{y_2 - y_1}{x_2 - x_1} = \frac{4 - (-3)}{2 - 2} = \frac{7}{0} \qquad \text{undefined}$$

The slope is undefined because 0 cannot be a denominator. This is true for any vertical line. The graphs of equations of the form $x = a$ are vertical lines with no slope (or with slope undefined). Note very carefully the distinction that **horizontal lines (y = b) have 0 slope and vertical lines (x = a) have no slope.** Both the equations for horizontal lines ($y = b$) and the equations for vertical lines ($x = a$) are special cases of the general linear equation in standard form, $Ax + By = C$, discussed in Section 8.2.

EXAMPLES

For each of the following linear equations, find the slope of the line, if it exists, and graph the line.

1. $2y = 5$

Solution:

$y = \frac{5}{2}$

$m = 0$

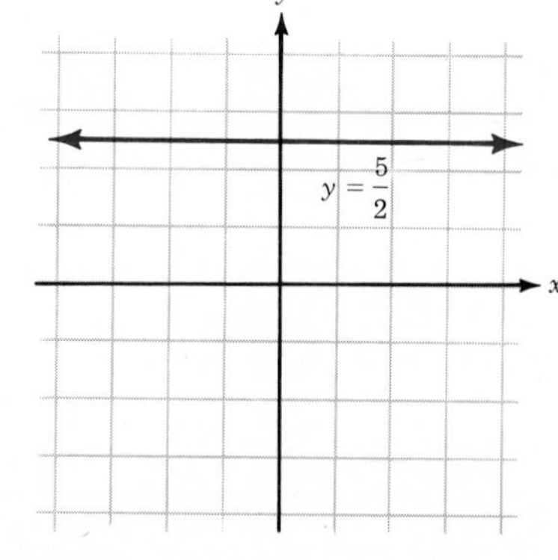

2. $x + 6 = 0$

Solution:

$x = -6$

no slope

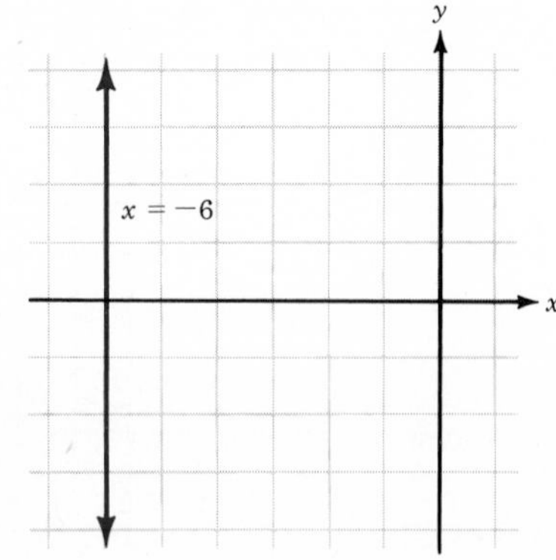

3. $x + y = 0$

Solution:

$y = -x$

$m = -1$

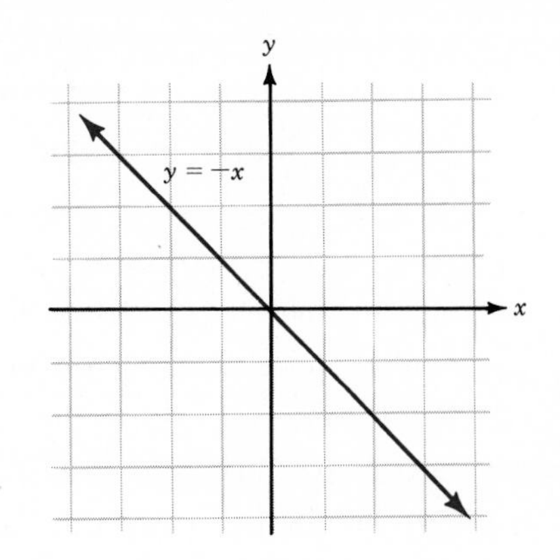

EXERCISES 8.5

Find the slope of the line, if it exists, and graph the line for Exercises 1–20.

1. $4y = 3x - 8$

2. $3y = 2x + 9$

3. $3y = 12$

4. $4x - y = 0$

5. $3x - 5y = 0$

6. $8 - 3y = 0$

7. $2x - 7 = 0$

8. $\frac{3}{2}x - y = 1$

9. $5y = 0$

10. $9 - 4x = 0$

11. $2y = \frac{x}{3} + 4$

12. $\frac{4x}{3} + y = 2$

13. $3x - y = 0$

14. $\frac{3}{4}x - 1 = 0$

15. $\frac{-3x}{5} + 2 = y$

16. $2y - 3 = 0$

17. $-\frac{x}{4} = 1$

18. $\frac{x}{2} - 2y = 4$

19. $\frac{2x}{3} + 2y = 9$

20. $3 - \frac{y}{2} = 0$

Write an equation for the line with the given slope through the given point in Exercises 21–30.

21. $(2, -3)$, $m = 0$

22. $(-1, 2)$, $m = -\frac{1}{4}$

23. $(1, 0)$, $m = \frac{3}{4}$

24. $(4, 6)$, no slope

25. $(-1, -4)$, $m = -\dfrac{2}{3}$

26. $\left(\dfrac{2}{3}, 6\right)$, $m = 0$

27. $(-5, 1)$, no slope

28. $(-2, 4)$, $m = -1$

29. $(3, -1)$, $m = 0$

30. $(0, 3)$, no slope

31. Write an equation for the horizontal line through point $(-2, 5)$.

32. Write an equation for the line parallel to the x-axis through point $(6, -2)$.

33. Write an equation for the line parallel to the y-axis through point $(-1, -3)$.

34. Write an equation for the vertical line through point $(-1, -1)$.

35. Write an equation for the line parallel to the line $3x - y = 4$ through the origin.

36. Write an equation for the line parallel to the line $2x = 5$ through point $(0, 3)$.

37. Write an equation for the line through point $\left(\dfrac{3}{2}, 1\right)$ and parallel to the line $4y - 6 = 0$.

38. Write an equation for the horizontal line through point $(-1, -6)$.

39. Find an equation for the line parallel to $2x - y = 7$ having the same y-intercept as $x - 3y = 6$.

40. Find an equation for the line parallel to $3x - 2y = 4$ having the same y-intercept as $5x + 4y = 12$.

8.6 LINEAR INEQUALITIES (OPTIONAL)

The points that satisfy the equation $y = -2x + 3$ lie on a straight line. For each value of x, the corresponding y-value is $-2x + 3$. Each point on the line is of the form $(x, -2x + 3)$. For example, if $x = 2$, then $y = -2 \cdot 2 + 3 = -4 + 3 = -1$; so the point $(2, -1)$ or $(2, -2 \cdot 2 + 3)$ is on the line (Figure 8.15).

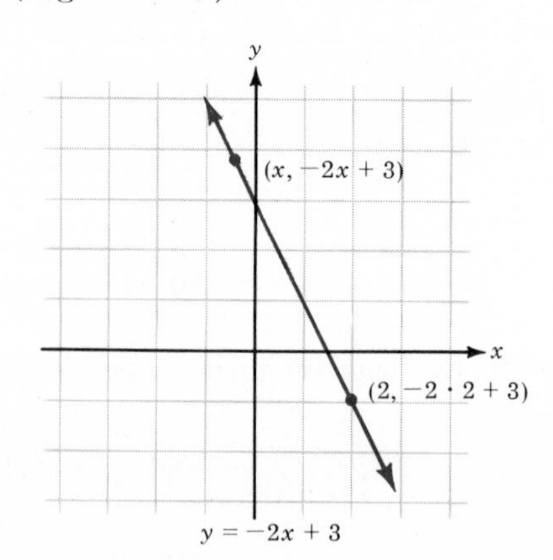

FIGURE 8.15

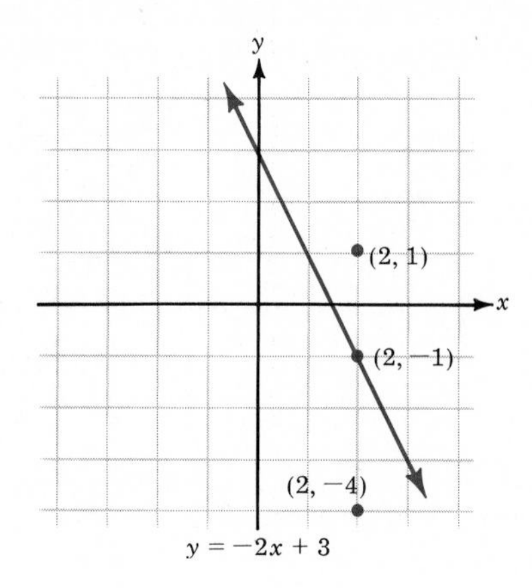

FIGURE 8.16

Where are the points in the plane where $y < -2x + 3$? Where are the points where $y > -2x + 3$? We know the points where $y = -2x + 3$ are on a line. Pick two points, one above the line and one below the line. Any two points will do, just so they are obviously on opposite sides of the line. For example, the points $(2, 1)$ and $(2, -4)$ are shown in Figure 8.16. The point $(2, -4)$ is below the point $(2, -1)$ since $-4 < -1$. Therefore, at least for $x = 2$, $-4 < -2x + 3$.

What about other points below the line? Will the y-value be less than $-2x + 3$? Choose another point below the line, say, $(0, \frac{1}{2})$. For $x = 0$, $-2x + 3 = -2(0) + 3 = 3$. Thus, for the point $(0, \frac{1}{2})$ below the line, $\frac{1}{2} < -2x + 3$. In fact, this will always happen for points below the line. The set of points that satisfy the inequality $y < -2x + 3$ are below the line $y = -2x + 3$. (See Figure 8.17. Note that the line

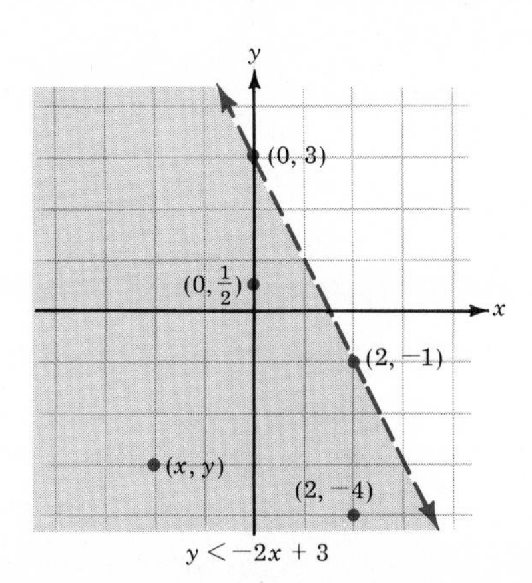

FIGURE 8.17

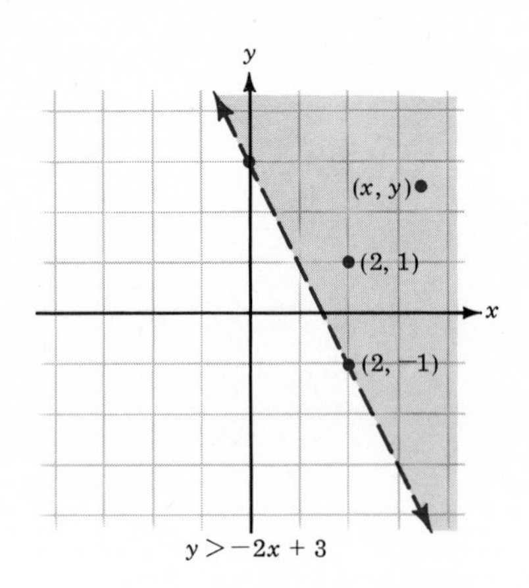

FIGURE 8.18

$y = -2x + 3$ is dotted, or broken, to show that the points on the line are **not** included in the graph.)

Now it should be clear that the points where $y > -2x + 3$ are above the line $y = -2x + 3$. The point (2, 1) above the line satisfies the relationship $1 > -2x + 3$. $y > -2(2) + 3$ (Figure 8.18).

EXAMPLES

1. Graph the points that satisfy the linear inequality $2x + 3y < 6$.

Solution: Solve for y.

$$2x + 3y < 6$$

$$3y < -2x + 6$$

$$y < \frac{-2x + 6}{3}$$

$$y < -\frac{2}{3}x + 2$$ The points are **below** the line $y = -\frac{2}{3}x + 2$.

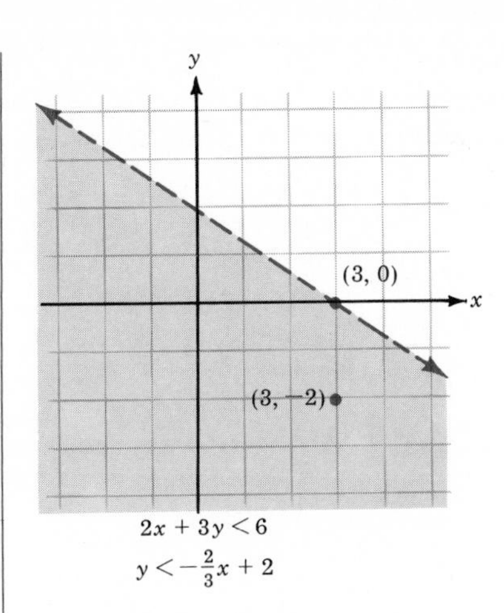

Checking one point:

For $(3, -2)$, $-2 < -\frac{2}{3}(3) + 2$ or $-2 < 0$.
Or $2(3) + 3(-2) = 6 - 6 = 0 < 6$.

2. Graph the points that satisfy the linear inequality $5x - y \leq 4$.

Solution: Solving for y, we have

$5x - y \leq 4$

$5x - 4 \leq y$

or $\quad y \geq 5x - 4 \quad$ The points are **above** and **on** the line $y = 5x - 4$.

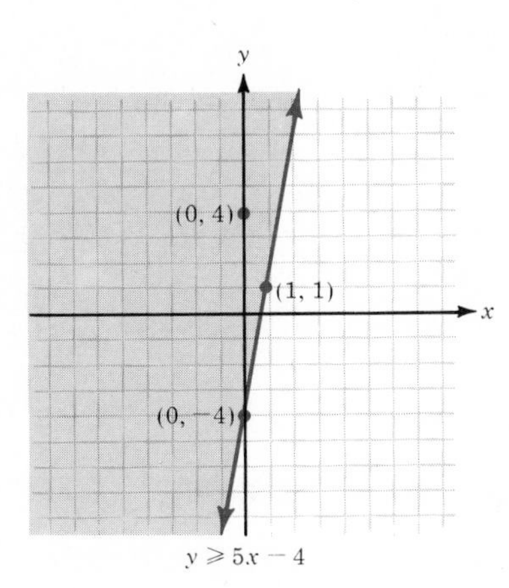

Checking one point:
For $(0, 4)$, $4 \geq 5 \cdot 0 - 4$ or $4 \geq -4$.

Or $5 \cdot 0 - 4 = -4 \leq 4$.

To see that checking one point is sufficient, try a point below the line, say, $(1, -2)$. Is $-2 \geq 5 \cdot 1 - 4$? Since $5 \cdot 1 - 4 = 1$ and $-2 \not\geq 1$, the points that satisfy the inequality $y \geq 5x - 4$ or $5x - y \leq 4$ are above and on the line.

3. Graph the points that satisfy both inequalities $x \leq 2$ and $y \geq -x + 1$.

Solution: For $x \leq 2$, the points are to the left of and on the line $x = 2$. For $y \geq -x + 1$, the points are above and on the line $y = -x + 1$. The points that satisfy both conditions are shown in the shaded portion and solid lines of the graph.

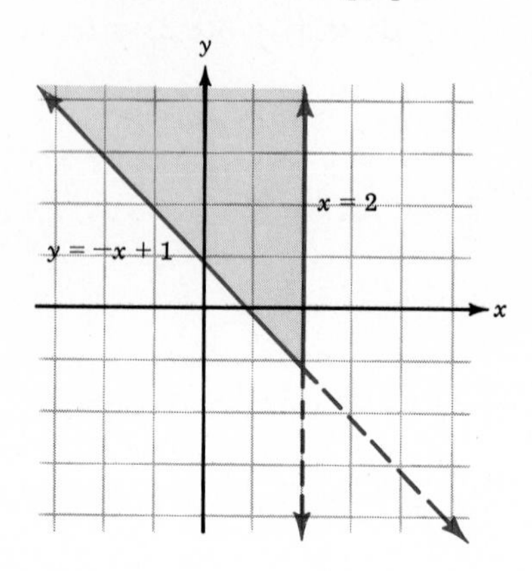

$x \leq 2$ and $y \geq -x + 1$

EXERCISES 8.6

Graph the linear inequalities in Exercises 1–30.

1. $y > 3x$

2. $x - y > 0$

3. $y \leq 2x - 1$

4. $y < 4 - x$

5. $y > -7$

6. $2y - x \geq 2$

7. $2x - 3 \leq 0$

8. $4x + y < 2$

9. $5x - 2y \geq 4$

10. $3y - 8 \leq 2$

11. $y < -\dfrac{x}{4}$

12. $2y - 3x \leq 4$

13. $2x - y < 1$

14. $4x + 3y \leq 6$

15. $2x - 3y \geq -3$

16. $2y + 5x > 0$

17. $2x + 5y \geq 10$

18. $4x + 2y \leq 10$

19. $3x < 4 - y$

20. $x < 2y + 3$

21. $2x - 3y < 9$

22. $x \leq 4 - 3y$

23. $5x - 2y \geq 4$

24. $3x + 6 \geq 2y$

25. $4y \geq 3x - 8$

26. $\dfrac{1}{2}x + y < 3$

27. $2x + \dfrac{1}{3}y < 2$

28. $2x - \dfrac{1}{2}y > -1$

29. $\dfrac{1}{2}x + \dfrac{2}{3}y \geq \dfrac{5}{6}$

30. $\dfrac{1}{4}x - \dfrac{1}{2}y \leq \dfrac{3}{4}$

Graph the points that satisfy both inequalities in Exercises 31–40.

31. $y < 2$ and $x \geq -3$

32. $2x + 5 < 0$ and $y \geq 2$

33. $x \geq -4$ and $x < 3$

34. $y \leq 5$ and $y > -2$

35. $x \leq 3$ and $2x + y > 7$

36. $2x - y > 4$ and $y < -1$

37. $x - 3y \leq 3$ and $x < 5$

38. $3x - 2y \geq 8$ and $y \geq 0$

39. $x - y \geq 0$ and $3x - 2y \geq 4$

40. $y \geq -x - 2$ and $x + y \geq -2$

CHAPTER 8 SUMMARY

In an **ordered pair** (x, y), x is called the **first component** and y is called the second component. The variable assigned to the first component is also called the **independent variable,** and the variable assigned to the second component is called the **dependent variable.**

The symbol $f(x)$, read "f of x", is called **function notation** and represents the value of the dependent variable related to x.

In the **Cartesian coordinate system,** two number lines intersect at right angles and separate the plane into four **quadrants.** The point of intersection $(0, 0)$ is called the **origin.** The horizontal number line is called the **horizontal axis** or **x-axis.** The vertical number line is called the **vertical axis** or **y-axis.**

There is a one-to-one correspondence between points in a plane and ordered pairs of real numbers.

The **standard form** of a **linear equation** is

$$Ax + By = C$$

The point where a line crosses the y-axis is called the **y-intercept.** The point where a line crosses the x-axis is called the **x-intercept.**

Given two points on a line, (x_1, y_1) and (x_2, y_2),

$$\text{slope} = \frac{\text{rise}}{\text{run}} = \frac{y_2 - y_1}{x_2 - x_1} = \frac{y_1 - y_2}{x_1 - x_2}$$

Definition: $y = mx + b$ is called the **slope-intercept form** for the equation of a line.

m is the slope, and b is the y-intercept.

Definition: $y - y_1 = m(x - x_1)$ is called the **point-slope form** for the equation of a line.

m is the slope, and (x_1, y_1) is the fixed point.

Any equation of the form $y = b$ (or $y = 0x + b$) is a horizontal line with slope 0. Any equation of the form $x = a$ is a vertical line and has no slope.

A **linear inequality** is an inequality of the form

$$ax + by < c$$

($\leq$, $>$, and $\geq$ can also be used.)

CHAPTER 8 REVIEW

List the set of ordered pairs corresponding to the graphs in Exercises 1–5.

1.

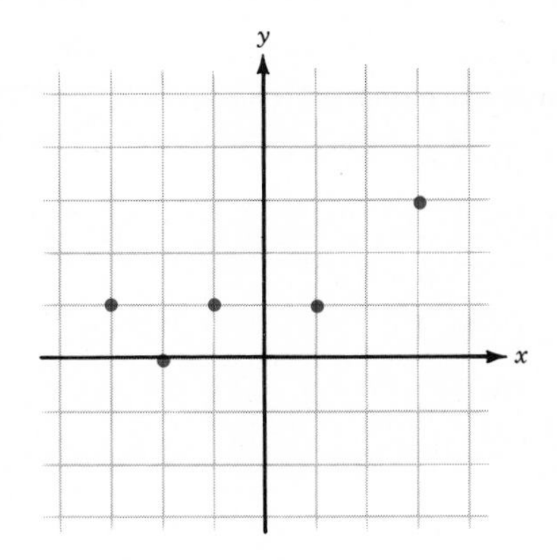

2.

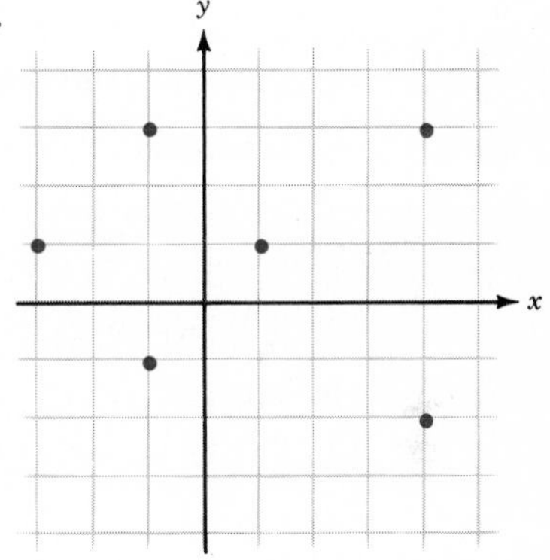

3.

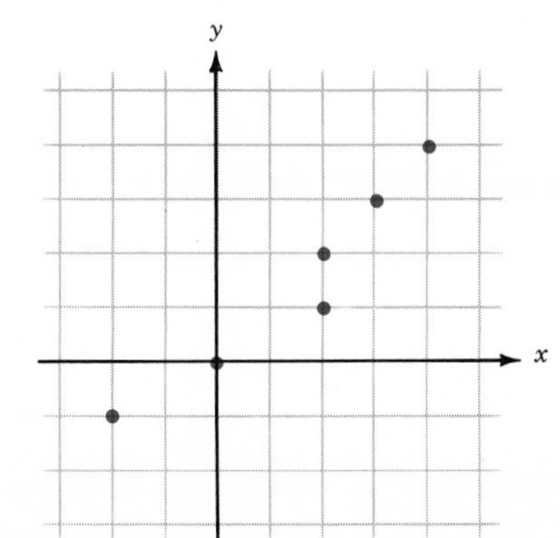

4.

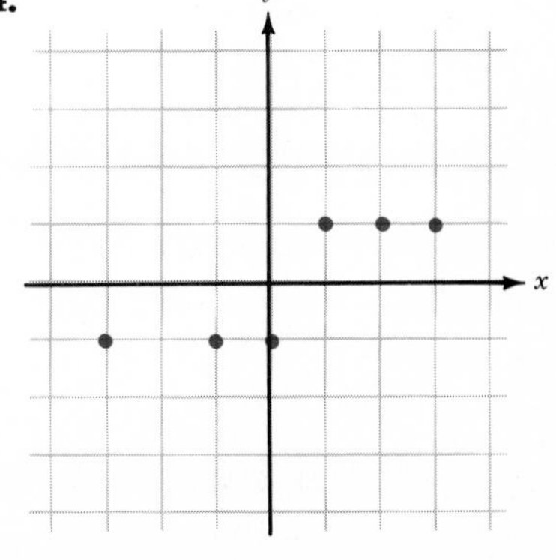

5.

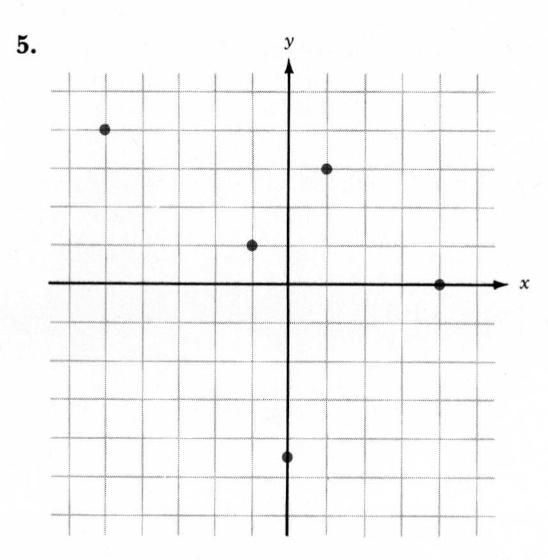

Graph each of the sets of ordered pairs in Exercises 6–13.

6. $\{(-1, -5), (0, 2), (2, 4), (3, 7)\}$

7. $\{(0, 4), (2, 2), (3, 1), (-2, 6)\}$

8. $\{(1, 3), (2, 7), (4, -1), (0, 0), (0, 2)\}$

9. $\{(1, -3), (2, 4), (-1, 1), (6, 2), (3, 1)\}$

10. $\{(0, 4), (-4, 0), (-5, 2), (3, 4)\}$

11. $\{(2, 4), (1, -4), (0, 4), (5, -4)\}$

12. $\{(2, \frac{1}{2}), (-1, 3), (\frac{2}{3}, -2), (1, 1)\}$

13. $\{(0, 1.4), (2.5, 6), (3, -1.75), (-2.25, 0)\}$

14. If $f(x) = 4x - 3$, find

a. $f(2)$
b. $f(-1)$
c. $f(0)$
d. $f(4)$

15. If $f(x) = \dfrac{x}{2} + 5$, find

a. $f(2)$
b. $f(-4)$
c. $f(0)$
d. $f(-1)$

16. If $g(x) = x^2 - 9$, find

a. $g(0)$
b. $g(-1)$
c. $g(-4)$
d. $g(2)$

17. If $f(x) = x^2 + 2$, find

a. $f(0)$
b. $f(-3)$
c. $f(-4)$
d. $f(1)$

18. If $f(x) = 2 + x - 2x^2$, find

a. $f(0)$
b. $f(1)$
c. $f(\frac{1}{2})$
d. $f(-2)$

Graph the linear equations in Exercises 19–23.

19. $y = -4x$ **20.** $x + 2y = 4$ **21.** $x + 4 = 0$

22. $x - 3y = -2$ **23.** $4y + x = 6$

Graph the linear equations in Exercises 24–28 by locating the x-intercept and the y-intercept.

24. $2x + y = 4$ **25.** $3x + y = -6$ **26.** $5x - y = 1$

27. $3y - 9 = 0$ **28.** $3x - 4y = 6$

For Exercises 29–33, determine the slope, m, and the y-intercept, b. Then graph the line.

29. $y - 2x = 3$ **30.** $2x + 5y = 10$ **31.** $4x - 8 = 0$

32. $4x - 3y = 7$ **33.** $2x + 3y = 5$

Find the slope of the line determined by each pair of points in Exercises 34–37.

34. $(2, 4), (-1, 6)$ **35.** $(6, -2), (3, 2)$ **36.** $(5, -4), (\frac{3}{2}, 3)$

37. $(\frac{3}{4}, 5), (2, 3)$

In Exercises 38–45, graph the line and then write an equation (in slope-intercept form or standard form) for it.

38. $(3, 7), (5, 2)$ **39.** $(-4, 2), (3, 2)$ **40.** $(2, -3), (-1, -5)$

41. $(6, -1), m = 0$ **42.** $(3, \frac{5}{2}), m = -\frac{1}{4}$ **43.** $(-3, 4)$, no slope

44. $(5, 0), m = -\frac{4}{3}$ **45.** $(4, -5), m = \frac{3}{2}$

46. Write an equation for the line parallel to the y-axis passing through $(-1, 5)$.

47. Write an equation for the line parallel to $2x + y = 5$ passing through $(1, -2)$.

48. Write an equation for the line parallel to $x - 2y = 5$ having the same y-intercept as $5x + 3y = 9$.

49. In a biology experiment, Oraib observed that the bacteria count in a culture was approximately 1000 at the end of 1 hour. At the end of 3 hours, the bacteria count was about 2000. Write a linear equation describing the bacteria count, N, in terms of the time, t.

50. A television repairman charges $20 for the house call plus $17 per hour. Write a linear equation for the total charges, C, in terms of time, t.

Optional

Graph the linear inequalities in Exercises 51–53.

51. $y < 5x$ **52.** $3x + y > 2$ **53.** $\frac{1}{2}x + \frac{1}{4}y \leq 1$

CHAPTER 8 TEST

1. List the ordered pairs corresponding to the points on the graph.

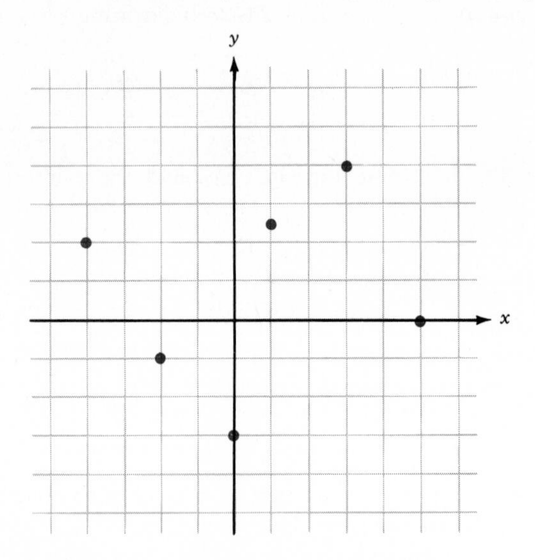

2. Graph the following set of ordered pairs:

$$\{(0, 2), (4, -1), (-3, 2), (-1, -5), (2, \tfrac{3}{2})\}$$

3. If $f(x) = 4 - 3x$, find

a. $f(-1)$
b. $f(0)$
c. $f(2)$
d. $f(\frac{1}{2})$

4. If $f(x) = \dfrac{x}{3} - 5$, find

a. $f(0)$
b. $f(-3)$
c. $f(3)$
d. $f(\frac{1}{2})$

5. Graph $x + y = 3$.

6. Graph $2x - y = 4$.

7. Graph the equation $3x + 4y = 9$ by locating the x-intercept and y-intercept.

8. What is the slope of the line determined by the points $(4, -3)$ and $(2, \frac{1}{2})$?

9. Graph the line passing through the point $(1, -5)$ with the slope $m = \frac{5}{2}$.

10. For $3x + 5y = -15$, determine the slope, m, and the y-intercept, b. Then graph the line.

11. For $2x + 4y = 7$, determine the slope, m, and the y-intercept, b. Then graph the line.

12. Graph the line passing through the point $(-2, 7)$ with no slope

In Exercises 13–19, write answers in slope-intercept form or in standard form.

13. Write an equation for the line passing through $(-3, 2)$ with the slope $m = \frac{1}{4}$.

14. Write an equation for the vertical line passing through $(4, -6)$.

15. Write an equation for the line passing through the points $(3, 1)$ and $(-2, 6)$.

16. Write an equation for the line passing through the points $(2, 5)$ and $(\frac{1}{2}, 5)$.

17. Write an equation for the line parallel to the x-axis through the point $(3, -2)$.

18. Write an equation for the line parallel to $3x - 2y = 5$ having the y-intercept $b = 2$.

19. Write an equation for the line parallel to the line $x - 3y = 1$ and passing through the point $(1, 2)$.

20. To rent a carpet cleaning machine, it costs a flat rate of \$9.00 plus \$4.00 per hour. Write a linear equation representing the cost, C, in terms of the hours used, t.

USING THE COMPUTER

```
10   REM
20   REM This program will find a linear equation given two
30   REM points on the line.
40   REM
50   '***************************************************
60   '*  X1,Y1,X2,Y2 : The coordinates of points          *
70   '*  DELTA.X     : The run                            *
80   '*  DELTA.Y     : The rise                           *
90   '*  SLOPE       : rise over run                      *
100  '*  CONSTANT    : The constant of equation           *
110  '***************************************************
120 PRINT
    "*Remark : First coordinate must be followed by a comma."
130 INPUT "Type in first point (a,b) example) 3,4 :";X1,Y1
140 INPUT "Type in second point (c,d)              :";X2,Y2
150 LET DELTA.X = X2 - X1                    'Compute the run
160 LET DELTA.Y = Y2 - Y1                    'Compute the rise
170 IF DELTA.X = 0 THEN PRINT "Equation is : x =";X1
180 IF DELTA.X = 0 GOTO 220
190 LET SLOPE = DELTA.Y /DELTA.X             'Compute the slope
200 LET CONSTANT = Y1 - SLOPE * X1           'Compute the constant
210 PRINT "The linear equation is : y = ";SLOPE;"x + ";CONSTANT
220 END                                      'End program
```

```
*Remark : First coordinate must be followed by a comma.
Type in first point (a,b) ex)3,4  : 1 , 2.5
Type in second point (c,d) ex)5,6 : 2 , 3
The linear equation is : y =  .5 x +  2
```

```
*Remark : First coordinate must be followed by a comma.
Type in first point (a,b) ex)3,4  : 1 , 9
Type in second point (c,d) ex)5,6 : 3 , 7
The linear equation is : y = -1 x +  10
```

```
*Remark : First coordinate must be followed by a comma.
Type in first point (a,b) ex)3,4  : 2 , 3
Type in second point (c,d) ex)5,6 : 2 , 5
Equation is : x = 2
```

```
*Remark : First coordinate must be followed by a comma.
Type in first point (a,b) ex)3,4  : 3 , 2
Type in second point (c,d) ex)5,6 : 8 , 2
The linear equation is : y =  0 x +  2
```

A scientist worthy of the name, above all a mathematician, experiences in his work the same impression as an artist, his pleasure is as great and of the same nature.

JULES HENRI POINCARÉ 1854–1912

DID YOU KNOW?

Most people recognize that mathematics has many applications in science and the social sciences. But what about the arts? Surely mathematics has had no impact on the fine arts—or has it?

As previously mentioned, the Pythagoreans investigated the relationship between rational numbers and harmony. For a long time after the Greek period, music was considered to be applied mathematics. The basic law of musical harmony is a mathematical equation stating that a musical tone is composed of a fundamental note and overtones whose frequencies are all integral multiples of the fundamental tone. Modern investigations into the physics of sound use mathematics heavily.

In painting, when styles demanded more realistic presentations of subjects, the science of projective geometry was developed by artist-mathematicians to add perspective to their paintings. Albrecht Dürer, the German artist, traveled to Italy in 1506 to learn the secret art of perspective which was being used by Italian painters in the early Renaissance period. In modern art, the Dutch artist Mauritis Escher has done a great deal of work investigating symmetry patterns in art. Symmetry patterns have algebraic interpretations and have also been investigated by mathematicians. Surrealism may have mathematical qualities because of the time-space transformations that distort size, shape, volume, and time. Optical art and computer-generated art also have a mathematical basis.

Stories, novels, and poems have been written using mathematical ideas for inspiration. William Wordsworth (1770–1850) wrote a poem entitled "Geometry." Science fiction writers often use mathematical ideas such as five-dimensional worlds, computers gone wild, probability, and geometry as themes for their short stories or novels. Arthur C. Clark, the author of *2001: A Space Odyssey*, has written several short stories involving computer technology. You will find that mathematics is a discipline with applications in all fields, arts as well as science.

9 Systems of Linear Equations

9.1 SOLUTIONS BY GRAPHING

Visualize two straight lines on the same graph. What two lines did you see? No matter what specific lines you envisioned, there are only three basic positions for the lines **relative to each other.** What do you think they are? The following discussion analyzes these three basic possibilities.

Do the lines $y = -x + 4$ and $y = 2x + 1$ intersect (cross each other)? If so, where do they **intersect?** If not, why not? We can answer these questions by graphing both equations on the same graph, as shown in Figure 9.1.

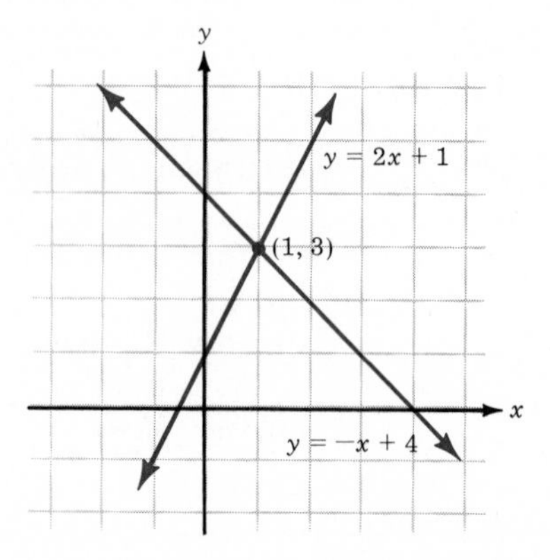

FIGURE 9.1

The lines appear to intersect at the point $(1, 3)$, or where $x = 1$ and $y = 3$. We can check this intersection by substituting $x = 1$ and $y = 3$ into **both** equations:

$$y = 2x + 1 \qquad\qquad y = -x + 4$$

$$3 \stackrel{?}{=} 2 \cdot 1 + 1 \qquad\qquad 3 \stackrel{?}{=} -1 + 4$$

$$3 = 3 \qquad\qquad 3 = 3$$

The lines do intersect at $(1, 3)$.

Do the lines $y = -x + 4$ and $x + y = 2$ intersect? If so, where do they intersect? If not, why not? The answers are in the graphs in Figure 9.2.

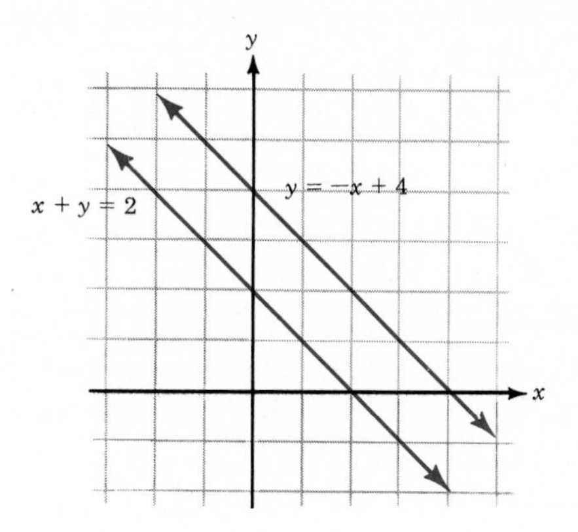

FIGURE 9.2

The lines do not intersect because they are **parallel.** In this case, we could have anticipated the result by writing both equations in the slope-intercept form and noted that both lines have the same slope, -1, and different intercepts.

$$y = -x + 2 \qquad m = -1$$

$$y = -x + 4 \qquad m = -1$$

Do the lines $y = -x + 4$ and $2y + 2x = 8$ intersect? If so, where do they intersect? If not, why not? The graphs of both lines are shown in Figure 9.3. Why is there just one line?

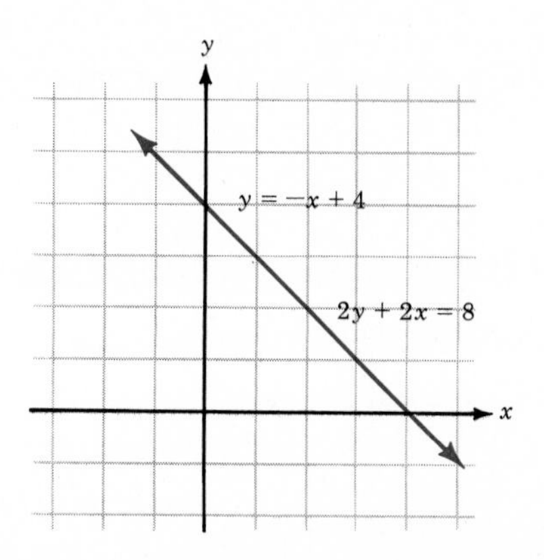

FIGURE 9.3

The reason there is just one line is that both equations represent the same line. The lines not only intersect; they are the same line. That is, they **coincide.** Any point that satisfies one equation will also satisfy the other. We can easily see this by putting both equations in the slope-intercept form:

$$y = -x + 4 \qquad\qquad \begin{aligned} 2y + 2x &= 8 \\ 2y &= -2x + 8 \\ \frac{2y}{2} &= \frac{-2x + 8}{2} \\ y &= -x + 4 \end{aligned}$$

Both equations are identical when written in the same form.

These three examples constitute all three possible situations involving the graphs of two linear equations. When two linear equations are considered together, they are called a **system of linear equations,** or a **set of simultaneous equations.** The term **simultaneous** is frequently used to emphasize the idea that the solution of a system is the point that satisfies both equations at the same time, or simultaneously.

If a system has a unique solution (one point of intersection), the system is **consistent.** If a system has no solution (the lines are parallel with no point of intersection), the system is **inconsistent.** If a system has an infinite number of solutions (the lines coincide), the system is **dependent.**

The following table summarizes the basic ideas and terminology.

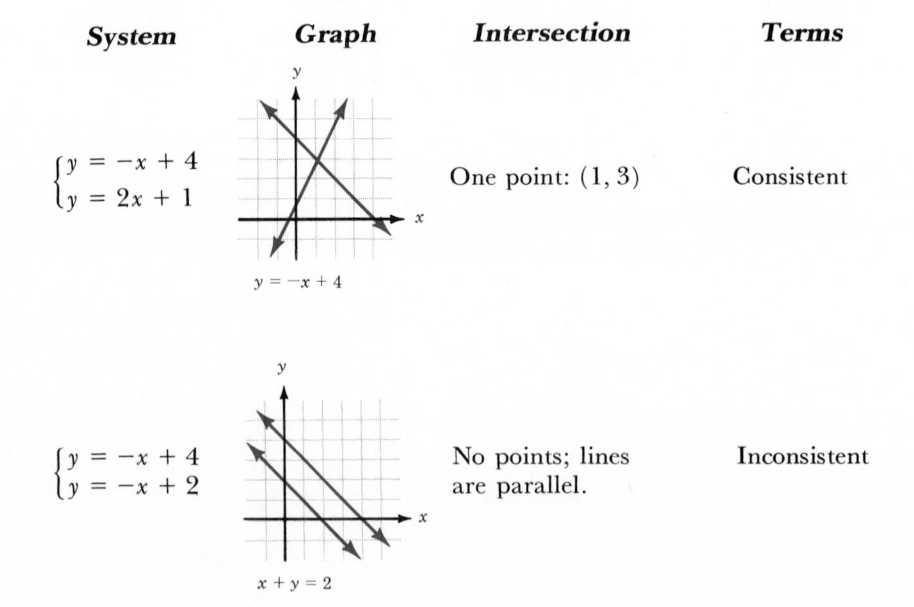

System	***Graph***	***Intersection***	***Terms***
$\begin{cases} y = -x + 4 \\ y = 2x + 1 \end{cases}$		One point: (1, 3)	Consistent
$\begin{cases} y = -x + 4 \\ y = -x + 2 \end{cases}$		No points; lines are parallel.	Inconsistent

System	***Graph***	***Intersection***	***Terms***
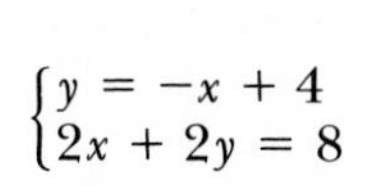 $\begin{cases} y = -x + 4 \\ 2x + 2y = 8 \end{cases}$	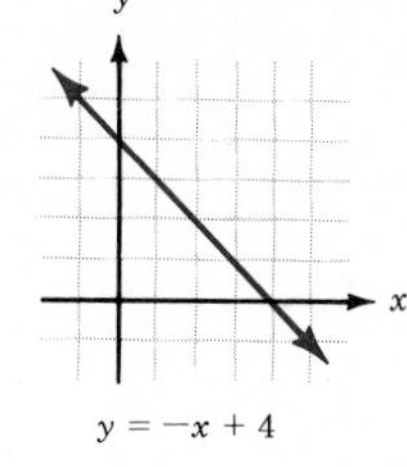	Infinite number of points; lines are the same (they coincide).	Dependent

EXAMPLES

Determine graphically whether the following systems are (a) consistent, (b) inconsistent, or (c) dependent. If the system is consistent, find (or estimate) the point of intersection.

1. $\begin{cases} x - y = 0 \\ 2x + y = 3 \end{cases}$

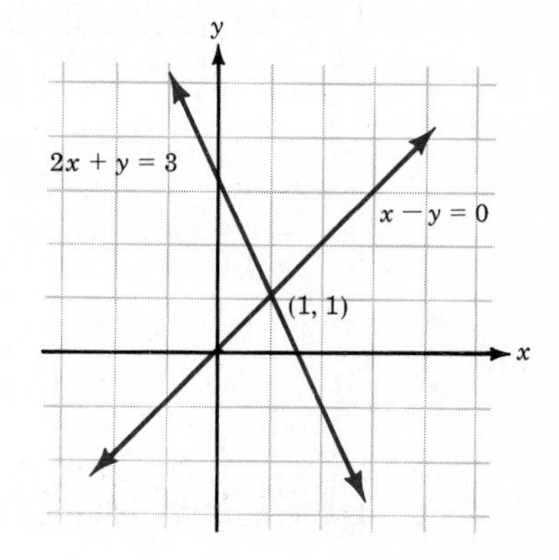

Solution: The system is consistent. The point of intersection is (1, 1).

Check: $x = 1,\ y = 1$

$1 - 1 = 0$

$2 \cdot 1 + 1 = 3$

2. $\begin{cases} x - 2y = 1 \\ x + 3y = 0 \end{cases}$

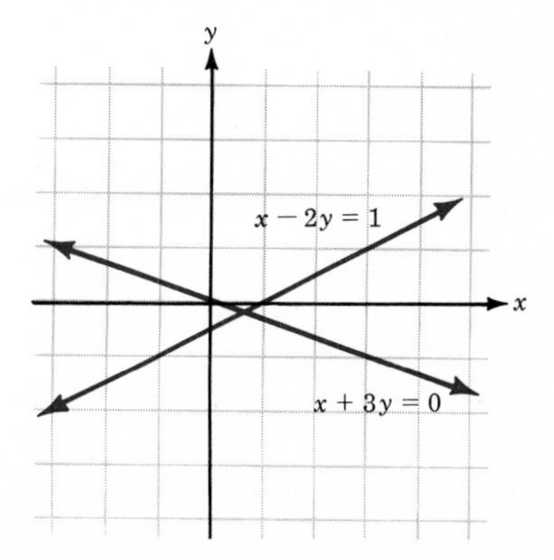

Solution: The system is consistent, but the point of intersection can only be estimated. This example points out the main weakness in solving a system graphically. In such cases, any reasonable estimate will be acceptable. For example, if you estimated $(\frac{1}{2}, -\frac{1}{4})$, or $(\frac{3}{4}, -\frac{1}{3})$, or some such point, your answer is acceptable. The actual point of intersection is $(\frac{3}{5}, -\frac{1}{5})$. We will be able to locate this point precisely using the techniques of the next section.

3. $\begin{cases} y = 3x \\ y - 3x = -4 \end{cases}$

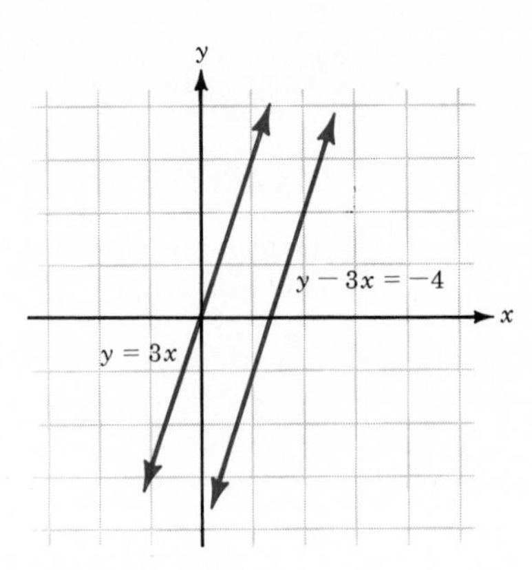

Solution: The system is inconsistent. The lines are parallel with the same slope, 3, and there are no points of intersection.

4. $\begin{cases} x + 2y = 6 \\ \quad\;\; y = -\frac{1}{2}x + 3 \end{cases}$

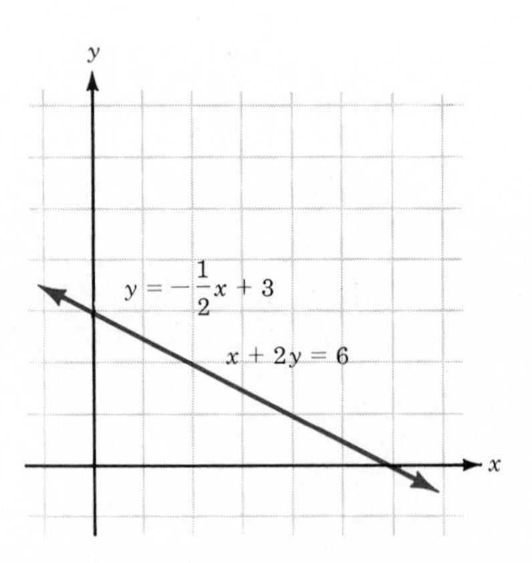

Solution: The system is dependent. All points that lie on one line also lie on the other line. For example, $(4, 1)$ is a point on the line $x + 2y = 6$ since $4 + 2(1) = 6$. The point $(4, 1)$ is also on the line $y = -\frac{1}{2}x + 3$ since $1 = -\frac{1}{2}(4) + 3$.

EXERCISES 9.1

Determine which of the given points lie on the line determined by the given equations in Exercises 1–4.

1. $2x - y = 4$
a. $(1, 1)$
b. $(2, 0)$
c. $(1, -2)$
d. $(3, 2)$

2. $x + 2y = -1$
a. $(1, -1)$
b. $(1, 0)$
c. $(2, 1)$
d. $(3, -2)$

3. $4x + y = 5$
a. $(\frac{3}{4}, 2)$
b. $(4, 0)$
c. $(1, 1)$
d. $(0, 3)$

4. $2x - 3y = 7$
a. $(1, 3)$
b. $(\frac{1}{2}, -2)$
c. $(\frac{7}{2}, 0)$
d. $(2, 1)$

Determine which of the given points lie on both of the lines determined by the given systems of equations in Exercises 5–8.

5. $\begin{cases} x - y = 6 \\ 2x + y = 0 \end{cases}$
a. $(1, -2)$
b. $(4, -2)$
c. $(2, -4)$
d. $(-1, 2)$

6. $\begin{cases} x + 3y = 5 \\ 3y = 4 - x \end{cases}$
a. $(2, 1)$
b. $(2, -2)$
c. $(-1, 2)$
d. $(4, 0)$

7. $\begin{cases} 2x + 4y - 6 = 0 \\ 3x + 6y - 9 = 0 \end{cases}$
a. $(1, 1)$
b. $(2, 0)$
c. $(0, \frac{3}{2})$
d. $(-1, 3)$

8. $\begin{cases} 5x - 2y - 5 = 0 \\ 5x = -3y \end{cases}$
a. $(1, 0)$
b. $(\frac{3}{5}, -1)$
c. $(0, 0)$
d. $(1, 4)$

Graph each of the systems in Exercises 9–35 and determine whether it is (a) consistent, (b) inconsistent, or (c) dependent. Estimate the coordinates of the intersection if the system is consistent.

9. $2x - y = 3$
$x + 3y = 5$

10. $x + y - 5 = 0$
$x - 4y = 5$

11. $3x - y = 6$
$y = 3x$

12. $x - y = 5$
$x = -3$

13. $x + 2y = 7$
$2x - y = -1$

14. $5x - 4y = 5$
$8y = 10x - 10$

15. $4x - 2y = 10$
$-6x + 3y = -15$

16. $y = 2x + 5$
$4x - 2y = 7$

17. $\frac{1}{2}x + 2y = 7$
$2x = 4 - 8y$

18. $4x + 3y + 7 = 0$
$5x - 2y + 3 = 0$

19. $2x + 3y = 4$
$4x - y = 1$

20. $7x - 2y = 1$
$y = 3$

21. $2x - 5y = 6$
$y = \frac{2}{5}x + 1$

22. $y = 4x - 3$
$x = 2y - 8$

23. $y = \frac{1}{2}x + 2$
$x - 2y + 4 = 0$

24. $2x + 3y = 5$
$3x - 2y = 1$

25. $\frac{2}{3}x + y = 2$
$x - 4y = 3$

26. $x + y = 8$
$5y = 2x + 5$

27. $y = 2x$
$2x + y = 4$

28. $x - y = 4$
$2y = 2x - 4$

29. $y = x + 1$
$y + x = -5$

30. $x + y = 4$
$2x - 3y = 3$

31. $2x + y + 1 = 0$
$3x + 4y - 1 = 0$

32. $4x + y = 6$
$2x + \frac{1}{2}y = 3$

33. $\frac{1}{2}x + \frac{1}{3}y = \frac{1}{6}$
$\frac{1}{4}x + \frac{1}{4}y = 0$

34. $\frac{1}{4}x - y = \frac{13}{4}$
$\frac{1}{3}x + \frac{1}{6}y = -\frac{1}{6}$

35. $\frac{1}{2}x - \frac{2}{3}y = -1$
$x - \frac{1}{5}y = \frac{7}{5}$

9.2 SOLUTIONS BY SUBSTITUTION

Solving a system of linear equations graphically can be time-consuming and does not always yield accurate results. When the system has one equation that can easily be solved for one of the variables, the algebraic technique of **substitution** is convenient. We simply substitute an expression for a variable from one equation into the other. For example, consider the system

$$\begin{cases} y = -2x + 5 \\ x + 2y = 1 \end{cases}$$

How would you substitute? The first equation is already solved for y. Would you put $-2x + 5$ for y in the second equation? Try this and see what happens.

$$x + 2(-2x + 5) = 1$$

We now have one equation in only one variable, x. We have reduced the problem of solving two equations in two variables to solving one equation in one variable. Solve the equation for x. Then find the corresponding y-value from **either of the two original equations.**

$$\begin{aligned} x + 2(-2x + 5) &= 1 \\ x - 4x + 10 &= 1 \\ -3x &= -9 \\ x &= 3 \end{aligned}$$

Substituting $x = 3$ into $y = -2x + 5$ gives

$$\begin{aligned} y &= -2 \cdot 3 + 5 \\ &= -6 + 5 = -1 \end{aligned}$$

The solution is $x = 3$ and $y = -1$, or the point $(3, -1)$.

Substitution is not the only algebraic technique for solving a system of linear equations. It does work in all cases but is generally used only when one equation is easily solved for one variable.

Solve the following system using the technique of substitution:

$$\begin{cases} 3x + y = 1 \\ 6x + 2y = 3 \end{cases}$$

Solving for y, we have

$$\begin{aligned} 3x + y &= 1 \\ y &= -3x + 1 \end{aligned}$$

Substituting yields

$$\begin{aligned} 6x + 2y &= 3 \\ 6x + 2(-3x + 1) &= 3 \\ 6x - 6x + 2 &= 3 \\ 2 &= 3 \end{aligned}$$

This last equation, $2 = 3$, is never true; and this tells us that the system is inconsistent. The lines are parallel and there is no point of intersection.

Solve the following system using the technique of substitution:

$$\begin{cases} x - 2y = 1 \\ 3x - 6y = 3 \end{cases}$$

Solving for x, we have

$$\begin{aligned} x - 2y &= 1, \\ x &= 2y + 1 \end{aligned}$$

Substituting yields

$$\begin{aligned} 3x - 6y &= 3 \\ 3(2y + 1) - 6y &= 3 \\ 6y + 3 - 6y &= 3 \\ 3 &= 3 \end{aligned}$$

This last equation, $3 = 3$, is always true; and this tells us that the system is dependent. The two lines are the same and there are an infinite number of points of intersection.

EXAMPLES

Solve the following systems of linear equations using the technique of substitution.

1. $\begin{cases} y = \dfrac{5}{6}x + 1 \\ 2x + 6y = 7 \end{cases}$

Solution: $y = \dfrac{5}{6}x + 1$ is already solved for y. Substituting, we find

$$2x + 6y = 7$$

$$2x + 6\left(\frac{5}{6}x + 1\right) = 7$$

$$2x + 5x + 6 = 7$$

$$7x = 1$$

$$x = \frac{1}{7}, \qquad y = \frac{5}{6} \cdot \frac{1}{7} + 1 = \frac{5}{42} + \frac{42}{42} = \frac{47}{42}$$

The solution is $\left(\dfrac{1}{7}, \dfrac{47}{42}\right)$, or $x = \dfrac{1}{7}$ and $y = \dfrac{47}{42}$.

2. $\begin{cases} x + 3y = 0 \\ -4x + 2y = 7 \end{cases}$

Solution: Solving for x, we have

$$x + 3y = 0$$

$$x = -3y$$

Substituting yields

$$-4x + 2y = 7$$

$$-4(-3y) + 2y = 7$$

$$12y + 2y = 7$$

$$14y = 7$$

$$y = \frac{1}{2}, \qquad x = -3\left(\frac{1}{2}\right) = -\frac{3}{2}$$

The solution is $\left(-\frac{3}{2}, \frac{1}{2}\right)$, or $x = -\frac{3}{2}$ and $y = \frac{1}{2}$.

3. $\begin{cases} x = -5 \\ y = 2x + 9 \end{cases}$

Solution: $x = -5$ is already solved for x. The equation represents a vertical line. Substituting, we have

$$y = 2x + 9$$

$$y = 2(-5) + 9 = -10 + 9 = -1$$

The solution is $x = -5$ and $y = -1$, or $(-5, -1)$.

4. $\begin{cases} x + y = 5 \\ 0.2x + 0.3y = 0.9 \end{cases}$

Solution: Solving the first equation for y, we have

$$x + y = 5$$

$$y = 5 - x$$

Substituting yields

$$0.2x + 0.3y = 0.9$$

$$0.2x + 0.3(5 - x) = 0.9$$

$$0.2x + 1.5 - 0.3x = 0.9$$

$$-0.1x = -0.6$$

$$\frac{-0.1x}{-0.1} = \frac{-0.6}{-0.1}$$

$$x = 6, \qquad y = 5 - x = 5 - 6 = -1$$

You could multiply each term by 10 to clear decimals at any step. The answer will be the same.

The solution is $x = 6$ and $y = -1$, or $(6, -1)$.

EXERCISES 9.2

Solve the following systems using the technique of substitution. If the system is inconsistent or dependent, say so in your answer.

1. $x + y = 6$, $y = 2x$

2. $5x + 2y = 21$, $x = y$

3. $x - 7 = 3y$, $y = 2x - 4$

4. $y = 3x + 4$, $2y = 3x + 5$

5. $x = 3y$, $3y - 2x = 6$

6. $4x = y$, $4x - y = 7$

7. $x - 5y + 1 = 0$
$x = 7 - 3y$

8. $2x + 5y = 15$
$x = y - 3$

9. $7x + y = 9$
$y = 4 - 7x$

10. $3y + 5x = 5$
$y = 3 - 2x$

11. $3x - y = 7$
$x + y = 5$

12. $4x - 2y = 5$
$y = 2x + 3$

13. $6x + 3y = 9$
$y = 3 - 2x$

14. $x - y = 5$
$2x + 3y = 0$

15. $4x = 8$
$3x + y = 8$

16. $3x - 9y = 6$
$y = \frac{1}{3}x - \frac{2}{3}$

17. $x + y = 8$
$3x + 2y = 8$

18. $y = 2x - 5$
$2x + y = -3$

19. $2x + 3y = 5$
$x - 6y = 0$

20. $2y = 5$
$3x - 4y = -4$

21. $x + 5y = 1$
$x - 3y = 5$

22. $3x + 8y = -2$
$x + 2y = -1$

23. $y = 7 - 3x$
$6x + 2y = 14$

24. $4x - 5y = 9$
$3x + y = 2$

25. $5x + 2y = -10$
$7x = 4 - y$

26. $x - 2y = -4$
$3x + y = -5$

27. $x + 4y = 3$
$3x - 4y = -23$

28. $3x - y = -1$
$7x - 4y = 0$

29. $4x + 2y = 9$
$x + \frac{1}{2}y = 2$

30. $x + 5y = -1$
$2x + 7y = 1$

31. $x + 3y = 5$
$3x + 2y = 7$

32. $3x - 2y = 4$
$x = \frac{2}{3}y + 2$

33. $3x - 4y - 39 = 0$
$2x - y - 13 = 0$

34. $\frac{x}{3} + \frac{y}{5} = 1$
$x + 6y = 12$

35. $\frac{x}{5} + \frac{y}{4} - 3 = 0$
$\frac{x}{10} - \frac{y}{2} + 1 = 0$

36. $6x - y = 15$
$0.2x + 0.5y = 2.1$

37. $x + 2y = 3$
$0.4x + y = 0.6$

38. $0.2x - 0.1y = 0$
$y = x + 10$

39. $0.1x - 0.2y = 1.4$
$3x + y = 14$

40. $0.5x + y = 4$
$0.75x + 0.5y = 4$

9.3 SOLUTIONS BY ADDITION

The system $\begin{cases} x + y = 4 \\ x - y = 6 \end{cases}$ can be solved using the algebraic technique of substitution discussed in Section 9.2. The technique of addition is also convenient and, in this case, probably easier than substitution. Put both equations in the standard form $Ax + By = C$ and set up one under the other so that like terms are aligned. Then **add like terms.**

$$\begin{aligned} x + y &= 4 \\ x - y &= 6 \\ \hline 2x \quad &= 10 \end{aligned}$$

Since $+y$ and $-y$ have opposite coefficients, the y-terms are eliminated, and the resulting equation, $2x = 10$, has only one variable. Just as with the technique of substitution, the solution of the system is reduced to solving one equation in one variable.

$$\begin{array}{rcl} x + y &=& 4 \\ x - y &=& 6 \\ \hline 2x \quad &=& 10 \\ x &=& 5 \end{array} \qquad \begin{array}{rcl} x + y &=& 4 \\ 5 + y &=& 4 \\ y &=& -1 \end{array}$$

The solution is $x = 5$ and $y = -1$, or $(5, -1)$.

If the two coefficients of one variable in the system are not opposites, then we multiply each equation by some nonzero constant so that either the two x-coefficients are opposites or the two y-coefficients are opposites. For example, consider the system

$$\begin{cases} 4x - 3y = 1 \\ 3x - 2y = 4 \end{cases}$$

Multiplying each term of the first equation by 2 and each term of the second equation by -3 will result in the y-coefficients being opposites. Or multiplying the terms of the first equation by 3 and the second equation by -4 will give opposite coefficients for the x-terms. Both cases yield the same solution and are illustrated below. The number used to multiply the terms of each equation is in brackets.

Method 1: Eliminate y-terms.

$$\begin{cases} [2] & 4x - 3y = 1 \\ [-3] & 3x - 2y = 4 \end{cases} \qquad \begin{array}{rcl} 8x - 6y &=& 2 \\ -9x + 6y &=& -12 \\ \hline -x \quad &=& -10 \\ x &=& 10 \end{array}$$

$$\begin{array}{rcl} 4x - 3y &=& 1 \\ 4 \cdot 10 - 3y &=& 1 \\ 40 - 3y &=& 1 \\ -3y &=& -39 \\ y &=& 13 \end{array}$$

The solution is $x = 10$ and $y = 13$, or $(10, 13)$.

***Method 2:* Eliminate x-terms.**

$$\begin{cases} [3] & 4x - 3y = 1 \\ [-4] & 3x - 2y = 4 \end{cases} \qquad \begin{aligned} 12x - 9y &= 3 \\ -12x + 8y &= -16 \\ \hline -y &= -13 \\ y &= 13 \end{aligned}$$

$$\begin{aligned} 4x - 3y &= 1 \\ 4x - 3 \cdot 13 &= 1 \\ 4x - 39 &= 1 \\ 4x &= 40 \\ x &= 10 \end{aligned}$$

The solution is $x = 10$ and $y = 13$, or $(10, 13)$.

As illustrated in Method 1 and Method 2, eliminating either the y-terms or the x-terms yields the same results. What if both the x- and y-terms are eliminated? In this situation, the system is either inconsistent or dependent. Just as with the technique of substitution, the resulting equation involving only constants tells which case is under consideration.

For the system

$$\begin{cases} 2x - y = 6 \\ 4x - 2y = 1 \end{cases}$$

multiplying the terms in the first equation by -2 and then adding gives

$$\begin{cases} [-2] & 2x - y = 6 \\ & 4x - 2y = 1 \end{cases} \qquad \begin{aligned} -4x + 2y &= -12 \\ 4x - 2y &= 1 \\ \hline 0 &= -11 \end{aligned}$$

Since the equation $0 = -11$ is not true, the system is inconsistent. There is no solution.

For the system

$$\begin{cases} 2x - 2y = 1 \\ 3x - 3y = \frac{3}{2} \end{cases}$$

multiplying the terms in the first equation by 3 and the terms in the second equation by -2 and then adding gives

$$\begin{cases} [3] & 2x - 2y = 1 \\ [-2] & 3x - 3y = \frac{3}{2} \end{cases} \qquad \begin{aligned} 6x - 6y &= 3 \\ -6x + 6y &= -3 \\ \hline 0 &= 0 \end{aligned}$$

Since the equation $0 = 0$ is always true, the system is dependent.

In solving a system by addition, find the constant for multiplying the terms of each equation by trying to get coefficients of like terms to be opposites. There are many possible choices. One approach is to find the least common multiple of the two coefficients already there, and then multiply so that one coefficient will be the LCM and the other coefficient, its opposite.

Thus, for the system

$$\begin{cases} 4x - 3y = 1 \\ 3x - 2y = 4 \end{cases}$$

in Method 1 the coefficients for the y-terms ended up being -6 and $+6$. The LCM for 3 and 2 is 6. In Method 2, the coefficients for the x-terms ended up being $+12$ and -12. The LCM for 4 and 3 is 12.

EXAMPLES

Solve the following systems of equations using the technique of addition.

1. $\begin{cases} x = 4 - 3y \\ y = 2x - 1 \end{cases}$

Solution: Rearrange the equations in standard form. $\begin{cases} x + 3y = 4 \\ 2x - y = 1 \end{cases}$

$$\begin{cases} & x + 3y = 4 \\ [3] & 2x - y = 1 \end{cases} \qquad \begin{aligned} x + 3y &= 4 \\ 6x - 3y &= 3 \\ \hline 7x \qquad &= 7 \\ x &= 1 \end{aligned} \qquad \begin{aligned} x + 3y &= 4 \\ 1 + 3y &= 4 \\ 3y &= 3 \\ y &= 1 \end{aligned}$$

The solution is $x = 1$ and $y = 1$, or $(1, 1)$.
(In this example, multiplying the terms of the first equation by -2 and eliminating the x-terms would have given the same results.)

2. $\begin{cases} 5x + 3y = -3 \\ 2x - 7x = 7 \end{cases}$

Solution:

$$\begin{cases} [-2] & 5x + 3y = -3 \\ [5] & 2x - 7y = 7 \end{cases} \qquad \begin{aligned} -10x - 6y &= 6 \\ 10x - 35y &= 35 \\ \hline -41y &= 41 \\ y &= -1 \end{aligned}$$

$$\begin{aligned} 5x + 3y &= -3 \\ 5x + 3(-1) &= -3 \\ 5x - 3 &= -3 \\ 5x &= 0 \\ x &= 0 \end{aligned}$$

The solution is $x = 0$ and $y = -1$, or $(0, -1)$.

3. $\begin{cases} x + y = 5 \\ x + y = 6 \end{cases}$

Solution:

$$\begin{aligned} & x + y = 5 \\ [-1] \quad & x + y = 6 \end{aligned} \qquad \begin{aligned} x + y &= 5 \\ -x - y &= -6 \\ \hline 0 &= -1 \end{aligned}$$

There is no solution. The system is inconsistent.

4. $\begin{cases} y = -2 + 4x \\ 8x - 2y = 4 \end{cases}$

Solution: Rearranging gives $\begin{cases} -4x + y = -2 \\ 8x - 2y = 4 \end{cases}$

$$\begin{cases} & -4x + y = -2 \\ [\tfrac{1}{2}] & 8x - 2y = 4 \end{cases} \qquad \begin{aligned} -4x + y &= -2 \\ 4x - y &= 2 \\ \hline 0 &= 0 \end{aligned}$$

The system is dependent. The solution is the set of all points that satisfy the equation $y = -2 + 4x$.

5. $\begin{cases} x + 0.4y = 3.08 \\ 0.1x - y = 0.1 \end{cases}$

Solution:

$$\begin{cases} x + 0.4y = 3.08 \\ [-10]\,0.1x - y = 0.1 \end{cases}$$

$$\begin{aligned} x + 0.4y &= 3.08 \\ -1.0x + 10.0y &= -1.0 \\ \hline 10.4y &= 2.08 \\ y &= 0.2 \end{aligned}$$

$$\begin{aligned} x + 0.4y &= 3.08 \\ x + 0.4(0.2) &= 3.08 \\ x + 0.08 &= 3.08 \\ x &= 3 \end{aligned}$$

The solution is $x = 3$ and $y = 0.2$, or $(3, 0.2)$.

6. Using the formula $y = mx + b$, find the equation of the line determined by the two points $(3, 5)$ and $(-6, 2)$.

Solution: Write two equations in m and b by substituting the coordinates of the points for x and y.

$$\begin{aligned} 5 &= 3m + b \\ 2 &= -6m + b \end{aligned}$$

$$\begin{aligned} 5 &= 3m + b \\ -2 &= 6m - b \\ \hline 3 &= 9m \\ \tfrac{1}{3} &= m \end{aligned}$$

$$\begin{aligned} 5 &= 3m + b \\ 5 &= 3 \cdot \tfrac{1}{3} + b \\ 5 &= 1 + b \\ 4 &= b \end{aligned}$$

The equation is $y = \frac{1}{3}x + 4$.

EXERCISES 9.3

Solve the systems in Exercises 1–40 using the technique of addition. If the system is inconsistent or dependent, say so in your answer.

1. $2x - y = 7$
$x + y = 2$

2. $x + 3y = 9$
$x - 7y = -1$

3. $3x + 2y = 0$
$5x - 2y = 8$

4. $4x - y = 7$
$4x + y = -3$

5. $2x + 2y = 5$
$x + y = 3$

6. $y = 2x + 14$
$x = 14 - 3y$

7. $x = 11 + 2y$
$2x - 3y = 17$

8. $6x - 3y = 6$
$y = 2x - 2$

9. $x - 2y = 4$
$y = \frac{1}{2}x - 2$

10. $x = 3y + 4$
$y = 6 - 2x$

11. $8x - y = 29$
$2x + y = 11$

12. $7x - y = 16$
$2y = 2 - 3x$

13. $3x + y = -10$
$2y - 1 = x$

14. $3x + 3y = 18$
$4x + 4y = 32$

15. $3x + 2y = 4$
$x + 5y = -3$

16. $x + 2y = 0$
$2x = 4y$

17. $2x + y = 4$
$5x + 3y = 12$

18. $y = 4x - 2$
$8x + 5y + 3 = 0$

19. $\frac{1}{2}x + y = -4$
$3x - 4y = 6$

20. $x + y = 1$
$x - \frac{1}{3}y = \frac{11}{3}$

21. $\frac{1}{4}x + y = 2$
$x = 8 - 4y$

22. $4x + 3y = 2$
$3x + 2y = 3$

23. $5x - 2y = 17$
$2x - 3y = 9$

24. $\frac{1}{2}x + 2y = 9$
$2x - 3y = 14$

25. $5x + 2y = -9$
$-3x + y = 1$

26. $3x + 2y = 14$
$7x + 3y = 26$

27. $4x + 3y = 28$
$5x + 2y = 35$

28. $2x + 7y = 2$
$5x + 3y = -24$

29. $7x - 6y = -1$
$5x + 2y = 37$

30. $9x + 2y = -42$
$5x - 6y = -2$

31. $7x + 4y = 7$
$6x + 7y = 31$

32. $6x - 5y = -40$
$8x - 7y = -54$

33. $\frac{3}{4}x - \frac{1}{2}y = -2$
$\frac{1}{3}x - \frac{7}{6}y = 1$

34. $\frac{4}{5}x - y = 2$
$\frac{1}{4}x + \frac{1}{3}y = \frac{23}{12}$

35. $x + y = 12$
$0.05x + 0.25y = 1.6$

36. $x + 0.5y = 8$
$0.1x + 0.01y = 0.64$

37. $0.5x - 0.3y = 7$
$0.3x - 0.4y = 2$

38. $0.6x + 0.5y = 5.9$
$0.8x + 0.4y = 6$

39. $2.5x + 1.8y = 7$
$3.5x - 2.7y = 4$

40. $2.1x + 3.4y = 0.8$
$0.9x - 1.7y = 3.5$

In Exercises 41–46, write an equation for the line determined by the two given points, using the formula $y = mx + b$ to set up a system of equations with m and b as the unknowns.

41. $(2, 3), (1, -2)$

42. $(4, 7), (-3, 2)$

43. $(-4, 1), (5, 2)$

44. $(0, 6), (-3, -3)$

45. $(3, -4), (7, 7)$

46. $(1, -3), (5, -3)$

9.4 APPLICATIONS

We have solved word problems in several previous sections, but those problems were all solved with one equation in one variable. Many of these same problems can be solved using two equations in two variables. The following example is Example 3 from Section 7.3, repeated here exactly as solved in that section.

EXAMPLE: Geometry

1. A rectangle with a perimeter of 140 meters has a length that is 20 meters less than twice the width. Find the dimensions of the rectangle.

Solution: Draw a diagram and use the formula $P = 2l + 2w$.

Let $w = \text{width}$

$$2w - 20 = \text{length}$$

$$2(w) + 2(2w - 20) = 140$$

$$2w + 4w - 40 = 140$$

$$6w = 180$$

$$w = 30 \text{ meters}$$

$$2w - 20 = 40 \text{ meters}$$

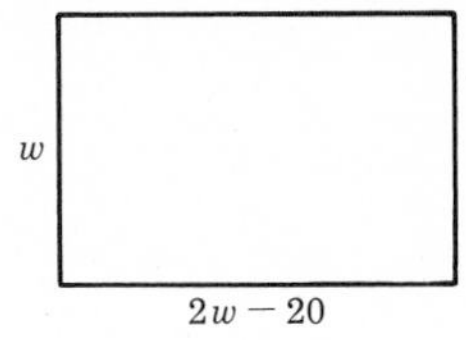

w

$2w - 20$

The rectangle is 30 meters wide and 40 meters long.

How would you solve this same problem using two equations in two variables? If you are going to use two variables, each variable must represent a different quantity. What are the two quantities in this case?

Let $x = \text{width}$

$y = \text{length}$

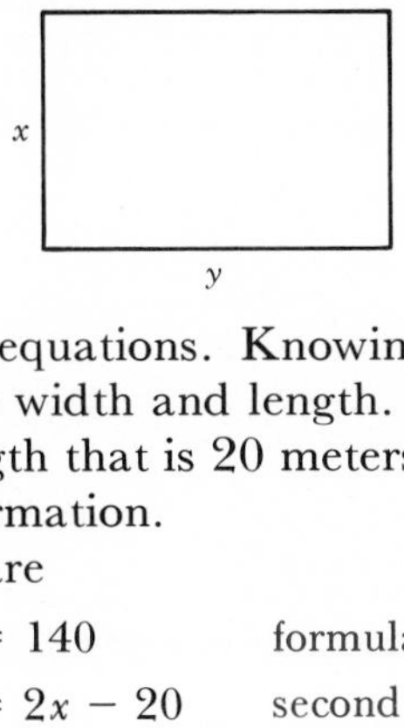

x

y

Now we need **two** equations. Knowing the perimeter gives one relationship between the width and length. What is the other relationship? The phrase "a length that is 20 meters less than twice the width" gives the necessary information.

The two equations are

$$\begin{cases} 2x + 2y = 140 & \text{formula } 2l + 2w = P \\ y = 2x - 20 & \text{second relationship} \end{cases}$$

Solving by substitution gives

$$2x + 2(2x - 20) = 140$$

the same equation obtained using one variable, w

$$2x + 4x - 40 = 140$$

$$6x = 180$$

$$x = 30 \text{ meters}$$

$$y = 2(30) - 20 = 40 \text{ meters}$$

Obviously, we have the same solution as before. The basic difference is that the thinking is easier. The technique can also be expanded to handle more difficult problems later on.

Consider the following solution to Example 2 of Section 7.4.

EXAMPLE: Interest

2. A woman has $7000. She decides to separate her funds into two investments. One yields an interest of 6% and the other, 10%. If she wants an annual income from the investments to be $580, how should she split the money?

Solution: Let x = amount invested at 6%
y = amount invested at 10%

$$x + y = 7000 \quad \text{Total invested is \$7000.}$$
$$0.06x + 0.10y = 580 \quad \text{Total interest is \$580.}$$

$$\begin{array}{ll} [-6] & x + y = 7000 \\ [100] & 0.06x + 0.10y = 580 \end{array}$$

$$\begin{aligned} -6x - 6y &= -42{,}000 \\ 6x + 10y &= 58{,}000 \\ \hline 4y &= 16{,}000 \\ y &= 4{,}000 \end{aligned}$$

$$\begin{aligned} x + y &= 7000 \\ x + 4000 &= 7000 \\ x &= 3000 \end{aligned}$$

She should invest $3000 at 6% and $4000 at 10%.

EXAMPLES: Mixture

3. How many ounces each of a 10% salt solution and a 15% salt solution must be used to produce 50 ounces of a 12% salt solution?

Solution: Let x = amount 10% solution
y = amount of 15% solution

	amount of solution	· percent of salt	= amount of salt
10% solution	x	0.10	$0.10x$
15% solution	y	0.15	$0.15y$
12% solution	50	0.12	0.12(50)

$$x + y = 50 \qquad\qquad x + y = 50$$
$$0.10x + 0.15y = 0.12(50) \qquad\qquad 10x + 15y = 12(50)$$

$$\begin{cases} [-10] & x + y = 50 \\ & 10x + 15y = 600 \end{cases} \qquad \begin{aligned} -10x - 10y &= -500 \\ 10x + 15y &= 600 \\ \hline 5y &= 100 \\ y &= 20 \end{aligned} \qquad \begin{aligned} x + 20 &= 50 \\ x &= 30 \end{aligned}$$

Use 30 ounces of the 10% solution and 20 ounces of the 15% solution.

4. One number is three less than twice the other, and their product is 83 more than their sum. Find the numbers.

Solution: Let x = one number
y = second number

$$\begin{cases} y = 2x - 3 \\ xy = x + y + 83 \end{cases}$$

Solve by substitution. The elimination method does not work here because a product is involved. The resulting equation is quadratic.

$$x(2x - 3) = x + (2x - 3) + 83$$
$$2x^2 - 3x = 3x + 80$$
$$2x^2 - 6x - 80 = 0$$
$$x^2 - 3x - 40 = 0$$
$$(x - 8)(x + 5) = 0$$

$$\begin{aligned} x - 8 &= 0 \\ x &= 8 \\ y &= 2 \cdot 8 - 3 \\ y &= 13 \end{aligned} \qquad \text{or} \qquad \begin{aligned} x + 5 &= 0 \\ x &= -5 \\ y &= 2(-5) - 3 \\ y &= -13 \end{aligned}$$

There are two sets of solutions: 8 and 13, or -5 and -13.

5. A small plane flew 300 miles in 2 hours. Then on the return trip, flying against the wind, it traveled only 200 miles in 2 hours. What were the wind velocity and the speed of the plane?

Solution: Let s = speed of plane
v = wind velocity

rate · time = distance

	rate	time	distance
with the wind	$s + v$	2	$2(s + v)$
against the wind	$s - v$	2	$2(s - v)$

$$\begin{cases} 2(s + v) = 300 \\ 2(s - v) = 200 \end{cases}$$

$$\begin{aligned} 2s + 2v &= 300 \\ 2s - 2v &= 200 \\ \hline 4s \quad\quad &= 500 \\ s &= 125 \end{aligned}$$

$$\begin{aligned} 2(125 + v) &= 300 \\ 125 + v &= 150 \\ v &= 25 \end{aligned}$$

The speed of the plane was 125 mph, and the wind velocity was 25 mph.

EXERCISES 9.4

Solve each of the problems in Exercises 1–25 by setting up a system of two equations in two unknowns.

1. The sum of two numbers is 56. Their difference is 10. Find the numbers.

2. The sum of two numbers is 40. The sum of twice the larger and 4 times the smaller is 108. Find the numbers.

3. An 18-foot board is cut into two parts so that one part is 2 feet longer than the other. How long is each part?

4. A couple recently bought a lot and had a house built. The cost of constructing the house was $25,000 more than the cost of the lot. The total cost for both was $80,000. What was the cost of each?

5. The perimeter of a rectangle is 242 feet. The length is 1 foot more than twice the width. Find the length and width of the rectangle.

6. Two sides of a triangle are the same length. The third side is 2 feet less than the sum of the other two. If the perimeter is 22 feet, find the length of each side.

7. H.V.C. High School played 32 games during the season. They won 5 games more than twice the number of games they lost. What was their won-lost record?

8. Ken, in his motor boat, makes the 4-mile trip downstream in 20 minutes ($\frac{1}{3}$ hr). The return trip takes 30 minutes ($\frac{1}{2}$ hr). Find the rate of the boat in still water and the rate of the current.

9. Mr. McKelvey finds that flying with the wind he can travel 1188 miles in 6 hours. However, when flying against the wind, he travels only $\frac{2}{3}$ of the distance in the same amount of time. Find the speed of the plane in still air and the wind speed.

10. Carmen invested $9000, part in a 6% passbook account and the rest in a 10% certificate account. If her annual interest was $680, how much did she invest at each rate?

11. A metallurgist has an alloy containing 20% copper and some containing 70% copper. How many pounds of each alloy must he use to make 50 pounds of an alloy containing 50% copper?

12. Frank bought 2 shirts and 1 pair of slacks for a total of $55. If he had bought 1 shirt and 2 pairs of slacks, he would have paid $68. What was the price of each shirt and each pair of slacks?

13. Four hamburgers and three orders of French fries cost $5.15. Three hamburgers and five orders of fries cost $5.10. What would one hamburger and one order of fries cost?

14. The length of a rectangle is twice the width. If the perimeter is 96 centimeters, what are the dimensions of the rectangle?

15. An airliner's average speed is $3\frac{1}{2}$ times the average speed of a private plane. Two hours after they leave the same airport at the same time, traveling in the same direction, they are 580 miles apart. What is the average speed of each plane?

16. The length of a rectangle is 1 millimeter less than twice the width. If each side is increased by 4 mm, the perimeter will be 116 mm. Find the length and the width of the original rectangle.

17. Mildred has money in two savings accounts. One account draws interest at a rate of 5% and the other, 6%. If she has $200 more in the 6% account, how much has she invested at 5% if her total interest is $61.50?

18. Mrs. Brown has $12,000 invested. Part is invested at 6% and the remainder at 8%. If the interest from the 6% investment exceeds the interest from the 8% investment by $230, how much is invested at each rate?

19. A manufacturer has received an order for 24 tons of a 60% copper alloy. His stock contains only 80% copper alloy and 50% copper alloy. How much of each will he need to fill the order?

20. A farmer has 160 meters of fencing to build a rectangular corral. If he uses the barn for one of the longer sides, what will be the dimensions of the corral if the length is 5 m less than three times the width?

21. Mr. Snyder had a sales meeting scheduled 80 miles from his home. Freeway traffic slowed him down for the first 30 miles, and he found that in order to make the meeting on time he had to increase his speed by 25 mph. If he traveled the same length of time at each rate, find the two rates. $\left(\textbf{Hint: } t = \frac{d}{r}.\right)$

22. A tobacco shop wants 50 ounces of tobacco that is 24% rare Turkish blend. How much each of a 30% Turkish blend and a 20% Turkish blend will be needed?

23. On two investments totaling $9500, Bill lost 3% on one and earned 6% on the other. If his net annual receipts were $282, how much was each investment?

24. How many liters each of a 40% acid solution and a 55% acid solution must be used to produce 60 liters of a 45% acid solution?

25. Sue's secretary bought 20 stamps, some 22¢ ones and some 40¢ ones. If she spent $5.30, how many of each kind did she buy?

Set up a system of two equations in two unknowns for Exercises 26–30, and then solve. (**Hint:** After substituting and simplifying, you will have a quadratic equation.)

26. The area of a rectangular field is 198 square meters. If it takes 58 meters of fencing to enclose the field, what are the dimensions of the field? (**Hint:** The length plus the width is 29 meters.)

27. The length of a rectangle is 4 meters more than its width. If the square of the length plus the square of the width is 400 meters, what are the dimensions of the rectangle? (**Hint:** Use the Pythagorean Theorem.)

28. Mr. Green traveled to a city 200 miles from his home to attend a meeting. Due to car trouble, his average speed returning was 10 mph less than his speed going. If the total driving time for the round trip was 9 hours, at what rate of speed did he travel to the city?

29. A man and his son can paint their cabin in 3 hours. Working alone, it would take the son 8 hours longer than it would the father. How long would it take the father to paint the cabin alone?

30. A farmer and his son can plow a field with two tractors in 4 hours. If it would take the son 6 hours longer than the father to plow the field alone, how long would it take each if they worked alone?

9.5 ADDITIONAL APPLICATIONS

Systems of equations do occur in practical situations related to supply and demand in economics and marketing as well as in "fun type" reasoning problems involving such topics as coins and people's ages. The following problems will acquaint you with some of these applications and provide more practice in solving systems of equations.

EXAMPLES

1. Mike has $1.05 worth of change in nickels and quarters. If he has twice as many nickels as quarters, how many of each type of coin does he have?

Solution: Let n = number of nickels
q = number of quarters

$n = 2q$ This equation relates the number of coins.

$0.05n + 0.25q = 1.05$ This equation relates the values of the coins

By substitution, we get

$$0.05(2q) + 0.25q = 1.05$$

$$0.10q + 0.25q = 1.05$$

$$10q + 25q = 105$$

$$35q = 105$$

$$q = 3 \text{ quarters}$$

$$n = 2q = 6 \text{ nickels}$$

2. Pat is 6 years older than her sister Sue. In 3 years, she will be twice as old as Sue. How old is each girl now?

Solution: Let P = Pat's age now
S = Sue's age now

$P - S = 6$ The difference in their age is 6.

$P + 3 = 2(S + 3)$ Each age is increased by 3.

Simplify the second equation and solve by addition.

$$P + 3 = 2(S + 3)$$

$$P + 3 = 2S + 6$$

$$P - 2S = 3$$

$$\begin{cases} P - S = 6 \\ [-1]P - 2S = 3 \end{cases} \qquad \begin{array}{r} P - S = 6 \\ -P + 2S = -3 \\ \hline S = 3 \end{array} \qquad \begin{array}{l} \\ \\ P - 3 = 6 \\ P = 9 \end{array}$$

Pat is 9 years old; Sue is 3 years old.

EXERCISES 9.5

In economics, an important application is the law of supply and demand. Let S be the price at which the manufacturer is willing to supply x units of a product and D be the price at which the consumer is willing to buy x units of the product. The equilibrium price is the price when supply equals demand.

1. Determine the equilibrium price for a product if the supply equation is $S = \frac{1}{2}x + 10$ dollars and the demand equation is $D = 40 - x$ dollars.

2. Determine the equilibrium price of a product if the supply equation is $S = 10 + \frac{1}{4}x$ dollars and the demand equation is $D = 16 - \frac{1}{2}x$ dollars.

3. A wholesaler will supply x hair dryers at a price of $S = 14 + \frac{1}{6}x$ dollars each. The local consumers will buy x hair dryers if the price is $D = 32 - \frac{1}{3}x$ dollars. Find the equilibrium price for the hair dryer.

In economics, the break-even point is the number of items sold so that the total revenue received, R, is equal to the total costs incurred, C.

4. Mr. Catz has invented a new mouse trap. He plans to sell it in a small shop where the rent is \$400 per month. The mouse traps cost him 15 cents each to produce, and he plans to sell them for 95 cents each.
a. Write equations for the cost, C, and revenue, R, if he sells x mouse traps. (**Hint:** The cost is the sum of the rent and $0.15x$. The revenue is $0.95x$.)
b. Find the break-even point.

5. A company is considering a new product for production. It estimates that setting up for production will cost \$9000. The material and labor are predicted to cost \$2.40 per item. The company expects to be able to sell each item for \$4.20.
a. Write equations for the cost, C, and the revenue, R.
b. Find the break-even point.

6. The local women's club plans to set up a hot dog stand at the county fair. The rental charge for the booth is \$520. The wholesale cost of the hot dogs is 40 cents each, and the women plan to sell them at 80 cents each.
a. Write equations for the cost, C, and the revenue, R.
b. Find the break-even point.

7. A furniture manufacturer sells lamp tables for \$80 each. His cost includes a fixed overhead of \$8000 plus a production cost of \$30 per table.
a. Write an equation for the cost, C, and the revenue, R.
b. Find the break-even point.

8. Sonja has some nickels and dimes. If she has 30 coins worth a total of \$2.00, how many of each type of coin does she have?

9. Louis has 27 coins consisting of quarters and dimes. The total value of the coins is \$5.40. How many of each type of coin does he have?

10. Jill is 8 years older than her brother Curt. Four years from now, Jill will be twice as old as Curt. How old is each at the present time?

11. Two years ago Anna was half as old as Beth. Eight years from now she will be two-thirds as old as Beth. How old are they now?

12. Seventy children and 160 adults attended the movie theater. The total receipts were \$620. One adult ticket and 2 children's tickets cost \$7. Find the price of each type of ticket.

13. Morton took some old newspapers and aluminium cans to the recycling center. Their total weight was 180 pounds. He received 1.5¢ per pound for the newspapers and 30¢ per pound for the cans. The total received was \$14.10. How many pounds of each did Morton have?

14. A dairyman wants to mix a 35% protein supplement and a standard 15% protein ratio to make 1800 pounds of a high-grade 20% protein ration. How many pounds of each should he use?

15. An insect spray can be purchased in an 80% strength solution. How many gallons must be used with water to fill a 300-gallon tank on a spray truck with a 2% spray?

16. A small manufacturer produces two kinds of radios, model X and model Y. Model X takes 4 hours to produce, and each costs \$8. Model Y takes 3 hours to produce, and each costs \$7. If the manufacturer decides to allot a total of 58 hours and \$126 each week, how many of each model will be produced?

17. A large cattle feed lot mixes two rations, I and II, to obtain the ration to be used. Ration I contains 20% crude protein and costs 45 cents per pound. Ration II contains 16% crude protein and costs 60 cents per pound. The daily ration contains a total of 300 pounds of protein and costs \$885. How many pounds of each ration is used?

18. A furniture shop refinishes chairs. Employees use two methods to refinish a chair. Method I takes 1 hour and the material costs \$3. Method II takes $1\frac{1}{2}$ hours and the material costs \$1.50. Last week, they took 36 hours and spent \$60 refinishing chairs. How many did they refinish with each method?

19. The Chopping Block sells two grades of hamburger. Grade A is 80% lean beef, while Grade B is 50% lean beef. Grade A sells for \$2.40 per pound and Grade B sells for \$2.00 per pound. How many pounds of each are sold if 20 pounds of lean beef is used and the sales amounted to \$64.00?

20. Larry sells sound systems. Brand X takes 2 hours to install and $\frac{1}{2}$ hour to adjust. Brand Y takes $1\frac{1}{2}$ hours to install and $\frac{3}{4}$ hour to adjust. If in one week, he spent 24 hours for installations and 6 hours for adjustments, how many of each brand did Larry sell?

CHAPTER 9 SUMMARY

When two linear equations are considered together, they are called a **system of linear equations** or a **set of simultaneous equations.**

If a system has a unique solution (one point of intersection), the system is **consistent.** If a system has no solution (the lines are parallel with no point of intersection), the system is **inconsistent.** If a system has an infinite number of solutions (the lines coincide), the system is **dependent.**

To solve a system by **substitution,** substitute the expression equal to one variable found in one of the equations for that variable in the other equation.

To solve a system by **addition,** write each equation in standard form. Then multiply each equation by some nonzero constant so that either the two x-coefficients or the two y-coefficients are opposites. Finally, **add like terms** so that one equation in one variable is formed.

CHAPTER 9 REVIEW

Determine which of the given points lie on the line determined by the equations in Exercises 1 and 2.

1. $4x - y = 7$

a. $(2, 1)$

b. $(3, 5)$

c. $(1, -3)$

d. $(0, 7)$

2. $2x + 5y = 6$

a. $\left(\frac{1}{2}, 1\right)$

b. $(3, 0)$

c. $\left(0, \frac{6}{5}\right)$

d. $(-2, 2)$

Which of the given points in Exercises 3 and 4 lie on both of the lines determined by the given equations?

3. $x - 2y = 7$

$2x - 3y = 5$

a. $(0, 3)$

b. $(7, 0)$

c. $(1, -1)$

d. $(-11, -9)$

4. $x - 2y = 6$

$y = \frac{1}{2}x - 3$

a. $(0, 6)$

b. $(6, 0)$

c. $(2, -2)$

d. $\left(3, -\frac{3}{2}\right)$

Graph each of the systems of equations in Exercises 5–10, and determine if the system is (a) consistent, (b) inconsistent, or (c) dependent.

5. $3x + y = 3$
$x - 3y = -9$

6. $y = 4x - 6$
$8x - 2y = -4$

7. $\frac{1}{2}x + \frac{1}{3}y = 2$
$x = -\frac{2}{3}y + 4$

8. $x + y = 4$
$2x + 7y = -2$

9. $6x + 2y = 8$
$y = -3x$

10. $x + y = 9$
$y = -\frac{1}{2}x + 1$

Solve the systems in Exercises 11–16 using substitution.

11. $x + y = -4$
$2x + 7y = 2$

12. $x = 2y$
$y = \frac{1}{2}x + 9$

13. $4x + 3y = 8$
$x + \frac{3}{4}y = 2$

14. $2x + y = 0$
$7x + 6y = -10$

15. $2x + 3y = 0$
$2x - y = 0$

16. $x = -\frac{1}{3}y$
$x + \frac{2}{3}y = -\frac{1}{3}$

Solve the systems in Exercises 17–22 using the technique of addition.

17. $2x + y = 7$
$2x - y = 1$

18. $3x - 2y = 9$
$x - 2y = 11$

19. $2x + 4y = 9$
$3x + 6y = 8$

20. $x + 5y = 10$
$y = 2 - \frac{1}{5}x$

21. $2x - 5y = 1$
$2x + 3y = -7$

22. $2x - \frac{5}{2}y = -\frac{1}{2}$
$3x - 2y = 1$

Solve the systems in Exercises 23–28.

23. $x + 3y = 7$
$5x - 2y = 1$

24. $3x - 5y = 17$
$x + 2y = 4$

25. $9x - 3y = 15$
$6x = 2y + 10$

26. $9x - 2y = -4$
$3x + 4y = 1$

27. $-3x + y = 4$
$2x - 4y = 5$

28. $\frac{1}{2}x + \frac{1}{3}y = 1$
$\frac{2}{3}x - \frac{1}{4}y = \frac{1}{12}$

Write an equation for the line determined by the two given points, using the formula $y = mx + b$ to set up a system of equations with m and b as the unknowns.

29. $(3, -1), (2, 6)$

30. $(-2, 5), (4, -3)$

Set up a system of two equations in two unknowns in Exercises 31–35. Then solve.

31. Two trains leave Kansas City at the same time. One train travels east and the other, west. The speed of the west-bound train is 5 mph greater than the speed of the east-bound train. After 6 hours, they are 510 miles apart. Find the rate of each train. (Assume that the trains travel in a straight line in directly opposite directions.)

32. To meet the government's specifications, an alloy must be 65% aluminum. How many pounds each of a 70% aluminum alloy and a 54% aluminum alloy will be needed to produce 640 pounds of the 65% aluminum alloy?

33. The Ski Club is planning to charter a bus to a ski resort. The cost will be \$900, and each member will share the cost equally. If the club had 15 more members, the cost per person would be \$10 less. How many are in the club now? (**Hint:** Substitution will result in a single quadratic equation.)

34. Pete's boat can travel 48 miles upstream in 4 hours. The return trip takes 3 hours. Find the speed of the boat in still water and the speed of the current.

35. The perimeter of a rectangle is 50 meters. The area of the rectangle is 154 square meters. Find the dimensions of the rectangle.

Optional

36. A company produces baseball gloves. Overhead costs are \$2400. Material and labor costs are \$10 per glove. The baseball gloves sell for \$22 each. Write an equation for total cost and an equation for total revenue, and find the break-even point.

37. A company manufactures two kinds of dresses, Model A and Model B. Each Model A dress takes 4 hours to produce and each costs \$18. Each Model B takes 2 hours to produce and each costs \$7. If during a week, there were 52 hours of production and costs of \$198, how many of each model were produced?

CHAPTER 9 TEST

1. Which of the following points lie on the line determined by the equation $x - 4y = 7$?

a. $(2, -2)$
b. $(-1, -2)$
c. $\left(5, -\frac{1}{2}\right)$
d. $(0, 7)$

2. Which of the following points lie on both of the lines determined by the equations $\begin{cases} 3x - 7y = 5 \\ 5x - 2y = -11 \end{cases}$

a. $(1, 8)$
b. $(4, 1)$
c. $(-3, -2)$
d. $(13, 10)$

Graph each of the systems in Exercises 3–6 and determine if the system is (a) consistent, (b) inconsistent, or (c) dependent.

3. $y = 2 - 5x$
$x - y = 6$

4. $x - y = 3$
$2x + 3y = 11$

5. $3x + y = 10$
$6x + 2y = 5$

6. $2x + 3y = 4$
$y = 4$

Solve the systems in Exercises 7 and 8 using substitution.

7. $5x - 2y = 0$
$y = 3x + 4$

8. $x = \frac{1}{3}y - 4$
$2x + \frac{3}{2}y = 5$

Solve the systems in Exercises 9 and 10 using the method of addition.

9. $x + 2y = 5$
$2x + 4y = 10$

10. $-2x + 3y = 6$
$4x + y = 1$

Solve the systems in Exercises 11–16.

11. $x + y = 2$
$y = -2x - 1$

12. $6x + 2y - 8 = 0$
$y = -3x$

13. $3y - x = 7$
$x + 2y = -2$

14. $7x + 5y = -9$
$6x + 2y = 6$

15. $y = 4x - 3$
$2x - \frac{1}{2}y = \frac{3}{2}$

16. $x - \frac{2}{5}y = \frac{4}{5}$
$\frac{3}{4}x + \frac{3}{4}y = \frac{5}{4}$

17. Write an equation for the line determined by the points (3, 1) and (5, 8), using the formula $y = mx + b$ and a system of equations.

Set up a system of two equations in two unknowns in Exercises 18–20. Then solve.

18. A boat can travel 24 miles downstream in 2 hours. The return trip takes 3 hours. Find the speed of the boat in still water and the speed of the current.

19. Eight pencils and two pens cost $2.22. Three pens and four pencils cost $2.69. What is the price of each pen and each pencil?

20. The perimeter of a rectangle is 60 inches. The area of the rectangle is 221 square inches. Find the length and width of the rectangle.

USING THE COMPUTER

```
10   REM
20   REM This program will estimate the coordinates of a point
30   REM where two lines intersect.
40   REM
50   '*******************************************************
60   '* A1 : The coefficient of X in first equation.        *
70   '* B1 : The coefficient of Y in first equation.        *
80   '* C1 : The constant in first equation.                *
90   '* A2,B2,C2 : same as above in second equation.        *
100  '* SLOPE1,SLOPE2 : The slope of each line.             *
110  '* Z   : A1 * B2 - B1 * A2 (from Cramer's rule).       *
120  '* (X,Y)  : Point of intersection.                     *
130  '*******************************************************
140 PRINT "Remark : aX + bY = c   (0,0,0 to quit.)"
150 INPUT
"Type in the value of coefficients a,b,c for first equation :"
,A1,B1,C1
160 IF A1 = 0 THEN GOTO 260
170 INPUT
"Type in the value of coefficients a,b,c for second equation:"
,A2,B2,C2
180 IF A2 = 0 THEN GOTO 260
190 LET Z = A1*B2 - B1*A2
200 IF Z = 0 THEN
    PRINT "Two lines are parallel." : PRINT : GOTO 140
210 LET X = (C1*B2 - B1*C2) / Z            'find first component
220 LET Y = (A1*C2 - C1*A2) / Z            'find second component
230 PRINT
"The lines appear to intersect at the point ("X","Y")."
240 PRINT : GOTO 140                       'goto 140 to continue
260 PRINT "Thank you for using this program..."
270 END                                    'end program
```

```
Remark : 0,0,0 to Quit.
Type in the value of coefficients a,b,c for first equation (aX+bY=c) :
 1 , 1 , 5
Type in the value of coefficients a,b,c for second equation (aX+bY=c):
 1 ,-4 , 5
The lines appear to intersect at the point ( 5 , 0 ).

Remark : 0,0,0 to Quit.
Type in the value of coefficients a,b,c for first equation (aX+bY=c) :
 3 ,-1 , 6
Type in the value of coefficients a,b,c for second equation (aX+bY=c):
-3 , 1 , 0
Two lines are parallel.

Remark : 0,0,0 to Quit.
Type in the value of coefficients a,b,c for first equation (aX+bY=c) :
 1 , 1 , 0
Type in the value of coefficients a,b                         X+bY=c):
 1 ,-1 , 0
The lines appear to intersect at the

Remark : 0,0,0 to Quit.
Type in the value of coefficients a,b                         +bY=c) :
 0 , 0 , 0
Thank you for using this program.....
```

The number of grains of sand on the beach at Coney Island is much less than a googol—10,000,000, 000,000,000,000,000,000,000,000,000,000,000,000, 000,000,000,000,000,000,000,000,000,000,000,000, 000,000,000,000,000,000.

EDWARD KASNER

DID YOU KNOW?

An important method of reasoning related to mathematical proofs is proof by contradiction. See if you can follow the reasoning in the following "proof" that $\sqrt{2}$ is an irrational number.

We need the following two statements (which can be proven algebraically):

1. The square of an even integer is an even integer.
2. The square of an odd integer is an odd integer.

Proof: $\sqrt{2}$ is either an irrational number or a rational number. Suppose that $\sqrt{2}$ is a rational number and

	$\frac{a}{b} = \sqrt{2}$	where a and b are integers and $\frac{a}{b}$ is reduced. (± 1 are the only factors common to a and b.)
	$\frac{a^2}{b^2} = 2$	Square both sides
	$a^2 = 2b^2$	This means a^2 is an even integer.
So,	$a = 2n$	Since a^2 is even, a must be even.
	$a^2 = 4n^2$	Square both sides.
	$a^2 = 4n^2 = 2b^2$	Substituting
	$2n^2 = b^2$	This means b^2 is an even integer.

Therefore, b is an even integer.
But if a and b are both even, 2 is a common factor.

This contradicts the statement that $\frac{a}{b}$ is reduced.

Thus, our original supposition that $\sqrt{2}$ is rational is false, and $\sqrt{2}$ is an irrational number.

10 Real Numbers and Radicals

10.1 REAL NUMBERS AND SIMPLIFYING RADICALS

We have discussed the **integers,** which include the whole numbers and their opposites,

$$\ldots, -4, -3, -2, -1, 0, 1, 2, 3, 4, \ldots \qquad \text{integers}$$

and fractions formed using integers in the numerator and denominator with no denominator equal to 0. The formal name for such fractions, including the integers, is **rational numbers.** A **rational number** is any number that can be written in the form

$$\frac{a}{b} \qquad \text{where } a \text{ and } b \text{ are integers and } b \neq 0$$

In decimal form, all rational numbers can be written as **repeating decimals.** For example,

$$\tfrac{1}{3} = 0.33333\ldots \qquad = 0.\bar{3}$$

$$\tfrac{1}{4} = 0.2500000\ldots \qquad = 0.25\bar{0} \qquad \text{repeat 0's}$$

$$\tfrac{1}{7} = 0.142857142857142857\ldots = 0.\overline{142857}$$

$$\tfrac{5}{3} = 1.66666\ldots \qquad = 1.\bar{6}$$

The bar over the digits indicates that that pattern of digits is to be repeated without end.

Other forms of rational numbers are **square roots** of perfect square numbers. For example, since $5^2 = 25$, 25 is a **perfect square integer,** and 5 is the **square root** of 25. We write $\sqrt{25} = 5$. The symbol $\sqrt{\ }$ is called a **radical sign,** and the number under the radical sign is called the **radicand.** For example,

$$\sqrt{49} = 7 \qquad \text{since} \qquad 7^2 = 49$$

and, in the expression $\sqrt{49}$, 49 is the radicand.

EXAMPLES

1. $\sqrt{4} = 2$ since $2^2 = 4$.

2. $\sqrt{36} = 6$ since $6^2 = 36$.

3. $\sqrt{\frac{4}{9}} = \frac{2}{3}$ since $\left(\frac{2}{3}\right)^2 = \frac{4}{9}$.

4. $\sqrt{81} = 9$ since $9^2 = 81$.

We also know that $(-7)^2 = 49$ and $(-6)^2 = 36$. To indicate that we want the negative number for a square root, we must put a negative sign in front of the radical. **No sign in front of a radical indicates a positive root.** This positive root is called the **principal square root.**

EXAMPLES

5. 49 has two square roots, one negative and one positive:

$-\sqrt{49} = -7$ and $\sqrt{49} = 7$ the principal square root

6. $-\sqrt{36} = -6$ and $\sqrt{36} = 6$ the principal square root

7. $-\sqrt{\frac{4}{9}} = -\frac{2}{3}$

8. $-\sqrt{\frac{25}{121}} = -\frac{5}{11}$

Not all roots are integers or rational numbers. Numbers such as $\sqrt{5}$, $\sqrt{7}$, $\sqrt{39}$, and $-\sqrt{10}$ are called **irrational numbers.** In decimal form, irrational numbers can be written as **nonrepeating decimals.**

EXAMPLES The following numbers are all irrational numbers.

9. $\sqrt{2} = 1.4142136\ldots$

10. $\sqrt[3]{4} = 1.5874011\ldots$ cube root of 4

11. $\pi = 3.14159265358979\ldots$ pi, the ratio of circumference to diameter of a circle

(You might be interested to know that computers have been used to find the calculation of pi accurate to 200,000 decimal places.)

12. $e = 2.718281828459045\ldots$ base of natural logarithms

Irrational numbers are just as important as rational numbers and just as useful in solving equations. Number lines have points corresponding to irrational numbers as well as rational numbers. Remember, in decimal form,

$$-\frac{1}{3} = -0.33333\ldots \qquad \text{repeating but never ending}$$

and $$\sqrt{2} = 1.4142136\ldots \qquad \text{nonrepeating and never ending}$$

However, both numbers correspond to just one point on a line (Figure 10.1).

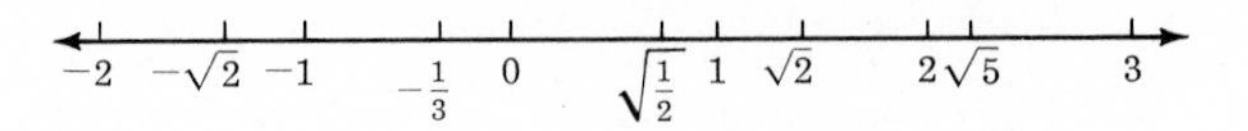

FIGURE 10.1

To further emphasize the idea that irrational numbers correspond to points on a number line, consider a circle with a diameter of 1 unit rolling on a line. If the circle touches the line at the point 0, at what point on the line will the same point on the circle again touch the line?

The point will be at π on the number line because π is the circumference of the circle (Figure 10.2).

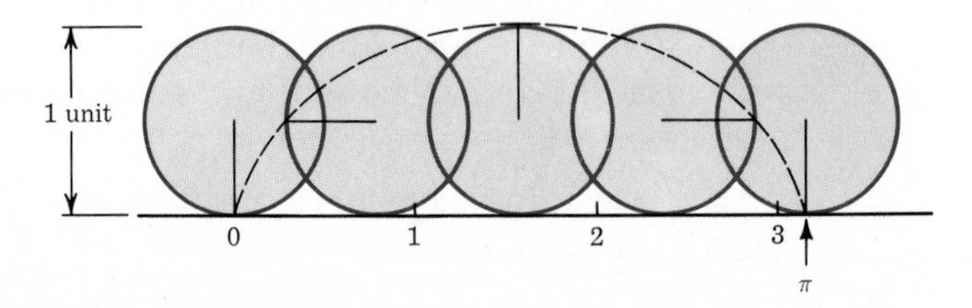

FIGURE 10.2

Real numbers are numbers that are either rational or irrational. That is, every rational number and every irrational number is also a real number, just as every integer is also a rational number. (Of course, every integer is a real number, too.)

You should now have a better understanding of the term **real number line** that was discussed in Chapter 7. When an inequality such as $2x + 1 < 7$ is solved and the solution is graphed, the solution and graph indicate real numbers.

EXAMPLE

13.
$$2x + 1 < 7$$
$$2x + 1 - 1 < 7 - 1$$
$$2x < 6$$
$$\frac{2x}{2} < \frac{6}{2}$$
$$x < 3$$

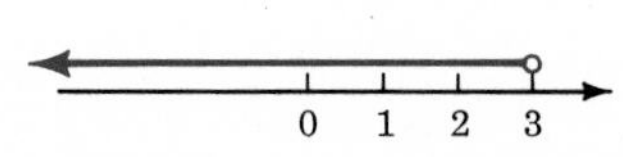

The shading indicates all real numbers less than 3. The open circle around 3 indicates that 3 is not included in the graph.

We are now interested in developing techniques that will aid in simplifying expressions that contain radicals. In this text, we will simplify only radicals that are square roots. Other radical expressions, such as cube roots ($\sqrt[3]{}$) and fourth roots ($\sqrt[4]{}$), will be discussed in later algebra courses.

The following two properties of radicals are basic to the discussion.

> If a and b are positive real numbers, then
>
> 1. $\sqrt{ab} = \sqrt{a}\sqrt{b}$ and
>
> 2. $\sqrt{\dfrac{a}{b}} = \dfrac{\sqrt{a}}{\sqrt{b}}$

Thus,

$$\sqrt{144} = \sqrt{36} \cdot \sqrt{4} = 6 \cdot 2 = 12$$

and
$$\sqrt{\frac{9}{25}} = \frac{\sqrt{9}}{\sqrt{25}} = \frac{3}{5}$$

To simplify $\sqrt{450}$, we can write

$$\sqrt{450} = \sqrt{25 \cdot 18} = \sqrt{25}\sqrt{18} = 5\sqrt{18}$$

Is $5\sqrt{18}$ the simplest form of $\sqrt{450}$? The answer is no, because $\sqrt{18}$ has a square number factor, 9, and $\sqrt{18} = \sqrt{9}\sqrt{2} = 3\sqrt{2}$.

We can write

$$\sqrt{450} = \sqrt{25 \cdot 18} = \sqrt{25} \cdot \sqrt{9} \cdot \sqrt{2} = 5 \cdot 3 \cdot \sqrt{2} = 15\sqrt{2}$$

or

$$\sqrt{450} = \sqrt{225 \cdot 2} = \sqrt{225} \cdot \sqrt{2} = 15\sqrt{2}$$

In simplifying a radical, try to find the largest square factor of the radicand. **A radical is considered to be in simplest form when the radicand has no square number factor.**

EXAMPLES

Simplify the following radicals.

14. $\sqrt{24}$

Solution: Factor 24 so that one factor is a square number.

$$\sqrt{24} = \sqrt{4 \cdot 6} = \sqrt{4} \cdot \sqrt{6}$$
$$= 2\sqrt{6}$$

15. $\sqrt{72}$

Solution: Find the largest square factor you can before simplifying.

$$\sqrt{72} = \sqrt{36 \cdot 2} = \sqrt{36} \cdot \sqrt{2}$$
$$= 6\sqrt{2}$$

Or, if you did not notice 36 as a factor, you could write

$$\sqrt{72} = \sqrt{9 \cdot 8} = \sqrt{9} \cdot \sqrt{8} = 3\sqrt{4 \cdot 2}$$
$$= 3 \cdot \sqrt{4} \cdot \sqrt{2} = 3 \cdot 2 \cdot \sqrt{2} = 6\sqrt{2}$$

16. $-\sqrt{288}$

Solution: $-\sqrt{288} = -\sqrt{144 \cdot 2} = -\sqrt{144} \cdot \sqrt{2}$

$$= -12\sqrt{2}$$

17. $\sqrt{\frac{75}{4}}$

Solution: $\sqrt{\frac{75}{4}} = \frac{\sqrt{75}}{\sqrt{4}} = \frac{\sqrt{25 \cdot 3}}{2} = \frac{\sqrt{25} \cdot \sqrt{3}}{2}$

$$= \frac{5\sqrt{3}}{2}$$

18. $\dfrac{6 + 3\sqrt{2}}{3}$

Solution: $\dfrac{6 + 3\sqrt{2}}{3} = \dfrac{6}{3} + \dfrac{3\sqrt{2}}{3}$

$= 2 + \sqrt{2}$

19. $\dfrac{10 + 15\sqrt{3}}{5}$

Solution: $\dfrac{10 + 15\sqrt{3}}{5} = \dfrac{10}{5} + \dfrac{15\sqrt{3}}{5}$

$= 2 + 3\sqrt{3}$

20. $\dfrac{3 + \sqrt{18}}{3}$

Solution: $\dfrac{3 + \sqrt{18}}{3} = \dfrac{3 + \sqrt{9 \cdot 2}}{3} = \dfrac{3 + \sqrt{9} \cdot \sqrt{2}}{3} = \dfrac{3 + 3\sqrt{2}}{3}$

$= \dfrac{3}{3} + \dfrac{3\sqrt{2}}{3} = 1 + \sqrt{2}$

Important Note: The square roots of negative numbers are undefined in this text. Thus, $\sqrt{-4}$ is undefined at this time. Such numbers are defined and discussed in detail in later courses in mathematics.

PRACTICE QUIZ

Questions	**Answers**
Simplify the following radical expressions.	
1. $\sqrt{\dfrac{3}{4}}$	1. $\dfrac{\sqrt{3}}{2}$
2. $\sqrt{32}$	2. $4\sqrt{2}$
3. $\sqrt{250}$	3. $5\sqrt{10}$
4. $\sqrt{\dfrac{54}{25}}$	4. $\dfrac{3\sqrt{6}}{5}$
5. $\dfrac{4 - \sqrt{12}}{2}$	5. $2 - \sqrt{3}$

EXERCISES 10.1

Identify the rational numbers and the irrational numbers in Exercises 1–20.

1. $\sqrt{64}$ **2.** $\sqrt{17}$ **3.** $-\sqrt{24}$ **4.** $\sqrt{81}$

5. $-\sqrt{12}$ **6.** $-\sqrt{49}$ **7.** $\sqrt{\frac{4}{9}}$ **8.** $\sqrt{\frac{7}{16}}$

9. $-\sqrt{\frac{12}{25}}$ **10.** $-\sqrt{\frac{25}{4}}$ **11.** $\sqrt{\frac{12}{75}}$ **12.** $\sqrt{\frac{18}{32}}$

13. 0.1010010001 . . . (nonrepeating)

14. 2.1632147985 . . . (nonrepeating)

15. 4.232323 . . . (repeating)

16. 1.9347193471 . . . (repeating)

17. −6.051051051 . . . (repeating)

18. −3.1629341251 . . . (nonrepeating)

19. 2.1350014728 . . . (nonrepeating)

20. −1.6936934725 . . . (nonrepeating)

Simplify each of the radicals in Exercises 21–60.

21. $\sqrt{36}$ **22.** $\sqrt{49}$ **23.** $\sqrt{\frac{1}{4}}$ **24.** $-\sqrt{121}$

25. $\sqrt{8}$ **26.** $\sqrt{18}$ **27.** $-\sqrt{56}$ **28.** $\sqrt{162}$

29. $\sqrt{\frac{5}{9}}$ **30.** $-\sqrt{\frac{7}{16}}$ **31.** $\sqrt{12}$ **32.** $-\sqrt{45}$

33. $\sqrt{288}$ **34.** $-\sqrt{63}$ **35.** $-\sqrt{72}$ **36.** $\sqrt{98}$

37. $-\sqrt{125}$ **38.** $\sqrt{\frac{32}{49}}$ **39.** $-\sqrt{\frac{11}{64}}$ **40.** $-\sqrt{\frac{125}{100}}$

41. $\sqrt{\frac{28}{25}}$ **42.** $\sqrt{\frac{147}{100}}$ **43.** $\sqrt{\frac{32}{81}}$ **44.** $\sqrt{\frac{75}{121}}$

45. $\frac{\sqrt{8}}{4}$ **46.** $\frac{-\sqrt{50}}{10}$ **47.** $\frac{-\sqrt{18}}{12}$ **48.** $\frac{\sqrt{28}}{6}$

49. $\frac{2 - 2\sqrt{3}}{4}$ **50.** $\frac{4 + 2\sqrt{6}}{2}$ **51.** $\frac{3 + 3\sqrt{6}}{6}$ **52.** $\frac{6 - 2\sqrt{3}}{8}$

53. $\frac{12 + \sqrt{45}}{15}$ **54.** $\frac{7 - \sqrt{98}}{14}$ **55.** $\frac{10 - \sqrt{108}}{4}$ **56.** $\frac{4 + \sqrt{288}}{20}$

57. $\frac{16 - \sqrt{60}}{12}$ **58.** $\frac{12 - \sqrt{18}}{21}$ **59.** $\frac{12 - \sqrt{192}}{28}$ **60.** $\frac{14 + \sqrt{147}}{35}$

10.2 ADDITION AND MULTIPLICATION WITH RADICALS

Like radicals are terms that either have the same radicals or can be simplified so that the radicals are the same. For example,

a. $7\sqrt{2}$ and $3\sqrt{2}$ are like radicals.

b. $2\sqrt{5}$ and $2\sqrt{3}$ are **not** like radicals.

c. $\sqrt{75}$ and $4\sqrt{3}$ are like radicals because $\sqrt{75}$ simplifies to $5\sqrt{3}$, and $5\sqrt{3}$ and $4\sqrt{3}$ are like radicals.

To find a sum of like radicals, such as $7\sqrt{2} + 3\sqrt{2}$, proceed just as when combining like terms. Thus,

$$7x + 3x = (7 + 3)x = 10x$$

and

$$7\sqrt{2} + 3\sqrt{2} = (7 + 3)\sqrt{2} = 10\sqrt{2}$$

Similarly,

$$10\sqrt{3} - 4\sqrt{3} + \sqrt{5} + 2\sqrt{5} = (10 - 4)\sqrt{3} + (1 + 2)\sqrt{5} = 6\sqrt{3} + 3\sqrt{5}$$

To add $\sqrt{75} + \sqrt{27}$, first simplify each radical. Then combine like radicals if possible.

$$\sqrt{75} + \sqrt{27} = \sqrt{25 \cdot 3} + \sqrt{9 \cdot 3} = 5\sqrt{3} + 3\sqrt{3} = 8\sqrt{3}$$

The procedure is the same if the radical contains a variable.

$$\sqrt{16x} + \sqrt{4x} = \sqrt{16 \cdot x} + \sqrt{4 \cdot x} = 4\sqrt{x} + 2\sqrt{x} = 6\sqrt{x}$$

To Find the Sum of Radicals

1. Simplify each radical expression.
2. Use the distributive property to combine any like radicals.

EXAMPLES

Simplify the following radical expressions.

1. $5\sqrt{2} - \sqrt{2} = (5 - 1)\sqrt{2} = 4\sqrt{2}$

2. $\sqrt{18} + \sqrt{8} = \sqrt{9}\sqrt{2} + \sqrt{4}\sqrt{2} = 3\sqrt{2} + 2\sqrt{2} = 5\sqrt{2}$

3. $\sqrt{5} - \sqrt{3} + \sqrt{12} = \sqrt{5} - \sqrt{3} + \sqrt{4}\sqrt{3} = \sqrt{5} - \sqrt{3} + 2\sqrt{3} = \sqrt{5} + \sqrt{3}$

4. $x\sqrt{x} + 7x\sqrt{x} = 8x\sqrt{x}$

5. $4\sqrt{a} - 3\sqrt{b} + 7\sqrt{a} + \sqrt{b} = 11\sqrt{a} - 2\sqrt{b}$

To find the product of radicals, we proceed just as in multiplying polynomials, as the following examples illustrate.

$$5(x + y) = 5x + 5y$$

$$\sqrt{2}(\sqrt{7} + \sqrt{3}) = \sqrt{2}\sqrt{7} + \sqrt{2}\sqrt{3} = \sqrt{14} + \sqrt{6}$$

And, with binomials,

$$(x + 5)(x - 3) = x^2 - 3x + 5x - 15 = x^2 + 2x - 15$$

$$(\sqrt{2} + 5)(\sqrt{2} - 3) = \sqrt{2}\sqrt{2} - 3\sqrt{2} + 5\sqrt{2} - 15$$

$$= 2 + 2\sqrt{2} - 15 = -13 + 2\sqrt{2}$$

(**Note:** $\sqrt{2}\sqrt{2} = \sqrt{4} = 2$. In general, $\sqrt{a}\sqrt{a} = a$ if a is positive.)

EXAMPLES

Find the following products and simplify.

6. $\sqrt{7}(\sqrt{7} - \sqrt{14}) = \sqrt{7}\sqrt{7} - \sqrt{7}\sqrt{14} = 7 - \sqrt{98} = 7 - \sqrt{49 \cdot 2}$
$= 7 - \sqrt{49}\sqrt{2} = 7 - 7\sqrt{2}$

7. $(\sqrt{2} + 4)(\sqrt{2} - 4) = (\sqrt{2})^2 - 4^2 = 2 - 16 = -14$

8. $(\sqrt{5} + \sqrt{3})(\sqrt{5} + \sqrt{3}) = \sqrt{5}\sqrt{5} + 2\sqrt{5}\sqrt{3} + \sqrt{3}\sqrt{3}$
$= 5 + 2\sqrt{15} + 3 = 8 + 2\sqrt{15}$

PRACTICE QUIZ

Questions	Answers
Simplify the following radical expressions.	
1. $-2(\sqrt{8} + \sqrt{2})$	1. $-6\sqrt{2}$
2. $\sqrt{75} - \sqrt{27} + \sqrt{20}$	2. $2\sqrt{3} + 2\sqrt{5}$
3. $(\sqrt{3} + \sqrt{8})(\sqrt{2} - \sqrt{3})$	3. $1 - \sqrt{6}$

EXERCISES 10.2

Simplify the radical expressions in Exercises 1–40.

1. $3\sqrt{2} + 5\sqrt{2}$

2. $7\sqrt{3} - 2\sqrt{3}$

3. $6\sqrt{5} + \sqrt{5}$

4. $4\sqrt{11} - 3\sqrt{11}$

5. $8\sqrt{10} - 11\sqrt{10}$

6. $6\sqrt{17} - 9\sqrt{17}$

7. $\sqrt{5} + 3\sqrt{5} - 6\sqrt{5}$

8. $14\sqrt{3} + 6\sqrt{3} - 20\sqrt{3}$

9. $6\sqrt{11} - 5\sqrt{11} - 2\sqrt{11}$

10. $\sqrt{7} + 6\sqrt{7} - 2\sqrt{7}$

11. $\sqrt{a} + 4\sqrt{a} - 2\sqrt{a}$

12. $2\sqrt{x} - 3\sqrt{x} + 7\sqrt{x}$

13. $5\sqrt{x} + 3\sqrt{x} - \sqrt{x}$

14. $6\sqrt{xy} - 10\sqrt{xy} + \sqrt{xy}$

15. $3\sqrt{2} + 5\sqrt{3} - 2\sqrt{3} + \sqrt{2}$

16. $\sqrt{5} + \sqrt{4} - 2\sqrt{5} + 6$

17. $\sqrt{5} + 2\sqrt{2} - 3\sqrt{5} + \sqrt{2}$

18. $2\sqrt{7} - \sqrt{3} - 5\sqrt{7} + 6\sqrt{3}$

19. $2\sqrt{a} + 7\sqrt{b} - 6\sqrt{a} + \sqrt{b}$

20. $4\sqrt{x} - 3\sqrt{x} + 2\sqrt{y} + 2\sqrt{x}$

21. $\sqrt{12} + \sqrt{27}$

22. $\sqrt{32} - \sqrt{18}$

23. $3\sqrt{5} - \sqrt{45}$

24. $2\sqrt{7} + 5\sqrt{28}$

25. $\sqrt{12} + \sqrt{75} - \sqrt{48}$

26. $\sqrt{20} + 2\sqrt{45} - \sqrt{80}$

27. $\sqrt{50} + \sqrt{18} - 3\sqrt{12}$

28. $2\sqrt{48} - \sqrt{54} + \sqrt{27}$

29. $2\sqrt{20} - \sqrt{45} - \sqrt{36}$

30. $\sqrt{18} - 2\sqrt{12} + 5\sqrt{2}$

31. $\sqrt{8} - 2\sqrt{3} + \sqrt{27} - \sqrt{72}$

32. $\sqrt{80} + \sqrt{8} - \sqrt{45} + \sqrt{50}$

33. $\sqrt{28} - \sqrt{75} + \sqrt{112} + \sqrt{48}$

34. $-\sqrt{162} - \sqrt{125} + \sqrt{288} - \sqrt{180}$

35. $6\sqrt{2x} - \sqrt{8x}$

36. $5\sqrt{3x} + 2\sqrt{12x}$

37. $5y\sqrt{2y} - y\sqrt{18y}$

38. $9x\sqrt{xy} - x\sqrt{16xy}$

39. $4x\sqrt{3xy} - x\sqrt{12xy} - 2x\sqrt{27xy}$

40. $x\sqrt{32x} - x\sqrt{50x} + 2x\sqrt{18x}$

Multiply the expressions in Exercises 41–60.

41. $\sqrt{2}(3 - 4\sqrt{2})$

42. $2\sqrt{7}(\sqrt{7} + 3\sqrt{2})$

43. $3\sqrt{18} \cdot \sqrt{2}$

44. $2\sqrt{10} \cdot \sqrt{5}$

45. $-2\sqrt{6} \cdot \sqrt{8}$

46. $2\sqrt{15} \cdot 5\sqrt{6}$

47. $\sqrt{3}(\sqrt{2} + 2\sqrt{12})$

48. $\sqrt{2}(\sqrt{3} - \sqrt{6})$

49. $\sqrt{y}(\sqrt{x} + 2\sqrt{y})$

50. $\sqrt{x}(\sqrt{x} - 3\sqrt{xy})$

51. $(5 + \sqrt{2})(3 - \sqrt{2})$

52. $(2\sqrt{3} + 1)(\sqrt{3} - 3)$

53. $(4\sqrt{3} + \sqrt{2})(\sqrt{3} - 2\sqrt{2})$

54. $(\sqrt{5} - \sqrt{3})(2\sqrt{5} + 3\sqrt{3})$

55. $(\sqrt{x} + 3)(\sqrt{x} - 3)$

56. $(\sqrt{a} + b)(\sqrt{a} + b)$

57. $(\sqrt{2} + \sqrt{7})(\sqrt{2} - \sqrt{7})$

58. $(\sqrt{x} + \sqrt{y})(\sqrt{x} - \sqrt{y})$

59. $(\sqrt{x} + 5)(\sqrt{x} - 3)$

60. $(\sqrt{3} + 7)(\sqrt{3} + 2\sqrt{7})$

10.3 RATIONALIZING DENOMINATORS

Each of the following expressions

$$\frac{5}{\sqrt{3}}, \quad \frac{\sqrt{7}}{\sqrt{8}}, \quad \text{and} \quad \frac{2}{3 - \sqrt{2}}$$

contains a radical in the denominator that is an irrational number. Such expressions are not considered in simplest form. The objective is to find an equal fraction that has a rational number for a denominator.

That is, we want to simplify the expression by **rationalizing the denominator.** The following examples illustrate the method:

a. $\frac{5}{\sqrt{3}} = \frac{5 \cdot \sqrt{3}}{\sqrt{3} \cdot \sqrt{3}} = \frac{5\sqrt{3}}{3}$ Multiply the numerator and the denominator by $\sqrt{3}$ because $\sqrt{3} \cdot \sqrt{3}$ gives a rational number.

b. $\frac{4}{\sqrt{x}} = \frac{4 \cdot \sqrt{x}}{\sqrt{x} \cdot \sqrt{x}} = \frac{4\sqrt{x}}{x}$ Multiply the numerator and the denominator by $\sqrt{x}$ because $\sqrt{x} \cdot \sqrt{x} = x$. We have no guarantee that x is rational, but the radical sign does not appear in the denominator of the expression. (Assume that $x > 0$.)

c. $\frac{\sqrt{7}}{\sqrt{8}} = \frac{\sqrt{7} \cdot \sqrt{2}}{\sqrt{8} \cdot \sqrt{2}} = \frac{\sqrt{14}}{4}$ Multiply the numerator and denominator by $\sqrt{2}$ because $\sqrt{8} \cdot \sqrt{2} = \sqrt{16}$ and 16 is a perfect square number.

If we had multiplied by $\sqrt{8}$, the results would have been the same, but the fraction would have to be reduced.

$$\frac{\sqrt{7}}{\sqrt{8}} = \frac{\sqrt{7} \cdot \sqrt{8}}{\sqrt{8} \cdot \sqrt{8}} = \frac{\sqrt{56}}{8} = \frac{\sqrt{4} \cdot \sqrt{14}}{8} = \frac{2\sqrt{14}}{8} = \frac{\sqrt{14}}{4}$$

Before trying to rationalize the denominator for $\frac{2}{3 - \sqrt{2}}$, recall that the product $(a + b)(a - b)$ results in the difference of two squares:

$$(a + b)(a - b) = a^2 - b^2$$

As long as a and b are real numbers, $a + b$ and $a - b$ are called **conjugate surds** of each other. Therefore, if the numerator and the denominator of a fraction are multiplied by the conjugate surd of the denominator, the denominator will be the difference of two squares and will be a rational number.

$$\frac{2}{3-\sqrt{2}} = \frac{2(3+\sqrt{2})}{(3-\sqrt{2})(3+\sqrt{2})}$$ $3+\sqrt{2}$ is the conjugate surd of $3-\sqrt{2}$.

$$= \frac{2(3+\sqrt{2})}{3^2-(\sqrt{2})^2}$$ The denominator is the difference of two squares.

$$= \frac{2(3+\sqrt{2})}{9-2}$$ The denominator is a rational number.

$$= \frac{2(3+\sqrt{2})}{7}$$

EXAMPLES

Rationalize the denominator of each of the following expressions and simplify.

1. $\sqrt{\frac{1}{2}} = \frac{1}{\sqrt{2}} = \frac{1\cdot\sqrt{2}}{\sqrt{2}\cdot\sqrt{2}} = \frac{\sqrt{2}}{2}$

2. $\frac{5}{\sqrt{5}} = \frac{5\cdot\sqrt{5}}{\sqrt{5}\cdot\sqrt{5}} = \frac{5\sqrt{5}}{5} = \sqrt{5}$

3. $\frac{6}{\sqrt{27}} = \frac{6\cdot\sqrt{3}}{\sqrt{27}\cdot\sqrt{3}} = \frac{6\cdot\sqrt{3}}{\sqrt{81}} = \frac{6\sqrt{3}}{9} = \frac{2\sqrt{3}}{3}$

4. $\frac{-3}{\sqrt{xy}} = \frac{-3\cdot\sqrt{xy}}{\sqrt{xy}\cdot\sqrt{xy}} = \frac{-3\sqrt{xy}}{xy}$

5. $\frac{3}{\sqrt{5}+\sqrt{2}} = \frac{3(\sqrt{5}-\sqrt{2})}{(\sqrt{5}+\sqrt{2})(\sqrt{5}-\sqrt{2})} = \frac{3(\sqrt{5}-\sqrt{2})}{(\sqrt{5})^2-(\sqrt{2})^2}$

$$= \frac{3(\sqrt{5}-\sqrt{2})}{5-2} = \frac{3(\sqrt{5}-\sqrt{2})}{3} = \sqrt{5}-\sqrt{2}$$

6. $\frac{x}{\sqrt{x}-3} = \frac{x(\sqrt{x}+3)}{(\sqrt{x}-3)(\sqrt{x}+3)} = \frac{x(\sqrt{x}+3)}{(\sqrt{x})^2-(3)^2} = \frac{x(\sqrt{x}+3)}{x-9}$

PRACTICE QUIZ

Questions	Answers
Rationalize each denominator and simplify.	
1. $\sqrt{\frac{5}{2}}$	1. $\frac{\sqrt{10}}{2}$
2. $\frac{\sqrt{7}}{\sqrt{18}}$	2. $\frac{\sqrt{14}}{6}$
3. $\frac{4}{\sqrt{7} + \sqrt{3}}$	3. $\sqrt{7} - \sqrt{3}$
4. $\frac{5}{\sqrt{x} + 2}$	4. $\frac{5(\sqrt{x} - 2)}{x - 4}$

EXERCISES 10.3

Rationalize each denominator and simplify.

1. $\frac{5}{\sqrt{2}}$ **2.** $\frac{7}{\sqrt{5}}$ **3.** $\frac{-3}{\sqrt{7}}$

4. $\frac{-10}{\sqrt{2}}$ **5.** $\frac{6}{\sqrt{3}}$ **6.** $\frac{8}{\sqrt{2}}$

7. $\frac{\sqrt{18}}{\sqrt{2}}$ **8.** $\frac{\sqrt{25}}{\sqrt{3}}$ **9.** $\frac{\sqrt{27x}}{\sqrt{3x}}$

10. $\frac{\sqrt{45y}}{\sqrt{5y}}$ **11.** $\frac{\sqrt{ab}}{\sqrt{9ab}}$ **12.** $\frac{\sqrt{5}}{\sqrt{12}}$

13. $\frac{\sqrt{4}}{\sqrt{3}}$ **14.** $\sqrt{\frac{3}{8}}$ **15.** $\sqrt{\frac{9}{2}}$

16. $\sqrt{\frac{3}{5}}$ **17.** $\sqrt{\frac{1}{x}}$ **18.** $\sqrt{\frac{x}{y}}$

19. $\sqrt{\frac{2x}{y}}$ **20.** $\sqrt{\frac{x}{4y}}$ **21.** $\frac{2}{\sqrt{2y}}$

22. $\frac{-10}{3\sqrt{5}}$ **23.** $\frac{21}{5\sqrt{7}}$ **24.** $\frac{x}{5\sqrt{x}}$

25. $\dfrac{-2y}{5\sqrt{2y}}$ **26.** $\dfrac{3}{1+\sqrt{2}}$ **27.** $\dfrac{2}{\sqrt{6}-2}$

28. $\dfrac{-11}{\sqrt{3}-4}$ **29.** $\dfrac{1}{\sqrt{5}-3}$ **30.** $\dfrac{7}{3-2\sqrt{2}}$

31. $\dfrac{-6}{5-3\sqrt{2}}$ **32.** $\dfrac{11}{2\sqrt{3}+1}$ **33.** $\dfrac{-\sqrt{3}}{\sqrt{2}+5}$

34. $\dfrac{\sqrt{2}}{\sqrt{7}+4}$ **35.** $\dfrac{\sqrt{7}}{1-3\sqrt{5}}$ **36.** $\dfrac{-3\sqrt{3}}{6+\sqrt{3}}$

37. $\dfrac{1}{\sqrt{3}-\sqrt{5}}$ **38.** $\dfrac{-4}{\sqrt{7}-\sqrt{3}}$ **39.** $\dfrac{-5}{\sqrt{2}+\sqrt{3}}$

40. $\dfrac{7}{\sqrt{2}+\sqrt{5}}$ **41.** $\dfrac{4}{\sqrt{x}+1}$ **42.** $\dfrac{-7}{\sqrt{x}-3}$

43. $\dfrac{5}{6+\sqrt{y}}$ **44.** $\dfrac{x}{\sqrt{x}+2}$ **45.** $\dfrac{8}{2\sqrt{x}+3}$

46. $\dfrac{3\sqrt{x}}{\sqrt{2x}-5}$ **47.** $\dfrac{\sqrt{4y}}{\sqrt{5y}-\sqrt{3}}$ **48.** $\dfrac{\sqrt{3x}}{\sqrt{2}+\sqrt{3x}}$

49. $\dfrac{3}{\sqrt{x}-\sqrt{y}}$ **50.** $\dfrac{4}{2\sqrt{x}+\sqrt{y}}$ **51.** $\dfrac{x}{\sqrt{x}+2\sqrt{y}}$

52. $\dfrac{y}{\sqrt{x}-\sqrt{3y}}$ **53.** $\dfrac{\sqrt{3}+1}{\sqrt{3}-2}$ **54.** $\dfrac{\sqrt{2}+4}{5-\sqrt{2}}$

55. $\dfrac{\sqrt{5}-2}{\sqrt{5}+3}$ **56.** $\dfrac{1+\sqrt{3}}{3-\sqrt{3}}$ **57.** $\dfrac{\sqrt{x}+1}{\sqrt{x}-1}$

58. $\dfrac{\sqrt{x}-4}{\sqrt{x}+3}$ **59.** $\dfrac{\sqrt{x}+2}{\sqrt{3x}+y}$ **60.** $\dfrac{3-\sqrt{x}}{2\sqrt{x}+y}$

10.4 DISTANCE BETWEEN TWO POINTS: $d = \sqrt{(x_2 - x_1)^2 + (y_2 - y_1)^2}$

What is the distance between the two points $P_1(2, 3)$ and $P_2(6, 3)$? What is the distance between the points $P_3(-1, -4)$ and $P_4(-1, 1)$? (See Figure 10.3.)
Did you find

$$\text{distance } (P_1 \text{ to } P_2) = 6 - 2 = 4?$$

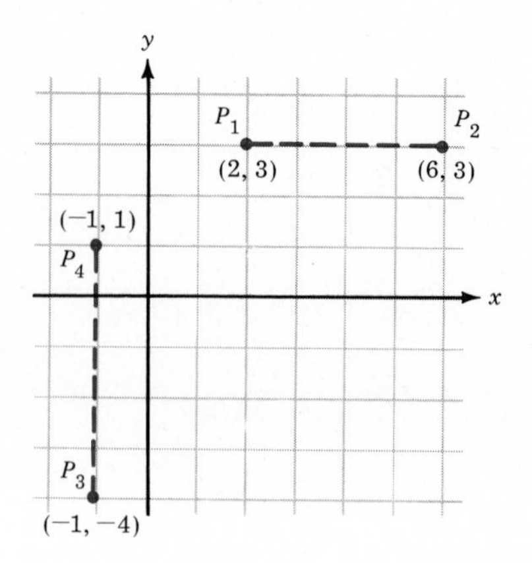

FIGURE 10.3

This is correct since P_1 and P_2 lie on a horizontal line and have the same y-coordinate. Did you find

$$\text{distance } (P_3 \text{ to } P_4) = 1 - (-4) = 5?$$

This is correct since P_3 and P_4 lie on a vertical line and have the same x-coordinate.

For the distance from P_1 to P_2, why not take

$$\text{distance } (P_1 \text{ to } P_2) = 2 - 6 = -4?$$

The reason is that the term **distance** is taken to mean a positive number. How can a positive number be guaranteed after we subtract the coordinates? The answer is to take the absolute value of the difference of the coordinates. The distance between two points is represented by d.

For $P_1(x_1, y_1)$ and $P_2(x_2, y_1)$ on a horizontal line,

$$d = |x_2 - x_1| \text{ (or } d = |x_1 - x_2|\text{)}$$

For $P_1(x_1, y_1)$ and $P_2(x_1, y_2)$ on a vertical line,

$$d = |y_2 - y_1| \text{ (or } d = |y_1 - y_2|\text{)}$$

EXAMPLES

1. Find the distance, d, between the two points $(5, 7)$ and $(-3, 7)$. Since the points are on a horizontal line (they have the same y-coordinate),

$$d = |-3 - 5| = |-8| = 8$$

or

$$d = |5 - (-3)| = |8| = 8$$

2. Find the distance, d, between the two points $(2, 8)$ and $(2, -\frac{1}{2})$. Since the points are on a vertical line (they have the same x-coordinate),

$$d = |8 - (-\tfrac{1}{2})| = |8\tfrac{1}{2}| = 8\tfrac{1}{2}$$

or

$$d = |-\tfrac{1}{2} - 8| = |-8\tfrac{1}{2}| = 8\tfrac{1}{2}$$

What if the points do not lie on a vertical line or on a horizontal line? Indeed, this is the more general case. We need the Pythagorean Theorem. This theorem and related applications will be discussed again in Chapter 11, so study it carefully.

A right triangle is a triangle in which one angle is a right angle (measures 90°). Two sides are perpendicular. The **hypotenuse** is the longest side and is opposite the 90° angle.

Pythagorean Theorem

In a right triangle, the square of the hypotenuse is equal to the sum of the squares of the two sides.

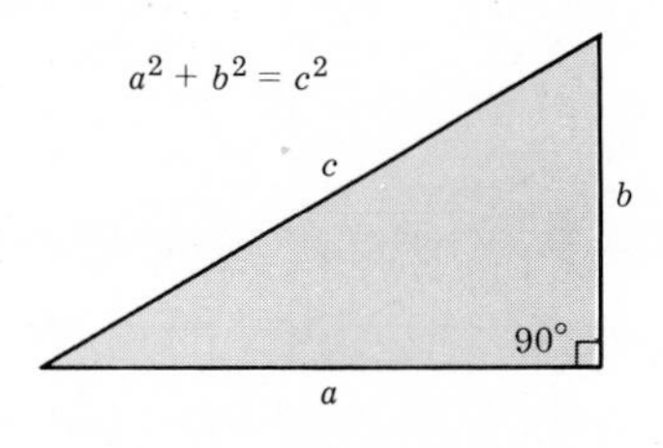

To find the distance between the two points $P_1(-1, 2)$ and $P_2(5, 4)$, apply the Pythagorean Theorem and calculate the length of the hypotenuse as shown in Figure 10.4.

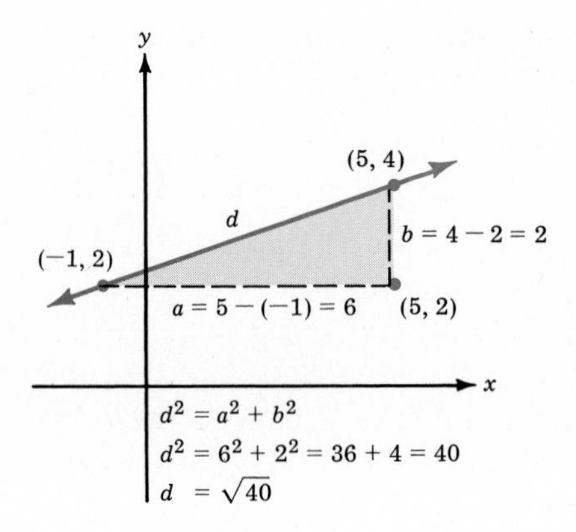

FIGURE 10.4

We could write directly $d = \sqrt{a^2 + b^2}$. Carrying this idea one step further, we can write a formula for d involving the coordinates of two general points $P_1(x_1, y_1)$ and $P_2(x_2, y_2)$.

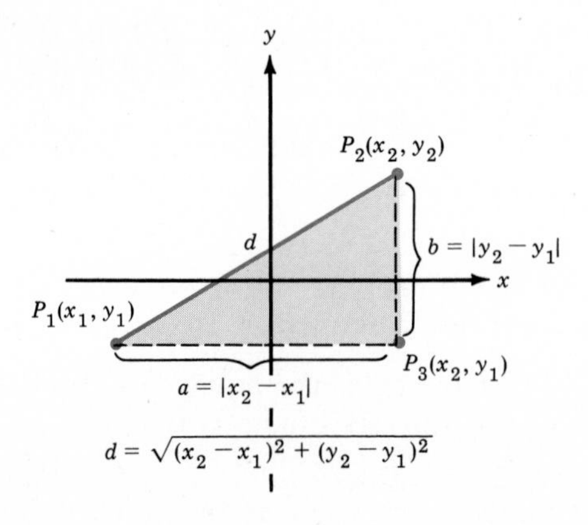

The formula for the distance between two points $P_1(x_1, y_1)$ and $P_2(x_2, y_2)$ is

$$d = \sqrt{(x_2 - x_1)^2 + (y_2 - y_1)^2}$$

Do we need $|x_2 - x_1|$ and $|y_2 - y_1|$ in the formula? Evidently not, but why not? The answer is that even though $x_2 - x_1$ and $y_2 - y_1$ might be negative, both are being squared, so neither of the terms

$(x_2 - x_1)^2$ or $(y_2 - y_1)^2$ will ever be negative. **In the actual calculation of d, be sure to add the squares before taking the square root.**

EXAMPLES

3. Find the distance between the two points $(3, 4)$ and $(-2, 7)$.

Solution:

$$\begin{aligned} d &= \sqrt{(3 - (-2))^2 + (4 - 7)^2} \\ &= \sqrt{5^2 + (-3)^2} \\ &= \sqrt{25 + 9} \\ &= \sqrt{34} \end{aligned}$$

4. Find the distance between the two points $(\frac{1}{2}, \frac{2}{3})$ and $(\frac{3}{4}, \frac{5}{3})$.

Solution:

$$\begin{aligned} d &= \sqrt{(\tfrac{1}{2} - \tfrac{3}{4})^2 + (\tfrac{2}{3} - \tfrac{5}{3})^2} \\ &= \sqrt{(-\tfrac{1}{4})^2 + (-\tfrac{3}{3})^2} \\ &= \sqrt{\tfrac{1}{16} + 1} \\ &= \sqrt{\tfrac{1}{16} + \tfrac{16}{16}} \\ &= \sqrt{\tfrac{17}{16}} = \frac{\sqrt{17}}{4} \end{aligned}$$

5. Show that the triangle determined by the points $A(-2, 1)$, $B(3, 4)$, and $C(1, -4)$ is an isosceles triangle. An isosceles triangle has two equal sides.

Solution: The length of the line segment $\overline{AB}$ is the distance between the points A and B. We denote this distance by $|\overline{AB}|$. Thus, to show that the triangle ABC is isosceles, we need to show that $|\overline{AB}| = |\overline{AC}|$ or that $|\overline{AB}| = |\overline{BC}|$. If neither of these relationships is true, then the triangle does not have two equal sides and is not isosceles.

$$\begin{aligned} |\overline{AB}| &= \sqrt{(-2 - 3)^2 + (1 - 4)^2} = \sqrt{(-5)^2 + (-3)^2} \\ &= \sqrt{25 + 9} = \sqrt{34} \\ |\overline{AC}| &= \sqrt{(-2 - 1)^2 + [1 - (-4)]^2} = \sqrt{(-3)^2 + (5)^2} \\ &= \sqrt{9 + 25} = \sqrt{34} \end{aligned}$$

Since $|\overline{AB}| = |\overline{AC}|$, the triangle is isosceles.

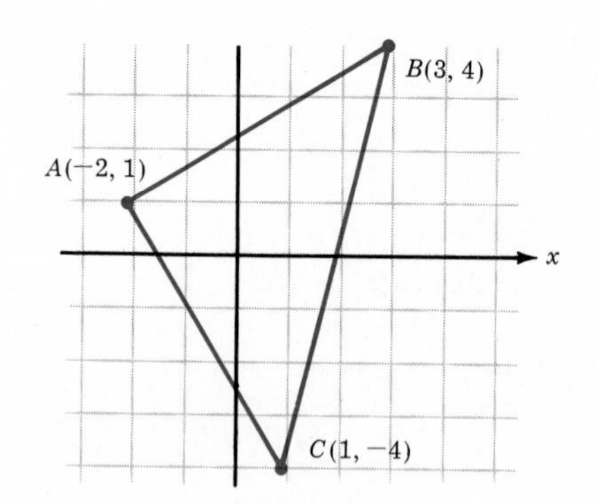

EXERCISES 10.4

Find the distances between the two given points in Exercises 1–24.

1. $(-3, 6), (-3, 2)$ **2.** $(5, 7), (-1, 7)$ **3.** $(4, -3), (7, -3)$

4. $(-1, 2), (5, 2)$ **5.** $(3, 1), \left(-\frac{1}{2}, 1\right)$ **6.** $\left(\frac{4}{3}, 7\right), \left(\frac{4}{3}, -\frac{2}{3}\right)$

7. $(3, 1), (2, 0)$ **8.** $(4, 6), (5, -2)$ **9.** $(1, 5), (-1, 2)$

10. $(0, 0), (-3, 4)$ **11.** $(2, -7), (-3, 5)$ **12.** $(5, -3), (7, -3)$

13. $\left(\frac{3}{7}, \frac{4}{7}\right), (0, 0)$ **14.** $(-5, 2), (1, 1)$ **15.** $(4, 1), (7, 5)$

16. $(10, 7), (1, 7)$ **17.** $(-10, 3), (2, -2)$ **18.** $\left(\frac{7}{3}, 2\right), \left(-\frac{2}{3}, 1\right)$

19. $(-3, 2), (3, -6)$ **20.** $\left(\frac{3}{4}, 6\right), \left(\frac{3}{4}, -2\right)$ **21.** $(4, 0), (0, -3)$

22. $(0, -2), (4, -3)$ **23.** $\left(\frac{4}{5}, \frac{2}{7}\right), \left(-\frac{6}{5}, \frac{2}{7}\right)$ **24.** $(6, 8), (2, 5)$

25. Use the distance formula and the Pythagorean Theorem to decide if the triangle determined by the points $A(1, -2)$, $B(7, 1)$, and $C(5, 5)$ is a right triangle.

26. Using the distance formula and the Pythagorean Theorem, decide if the triangle determined by the points $A(-5, -1)$, $B(2, 1)$, and $C(-1, 6)$ is a right triangle.

In Exercises 27 and 28, show that the triangle determined by the given points is an isosceles triangle (has two equal sides).

27. $A(1, 1), B(5, 9), C(9, 5)$ **28.** $A(1, -4), B(3, 2), C(9, 4)$

In Exercises 29 and 30, show that the triangle determined by the given points is an equilateral triangle (all sides equal). [**Hint:** $(\sqrt{a})^2 = a$.]

29. $A(1, 0), B(3, \sqrt{12}), C(5, 0)$

30. $A(0, 5), B(0, -3), C(\sqrt{48}, 1)$

In Exercises 31 and 32, show that the diagonals of the rectangle *ABCD* are equal.

31. $A(2, -2), B(2, 3), C(8, 3), D(8, -2)$

32. $A(-1, 1), B(-1, 4), C(4, 4), D(4, 1)$

In Exercises 33–35, find the perimeter of the triangle determined by the given points.

33. $A(-5, 0), B(3, 4), C(0, 0)$

34. $A(-6, -1), B(-3, 3), C(6, 4)$

35. $A(-2, 5), B(3, 1), C(2, -2)$

CHAPTER 10 SUMMARY

A **rational number** is any number that can be written in the form

$$\frac{a}{b} \quad \text{where } a \text{ and } b \text{ are integers and } b \neq 0$$

In decimal form, all **rational numbers** can be written as **repeating decimals.**

In decimal form, **irrational numbers** can be written as **nonrepeating decimals.**

Real numbers are numbers that are either rational or irrational.

The symbol $\sqrt{\quad}$ is called a **radical sign,** and the number under the radical sign is called the **radicand.**

A positive square root is called the **principal square root.**

If a and b are positive real numbers, then

1. $\sqrt{ab} = \sqrt{a}\sqrt{b}$ and
2. $\sqrt{\frac{a}{b}} = \frac{\sqrt{a}}{\sqrt{b}}$

To Find the Sum of Radicals

1. Simplify each radical expression.
2. Use the distributive property to combine any like radicals.

To **rationalize the denominator** of a fraction, multiply the numerator and denominator by a number that will give a denominator that is a rational number.

Pythagorean Theorem

In a right triangle, the square of the hypotenuse is equal to the sum of the squares of the two sides.

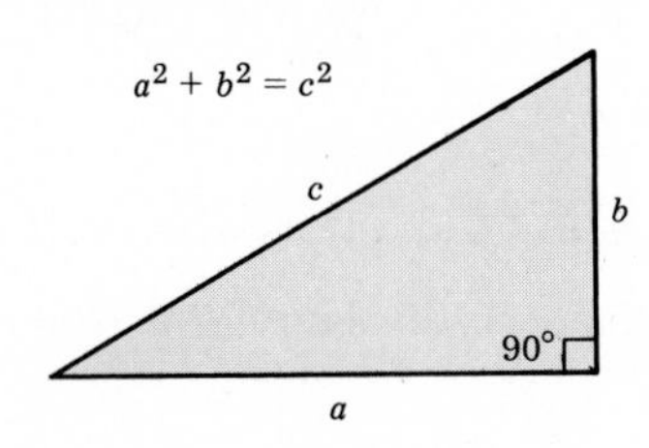

The formula for the distance between two points $P_1(x_1, y_1)$ and $P_2(x_2, y_2)$ is

$$d = \sqrt{(x_2 - x_1)^2 + (y_2 - y_1)^2}$$

CHAPTER 10 REVIEW

Identify the rational numbers and the irrational numbers in Exercises 1–8.

1. $-\sqrt{81}$ **2.** $\sqrt{12}$ **3.** $-\sqrt{\frac{18}{50}}$

4. $\sqrt{\frac{5}{16}}$ **5.** $\sqrt{\frac{4}{25}}$ **6.** $\sqrt{38}$

7. 0.15785136 . . . (nonrepeating) **8.** 0.1278312783 . . . (repeating)

Simplify the square roots in Exercises 9–20.

9. $\sqrt{169}$ **10.** $\sqrt{196}$ **11.** $-\sqrt{48}$

12. $\sqrt{54}$ **23.** $\sqrt{40}$ **14.** $\sqrt{150}$

15. $\sqrt{\frac{7}{196}}$ **16.** $-\sqrt{\frac{5}{121}}$ **17.** $-\sqrt{\frac{15}{48}}$

18. $\sqrt{\frac{75}{64}}$ **19.** $\sqrt{\frac{17}{16}}$ **20.** $\sqrt{\frac{80}{81}}$

Simplify the radical expressions in Exercises 21–35.

21. $\frac{8+\sqrt{12}}{2}$ **22.** $\frac{20-\sqrt{48}}{8}$ **23.** $\frac{18+\sqrt{45}}{30}$

24. $\frac{18+\sqrt{72}}{12}$ **25.** $\frac{\sqrt{12}-\sqrt{20}}{4}$ **26.** $\frac{\sqrt{27}+\sqrt{18}}{3}$

27. $11\sqrt{2}+\sqrt{50}$ **28.** $4\sqrt{20}-\sqrt{45}$ **29.** $\sqrt{18}+5\sqrt{2}$

30. $\sqrt{50x}-2\sqrt{8x}$ **31.** $\sqrt{12}-5\sqrt{3}+\sqrt{48}$ **32.** $4\sqrt{3}-\sqrt{8}+\sqrt{75}$

33. $\sqrt{49}-\sqrt{56}+2\sqrt{25}$ **34.** $\sqrt{72}+\sqrt{98}-3\sqrt{2}$ **35.** $\sqrt{12y}-3\sqrt{48y}+\sqrt{75y}$

Multiply each expression in Exercises 36–45 and simplify if possible.

36. $\sqrt{24}\cdot\sqrt{6}$ **37.** $2\sqrt{30}\cdot\sqrt{20}$ **38.** $2\sqrt{48}\cdot\sqrt{2}$

39. $\sqrt{14}\cdot\sqrt{21}$ **40.** $(\sqrt{3}-5)(\sqrt{3}+5)$ **41.** $(\sqrt{2}+3)(\sqrt{2}+5)$

42. $(\sqrt{6}+2\sqrt{3})(3\sqrt{6}-\sqrt{3})$ **43.** $(\sqrt{x}+\sqrt{2})(\sqrt{x}-\sqrt{2})$

44. $(5-\sqrt{3})(7+\sqrt{3})$ **45.** $(2\sqrt{x}+3)(\sqrt{x}-8)$

Rationalize each denominator in Exercises 46–60 and simplify if possible.

46. $\frac{9}{\sqrt{3}}$ **47.** $\sqrt{\frac{3}{20}}$ **48.** $\sqrt{\frac{5}{27}}$

49. $\frac{\sqrt{2}}{\sqrt{5}}$ **50.** $\frac{14}{\sqrt{7x}}$ **51.** $-\sqrt{\frac{9}{2y}}$

52. $\frac{2}{\sqrt{5y}}$ **53.** $\frac{2}{\sqrt{3}-5}$ **54.** $\frac{1}{\sqrt{2}+3}$

55. $\frac{6}{\sqrt{6}-3}$ **56.** $\frac{4}{\sqrt{5}-\sqrt{13}}$ **57.** $\frac{\sqrt{3}}{\sqrt{x}-7}$

58. $\frac{2}{\sqrt{2}+\sqrt{5}}$ **59.** $\frac{12}{\sqrt{7}+\sqrt{3}}$ **60.** $\frac{\sqrt{y}-1}{\sqrt{y}+3}$

Find the distance between the points in Exercises 61–66.

61. $(4, 7), (-3, 4)$ **62.** $(-2, 5), (3, 0)$

63. $(-4, -2), (-5, 1)$ **64.** $\left(\frac{1}{2}, 2\right), \left(-\frac{3}{2}, 4\right)$

65. $\left(-\frac{1}{2}, 3\right), \left(-\frac{1}{2}, 5\right)$ **66.** $\left(\frac{4}{3}, 2\right), \left(\frac{1}{3}, 2\right)$

67. Show that the triangle determined by points $A(-5, 1)$, $B(-2, 4)$, and $C(1, 1)$ is an isosceles triangle (two equal sides).

68. Show that the triangle determined by points $A(-4, 1)$, $B(4, 9)$, and $C(2, 3)$ is an isosceles triangle (two equal sides).

69. Find the perimeter of the triangle determined by the points $A(2, 0)$, $B(4, 0)$, and $C(7, 4)$.

70. Find the perimeter of the triangle determined by the points $A(-3, -1)$, $B(1, 2)$, and $C(0, 4)$.

CHAPTER 10 TEST

1. State whether $\sqrt{48}$ is a rational or an irrational number.

Simplify the square roots in Exercises 2–5.

2. $\sqrt{125}$

3. $\sqrt{144}$

4. $-\sqrt{\frac{16}{64}}$

5. $-\sqrt{\frac{24}{36}}$

Simplify the radical expressions in Exercises 6–10.

6. $\frac{\sqrt{96} + 12}{24}$

7. $\frac{21 - \sqrt{98}}{14}$

8. $5\sqrt{8} + 3\sqrt{2}$

9. $2\sqrt{27} + 5\sqrt{48} - 4\sqrt{3}$

10. $3\sqrt{2x} - 4\sqrt{50x} + 2\sqrt{72x}$

Multiply the expressions in Exercises 11–14 and simplify if possible.

11. $3\sqrt{5} \cdot 5\sqrt{5}$

12. $\sqrt{15} \cdot (\sqrt{3} + \sqrt{5})$

13. $(5 + \sqrt{11})(5 - \sqrt{11})$

14. $(6 + \sqrt{2})(5 - \sqrt{2})$

Rationalize each denominator in Exercises 15–18 and simplify if possible.

15. $\frac{3}{\sqrt{18}}$

16. $\sqrt{\frac{4}{75}}$

17. $\frac{2}{\sqrt{3} - 1}$

18. $\frac{\sqrt{2}}{\sqrt{3} + \sqrt{2}}$

19. Find the distance between the points $(1, -6)$ and $(4, 3)$.

20. What is the perimeter of the triangle determined by the points $A(1, -3)$, $B(4, 3)$, and $C(1, 5)$?

USING THE COMPUTER

```
10  REM
20  REM  This program uses the Pythagorean theorem to find the
30  REM  length of the hypotenuse of a right triangle given the
40  REM  other two sides.
50  REM
60  '*******************************************************
70  '* SUMSQUARES : The sum of adjacent square and         *
71  '*                opposite square                       *
80  '*******************************************************
90  REM
100 PRINT
"Type in the length of two sides in a right triangle."
110 PRINT
"Then, I'll give you the length of the hypotenuse.(0 to Quit)";
120 INPUT ADJACENT,OPPOSITE
130 IF (ADJACENT = 0) OR (OPPOSITE = 0) GOTO 190
140 LET SUMSQUARES = ADJACENT^2 + OPPOSITE^2
150 LET HYPOTENUSE = SUMSQUARES^.5
160 PRINT
"The hypotenuse of the right triangle is ",HYPOTENUSE
170 PRINT
180 GOTO 100
190 PRINT "Bye!"
200 END
```

```
Type in the length of two sides in a right triangle.
Then, I'll give you the length of the hypotenuse.(0 to Quit)
 3 , 4
The hypotenuse of the right triangle is  5

Type in the length of two sides in a right triangle.
Then, I'll give you the length of the hypotenuse.(0 to Quit)
 5 , 7
The hypotenuse of the right triangle is  8.60233

Type in the length of two sides in a right triangle.
Then, I'll give you the length of the hypotenuse.(0 to Quit)
 12 , 5
The hypotenuse of the right triangle is  13

Type in the length of two sides in a right triangle.
Then, I'll give you the length of the hypotenuse.(0 to Quit)
 6 , 9
The hypotenuse of the right triangle is  10.8167

Type in the length of two sides in a right triangle.
Then, I'll give you the length of the hypotenuse.(0 to Quit)
 0 , 0
Bye!
```

What is the difference between method and device?
A method is a device you can use twice.

GEORGE PÓLYA in *How to Solve It*

DID YOU KNOW?

The quadratic formula is a general method for solving second-degree equations of the form $ax^2 + bx + c = 0$, where a, b, and c can be any real numbers. The quadratic formula is a very old formula; it was known to Babylonian mathematicans around 2000 B.C. However, Babylonian, and later Greek, mathematicians always discarded negative solutions of quadratic equations because they felt that these solutions had no physical meaning. Greek mathematicians always tried to interpret their algebraic problems from a geometrical viewpoint and hence the development of the geometric method of "completing the square."

Consider the following equation: $x^2 + 6x = 7$. A geometric figure is constructed having areas x^2, $3x$ and $3x$:

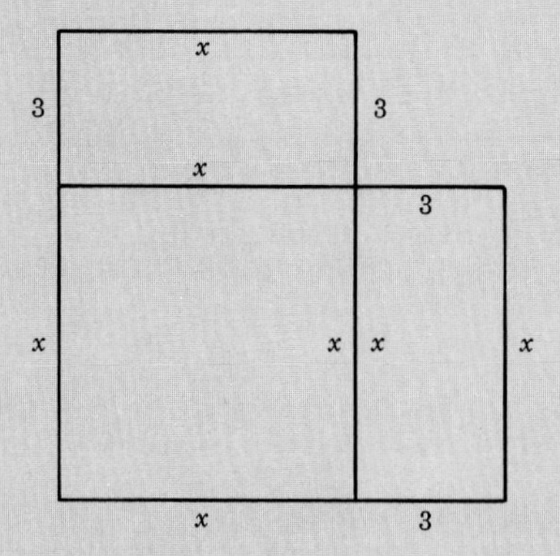

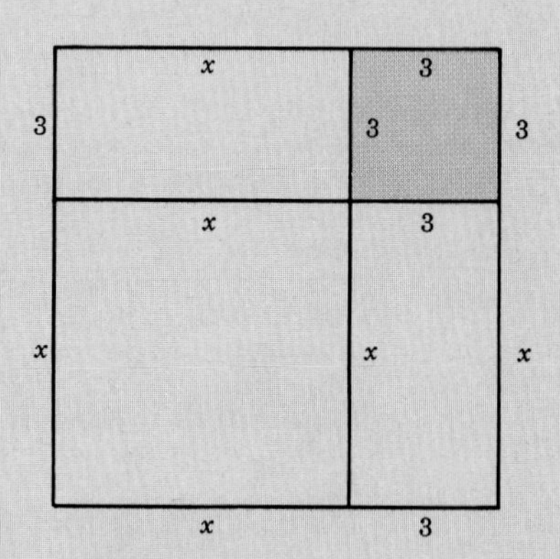

Note that, in order to make the figure a square, one must add a 3-by-3 section (area = 9). Thus, 9 must be added to both sides of the equation to restore equality. Therefore,

$$x^2 + 6x = 7$$

11 Quadratic Equations

and $$x^2 + 6x + 9 = 7 + 9$$
$$x^2 + 6x + 9 = 16$$

So, the square with side $x + 3$ now has an area of 16 square units. Therefore, the sides must be of length 4, which means $x + 3 = 4$. Hence, $x = 1$.

Note that actually the solution set of the original equation is $\{1, -7\}$, since $(-7)^2 + 6(-7) = 49 - 42 = 7$. Thus the Greek mathematicans "lost" the negative solution because of their strictly geometric interpretation of quadratic equations. There were, therefore, many quadratic equations that the Greek mathematicians could not solve because both solutions were negative numbers or complex numbers. Negative solutions to equations were almost completely ignored until the early 1500s during the Renaissance.

11.1 COMPLETING THE SQUARE

We can solve a quadratic equation such as $x^2 = 49$ by factoring as follows:

$$x^2 = 49$$

$$x^2 - 49 = 0 \qquad \text{Get 0 on one side.}$$

$$(x + 7)(x - 7) = 0 \qquad \text{Factor.}$$

$$x + 7 = 0 \quad \text{or} \quad x - 7 = 0 \qquad \text{Solve two equations.}$$

$$x = -7 \qquad\qquad x = 7 \qquad \text{Two solutions.}$$

If the constant is not a perfect square, we can use radicals and still factor as follows:

$$x^2 = 5$$

$$x^2 - 5 = 0 \qquad \text{Get 0 on one side.}$$

$$(x + \sqrt{5})(x - \sqrt{5}) = 0 \qquad \text{Use } \sqrt{5} \text{ since } (\sqrt{5})^2 = 5.$$

$$x + \sqrt{5} = 0 \quad \text{or} \quad x - \sqrt{5} = 0 \qquad \text{Solve two equations.}$$

$$x = -\sqrt{5} \qquad\qquad x = \sqrt{5} \qquad \text{Two solutions.}$$

An equation of the form $x^2 = a$ can be solved by going directly to the square roots, as the following statement indicates.

> If $x^2 = c$ $(c > 0)$, then $x = \sqrt{c}$ or $x = -\sqrt{c}$.
>
> We can write $x = \pm\sqrt{c}$.

EXAMPLES

Solve the following quadratic equations.

1. $x^2 = 17$

$$x = \pm\sqrt{17}$$

Keep in mind that the expression $x = \pm\sqrt{17}$ represents the two equations, $x = \sqrt{17}$ and $x = -\sqrt{17}$.

2. $x^2 = 39$

$$x = \pm\sqrt{39}$$

3. $(x + 4)^2 = 7$

$$x + 4 = \pm\sqrt{7}$$

$$x = -4 \pm \sqrt{7}$$

4. $2(x + 9)^2 = \dfrac{20}{9}$

$$(x + 9)^2 = \frac{10}{9} \quad \text{Divide both sides by 2.}$$

$$x + 9 = \pm\sqrt{\frac{10}{9}}$$

$$x = -9 \pm \frac{\sqrt{10}}{3}$$

The example could also be written as follows to emphasize that there are two solutions.

$$2(x + 9)^2 = \frac{20}{9}$$

$$(x + 9)^2 = \frac{10}{9}$$

$$x + 9 = \sqrt{\frac{10}{9}} \quad \text{or} \quad x + 9 = -\sqrt{\frac{10}{9}}$$

$$x = -9 + \frac{\sqrt{10}}{3} \qquad x = -9 - \frac{\sqrt{10}}{3}$$

There is no real number whose square is negative. So, to solve equations such as $x^2 = -4$ and $(x - 3)^2 = -10$, numbers called **complex numbers** are needed. These numbers will be discussed in the next course in algebra.

Expressions such as $x^2 + 6x + 9$, $x^6 - 4x^3 + 4$, and $a^2 + 2ab + b^2$ are perfect square trinomials since they are the squares of binomials.

$$x^2 + 6x + 9 = (x + 3)^2$$

$$x^6 - 4x^3 + 3 = (x^3 - 2)^2$$

$$a^2 + 2ab + b^2 = (a + b)^2$$

In Section 5.2, we discussed **completing the square** of a quadratic (second-degree) expression. For example, given

$$x^2 + 12x,$$

36 will complete the square since $\frac{1}{2}(12) = 6$ and $6^2 = 36$.

$$x^2 + 12x + \underline{36} = (x + 6)^2$$

Similarly, given

$$a^2 - 20a$$

100 will complete the square since $\frac{1}{2}(-20) = -10$ and $(-10)^2 = 100$.

$$a^2 - 20a + \underline{100} = (a - 10)^2$$

Now we want to use this idea of completing the square to solve quadratic equations.

To Solve Quadratic Equations by Completing the Square

1. Arrange terms with variables on one side and constants on the other.
2. Divide each term by the coefficient of x^2. (We want the coefficient of x^2 to be 1.)
3. Find the number that completes the square of the quadratic expression, and add this number to **both** sides of the equation.
4. Find the square roots of both sides.
5. Solve for x. Remember, there will usually be two solutions.

EXAMPLES

Solve the following quadratic equations by completing the squares.

5. $x^2 - 6x + 4 = 0$

$x^2 - 6x = -4$	Add -4 to both sides of the equation.
$x^2 - 6x + \mathbf{9} = -4 + \mathbf{9}$	Add 9 to both sides of the equation. The left side is now a **perfect square trinomial.** $\frac{1}{2}(-6) = -3$ and $(-3)^2 = 9$
$(x - 3)^2 = 5$	Simplify.
$x - 3 = \pm\sqrt{5}$	Find square roots.
$x = 3 \pm \sqrt{5}$	Solve for x.

6. $x^2 + 5x = 7$

$x^2 + 5x + \mathbf{\frac{25}{4}} = 7 + \mathbf{\frac{25}{4}}$	Complete the square on the left. $\frac{1}{2} \cdot 5 = \frac{5}{2}$ and $\left(\frac{5}{2}\right)^2 = \frac{25}{4}$
$\left(x + \frac{5}{2}\right)^2 = \frac{53}{4}$	Simplify.
$x + \frac{5}{2} = \pm\sqrt{\frac{53}{4}}$	Find square roots.
$x = -\frac{5}{2} \pm \sqrt{\frac{53}{4}}$	Solve for x.
$x = -\frac{5}{2} \pm \frac{\sqrt{53}}{2}$	Special property of square roots $\sqrt{\frac{a}{b}} = \frac{\sqrt{a}}{\sqrt{b}}$ for $a > 0$ and $b > 0$.
$x = \frac{-5 \pm \sqrt{53}}{2}$	Simplify.

7. $6x^2 + 12x - 9 = 0$

$6x^2 + 12x = 9$	Add 9 to both sides of the equation.
$\frac{6x^2}{6} + \frac{12x}{6} = \frac{9}{6}$	Divide each term by 6 **so that the leading coefficient will be 1.**
$x^2 + 2x = \frac{3}{2}$	The leading coefficient is 1.
$x^2 + 2x + \mathbf{1} = \frac{3}{2} + \mathbf{1}$	Complete the square, $\frac{1}{2} \cdot 2 = 1$ and $1^2 = 1$.
$(x + 1)^2 = \frac{5}{2}$	Simplify.
$x + 1 = \pm\sqrt{\frac{5}{2}}$	Find square roots.
$x = -1 \pm \sqrt{\frac{5}{2}}$	Solve for x.
$x = -1 \pm \frac{\sqrt{5}}{\sqrt{2}} \cdot \frac{\sqrt{2}}{\sqrt{2}}$	Rationalize the denominator.
$x = -1 \pm \frac{\sqrt{10}}{2}$	Simplify.
or $x = \frac{-2 \pm \sqrt{10}}{2}$	

8. $2x^2 + 5x - 8 = 0$

$2x^2 + 5x = 8$ — Add 8 to both sides of the equation.

$\frac{2x^2}{2} + \frac{5x}{2} = \frac{8}{2}$ — Divide each term by 2 **so that the leading coefficient will be 1.**

$x^2 + \frac{5}{2}x = 4$ — Simplify.

$x^2 + \frac{5}{2}x + \mathbf{\frac{25}{16}} = 4 + \mathbf{\frac{25}{16}}$ — Complete the square, $\frac{1}{2} \cdot \frac{5}{2} = \frac{5}{4}$ and $\left(\frac{5}{4}\right)^2 = \frac{25}{16}$.

$\left(x + \frac{5}{4}\right)^2 = \frac{89}{16}$ — Simplify.

$x + \frac{5}{4} = \pm\sqrt{\frac{89}{16}}$ — Find the square roots.

$x = -\frac{5}{4} \pm \frac{\sqrt{89}}{4}$ — Solve for x; also, $\sqrt{\frac{89}{16}} = \frac{\sqrt{89}}{\sqrt{16}} = \frac{\sqrt{89}}{4}$.

$x = \frac{-5 \pm \sqrt{89}}{4}$ — Simplify.

9. $x^2 + 8x = -4$

$x^2 + 8x + \mathbf{16} = -4 + \mathbf{16}$ — Complete the square.

$(x + 4)^2 = 12$ — Simplify.

$x + 4 = \pm\sqrt{12}$ — Find square roots.

$x = -4 \pm \sqrt{12}$ — Solve for x.

$x = -4 \pm 2\sqrt{3}$ — $\sqrt{12} = \sqrt{4} \cdot \sqrt{3} = 2\sqrt{3}$ by special property $\sqrt{ab} = \sqrt{a} \cdot \sqrt{b}$ for $a > 0$ and $b > 0$.

EXERCISES 11.1

Solve the quadratic equations in Exercises 1–12.

1. $x^2 = 121$

2. $x^2 - 35 = 0$

3. $x^2 - 42 = 0$

4. $x^2 - 62 = 0$

5. $(x - 1)^2 = 4$

6. $(x + 3)^2 = 9$

7. $(x + 1)^2 = \frac{1}{4}$

8. $(x - 9)^2 = \frac{9}{25}$

9. $2(x - 6)^2 = 18$

10. $3(x + 8)^2 = 75$

11. $2(x - 7)^2 = 24$

12. $(x + 11)^2 = \frac{3}{4}$

Add the correct constant in Exercises 13–22 in order to make the trinomial factorable as indicated. (Constant factors must be dealt with in Exercises 15, 16, 21, and 22.)

13. $x^2 + 12x + \_\_\_\_\_\_ = (\qquad)^2$

14. $x^2 - 14x + \_\_\_\_\_\_ = (\qquad)^2$

15. $2x^2 - 16x + \_\_\_\_\_\_ = 2(\qquad)^2$

16. $5x^2 + 10x + \_\_\_\_\_\_ = 5(\qquad)^2$

17. $x^2 - 3x + \_\_\_\_\_\_ = (\qquad)^2$

18. $x^2 + 5x + \_\_\_\_\_\_ = (\qquad)^2$

19. $x^2 + x + \_\_\_\_\_\_ = (\qquad)^2$

20. $x^2 - 7x + \_\_\_\_\_\_ = (\qquad)^2$

21. $2x^2 + 4x + \_\_\_\_\_\_ = 2(\qquad)^2$

22. $3x^2 + 18x + \_\_\_\_\_\_ = 3(\qquad)^2$

Solve the quadratic equations in Exercises 23–40 by completing the squares.

23. $x^2 + 6x - 7 = 0$

24. $x^2 + 8x + 12 = 0$

25. $x^2 - 4x - 45 = 0$

26. $x^2 - 10x + 21 = 0$

27. $x^2 - 3x - 40 = 0$

28. $x^2 + x - 42 = 0$

29. $3x^2 + x - 4 = 0$

30. $2x^2 + x - 6 = 0$

31. $4x^2 - 4x - 3 = 0$

32. $3x^2 + 11x + 10 = 0$

33. $x^2 + 6x + 3 = 0$

34. $x^2 - 4x - 3 = 0$

35. $x^2 + 2x - 5 = 0$

36. $x^2 + 8x + 2 = 0$

37. $x^2 + x - 3 = 0$

38. $x^2 + 5x + 1 = 0$

39. $2x^2 + 3x - 1 = 0$

40. $3x^2 - 4x - 2 = 0$

Solve the quadratic equations in Exercises 41–60 by factoring or completing the square.

41. $x^2 - 9x + 2 = 0$

42. $x^2 - 8x - 20 = 0$

43. $x^2 + 7x - 14 = 0$

44. $x^2 + 5x + 3 = 0$

45. $x^2 - 11x - 26 = 0$

46. $x^2 + 6x - 4 = 0$

47. $2x^2 + 5x + 2 = 0$

48. $6x^2 - 2x - 2 = 0$

49. $3x^2 - 6x + 3 = 0$

50. $5x^2 + 15x - 5 = 0$

51. $4x^2 + 20x - 8 = 0$

52. $2x^2 - 7x + 4 = 0$

53. $6x^2 - 8x + 1 = 0$

54. $5x^2 - 10x + 3 = 0$

55. $2x^2 + 7x + 4 = 0$

56. $3x^2 - 5x - 3 = 0$

57. $4x^2 - 2x - 3 = 0$

58. $2x^2 - 9x + 7 = 0$

59. $3x^2 + 8x + 5 = 0$

60. $5x^2 + 11x - 1 = 0$

11.2 THE QUADRATIC FORMULA: $\mathbf{x = \dfrac{-b \pm \sqrt{b^2 - 4ac}}{2a}}$

Now we are interested in developing a formula that will be useful in solving quadratic equations of any form. **This formula will always work,** but you should not forget the factoring and completing the square techniques because they can be easier to apply than the formula.

The **general quadratic equation** is

$$ax^2 + bx + c = 0, \qquad a \neq 0$$

We want to solve the general quadratic equation for x in terms of the coefficients a, b, and c. The technique is to **complete the square** (Section 11.1), treating a, b, and c as constants.

Development of the Quadratic Formula

$$ax^2 + bx + c = 0, \qquad a > 0$$

$$ax^2 + bx = -c$$

Add $-c$ to both sides.

$$\frac{ax^2}{a} + \frac{bx}{a} = \frac{-c}{a}$$

Divide each term by a.

$$x^2 + \frac{b}{a}x = \frac{-c}{a}$$

Simplify: $\dfrac{bx}{a} = \dfrac{b}{a}x$.

$$x^2 + \frac{b}{a}x + \left(\frac{b}{2a}\right)^2 = \left(\frac{b}{2a}\right)^2 + \frac{-c}{a}$$

Complete the square, $\dfrac{1}{2}\left(\dfrac{b}{a}\right) = \dfrac{b}{2a}$.

$$\left(x + \frac{b}{2a}\right)^2 = \frac{b^2}{4a^2} + \frac{-c}{a}$$ Simplify.

$$\left(x + \frac{b}{2a}\right)^2 = \frac{b^2}{4a^2} + \frac{-c \cdot 4a}{a \cdot 4a}$$ Common denominator is $4a^2$.

$$\left(x + \frac{b}{2a}\right)^2 = \frac{b^2 - 4ac}{4a^2}$$ Simplify.

$$x + \frac{b}{2a} = \pm\sqrt{\frac{b^2 - 4ac}{4a^2}}$$ Find the square roots.

$$x + \frac{b}{2a} = \frac{\pm\sqrt{b^2 - 4ac}}{\pm\sqrt{4a^2}}$$ Use the relationship $\sqrt{\frac{a}{b}} = \frac{\sqrt{a}}{\sqrt{b}}$ if $a, b > 0$.

$$x + \frac{b}{2a} = \pm\frac{\sqrt{b^2 - 4ac}}{2a}$$ Simplify.

$$x = \frac{-b}{2a} \pm \frac{\sqrt{b^2 - 4ac}}{2a}$$ Solve for x.

$$x = \frac{-b \pm \sqrt{b^2 - 4ac}}{2a}$$ **THE QUADRATIC FORMULA**

Special Note: The expression $b^2 - 4ac$ is called the **discriminant.** If $b^2 - 4ac < 0$, then there are no real number solutions because the square root of a negative number is not a real number. A discussion of negative discriminants is given in later courses in algebra.

Now the solutions to quadratic equations can be found by going directly to the formula.

EXAMPLES

Solve the following quadratic equations using the quadratic formula.

1. $2x^2 + x - 2 = 0$

   $a = 2$, $b = 1$, and $c = -2$

$$x = \frac{-1 \pm \sqrt{1^2 - 4(2)(-2)}}{2 \cdot 2} = \frac{-1 \pm \sqrt{1 + 16}}{4} = \frac{-1 \pm \sqrt{17}}{4}$$

2. $3x^2 - 5x + 1 = 0$

   $a = 3$, $b = -5$, and $c = 1$

$$x = \frac{-(-5) \pm \sqrt{(-5)^2 - 4(3)(1)}}{2 \cdot 3} = \frac{5 \pm \sqrt{25 - 12}}{6} = \frac{5 \pm \sqrt{13}}{6}$$

3. $\frac{1}{6}x^2 - x + \frac{1}{2} = 0$

$$6 \cdot \frac{1}{6}x^2 - 6 \cdot x + 6 \cdot \frac{1}{2} = 6 \cdot 0$$

Multiply each term by 6, the least common denominator.

$$x^2 - 6x + 3 = 0$$

$a = 1$, $b = -6$, and $c = 3$

$$x = \frac{-(-6) \pm \sqrt{(-6)^2 - 4(1)(3)}}{2 \cdot 1} = \frac{6 \pm \sqrt{36 - 12}}{2}$$

$$= \frac{6 \pm \sqrt{24}}{2} = \frac{6 \pm 2\sqrt{6}}{2}$$

$$= 3 \pm \sqrt{6}$$

$\left(\textbf{Note: } \frac{6 \pm \sqrt{4} \cdot \sqrt{6}}{2} = \frac{6 \pm 2\sqrt{6}}{2} = \frac{6}{2} \pm \frac{2\sqrt{6}}{2} = 3 \pm \sqrt{6}\right)$

4. $2x^2 - 25 = 0$

We could add 25 to both sides, divide by 2, and then take the square roots. We can arrive at the same result using the quadratic formula, but note that $b = 0$.

$2x^2 - 25 = 0$

$a = 2$, $b = 0$, and $c = -25$

$$x = \frac{-(0) \pm \sqrt{0^2 - 4(2)(-25)}}{2 \cdot 2} = \frac{\pm\sqrt{200}}{4} = \frac{\pm 10\sqrt{2}}{4}$$

$$= \frac{\pm 5\sqrt{2}}{2}$$

5. $-5x^2 + 3x = -2$

$-5x^2 + 3x + 2 = 0$ One side must be 0.

$a = -5$, $b = 3$, and $c = 2$

$$x = \frac{-3 \pm \sqrt{3^2 - 4(-5)(2)}}{2(-5)} = \frac{-3 \pm \sqrt{9 + 40}}{-10}$$

$$= \frac{-3 \pm \sqrt{49}}{-10} = \frac{-3 \pm 7}{-10}$$

$$x = \frac{-3 + 7}{-10} = \frac{4}{-10} = -\frac{2}{5} \quad \text{or} \quad x = \frac{-3 - 7}{-10} = \frac{-10}{-10} = 1$$

This example could also be solved by factoring:

$$-5x^2 + 3x + 2 = 0$$

$$5x^2 - 3x - 2 = 0 \qquad \text{Multiply both sides of the equation by } -1.$$

$$(5x + 2)(x - 1) = 0$$

$$5x + 2 = 0 \qquad \text{or} \qquad x - 1 = 0$$

$$5x = -2 \qquad\qquad x = 1$$

$$x = -\tfrac{2}{5}$$

Equations such as in Example 5 should be solved by factoring because factoring is generally easier than applying the quadratic formula. However, if you do not "see" the factors readily, the quadratic formula will give the solutions, as the following example illustrates.

EXAMPLES

6. $4x^2 + 12x + 9 = 0$

$a = 4,\ b = 12,\ c = 9$

$$x = \frac{-12 \pm \sqrt{12^2 - 4(4)(9)}}{2 \cdot 4} = \frac{-12 \pm \sqrt{144 - 144}}{8}$$

$$= \frac{-12 \pm \sqrt{0}}{8} \qquad (\sqrt{0} = 0)$$

$$= -\frac{12}{8} = -\frac{3}{2}$$

Note that when the discriminant is 0, there is only one solution. This equation could also be solved by factoring.

7. $x^2 + x + 1 = 0$

$a = 1,\ b = 1,\ c = 1$

$$x = \frac{-1 \pm \sqrt{1^2 - 4(1)(1)}}{2 \cdot 1} = \frac{-1 \pm \sqrt{1 - 4}}{2} = \frac{-1 \pm \sqrt{-3}}{2}$$

$$= \text{undefined}$$

There is no real solution. This example illustrates the fact that not every equation has real solutions. None of the exercises has this kind of answer.

PRACTICE QUIZ

Questions	Answers
Solve the equations using the quadratic formula.	
1. $x^2 + 5x + 1 = 0$	1. $x = \dfrac{-5 \pm \sqrt{21}}{2}$
2. $4x^2 - x - 1 = 0$	2. $x = \dfrac{1 \pm \sqrt{17}}{8}$
3. $x^2 - 4x - 4 = 0$	3. $x = 2 \pm 2\sqrt{2}$
4. $-2x^2 + 3 = 0$ (**Note:** Here $b = 0$.)	4. $x = \dfrac{0 \pm 2\sqrt{6}}{-4}$ or $x = \dfrac{\pm\sqrt{6}}{2}$
5. $5x^2 - x = 0$ (**Note:** Here $c = 0$.)	5. $x = \dfrac{1 \pm \sqrt{1}}{10}$ or $x = 0, x = \dfrac{1}{5}$ (This problem could be solved easily by factoring.)

EXERCISES 11.2

Write each of the quadratic equations in Exercises 1–12 in the form $ax^2 + bx + c = 0$ with $a > 0$ and a, b, c as integers; then identify the constants a, b, and c.

1. $x^2 - 3x = 2$

2. $x^2 + 2 = 5x$

3. $x = 2x^2 + 6$

4. $5x^2 = 3x - 1$

5. $4x + 3 = 7x^2$

6. $x = 4 - 3x^2$

7. $4 = 3x^2 - 9x$

8. $6x + 4 = 3x^2$

9. $x^2 + 5x = 3 - x^2$

10. $x^2 + 4x - 1 = 2x + 3x^2$

11. $\dfrac{2}{3}x^2 + x - 1 = 0$

12. $\dfrac{x^2}{4} + x - \dfrac{1}{2} = 0$

Solve the quadratic equations in Exercises 13–32 using the quadratic formula.

13. $x^2 + 4x - 1 = 0$

14. $x^2 - 3x + 1 = 0$

15. $x^2 - 3x - 4 = 0$

16. $x^2 + 5x - 2 = 0$

17. $-2x^2 + x + 1 = 0$

18. $-3x^2 + x + 1 = 0$

19. $5x^2 + 3x - 2 = 0$

20. $-2x^2 + 5x - 1 = 0$

21. $6x^2 - 3x - 1 = 0$

22. $4x^2 - 7x + 3 = 0$

23. $x^2 - 7 = 0$

24. $2x^2 + 5x - 3 = 0$

25. $x^2 + 4x = x - 2x^2$

26. $x^2 - 2x + 1 = 2 - 3x^2$

27. $3x^2 + 4x = 0$

28. $4x^2 - 10 = 0$

29. $\frac{2}{5}x^2 + x - 1 = 0$

30. $2x^2 + 3x - \frac{3}{4} = 0$

31. $-\frac{x^2}{2} - 2x + \frac{1}{3} = 0$

32. $\frac{x^2}{3} - x - \frac{1}{5} = 0$

Solve the quadratic equations in Exercises 33–50 using any method (factoring, completing the square, or quadratic formula).

33. $2x^2 + 7x + 3 = 0$

34. $5x^2 - x - 4 = 0$

35. $9x^2 - 6x - 1 = 0$

36. $3x^2 - 7x + 1 = 0$

37. $2x^2 - 2x - 1 = 0$

38. $10x^2 = x + 24$

39. $9x^2 + 12x + 2 = 0$

40. $3x^2 - 11x - 4 = 0$

41. $5x^2 = 7x + 5$

42. $-4x^2 + 11x - 5 = 0$

43. $-4x^2 + 12x - 9 = 0$

44. $10x^2 + 35x + 30 = 0$

45. $6x^2 + 2x = 20$

46. $25x^2 + 4 = 20$

47. $3x^2 - 4x + \frac{1}{3} = 0$

48. $\frac{3}{4}x^2 - 2x + \frac{1}{8} = 0$

49. $\frac{11}{2}x + 1 = 3x^2$

50. $\frac{3}{7}x^2 = \frac{1}{2}x + 1$

11.3 APPLICATIONS

Application problems are designed to teach you to read carefully, to think clearly, and to translate from English to algebraic expressions and equations. The problems do not tell you directly to add, subtract, multiply, divide, or square. You must decide on a method of attack from the wording of the problem, combined with your previous experience and knowledge. Study the following examples and read the explanations carefully. They are similar to some, but not all, of the problems in the exercises.

In a **right triangle,** one of the angles is a right angle (measures 90°), and the side opposite this angle (the longest side) is called **the hypotenuse.** Pythagoras was a famous Greek mathematician who is given credit for proving the following very important and useful theorem.

Pythagorean Theorem

In a right triangle, the square of the hypotenuse is equal to the sum of the squares of the two sides.

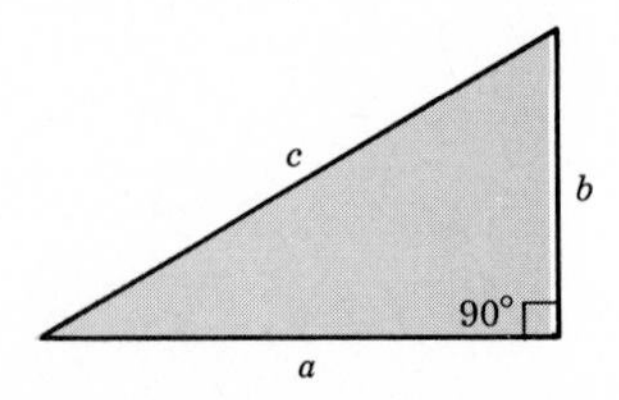

In Section 10.4, we used the Pythagorean Theorem to develop the formula for the distance between two points. Example 1 illustrates how the theorem can be used to solve word problems.

EXAMPLE 1: The Pythagorean Theorem

The length of a rectangle is 7 feet longer than the width. If one diagonal measures 13 feet, what are the dimensions of the rectangle?

Solution: Draw a diagram for problems involving geometric figures whenever possible.

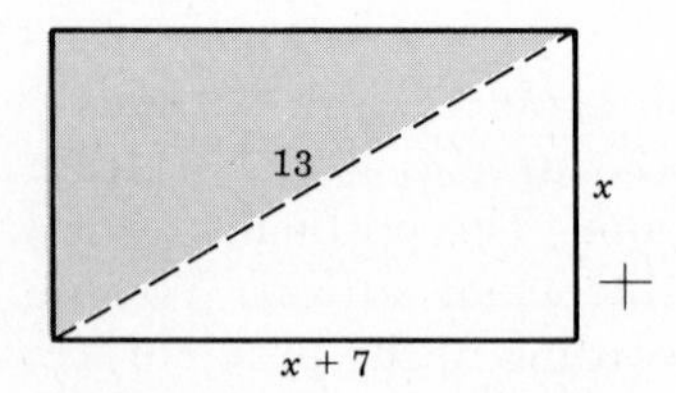

Let x = width of the rectangle
$x + 7$ = length of the rectangle

$$(x + 7)^2 + x^2 = 13^2 \quad \text{Use the Pythagorean Theorem.}$$

$$x^2 + 14x + 49 + x^2 = 169$$

$$2x^2 + 14x + 49 - 169 = 0$$

$$2x^2 + 14x - 120 = 0$$

$$2(x^2 + 7x - 60) = 0$$

$$2(x - 5)(x + 12) = 0$$

$$x - 5 = 0 \quad \text{or} \quad x + 12 = 0$$

$$x = 5 \quad \cancel{x = -12}$$

A negative number does not fit the conditions of the problem.

$$x = 5 \text{ ft width}$$

$$x + 7 = 12 \text{ ft length}$$

Check: $5^2 + 12^2 \stackrel{?}{=} 13^2$

$$25 + 144 \stackrel{?}{=} 169$$

$$169 = 169$$

The width is 5 ft and the length is 12 ft.

EXAMPLE 2: Work

Working for a janitorial service, a woman and her daughter clean a building in 5 hours. If the daughter were to do the job by herself, she would take 24 hours longer than her mother would take. How long would it take her mother to clean the building without the daughter's help?

Solution: This problem is similar to the work problems discussed in Section 7.4.

Let x = time for mother alone

$x + 24$ = time for daughter alone

	hours	part in 1 hour
mother	x	$\frac{1}{x}$
daughter	$x + 24$	$\frac{1}{x + 24}$
together	5	$\frac{1}{5}$

$$\underbrace{\begin{array}{c}\text{part done}\\ \text{by mother}\\ \text{in 1 hour}\end{array}}_{\frac{1}{x}} + \underbrace{\begin{array}{c}\text{part done by}\\ \text{daughter in}\\ \text{1 hour}\end{array}}_{\frac{1}{x+24}} = \underbrace{\begin{array}{c}\text{part done work-}\\ \text{ing together in}\\ \text{1 hour}\end{array}}_{\frac{1}{5}}$$

$$\frac{1}{x}\mathbf{(5x)(x+24)} + \frac{1}{x+24}\mathbf{(5x)(x+24)} = \frac{1}{5}\mathbf{(5x)(x+24)}$$ Multiply each term by the LCM of the denominators.

$$5(x+24) + 5x = x(x+24)$$

$$5x + 120 + 5x = x^2 + 24x$$

$$0 = x^2 + 24x - 10x - 120$$

$$0 = x^2 + 14x - 120$$

$$0 = (x-6)(x+20)$$

$$x - 6 = 0 \quad \text{or} \quad x + 20 = 0$$

$$x = 6 \qquad \cancel{x = -20}$$

The mother could do the job alone in 6 hours.

EXAMPLE 3: Distance, Rate, Time

An airplane travels at a top speed of 200 mph. The plane, flying at top speed, is clocked over a distance of 960 miles and takes 2 hours more on the return trip because of a head wind. (A tail wind was helping on the first leg of the flight.) What was the wind velocity?

Solution: The basic formula is $d = rt$ (distance = rate × time). Also, $t = \frac{d}{r}$ and $r = \frac{d}{t}$.

If we know or can represent any two of the three quantities, the formula using these two should be used.

Let

$$x = \text{wind velocity}$$

$$200 + x = \text{speed going}$$

$$200 - x = \text{returning}$$

$$960 = \text{distance each way}$$

We know distance and can represent rate (or speed), so the formula $t = \frac{d}{r}$ is appropriate.

	rate	time	distance
going	$200 + x$	$\frac{960}{200 + x}$	960
returning	$200 - x$	$\frac{960}{200 - x}$	960

$$\underbrace{\begin{matrix}\text{time}\\ \text{returning}\end{matrix}}_{\frac{960}{200 - x}} - \underbrace{\begin{matrix}\text{time}\\ \text{going}\end{matrix}}_{\frac{960}{200 + x}} = \underbrace{\begin{matrix}\text{difference}\\ \text{in time}\end{matrix}}_{2}$$

$$\frac{960}{200 - x}\mathbf{(200 - x)(200 + x)} - \frac{960}{200 + x}\mathbf{(200 - x)(200 + x)} = 2\mathbf{(200 - x)(200 + x)}$$

$$960(200 + x) - 960(200 - x) = 2(40{,}000 - x^2)$$

$$192{,}000 + 960x - 192{,}000 + 960x = 80{,}000 - 2x^2$$

$$2x^2 + 1920x - 80{,}000 = 0$$

$$x^2 + 960x - 40{,}000 = 0$$

$$(x - 40)(x + 1000) = 0$$

$$x - 40 = 0 \quad \text{or} \quad x + 1000 = 0$$

$$x = 40 \qquad \cancel{x = -1000}$$

The wind velocity was 40 mph.

Check: $\frac{960}{200 - 40} - \frac{960}{200 + 40} \stackrel{?}{=} 2$

$$\frac{960}{160} - \frac{960}{240} \stackrel{?}{=} 2$$

$$6 - 4 \stackrel{?}{=} 2$$

$$2 = 2$$

EXAMPLE 4: Geometry

A square piece of cardboard has a small square, 2 in. by 2 in., cut from each corner. The edges are then folded up to form a box with a volume of 5000 cu in. What are the dimensions of the box? (**Hint:** The volume is the product of the length, width, and height: $V = lwh$.)

Solution: Draw a diagram illustrating the information.

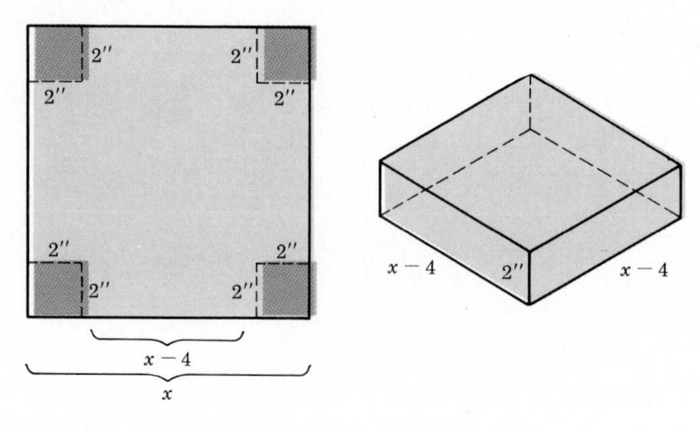

Let x = one side of the square.

$$2(x - 4)(x - 4) = 5000$$
$$2(x^2 - 8x + 16) = 5000$$
$$x^2 - 8x + 16 = 2500$$
$$x^2 - 8x - 2484 = 0$$
$$(x - 54)(x + 46) = 0$$
$$x - 54 = 0 \quad \text{or} \quad x + 46 = 0$$
$$x = 54 \qquad \cancel{x = -46}$$
$$x - 4 = 50$$

The dimensions of the box are 50 in. by 50 in. by 2 in.

EXERCISES 11.3

Determine a quadratic equation for each of the following problems. Then solve the equation.

1. The sum of a positive number and its square is 132. Find the number.

2. The dimensions of a rectangle can be represented by two consecutive even integers. The area of the rectangle is 528 square centimeters. Find the width and the length of the rectangle. (**Hint:** area = length · width.)

3. A rectangle has a length 5 meters less than twice its width. If the area is 63 square meters, find the dimensions of the rectangle. (**Hint:** area = length · width.)

4. The sum of a positive number and its square is 992. Find the number.

5. The area of a rectangular field is 198 square meters. If it takes 58 meters of fencing to enclose the field, what are the dimensions of the field? (**Hint:** The length plus the width is 29 meters.)

6. The length of a rectangle exceeds the width by 5 cm. If both were increased by 3 cm, the area would be increased by 96 sq cm. Find the dimensions of the original rectangle. (**Hint:** The original rectangle has area $w(w + 5)$.)

7. The Willsons have a rectangular swimming pool that is 10 ft longer than it is wide. The pool is completely surrounded by a concrete deck that is 6 ft wide. The total area of the pool and the deck is 1344 sq ft. Find the dimensions of the pool.

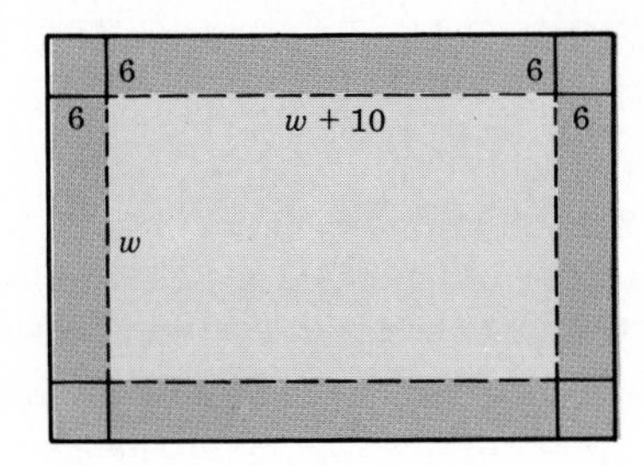

8. The product of two positive consecutive odd integers exceeds their sum by 287. Find the numbers.

9. The sum of the squares of two positive consecutive integers is 221. Find the positive integers.

10. The difference between two positive numbers is 9. If the small number is added to the square of the larger number, the result is 147. Find the numbers.

11. The difference between a positive number and 3 is four times the reciprocal of the number. Find the number. $\left(\textbf{Hint:} \text{ The reciprocal of } x \text{ is } \frac{1}{x}.\right)$

12. The sum of a positive number and 5 is fourteen times the reciprocal of the number. Find the number. $\left(\textbf{Hint:} \text{ The reciprocal of } x \text{ is } \frac{1}{x}.\right)$

13. Each side of a square is increased by 10 cm. The area of the resulting square is 9 times the area of the original square. Find the length of the sides of the original square.

14. If 5 meters are added to each side of a square, the area of the resulting square is four times the area of the original square. Find the length of the sides of the original square.

16. The diagonal of a rectangle is 13 meters. The length is 2 meters more than twice the width. Find the dimensions of the rectangle.

16. The length of a rectangle is 4 meters more than its width. If the diagonal is 20 meters, what are the dimensions of the rectangle?

17. A right triangle has two equal sides. If the hypotenuse is 12 cm, what is the length of the equal sides?

18. Mr. Prince owns a 15-unit apartment complex. The rent for each apartment is presently $200 per month and all units are rented. Each time the rent is increased $20, he will lose 1 tenant. What is the rental rate if he receives $3120 monthly in rent? (**Hint:** Let x = number of empty units.)

19. The Ski Club is planning to charter a bus to a ski resort. The cost will be $900, and each member will share the cost equally. If the club had 15 more members, the cost per person would be $10 less. How many are in the club now? $\left(\textbf{Hint: } \text{If } x = \text{number in club now, } \frac{900}{x} = \text{cost per person.}\right)$

20. A sporting goods store owner estimates that if he sells a certain model of basketball shoes for x dollars a pair, he will be able to sell $40 - x$ pairs. Find the price if his sales were $375. Is there more than one possible answer?

21. Mr. Green traveled to a city 200 miles from his home to attend a meeting. Due to car trouble, his average speed returning was 10 mi/hr less than his speed going. If the total driving time for the round trip was 9 hours, at what rate of speed did he travel to the city?

	rate	time	distance
going	x		200
returning	$x - 10$		200

22. A motor boat takes a total of 2 hours to travel 8 miles downstream and 4 miles back on a river that is flowing at a rate of 2 mph. Find the rate of the boat in still water.

23. A small motor boat travels 12 mph in still water. It takes 2 hours longer to travel 45 miles going upstream than it does going downstream. Find the rate of the current. (**Hint:** $12 + c$ = rate going downstream and $12 - c$ = rate going upstream.)

24. Recently Mr. and Mrs. Roberts spent their vacation in San Francisco, which is 540 miles from their home. Being a little reluctant to return home, the Roberts took 2 hours longer on their return trip and their average speed was 9 mi/hr slower than when they were going. What was their average rate of speed as they traveled from home to San Francisco?

25. The Blumin Garden Club planned to give their president a gift of appreciation costing $120 and to divide the cost evenly. In the meantime, 5 members dropped out of the club. If it now costs each of the remaining members $2 more than originally planned, how many members did the club initially have? $\left(\textbf{Hint:}\text{ If } x = \text{number in club initially, } \frac{120}{x} = \text{cost per member.}\right)$

26. A rectangular sheet of metal is 6 in. longer than it is wide. A box is to be made by cutting out 3-in. squares at each corner and folding up the sides. If the box has a volume of 336 cu in., what were the dimensions of the sheet metal? (See Example 4.)

27. A box is to be made out of a square piece of cardboard by cutting out 2-in. squares at each corner and folding up the sides. If the box has a volume of 162 cu in., how big was the piece of cardboard? (See Example 4.)

28. A man and his son can paint their cabin in 3 hours. Working alone, it would take the son 8 hours longer than it would the father. How long would it take the father to paint the cabin alone?

29. Two pipes can fill a tank in 8 minutes if both are turned on. If only one is used, it would take 30 minutes longer for the smaller pipe to fill the tank than the larger pipe. How long will it take this small pipe to fill the tank?

30. A farmer and his son can plow a field with two tractors in 4 hours. If it would take the son 6 hours longer than the father to plow the field alone, how long would it take each if they worked alone?

CHAPTER 11 SUMMARY

If $x^2 = c$ $(c > 0)$, then $x = \sqrt{c}$ or $x = -\sqrt{c}$.

We can write $x = \pm\sqrt{c}$.

To Solve Quadratic Equations by Completing the Square

1. Arrange terms with variables on one side and constants on the other.
2. Divide each term by the coefficient of x^2. (We want the coefficient of x^2 to be 1.)
3. Find the number that completes the square of the quadratic expression, and add this number to **both** sides of the equation.
4. Find the square roots of both sides.
5. Solve for x. Remember, there will usually be two solutions.

The **general quadratic equation is**

$$ax^2 + bx + c = 0, \qquad a \neq 0$$

The **QUADRATIC FORMULA** is

$$x = \frac{-b \pm \sqrt{b^2 - 4ac}}{2a}$$

The expression $b^2 - 4ac$ is called the **discriminant.**

CHAPTER 11 REVIEW

Solve the quadratic equation in Exercises 1–8.

1. $x^2 = 49$ **2.** $x^2 = 25$ **3.** $x^2 - \frac{1}{4} = 0$

4. $9x^2 - 162 = 0$ **5.** $(x + 2)^2 = 4$ **6.** $(x - 1)^2 = 9$

7. $(x - 5)^2 = 7$ **8.** $3(x - 4)^2 = 24$

Add the correct constant in Exercises 9–14 in order to make the trinomial factorable as indicated.

9. $x^2 - 4x +$ _____ $= (\qquad)^2$

10. $3x^2 + 18x +$ _____ $= 3(\qquad)^2$

11. $x^2 - \frac{2}{3}x +$ _____ $= (\qquad)^2$

12. $x^2 - x +$ _____ $= (\qquad)^2$

13. $x^2 + 9x +$ _____ $= (\qquad)^2$

14. $5x^2 - 10x +$ _____ $= 5(\qquad)^2$

Solve the quadratic equations in Exercises 15–20 by completing the square.

15. $x^2 - 6x - 7 = 0$

16. $x^2 + 4x + 2 = 0$

17. $x^2 + 2x - 4 = 0$

18. $2x^2 + 8x - 1 = 0$

19. $2x^2 + 2x - 3 = 0$

20. $3x^2 - 2x - 2 = 0$

Solve the quadratic equations in Exercises 21–28 by factoring or completing the square.

21. $x^2 + 6x - 16 = 0$

22. $x^2 + 8x - 48 = 0$

23. $2x^2 + 7x - 4 = 0$

24. $5x^2 - 6x = 0$

25. $5x^2 - 10x + 3 = 0$

26. $4x^2 + 2x - 3 = 0$

27. $3x^2 + 6x - 5 = 0$

28. $8x^2 + 24x = 0$

Write each of the quadratic equations in Exercises 29–34 in the form $ax^2 + bx + c = 0$. Identify the constants a, b, and c. Then solve using the quadratic formula.

29. $2x^2 + 3x - 2 = 0$

30. $3x^2 + x - 4 = 0$

31. $2x^2 = 2x + 1$

32. $2x^2 + 5x = 6$

33. $5x^2 = 1 - x$

34. $\frac{3}{4}x^2 = 2x - \frac{1}{8}$

Solve the quadratic equations in 35–42 using any method.

35. $2x^2 = 72$

36. $(x - 2)^2 = 17$

37. $x^2 - 6x + 4 = 0$

38. $5x^2 - 2x - 7 = 0$

39. $x^2 - 7x = 60$

40. $4x^2 - 1 = 3x + 3$

41. $3x^2 - 9x - 12 = 0$

42. $3x^2 - 4x - 2 = 0$

Determine a quadratic equation for each of the Exercises 43–50. Then solve.

43. The product of two consecutive positive even integers is 168. Find the integers.

44. The length of a rectangle is three centimeters greater than twice the width. If the area of the rectangle is 275 square centimeters, find the length and width.

45. The diagonal of a rectangle is 15 meters. The length of the rectangle is 6 meters less than twice the width. Find the length and width.

46. A square garden has a 3-foot walk surrounding it. If the walk is removed and the space is included in the garden, the area will then be four times the original area. Find the length of one side of the original square.

47. A boat travels 12 miles downstream and then returns. The current in the stream is moving at 2 mph. If the time for the total round trip is $4\frac{1}{2}$ hours, find the speed of the boat in still water.

48. A tank can be filled by two pipes in 4 hours. The larger pipe alone will fill the tank in 6 hours less than the smaller one. Find the time required by each pipe to fill the tank alone.

49. A store manager ordered $300 worth of shirts. If each shirt had cost $3 more, he would have obtained 5 fewer shirts for the same amount of money. How much did each shirt cost, and how many did he purchase?

50. A rectangular sheet of metal is three times as long as it is wide. A box is to be made by cutting out 2-inch squares at each corner and folding up the sides. If the volume of the box is 312 cubic inches, what are the dimensions of the sheet of metal?

CHAPTER 11 TEST

Solve the equations in Exercises 1 and 2.

1. $(x - 5)^2 = 49$

2. $8x^2 = 96$

Add the correct constant in Exercises 3–5 in order to make the trinomial factorable as indicated.

3. $x^2 - 24x +$ ______ $= (\qquad)^2$

4. $x^2 + 9x +$ ______ $= (\qquad)^2$

4. $3x^2 + 9x +$ ______ $= 3(\qquad)^2$

Solve the quadratic equations in Exercise 6–8 by factoring.

6. $x^2 - 3x + 2 = 0$

7. $5x^2 + 12x = 0$

8. $3x^2 - x - 10 = 0$

Solve the quadratic equations in Exercises 9–11 by completing the square.

9. $x^2 + 6x + 8 = 0$

10. $x^2 - 2x - 5 = 0$

11. $2x^2 + 3x - 3 = 0$

Solve the quadratic equations in Exercises 12–14 using the quadratic formula.

12. $3x^2 + 8x + 2 = 0$

13. $2x^2 = 4x + 3$

14. $\frac{1}{2}x^2 - x = 2$

Solve the quadratic equations in Exercises 15–17 using any method.

15. $x^2 + 6x = 2$

16. $3x^2 - 7x + 2 = 0$

17. $2x^2 - 3x = 1$

Determine a quadratic equation for each of the Exercises 18–20. Then solve.

18. The diagonal of a square is 36 inches. Find the length of each side.

19. A small boat travels 8 mph in still water. It takes 10 hours longer to travel 60 miles upstream than it takes to travel 60 miles downstream. Find the rate of the current.

20. A rectangular sheet of metal is 3 inches longer than it is wide. A box is to be made by cutting out 4-inch squares at each corner and folding up the sides. If the box has a volume of 720 cubic inches, what are the dimensions of the sheet of metal?

USING THE COMPUTER

```
10  REM
20  REM This program will figure out what kind of solutions a
30  REM quadratic equation has by looking at the discriminant,
40  REM B²-4AC.
50 '*********************************************************
60 '*   FIRST  : The coefficient of X^2                    *
70 '*   SECOND : The coefficient of X                      *
80 '*   THIRD  : The constant                              *
90 '*********************************************************
100 PRINT "Type in the value of a,b,c (ax^2+bx+c)";
110 INPUT FIRST,SECOND,THIRD          'Get the values
120 PRINT
"The given quadratic equation was ";FIRST;"x^2 +";SECOND;
130 PRINT "x +";THIRD;" = 0"          'Print the given equation
139 REM Compute the discriminant
140 LET DISC = SECOND^2 - 4*FIRST*THIRD
149 REM Case 1
150 IF DISC>0 THEN PRINT
"This equation has discriminant ";DISC;"and two real roots"
159 REM Case 2
160 IF DISC=0 THEN PRINT "This equation has a double root "
169 REM Case 3
170 IF DISC<0 THEN PRINT
"This equation has discriminant ";DISC;"and two complex roots."
180 END
```

```
Type in the value of a,b,c (ax^2+bx+c) 1 , 2 , 3
The given quadratic equation was  1 x^2 + 2 x + 3  = 0
This equation has discriminant -8 and two complex roots
```

```
Type in the value of a,b,c (ax^2+bx+c) 1 , 2 , 1
The given quadratic equation was  1 x^2 + 2 x + 1  = 0
This equation has a double root
```

```
Type in the value of a,b,c (ax^2+bx+c) 3 , 4 , 5
The given quadratic equation was  3 x^2 + 4 x + 5  = 0
This equation has discriminant -44 and two complex roots
```

```
Type in the value of a,b,c (ax^2+bx+c) 1 , 3 , 2
The given quadratic equation was  1 x^2 + 3 x + 2  = 0
This equation has discriminant  1 and two real roots
```

Answers

CHAPTER 1

EXERCISES 1.1, page 8

1. 186
3. 295
5. 2262
7. 57
9. 44
11. 15
13. 9
15. 286
17. 24
19. 35
21. 296
23. 1215
25. 26,751
27. 4386
29. 11,346
31. 75
33. 34
35. 14
37. 15
39. 34
41. $18 + 48 = 66$
43. $56 + 35 = 91$
45. $70 + 60 = 130$
47. $3 + 7$
49. $4 \cdot 19$
51. $6 \cdot 5 + 6 \cdot 8$
53. $(2 \cdot 3) \cdot x$
55. $(3 + x) + 7$
57. commutative property for multiplication
$6 \cdot 4 = 4 \cdot 6$
$24 = 24$
59. associative property for addition
$8 + (4 + 4) = (8 + 4) + 4$
$8 + 8 = 12 + 4$
$16 = 16$
61. distributive property
$5(4 + 18) = 5(22)$
$5 \cdot 4 + 90 = 20 + 90$
$110 = 110$
63. associative property for multiplication
$(6 \cdot 4) \cdot 9 = 6 \cdot (4 \cdot 9)$
$24 \cdot 9 = 6 \cdot 36$
$216 = 216$
65. commutative property for addition
$4 + 34 = 34 + 4$
$38 = 38$
67. 17
69. 25

EXERCISES 1.2, page 15

1. 9
3. 16
5. 64
7. 625
9. 144
11. 5^2
13. 7^4
15. 3^2a^3
17. $2^25^3a^2$
19. $7a^4b^3$
21. **a.** 1, 2, 3, 6
b. 6, 12, 18, 24, 30, 36
23. **a.** 1, 2, 5, 10
b. 10, 20, 30, 40, 50, 60
25. **a.** 1, 3, 5, 15
b. 15, 30, 45, 60, 75, 90
27. $2 \cdot 17$
29. $3 \cdot 3 \cdot 3 = 3^3$
31. $2 \cdot 2 \cdot 2 \cdot 7 = 2^3 \cdot 7$

33. $2 \cdot 2 \cdot 2 \cdot 3 \cdot 5 = 2^3 \cdot 3 \cdot 5$
35. $2 \cdot 2 \cdot 2 \cdot 3 \cdot 7 = 2^3 \cdot 3 \cdot 7$
37. prime
39. $2 \cdot 2 \cdot 7 \cdot 7 = 2^2 \cdot 7^2$
41. 105 **43.** 120 **45.** 120 **47.** 390
49. xyz **51.** $12x^2y^2$ **53.** $210x^2y^2$ **55.** $840x^2y$
57. 9 **59.** 33 **61.** 11 **63.** 5
65. 14 **67.** 39 **69.** 5

EXERCISES 1.3, page 21

1. $\frac{5}{6}$ **3.** $\frac{2}{3}$ **5.** $\frac{3}{7}$ **7.** 1
9. $\frac{3}{2}$ **11.** $\frac{3}{7}$ **13.** 20 **15.** 9
17. 0 **19.** 84 **21.** $66a$ **23.** $77x$
25. $\frac{1}{6}$ **27.** $\frac{45}{128}$ **29.** $\frac{5}{6}$ **31.** $\frac{5}{6}$
33. $\frac{7}{4}$ **35.** $\frac{1}{2}$ **37.** $\frac{25}{4}$ **39.** $\frac{7b}{6a}$
41. 0 **43.** $\frac{11}{15x}$ **45.** 1 **47.** undefined
49. $\frac{18}{35}$ **51.** $\frac{2}{5ab}$ **53.** $\frac{50}{9}$ **55.** $\frac{5}{9}$
57. $\frac{17x}{26}$ **59.** $\frac{3ax}{16}$

EXERCISES 1.4, page 26

1. $\frac{8}{7}$ **3.** $\frac{7}{3}$ **5.** $\frac{1}{5}$ **7.** $\frac{13}{19}$
9. 2 **11.** $\frac{5}{4}$ **13.** $\frac{10}{9}$ **15.** $\frac{1}{8}$
17. 1 **19.** $\frac{25}{32a}$ **21.** $\frac{5}{6}$ **23.** $\frac{1}{8}$
25. $\frac{7}{20}$ **27.** $\frac{9}{8}$ **29.** $\frac{7}{12}$ **31.** $\frac{11}{20}$
33. $\frac{7}{18a}$ **35.** $\frac{19}{32c}$ **37.** $\frac{29}{30}$ **39.** $\frac{15}{8}$
41. $\frac{35}{48}$ **43.** $\frac{49}{60}$ **45.** $\frac{39}{40}$ **47.** $\frac{31}{60}$

49. $\frac{33}{70}$ **51.** $\frac{3}{4x}$ **53.** $\frac{5}{24x}$ **55.** 0
57. $\frac{1}{24}$ **59.** $\frac{3}{10}$ **61.** $\frac{7}{4}$ **63.** $\frac{1}{8}$
65. $\frac{49}{40}$

EXERCISES 1.5, page 31

1. 37% **3.** 72.3% **5.** 3.2% **7.** 280%
9. 123% **11.** 0.43 **13.** 0.06 **15.** 0.113
17. 2.38 **19.** 2.00 **21.** 157.22 **23.** 317.23
25. 67.11 **27.** 263.51 **29.** 187.01 **31.** 34.04
33. 141.17 **35.** 8.18 **37.** 0.623 **39.** 153.92
41. 32.85 **43.** 3.7992 **45.** 11,889.28 **47.** 13.23
49. 28.8 **51.** 61.56 **53.** 694.4 **55.** 437.5
57. 101.558 **59.** 80.648 **61.** 21.4229 **63.** 20.1939
65. 93.1095 **67.** 4.05081 **69.** 43.239

EXERCISES 1.6, page 33

1. \$36,464; \$9116 **3.** 9720 ft **5.** 10.5 in. **7.** 9
9. 48 **11.** 344 yd **13.** $\frac{19}{6}$ **15.** $\frac{1}{3}$
17. $\frac{92}{19}$ **19.** 15 mpg **21.** \$757.80 **23.** \$706
25. \$30.85; \$4.63 **27.** 10.1875 in. **29.** 9 pieces

EXERCISES 1.7, page 38

1. 9 **3.** 4 **5.** 1 **7.** 1
9. 1 **11.** 10 **13.** 33 **15.** 39
17. 15 **19.** 7 **21.** $4\frac{1}{2}, 4\frac{1}{3}$ **23.** $6, 5\frac{2}{3}$
25. $\frac{5}{4}, \frac{3}{4}$ **27.** 5.8, 3.5 **29.** 15, 9.25 **31.** 9.9, 0.7
33. 6 is a solution. **35.** 3 is not a solution.
37. 4 is not a solution. **39.** 5 is a solution.
41. 1 is a solution. **43.** 3 is a solution.
45. $\frac{1}{2}$ is a solution. **47.** $\frac{1}{4}$ is a solution.

49. 0.2 is a solution.
51. 1.4 is not a solution.
53. 2.25 is a solution.
55. 2.36 is not a solution.
57. 2.08 is a solution.

EXERCISES 1.8, page 45

1. e
3. k
5. c
7. d
9. g
11. l
13. m
15. h
17. 70 cm; 300 cm^2
19. 54 mm; 126 mm^2
21. 31.4 in.; 78.5 in.^{2}
23. 1436.03 in.^{2}
25. 729 m^3
27. 56 in.
29. 63 cm^2
31. 6154.4 ft^3
33. 125 ft^3
35. 272 cm^2
37. 529.88 in.^{3}
39. 28 cm; 48 cm^2

CHAPTER 1 REVIEW, page 53

1. 36
2. 40
3. 9
4. 15
5. 126
6. 459
7. 39
8. 124
9. 840
10. 196
11. 3318
12. 24
13. $21 + 42 = 63$
14. $72 + 56 = 128$
15. 17
16. 26
17. 81
18. 121
19. 2^3a^2b
20. $3^25^2ab^3$
21. $2 \cdot 2 \cdot 3 \cdot 5 = 2^2 \cdot 3 \cdot 5$
22. $3 \cdot 3 \cdot 17 = 3^2 \cdot 17$
23. 108
24. $72xy^2$
25. 90
26. 12
27. $\frac{5}{11}$
28. $\frac{2}{3}$
29. 30
30. 52
31. commutative property for addition
32. associative property for addition
33. distributive property
34. $\frac{43}{40}$
35. $\frac{1}{36}$
36. $\frac{22}{15y}$
37. $\frac{1}{4a}$
38. $\frac{15}{2}$
39. $\frac{5}{6}$
40. $\frac{7}{5}$
41. $\frac{2}{9}$
42. $\frac{2}{9}$
43. 1
44. 0
45. $\frac{7}{16}$
46. 73%
47. 6.5%
48. 0.07
49. 0.125
50. 32.53
51. 4.69
52. 14.6
53. 58.19
54. 5.292
55. 10.8
56. 58.3
57. 43.7
58. 12
59. 35
60. $3\frac{1}{2}$
61. $\frac{1}{4}$
62. 1.7
63. 3.66
64. 2 is not a solution.

65. 3 is a solution. **66.** $\frac{1}{2}$ is a solution. **67.** 1.5 is a solution. **68.** 76 ft
69. 254.34 m^2 **70.** 693 $in.^3$ **71.** $3\frac{1}{3}$ cups **72.** $21
73. 20.8 in. **74.** $365 **75.** $68.01

CHAPTER 1 TEST, page 57

1. 1794 **2.** 366 **3.** 23 **4.** distributive property
5. 53 **6.** $2 \cdot 3^2xy^3$ **7.** $3 \cdot 3 \cdot 5 \cdot 7 = 3^2 \cdot 5 \cdot 7$ **8.** $90x^2y^2$
9. 88 **10.** $\frac{6}{13}$ **11.** 60 **12.** $\frac{3}{5}$
13. $\frac{3}{10}$ **14.** $\frac{29}{40}$ **15.** 21.06 **16.** 13.588
17. 42.5 **18.** 9 **19.** 102 ft^2 **20.** $32

CHAPTER 2

EXERCISES 2.1, page 66

1. > **3.** > **5.** < **7.** =
9. < **11.** < **13.** = **15.** <
17. true **19.** true **21.** false **23.** true
25. false **27.** true **29.** true **31.** false
33. true **35.** true

37. −3 −2 0 1 **39.** −3 −2 −1 4
41. −2 −1 $-\frac{1}{3}$ 2 **43.** −3.4 −2 0.5 1 $\frac{5}{3}$
45. −4 $-\frac{7}{3}$ −1 0.2 $\frac{5}{2}$ **47.** 1 2 3 4 5 6
49. −3 −2 −1 **51.** −1 0 1 2
53. 0 1 2 3 4 **55.** −1 0 1 2 . . .

EXERCISES 2.2, page 69

1. −5 1 2 3 4 **3.**

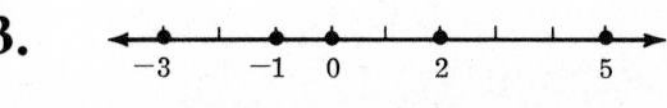

5. −8 0 6 8 **7.** −2 −1 $\frac{3}{2}$ 2 2.7

9. 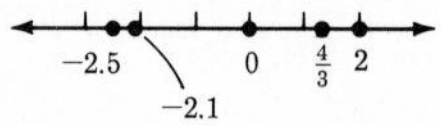

11. true

13. true
15. true
17. false
19. true
21. $x = 4$ or $x = -4$
23. $x = 9$ or $x = -9$
25. $y = 0$
27. no solution
29. $x = 4.7$ or $x = -4.7$
31. $y = 12$ or $y = -12$
33. $x = \frac{4}{7}$ or $x = -\frac{4}{7}$
35. $x = \frac{5}{4}$ or $x = -\frac{5}{4}$
37. no solution
39. no solution

EXERCISES 2.3, page 74

1. −11
3. 6
5. −4.7
7. 0
9. $\frac{5}{16}$
11. 13
13. −4
15. 0
17. 5
19. −13
21. −2
23. −10
25. 0
27. 17
29. −19
31. $-9\frac{5}{6}$
33. $-3\frac{7}{16}$
35. 22.02
37. −55.52
39 −7
41. −16
43. −26
45. 0
47. −32
49. 5
51. −83
53. 12
55. −32
57. −2 is a solution.
59. −4 is a solution.
61. −7 is a solution.
63. 2 is not a solution.
65. −1 is a solution.
67. sometimes
69. always
71. 85.179
73. −97.714

EXERCISES 2.4, page 79

1. 5
3. −10
5. 12
7. 3
9. $-6\frac{7}{8}$
11. 24.74
13. −15
15. 30
17. −57
19. 0
21. $-25\frac{7}{8}$
23. −31.04
25. −15
27. −1
29. −3
31. 7
33. $\frac{19}{30}$
35. −15.26
37. $<$
39. $<$
41. $=$
43. $>$
45. $>$
47. $<$
49. $=$
51. −8 is a solution.
53. −2 is a solution.
55. 4 is not a solution.
57. 5 is not a solution.
59. 1 is a solution.
61. 0.262
63. 20.405

EXERCISES 2.5, page 86

1. -12 **3.** 56 **5.** 57 **7.** 56
9. -32.4 **11.** 2.73 **13.** $-\frac{3}{5}$ **15.** $\frac{3}{4}$
17. -288 **19.** 0 **21.** 4 **23.** -6
25. -3 **27.** 13 **29.** 0 **31.** undefined
33. -1.06 **35.** $\frac{4}{7}$ **37.** negative **39.** negative
41. negative **43.** zero **45.** undefined **47.** true
49. true **51.** false **53.** false **55.** true
57. -12 is a solution. **59.** -72 is a solution.
61. -8 is not a solution. **63.** 5 is a solution.
65. -4 is a solution. **67.** -3.445911
69. -2.671

EXERCISES 2.6, page 88

1. $(-23) + 13 + (-6) = -16$ lb **3.** $4° + 3° + (-9°) = -2°$
5. $\$47 + (-\$22) + (\$8) + (-\$45) = -\$12$
7. $14° + (-6°) + (11°) + (-15°) = 4°$
9. $\$187 + (-\$241) + (\$82) + (\$26) = \$54$
11. 4 yd **13.** 8th floor **15.** \$40 **17.** -13
19. -10.8 **21.** $2\frac{1}{8}$ **23.** -15 **25.** -86
27. \$220 **29.** $-19°$ **31.** $-\$8$ **33.** $-14{,}777$ ft
35. 52 yr **37.** 5 under par **39.** $1\frac{7}{8}$

CHAPTER 2 REVIEW, page 92

1. $<$ **2.** $=$ **3.** $>$ **4.** $>$
5. $>$ **6.** $>$

7. 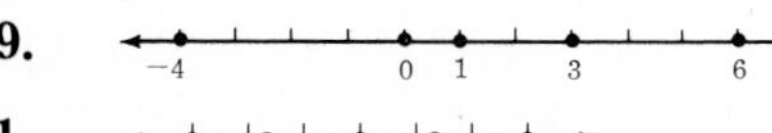

8. 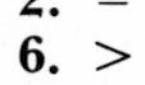

9. 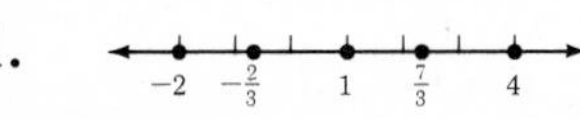

10. 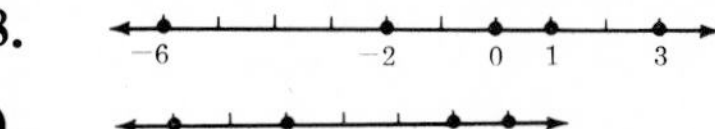

11. 

12. 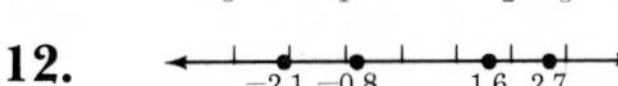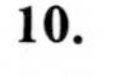

13. 

14. 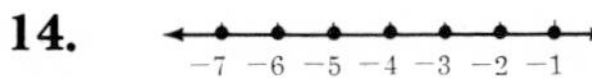

15. 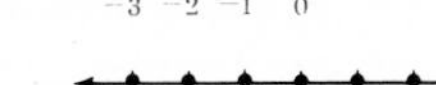

16. 

17. $x = 6$ or $x = -6$ **18.** $x = 10$ or $x = -10$
19. $y = \frac{3}{5}$ or $y = -\frac{3}{5}$ **20.** $y = 2.9$ or $y = -2.9$
21. no solution **22.** $y = 1.724$ or $y = -1.724$

23. −8 **24.** −12 **25.** −15 **26.** 1.8

27. $-\frac{2}{15}$ **28.** $-\frac{23}{12}$ **29.** −1.18 **30.** −114

31. 253 **32.** 160 **33.** $-\frac{3}{10}$ **34.** −3.9

35. 4 **36.** undefined **37.** −7 **38.** 0

39. −6.8 **40.** $\frac{10}{3}$ **41.** sometimes **42.** never

43. always **44.** sometimes **45.** sometimes **46.** 6 is a solution.

47. −2 is a solution. **48.** −5 is a solution.

49. −3 is not a solution. **50.** −16 is a solution.

51. 16 is a solution. **52.** $\frac{3}{4}$ is a solution.

53. $-\frac{2}{3}$ is a solution. **54.** −2.6 is a solution.

55. 1.3 is not a solution. **56.** $10

57. $88 **58.** 3 **59.** $1\frac{3}{16}$ **60.** 17.49

CHAPTER 2 TEST, page 95

1. a. < **b.** > **2.** 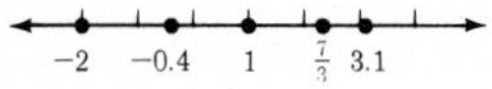

3. 

4. $y = 7$ or $y = -7$

5. $x = \frac{7}{8}$ or $x = -\frac{7}{8}$ **6.** 13 **7.** 6.68 **8.** $\frac{1}{8}$

9. 19 **10.** 11 **11.** 162 **12.** −4

13. −25.132 **14.** $-\frac{9}{4}$ **15.** never **16.** always

17. 5 is a solution. **18.** −1.5 is a solution.

19. −60 **20.** $15.47

CHAPTER 3

EXERCISES 3.1, page 103

1. $15x$ **3.** $3x$ **5.** $5y^2$ **7.** $12x^2$

9. $7x + 2$ **11.** $x - 3y$ **13.** $4x^2 + 3y$ **15.** $2x + 3$

17. $7x - 8y$ **19.** $8x + y$ **21.** $2x^2 - x$ **23.** $-2x^2 + 14x$

25. $3x^2 - xy + y^2$ **27.** $2x$ **29.** $-y$ **31.** 0

33. $10x$ **35.** x **37.** −12 **39.** −7

41. 0 **43.** 2 **45.** 0 **47.** $13x - 7$; 45

49. $9y + 2$; −25 **51.** $5x - 3y$; 13 **53.** $7x + 2y$; −10 **55.** $6y + 2xy$; 42

57. $5x$; −30 **59.** $6x$; 18

EXERCISES 3.2, page 109

1. $x = 6$ **3.** $x = -4$ **5.** $y = -4$ **7.** $y = -18$
9. $x = 34$ **11.** $x = \frac{1}{2}$ **13.** $y = \frac{11}{15}$ **15.** $x = 0.24$
17. $x = 11$ **19.** $x = -8$ **21.** $x = 4$ **23.** $x = 12$
25. $x = -9$ **27.** $x = 0$ **29.** $x = \frac{1}{4}$ **31.** $x = -\frac{3}{10}$
33. $x = 4.2$ **35.** $x = 0.84$ **37.** $x = -3$ **39.** $x = 12$
41. $x = 10$ **43.** $x = 0$ **45.** $9013.50 **47.** 29 cm
49. $3.70 **51.** -50.753 **53.** $y = 26.087$

EXERCISES 3.3, page 116

1. x **3.** x **5.** y **7.** $\frac{1}{8}$
9. 10 **11.** $-\frac{4}{9}$ **13.** $x = 4$ **15.** $x = -7$
17. $y = 6$ **19.** $x = -9$ **21.** $x = -14$ **23.** $x = -15$
25. $x = -16$ **27.** $y = 18$ **29.** $y = -21$ **31.** $x = -32$
33. $x = 12$ **35.** $x = \frac{2}{5}$ **37.** $x = \frac{9}{2}$ **39.** $x = -\frac{3}{2}$
41. $x = \frac{3}{4}$ **43.** $x = -\frac{1}{3}$ **45.** $x = \frac{2}{3}$ **47.** $y = 2$
49. $x = -3$ **51.** $x = -0.9$ **53.** $x = 10$ **55.** 7 hr
57. 3 yr **59.** 6 in. **61.** $x = 246$ **63.** $y = -498$

EXERCISES 3.4, page 122

1. $x = -11$ **3.** $x = 2$ **5.** $x = 10$ **7.** $x = -2$
9. $x = 4$ **11.** $x = 2$ **13.** $x = \frac{1}{5}$ **15.** $y = \frac{5}{2}$
17. $y = 1.8$ **19.** $x = -0.9$ **21.** $x = 18$ **23.** $x = 5$
25. $x = 18$ **27.** $x = \frac{4}{3}$ **29.** $x = -\frac{10}{3}$ **31.** $x = 6$
33. $x = 3$ **35.** $x = -6$ **37.** $x = \frac{1}{2}$ **39.** $x = -\frac{2}{7}$
41. $x = 21$ **43.** $x = -4$ **45.** $x = -3$ **47.** $x = 10$
49. $x = -5$ **51.** $x = 7$ **53.** $x = 2$ **55.** $x = -\frac{7}{2}$

57. $x = -\frac{7}{3}$ **59.** $6\frac{1}{2}$ cm **61.** 18 in. **63.** $490
65. $x = -9.29$

EXERCISES 3.5, page 126

1. the product of 4 and a number
3. 1 more than the product of 2 and a number
5. the difference between 7 times a number and 5.3
7. the product of −2 with the difference between a number and 8
9. 5 five times the quantity found by taking the sum of twice a number and 3
11. $\frac{2}{3}$ added to the product of 6 with the difference between a number and 1
13. 7 added to the product of 3 with a number; the product of 3 with the sum of a number and 7
15. the difference between 7 times a number and 3; the product of 7 with the difference between a number and 3
17. $x + 6$ **19.** $x - 4$ **21.** $3x - 5$
23. $\frac{x - 3}{7}$ **25.** $3(x - 8)$ **27.** $3x - 5$
29. $3x + (8 - 2x)$ **31.** $8(x - 6) + 4$ **33.** $(3x - 9) - 5x$
35. $4(x + 5) - 2x$ **37.** $0.11x$ dollars **39.** $(60h + 20)$ minutes
41. $0.20m + 20$ dollars **43.** $0.09x + 250$ dollars **45.** $2(2w - 3) + 2w$ cm

EXERCISES 3.6, page 131

1. 36 **3.** 4 **5.** 3 **7.** −6
9. 13 **11.** −35 **13.** −7 **15.** −4
17. 12 **19.** 22, 23, 24 **21.** 29, 31 **23.** −54, −52, −50
25. 83, 88 **27.** 37, 77 **29.** 11, 6, 22 **31.** 27 ft
33. $285 **35.** $910 **37.** $60 **39.** 11 yr, 33 yr
41. $8.50 **43.** $15.60 for calculator; $22.35 for book **45.** $14.75

EXERCISES 3.7, page 138

1. $b = P - a - c$ **3.** $m = \frac{f}{a}$ **5.** $w = \frac{A}{l}$ **7.** $n = \frac{R}{p}$
9. $p = A - i$ **11.** $m = 2A - n$ **13.** $s = \frac{p}{4}$ **15.** $t = \frac{d}{r}$

17. $t = \dfrac{I}{pr}$ **19.** $b = \dfrac{p - a}{2}$ **21.** $\beta = 180 - \alpha - \gamma$ **23.** $x = \dfrac{y - b}{m}$

25. $r^2 = \dfrac{A}{4\pi}$ **27.** $M = \dfrac{(IQ)C}{100}$ **29.** $h = \dfrac{3V}{\pi r^2}$ **31.** $I = \dfrac{E}{R}$

33. $L = \dfrac{R}{2A}$ **35.** $t = \dfrac{v - k}{g}$ **37.** $h = \dfrac{S - 2\pi r^2}{2\pi r}$ **39.** $r = \dfrac{s - a}{s}$

41. $C = \dfrac{5}{9}(F - 32)$ **43.** $m = \dfrac{2gk}{v^2}$ **45.** $y = x - 3$ **47.** $y = 8 - 2x$

49. $x = 5 - 2y$ **51.** $y = \dfrac{4x - 9}{3}$

53. $x = \dfrac{4 - 8y}{3}$ **55.** $y = \dfrac{3 - 1.2x}{1.5}$ or $y = \dfrac{30 - 12x}{15}$

EXERCISES 3.8, page 140

1. 48 ft per sec **3.** 2 sec **5.** 100 milligrams **7.** 5 yr

9. 16 studs **11.** \$1030 **13.** 230 calculators **15.** 5%

17. $C = 2\pi r$; $r = 17$ cm **19.** $p = b + 2a$; $a = 22$

21. $A = \pi r^2$; $A = 49\pi$ m^2 **23.** $R = \dfrac{E}{I}$; $R = 6$ ohms

25. $A = \dfrac{1}{2}h(a + b)$; $b = 10$ cm

CHAPTER 3 REVIEW, page 144

1. $7x + 1$ **2.** $-2x - 12$ **3.** 0 **4.** $4x$

5. $12x^2 + 29x$ **6.** 12 **7.** 15 **8.** 11

9. -22 **10.** 7 **11.** $x = -1.5$ **12.** $x = \dfrac{5}{4}$

13. $x = \dfrac{3}{2}$ **14.** $x = 4.1$ **15.** $x = -3$ **16.** $x = 2$

17. $x = 6$ **18.** $x = -4$ **19.** $x = -3$ **20.** $x = -4$

21. $x = 3$ **22.** $x = \dfrac{25}{4}$ **23.** $x = \dfrac{32}{3}$ **24.** $x = -40$

25. $x = -4$ **26.** the difference between 3 times a number and 1

27. the difference between 4 and the product of a number and 7 **28.** the product of 5 with the sum of a number and 1

29. twice the difference between 4 times a number and 1

30. 4 divided by the sum of a number and 7

31. $3x + 4$ **32.** $9 - 2x$ **33.** $11(x - 4)$ **34.** $24x + 5$ hr

35. $2x + 17$ points **36.** $m = \frac{E}{c^2}$ **37.** $b = \frac{Fd^2}{ka}$ **38.** $h = \frac{3V}{\pi r^2}$

39. $y = \frac{8 - 2x}{7}$ **40.** $x = \frac{-2y - 3}{5}$

41. $240 = 2(85) + 2w$; $w = 35$ in. **42.** $6400 = 150t + 1000$; $t = 36$ mo

43. $3x - 9 = 30$; $x = 13$ **44.** $2x + 3 = 3x - 8$; $x = 11$

45. $3x = 2x + 10$; $x = 10$ **46.** $n + (n + 2) = 84$; 41, 43

47. $5 - 2n = (n + 1) + 19$; $-5, -4$ **48.** $x + (x + 9.35) = 16.25$
$3.45 for ball; $12.80 for bat

49. $V = 7200 - (0.7200)(0.08)(7)$
$V = \$3168$

50. $147\pi = \frac{1}{3}\pi(7)^2h$; $h = 9$ in.

CHAPTER 3 TEST, page 147

1. $8x^2 - 17x$ **2.** $5x - 10$ **3.** -3 **4.** 17

5. $x = -3$ **6.** $x = 5$ **7.** $x = 0$ **8.** $x = 14$

9. $x = 20$ **10.** the product of 4 with the difference between a number and 2

11. 7 subtracted from the product of 3 with the sum of a number and 4 **12.** $3(2x + 5)$

13. $4x + 3$ qt **14.** $h = \frac{s}{2\pi r}$ **15.** $m = \frac{N - p}{rt}$ **16.** $y = \frac{5x - 7}{3}$

17. $-9, -13$ **18.** 16, 17 **19.** 24 in.

20. $0.95 for shake; $1.85 for hamburger

CHAPTER 4

EXERCISES 4.1, page 159

1. $3^3 = 27$ **3.** $8^3 = 512$ **5.** $\frac{1}{4^2} = \frac{1}{16}$ **7.** $\frac{1}{6^3} = \frac{1}{216}$

9. $(-4)^3 = -64$ **11.** 54 **13.** -54 **15.** $\frac{4}{9}$

17. $-\frac{5}{4}$ **19.** x^4 **21.** y^{11} **23.** $\frac{1}{y^2}$

25. $\frac{5}{y^4}$ **27.** $\frac{1}{x^2}$ **29.** y^3 **31.** $3^2 = 9$

33. $9^3 = 729$ **35.** $\frac{1}{10^3} = \frac{1}{1000}$ **37.** x^2 **39.** x^2
41. x^4 **43.** $\frac{1}{x^4}$ **45.** x^6 **47.** x^2
49. y^2 **51.** $3x^3$ **53.** $10x^5$ **55.** $36x^3$
57. $-14x^5$ **59.** $-12x^6$ **61.** $4y$ **63.** $3y^2$
65. $-2y^2$ **67.** $-7x^2$ **69.** $10^3 = 1000$ **71.** $\frac{1}{10^2} = \frac{1}{100}$
73. x^5 **75.** $y^0 = 1$

EXERCISES 4.2, page 167

1. $36x^6$ **3.** $-27x^6$ **5.** x^3y^6 **7.** $64a^8b^2$
9. -1 **11.** $\frac{n^6}{16m^4}$ **13.** $-\frac{2y^6}{27x^{15}}$ **15.** $\frac{16x^2}{y^4}$
17. $\frac{9x^4}{y^6}$ **19.** $\frac{y^2}{x^2}$ **21.** $\frac{y^{10}}{4x^2}$ **23.** $\frac{1}{64a^6b^9}$
25. $25x^2y^2$ **27.** $16a^4b^4$ **29.** $\frac{1}{8a^3b^6}$ **31.** $\frac{y^8}{16x^8}$
33. $\frac{8a^6}{b^9}$ **35.** $\frac{49y^4}{x^6}$ **37.** $\frac{3y^4}{2x^3}$ **39.** $\frac{4}{9x^{10}}$
41. 8.6×10^4 **43.** 3.62×10^{-2} **45.** 1.83×10^7 **47.** 2.17×10^{-4}
49. 4.5×10^6 **51.** 1.23×10^{10} **53.** 4.5×10^{-2} **55.** 7.5×10^{-9}
57. 1.3×10^6 **59.** 2.5×10^{-3} **61.** 6.0×10^3 **63.** 3.9×10^0
65. 2.4×10^{11} **67.** 5.0×10^{-2} **69.** 1.8×10^{12}; 1.08×10^{14}

EXERCISES 4.3, page 172

1. $4x$; first-degree **3.** $x^3 + 3x^2 - 2x$; third-degree
5. $-2x^2$; second-degree **7.** 0; no degree
9. $-x^5 - x^2$; fifth-degree **11.** $2x^3 + 4x$; third-degree
13. 4; zero-degree **15.** $3x^2 + x + 1$; second-degree.
17. $2x^2 + 2x + 2$ **19.** $4x^2 + x - 4$ **21.** $5x^2 + 6x - 10$ **23.** $4x^2 + 7x - 8$
25. $5x^2 + 14x - 2$ **27.** $2x^3 - 3x^2 - 3x - 6$ **29.** $3x^2 - 2x + 5$
31. $3x^3 + 2x^2 + 3x - 4$ **33.** $5x^3 + 11x^2 + 10x - 14$ **35.** x
37. $2x^2 + 5x - 11$ **39.** $x^2 - x - 12$ **41.** $3x^2 - 4x - 8$ **43.** $2x^2 + 13x + 9$
45. $-7x^2 - x + 9$ **47.** $4x^2 + 3x + 11$ **49.** $6x^2 - 7x + 18$ **51.** $14x^2 - 15$
53. $2x^3 + 4x^2 + 3x - 10$ **55.** $2x^3 - 5x^2 + 11x - 11$ **57.** $-12x + 22$
59. $-x - 15$ **61.** $-9x + 26$ **63.** $7x + 7$
65. $-x - 2$ **67.** $5x - 7$ **69.** $2x^2 - 3x$

EXERCISES 4.4, page 177

1. $6x^5$ **3.** $-20x^3$ **5.** $-3x^2 - 2x$
7. $-4x^3 - 4x$ **9.** $7x^4 + 14x^3 - 7x^2$ **11.** $-x^4 - 5x^2 + 4x$
13. $6x^2 - x - 2$ **15.** $9x^2 - 25$ **17.** $-10x^2 + 39x - 14$
19. $x^3 + 5x^2 + 8x + 4$ **21.** $x^2 + x - 12$ **23.** $x^2 - 2x - 48$
25. $x^2 - 3x + 2$ **27.** $3x^2 - 3x - 60$ **29.** $x^3 + 11x^2 + 24x$
31. $2x^2 - 7x - 4$ **33.** $6x^2 + 17x - 3$ **35.** $4x^2 - 9$
37. $16x^2 + 8x + 1$ **39.** $x^3 + 2x^2 + x + 12$ **41.** $3x^2 - 8x - 35$
43. $-16x^3 + 50x^2 + 25x - 14$ **45.** $12x^4 + 28x^3 + 47x^2 + 34x + 48$

47. $4x^2 + 17x - 42$
$$[2 + 6][4(2) - 7] = (8)(1) = 8$$
$$4(2)^2 + 17(2) - 42 = 8$$

49. $5x^2 + 17x - 12$
$$[5(2) - 3][2 + 4] = (7)(6) = 42$$
$$5(2)^2 + 17(2) - 12 = 42$$

51. $3x^2 + 2x - 16$
$$[2 - 2][3(2) + 8] = (0)(14) = 0$$
$$3(2)^2 + 2(2) - 16 = 0$$

53. $6x^2 - x - 35$
$$[3(2) + 7][2(2) - 5] = (13)(-1) = -13$$
$$6(2)^2 - (2) - 35 = -13$$

55. $25x^2 + 20x + 4$
$$[5(2) + 2][5(2) + 2] = (12)(12) = 144$$
$$25(2)^2 + 20(2) + 4 = 144$$

57. $x^3 - 4x^2 + 2x - 8$
$$[(2)^2 + 2][2 - 4] = (6)(-2) = -12$$
$$(2)^3 - 4(2)^2 + 2(2) - 8 = -12$$

59. $9x^2 - 16$
$$[3(2) - 4][3(2) + 4] = (2)(10) = 20$$
$$9(2)^2 - 16 = 20$$

61. $x^3 - 8$
$$[2 - 2][(2)^2 + 2(2) + 4] = (0)(12) = 0$$
$$(2)^3 - 8 = 0$$

63. $25x^2 - 60x + 36$
$$[5(2) - 6][5(2) - 6] = (4)(4) = 16$$
$$25(2)^2 - 60(2) + 36 = 16$$

65. $3x^3 - 2x^2 + 26x + 9$
$$[3(2) + 1][(2)^2 - (2) + 9] = (7)(11) = 77$$
$$3(2)^3 - 2(2)^2 + 26(2) + 9 = 77$$

EXERCISES 4.5, page 183

1. $x^2 - 9$; difference of squares **3.** $x^2 - 10x + 25$; perfect square trinomial
5. $x^2 - 36$; difference of squares **7.** $x^2 + 16x + 64$; perfect square trinomial
9. $2x^2 + x - 3$ **11.** $x^2 - c^2$; difference of squares
13. $4x^2 - 1$; difference of squares **15.** $9x^2 - 12x + 4$; perfect square trinomial
17. $9 + 6x + x^2$; perfect square trinomial **19.** $25 - 10x + x^2$; perfect square trinomial
21. $25x^2 - 81$; difference of squares **23.** $4x^2 + 28x + 49$; perfect square trinomial
25. $81x^2 - 4$; difference of squares **27.** $10x^2 - 11x - 6$
29. $1 + 14x + 49x^2$; perfect square trinomial
31. $x^2 + 3x + 2$ **33.** $x^2 + x - 12$ **35.** $x^2 - 13x + 42$ **37.** $25 + 10x + x^2$
39. $x^4 - 1$ **41.** $x^6 + 4x^3 + 4$ **43.** $x^6 - 12x^3 + 36$ **45.** $x^4 - 2x^2 - 15$
47. $x^6 - 11x^3 + 28$ **49.** $x^8 - 1$ **51.** $x^2 - \frac{4}{9}$ **53.** $x^2 - \frac{9}{16}$

55. $x^2 + \frac{6}{5}x + \frac{9}{25}$ **57.** $x^2 - \frac{5}{3}x + \frac{25}{36}$ **59.** $x^2 + 3x + \frac{9}{4}$ **61.** $x^2 - \frac{1}{4}x - \frac{1}{8}$

63. $x^2 + \frac{5}{6}x + \frac{1}{6}$ **65.** $x^2 - \frac{11}{10}x + \frac{6}{25}$

CHAPTER 4 REVIEW, page 187

1. $6^7 = 279{,}936$ **2.** $(-3)^5 = -243$ **3.** $\frac{1}{7}$ **4.** $\frac{1}{5}$

5. y^5 **6.** y **7.** $2x^4$ **8.** $\frac{2y^2}{3}$

9. $\frac{1}{x^7}$ **10.** $2x$ **11.** $64x^6y^3$ **12.** $\frac{49x^{10}}{y^4}$

13. $\frac{36x^4}{y^{10}}$ **14.** $\frac{1}{x^6y^2}$ **15.** $\frac{a^6}{b^4}$ **16.** $\frac{x}{3y^4}$

17. $\frac{y^3}{8x^9}$ **18.** 1 **19.** x^4y^2 **20.** $\frac{x^8}{9y^6}$

21. 4.27×10^6 **22.** 2.3×10^{-4} **23.** 6.3 **24.** 4.0×10^6

25. 1.2×10^{-1} **26.** $5x^2 + 1$; second-degree

27. $4x^2 + 9x$; second-degree **28.** $x^2 + 4x$; second-degree

29. $3x^2 + 8x$; second-degree **30.** $-x^3 + 3x^2 + x - 2$; third-degree

31. $-x^3 - 6x^2 + 2x - 4$ **32.** $7x^2 + 10x - 5$

33. $2x^2 - 8x + 12$ **34.** $x^3 - 2x^2 - 6x - 7$

35. $-x^2 + x - 5$ **36.** $3x^3 + 4x^2 - 7x - 3$

37. $-x^3 + 7x^2 - 6$ **38.** $6x^2 - 7x + 1$

39. $-x - 17$ **40.** $-2x^2 - 3x$ **41.** $-10x + 4$ **42.** $5x - 11$

43. $2x^3 + 5x^2 + 3x$ **44.** $-18x + 26$ **45.** $7x - 38$ **56.** $-3x^3 + 12x$

47. $x^3 - x^5$ **48.** $x^2 - 36$; difference of squares

49. $x^2 + x - 12$ **50.** $9x^2 + 42x + 49$; perfect square trinomial

51. $4x^2 - 1$; difference of squares **52.** $x^4 - 25$; difference of squares

53. $x^4 - 4x^2 + 4$; perfect square trinomial **54.** $x^2 - \frac{4}{25}$; difference of squares

55. $x^2 + \frac{5}{4}x + \frac{25}{64}$; perfect square trinomial

56. $-x^2 + 5x - 2$

$$-1[(3)^2 - 5(3) + 2] = -1(-4) = 4$$

$$-(3)^2 + 5(3) - 2 = 4$$

57. $3x^3 + 6x^2 - 3x$

$$3(3)[(3)^2 + 2(3) - 1] = (9)(14) = 126$$

$$3(3)^3 + 6(3)^2 - 3(3) = 126$$

58. $12x^2 - 13x - 14$

$$[3(3) + 2][4(3) - 7] = (11)(5) = 55$$

$$12(3)^2 - 13(3) - 14 = 55$$

59. $2x^2 - x - 36$

$$[2(3) - 9][3 + 4] = (-3)(7) = -21$$

$$2(3)^2 - (3) - 36 = -21$$

60. $5x^2 - 27x - 18$
$[3 - 6][5(3) + 3] = (-3)(18) = -54$
$5(3)^2 - 27(3) - 18 = -54$

61. $6x^2 + x - 12$
$[3(3) - 4][2(3) + 3] = (5)(9) = 45$
$6(3)^2 + 3 - 12 = 45$

62. $x^4 - 16$
$(3)^2 + 4(3)^2 - 4 = (13)(5) = 65$
$(3)^4 - 16 = 65$

63. $x^3 - 6x^2 + 8x - 48$
$3 - 6(3)^2 + 8 = (-3)(17) = -51$
$(3)^3 - 6(3)^2 + 8(3) - 48 = -51$

64. $4x^3 - 3x^2 - x$
$4(3) + 1(3)^2 - (3) = (13)(6) = 78$
$4(3)^3 - 3(3)^2 - 3 = 78$

65. $5x^3 + 3x^2 - 12x + 4$
$5(3) - 2(3)^2 + (3) - 2 = (13)(10) = 130$
$5(3)^3 + 3(3)^2 - 12(3) + 4 = 130$

CHAPTER 4 TEST, page 190

1. $-12x^6$
2. $2x^9$
3. $\frac{x^2}{4y^2}$
4. $4x^2y^4$
5. 1.15
6. 5.2×10^{-3}
7. $8x^2 + 3x$; second-degree
8. $-x^3 + 3x^2 + 3x - 1$; third-degree
9. $3x^2 + 2x - 7$
10. $2x^2 + 9x - 8$
11. $5x^3 - x^2 + 6x + 5$
12. $-2x^2 + x - 6$
13. $7x^2 + 14x + 4$
14. $3x^3 - 4x^2 - 5x + 8$
15. $-3x + 2$
16. $20x - 8$
17. $-15x^4 + 45x^3$
18. $25x^2 - 16$
19. $16 - 24x + 9x^2$
20. $6x^3 - 27x^2 - 105x$

CHAPTER 5

EXERCISES 5.1, page 198

1. x^2
3. x^4
5. $-4y$
7. $3x^3$
9. $2x^2y$
11. $3(2x - 7)$
13. $-9(x^2 - 4)$
15. $6(x + 2y)$
17. $x(x^2 - 9)$
19. $a^2(b - c)$
21. $-3x(x - 2)$
23. $7y(x - 2z)$
25. $12yz(2y - 1)$
27. $6ax(4x^2 - 9)$
29. $11xy^3(4y^2 - 11)$
31. $5y^3(16m^2 - 1)$
33. $11(x^2 - 2x + 1)$
35. $-3(x^2 - 2x + 3)$
37. $a(a - 1 + 2b)$
39. $x(x^2 - 4x + 6)$
41. $2a(7b + 3c - d)$
43. $2b(c^2 + 3c + 4)$
45. $-8y(y^2 + 2y - 3)$
47. $x^2(14x^2 + 27x + 9)$
49. $11x(10x^2 - 11x + 1)$
51. $x^4(15x^3 + 24x^2 - 32)$

53. $2ax(25b^2 - c^2)$
55. $-8xy(x^2 - 4y^2)$
57. $9axy(y^2 - 1)$
59. $4c^2(x^2 - 2xy - 3y^2)$
61. $-3xy(xy + 2x^2y^2 + 3)$
63. $abx^2(1 - b + a)$
65. $6y^3(2xy^2 - 3xy + 4)$
67. $x(x^4 - 3x^3 + 7x - 21)$
69. $2x(2x^3 - 3x^2 + 7x - 1)$

EXERCISES 5.2, page 204

1. $(x - 2)(x - 1)$
3. $(x + 1)(x - 1)$
5. $(x + 4)(x + 3)$
7. $(y - 5)(y + 2)$
9. $(t - 2)(t + 1)$
11. irreducible
13. $(y - 8)(y + 3)$
15. $(x + 5)(x + 5)$
17. $(c + d)(c - d)$
19. $(3a + 1)(3a - 1)$
21. $(x - 8)(x - 4)$
23. irreducible
25. $(x + 11)(x + 2)$
27. $(y - 10)(y - 10)$
29. $(5 + 3x)(5 - 3x)$
31. $(z - 14)(z + 1)$
33. $3(x + 3y)(x - 3y)$
35. $c(x + 4)(x - 1)$
37. $3a(4x^2 + 25y^2)$
39. $a(y + 1)(y + 1)$
41. $m(m + n)(m - n)$
43. $2a(y - 4)(y - 3)$
45. $5y(x^2 - 2x + 4)$
47. $3(3m + 4n)(3m - 4n)$
49. $(3 - x)(2 - x)$
51. $2a(y - 11)(y - 11)$
53. $(a^2 + 1)(a + 1)(a - 1)$
55. $-3(x - 7)(x - 5)$
57. $3(x + 2)(x + 2)(x - 2)(x - 2)$
59. $(a^3 + b^2)(a^3 - b^2)$
61. $x^2 - 6x + \underline{9} = (x - 3)^2$
63. $x^2 + 4x + \underline{4} = (x + 2)^2$
65. $x^2 - 8x + \underline{16} = (x - 4)^2$
67. $x^2 - 18x + \underline{81} = (x - 9)^2$
69. $x^2 - 16x + \underline{64} = (x - 8)^2$
71. $x^2 + x + \underline{\frac{1}{4}} = \left(x + \frac{1}{2}\right)^2$
73. $y^2 - 9y + \underline{\frac{81}{4}} = \left(y - \frac{9}{2}\right)^2$
75. $x^2 + 11x + \underline{\frac{121}{4}} = \left(x + \frac{11}{2}\right)^2$

EXERCISES 5.3, page 210

1. $(2x + 1)(x - 1)$
3. $(4t + 1)(t - 1)$
5. $(2a + 1)(a + 3)$
7. $(5a - 6)(a + 1)$
9. $(7x - 2)(x + 1)$
11. $(4x + 3)(x + 5)$
13. $(x + 8)(x - 2)$
15. $(3x - 2)(2x - 5)$
17. $(5y - 8)(2y + 3)$
19. $(3x - 1)(x - 2)$
21. $(3x - 1)^2$
23. irreducible
25. $(4y + 3)(3y - 4)$
27. $5(x^2 + 9)$
29. $(4a - 3)(a - 2)$
31. irreducible
33. $(4x + y)(4x - y)$
35. $(3x - 1)(4x - 7)$
37. $(8x - 3)^2$
39. $(5m + 2)(3m - 5)$
41. $2(3x - 5)(x + 2)$
43. $5(2x + 3)(x + 2)$
45. $2(3x + 2y)(3x - 2y)$
47. $x^2(7x^2 - 5x + 3)$
49. $-2(6m + 1)(m - 2)$
51. $3x(2x - 1)(x + 2)$
53. $9xy(x^2y^2 + 1)$
55. $3x(2x - 9)^2$
57. $4xy(4y - 3)(3y - 4)$
59. $7y^2(3y - 2)(y - 4)$
61. $(b + c)(x + 1)$
63. $(x^2 + 6)(x + 3)$
65. $(x + 6y)(x - 4)$
67. $(y - 4)(5x + z)$
69. $(8 - z)(3y - 2x)$ or $(z - 8)(2x - 3y)$

EXERCISES 5.4, page 216

1. $x = 3, x = -4$ **3.** $x = -11, x = 9$ **5.** $x = -6, x = 2$
7. $x = -3, x = 1$ **9.** $x = 0, x = -9$ **11.** $x = 7, x = -3$
13. $x = 0, x = 5$ **15.** $x = -3, x = 8$ **17.** $x = 0, x = -5, x = -3$
19. $x = -11, x = 5$ **21.** $x = 2, x = 3$ **23.** $x = -2, x = 7$
25. $x = 3$ **27.** $x = -5, x = 5$ **29.** $x = 0, x = 2$
31. $x = 0, x = 4, x = -1$ **33.** $x = 0, x = -3$ **35.** $x = -3, x = 4$
37. $x = 2, x = 4$ **39.** $x = -4, x = 3$ **41.** $x = -1, x = 2$
43. $x = -2, x = 2$ **45.** $x = -7, x = 2$ **47.** $x = 0, x = 6$
49. $x = -3, x = 1$ **51.** $x = 2$ **53.** $x = -3, x = 3$
55. $x = -8, x = 2$ **57.** $x = 3$ **59.** $x = -5, x = 10$

EXERCISES 5.5, page 221

1. 0, 7 **4.** 4 **5.** 6, 13 or −13, −6
7. 6 **9.** 3, 11 **11.** 8, 20
13. 8, 9 **15.** 6, 7 **17.** 8, 9
19. 10, 12 or −12, −10 **21.** 6 in., 12 in. **23.** 5 m, 17 m
25. 13 ft, 9 ft **27.** 12 m, 7 m **29.** 14 ft
31. 13 cm, 8 cm **33.** 8 m by 12 m **35.** 9 rows
37. 3 cm by 10 cm **39.** 10 ft by 30 ft or 15 ft by 20 ft

EXERCISES 5.6, page 223

1. 2260.8 cu in. **3.** 5 in. **5.** 800 lb **7.** 6 in. **9.** 10 in. **11.** 200 ft
13. 4 sec or 6 sec **15.** 12 amps or 20 amps **17.** \$16 or \$20 **19.** \$4

EXERCISES 5.7, page 225

1. $(m + 6)(m + 1)$ **3.** $(x + 9)(x + 2)$ **5.** $(x + 10)(x - 10)$
7. $(m - 3)(m + 2)$ **9.** irreducible **11.** $(8x + 1)(8x - 1)$
13. $(x + 5)^2$ **15.** $(x + 12)(x - 3)$ **17.** $(x + 9)(x + 4)$
19. $5(x - 6)(x - 8)$ **21.** $-4(x - 10)(x + 5)$ **23.** $3(x + 7)(x - 7)$
25. $3x(x + 2)(x + 3)$ **27.** $4x(2x + 5)(2x - 5)$ **29.** $(3x - 2)(x - 5)$
31. $2(x - 3)(2x - 1)$ **33.** $(2x - 5)(6x - 1)$ **35.** $(3x - 7)(2x + 5)$
37. $(2x + 5)(4x - 7)$ **39.** $(5x + 6)(4x - 9)$ **41.** $3(4x^2 - 20x - 25)$
43. $(3 - x)(8 + 3x)$ **45.** $(2x - 3)(4x - 5)$ **47.** $(4x + 5)(5x - 4)$
49. $(6x - 1)(3x - 2)$ **51.** $7x(6 + 5x)(6 - 5x)$ **53.** $x(7x - 2)(3x - 1)$
55. $3x(3x - 2)(4x + 5)$ **57.** $2x(4x - 11)(2x - 1)$ **59.** $-5(4x + 3)(6x - 5)$
61. $(y - 4)(x + 3)$ **63.** $(x - 6)(x + 2y)$ **65.** $(x^2 - 5)(x - 8)$

CHAPTER 5 REVIEW, page 229

1. x^3 **2.** x^3 **3.** $2x$
4. $-8x^2y$ **5.** $4x^2y$ **6.** $5(x - 2)$
7. $-4x(3x - 4)$ **8.** $8xy(2x - 3)$ **9.** $5x^2(2x^2 - 5x + 1)$
10. $(2x + 1)(2x - 1)$ **11.** $(y - 10)^2$ **12.** $(9x + 4y)(9x - 4y)$
13. $(2x - 3)^2$ **14.** $a(c^2 + ab^2 + d)$ **15.** $3(x + 4y)(x - 4y)$
16. $(x - 9)(x + 2)$ **17.** $5(x + 4)^2$ **18.** $(5x + 2)(x + 3)$
19. $(x + 6)(x - 5)$ **20.** irreducible **21.** $(5x + 2)^2$
22. $(3x + 2)(x + 1)$ **23.** $2x(x - 5)^2$ **24.** $4x(x^2 + z^2)$
25. $(3x - 2)(2x + 1)$ **26.** $2x(4x + 3)(x - 2)$ **27.** $(x^2 + 7)(x + 2)(x - 2)$
28. $(x^2 + 2)(x + 2)(x - 2)$ **29.** $(y + 3)(x + 2)$ **30.** $(a + b)(x - 2)$
31. $x^2 - 4x + \underline{4} = (x - 2)^2$ **32.** $x^2 + 18x + \underline{81} = (x + 9)^2$
33. $x^2 - 8x + \underline{16} = (x - 4)^2$ **34.** $x^2 - 10x + \underline{25} = (x - 5)^2$
35. $x^2 + 5x + \underline{\frac{25}{4}} = \left(x + \frac{5}{2}\right)^2$ **36.** $x = 7,\ x = -1$
37. $x = 0,\ x = -\frac{5}{3}$ **38.** $x = 0,\ x = -5,\ x = 2$ **39.** $x = 0,\ x = 7$
40. $x = -6,\ x = -2$ **41.** $x = 7,\ x = -4$ **42.** $x = -3,\ x = 3$
43. $x = 0,\ x = -6,\ x = 1$ **44.** $x = 11,\ x = -5$ **45.** $x = -10,\ x = 6$
46. $x = 1,\ x = -5$ **47.** $x = 0,\ x = -7$ **48.** $x = 0,\ x = -2$
49. $x = 0,\ x = 4$ **50.** $x = -8,\ x = 6$ **51.** 3, 12
52. 12 cm by 17 cm **53.** 8, 9 **54.** 13 ft, 6 ft
55. 22 in., 17 in. **56.** $2\frac{1}{2}$ sec or 3 sec

CHAPTER 5 TEST, page 231

1. $2x^3$ **2.** $-7y^2$ **3.** $x^3y(20y + 18 - 15x)$
4. $(x - 5)(x - 4)$ **5.** $(x + 7)^2$ **6.** $(6x + 1)(6x - 1)$
7. $(6x - 5)(x + 1)$ **8.** $(3x - 8)(x + 3)$ **9.** $(4x + 5y)(4x - 5y)$
10. $x(2x - 3)(x + 1)$ **11.** $x^2 - 10x + \underline{25} = (x - 5)^2$ **12.** $x^2 + 16x + \underline{64} = (x + 8)^2$
13. $x = -2,\ x = \frac{5}{3}$ **14.** $x = 8,\ x = -1$ **15.** $x = 0,\ x = -6$
16. $x = 5,\ x = -3$ **17.** $x = \frac{3}{4},\ x = -5$ **18.** 6, 20 or -4, -30
19. 18, 19 **20.** 15 cm, 11 cm

CHAPTER 6

EXERCISES 6.1, page 241

1. $x^2 + 2x + \frac{3}{4}$ **3.** $2x^2 - 3x - \frac{3}{5}$ **5.** $2x + 5$

7. $x + 6 - \frac{3}{x}$ **9.** $2x + 3 - \frac{3}{2x}$ **11.** $2x - 3 - \frac{1}{x}$

13. $x + 3y - \frac{11y}{7x}$ **15.** $\frac{3x}{4} - 2y - \frac{y^2}{x}$ **17.** $\frac{5x}{8} - 1 + \frac{2}{y}$

19. $\frac{8x^2}{9} - xy + \frac{5}{9y}$ **21.** $12\frac{2}{23}$ **23.** $7\frac{24}{59}$

25. $x - 8 + \frac{18}{x + 3}$ **27.** $a + \frac{-15}{a - 2}$ **29.** $y + 3 + \frac{-30}{y - 4}$

31. $5y - 11 + \frac{48}{y + 5}$ **33.** $4c - 5 + \frac{1}{2c + 3}$ **35.** $2m + 1 + \frac{-3}{5m - 3}$

37. $x - 2 + \frac{10}{x + 5}$ **39.** $y^2 - 7y + 12$ **41.** $3a^2 + 1 + \frac{2}{4a - 1}$

43. $x^2 + 7x + 35 + \frac{170}{x - 5}$ **45.** $x^2 - 3x + 9$

EXERCISES 6.2, page 247

1. $\frac{x - 2}{2x + 1}; x \neq -\frac{1}{2}$ **3.** $\frac{1}{x - 4}; x \neq 4, 0$ **5.** $7; x \neq -2$

7. $1; x \neq -\frac{5}{3}$ **9.** $-1; x \neq 1, -1$ **11.** $\frac{2}{x + 3}; x \neq -3$

13. $\frac{x + 2}{x - 5}; x \neq 5, -5$ **15.** $\frac{x - 6}{x + 3}; x \neq -3$

17. $\frac{-(x - 3)}{2x(2 + x)}$ or $\frac{3 - x}{2x(2 + x)}; x \neq 0, 2, -2$ **19.** $\frac{4x + 5}{x + 2}; x \neq -2, -\frac{5}{4}$

21. $\frac{2x}{x + 2}$ **23.** $\frac{2x(x - 3)}{x - 1}$ **25.** $3(x - 2)$ **27.** $\frac{2(x - 1)}{x + 1}$

29. $\frac{5x}{6}$ **31.** $\frac{x(x - 2)}{(x + 1)(x - 1)}$ **33.** $\frac{x + 3}{x + 4}$ **35.** $\frac{3x + 1}{x + 1}$

37. $\frac{-(x - 5)}{(x - 2)(x + 7)}$ or $\frac{5 - x}{(x - 2)(x + 7)}$ **39.** -1

41. $\frac{x(2x - 1)}{x + 4}$ **43.** $\frac{4x}{x - 3}$ **45.** $\frac{x - 3}{2x - 3}$ **47.** $\frac{3x - 2}{3x + 2}$

49. $\frac{(x + 3)(x - 4)}{(2x + 1)(x - 1)}$ **51.** $2x^2$ **53.** $\frac{x(x - 3)}{(x + 1)^2}$ **55.** $\frac{x(x + 5)}{2x + 1}$

57. $\frac{x - 1}{x + 1}$ **59.** $\frac{2x - 1}{(6x + 1)(2x + 1)}$

EXERCISES 6.3, page 253

1. 4
3. 2
5. $\frac{x-3}{x+1}$
7. $\frac{1}{x-1}$
9. $\frac{x-2}{x+2}$
11. $\frac{x}{2x-1}$
13. $\frac{8}{7x^2y^3}$
15. $\frac{2}{x-3}$
17. $\frac{-2}{5(x-2)}$
19. $\frac{3x+8}{x(x+4)}$
21. $\frac{2x^2}{(x+4)(x-4)}$
23. $\frac{3}{x-2}$
25. $\frac{x+3}{x-5}$
27. $\frac{x^2+x+1}{(x-1)(x+2)}$
29. $\frac{-x^2-11x-4}{(x+4)(x-4)}$
31. $\frac{4x+24}{(x+2)(x-2)}$
33. $\frac{-9x-4}{12(x-2)}$
35. $\frac{-2x-17}{(x+5)(x-1)}$
37. $\frac{3x^2-20x}{(x+6)(x-6)}$
39. $\frac{2x^2+2x-1}{(x-1)(x-7)}$
41. $\frac{x^2+4x+2}{2(x+2)(x+3)}$
43. $\frac{x^2+8x+2}{2(x+4)(x-4)}$
45. $\frac{-1}{(x+2)(x+1)(x-1)}$
47. $\frac{5x^2+18x}{(x+4)(x+3)(x-3)}$
49. $\frac{-(2x^2+8x+3)}{(x+2)(x+2)(x+1)}$
51. $\frac{-2x^2+13x}{(x+4)(x-4)(x+1)}$
53. $\frac{3x^2-5x-7}{(2x-1)(x+3)(x-4)}$
55. $\frac{-x(2x+13)}{(x+5)(x-2)(x+2)}$
57. $\frac{10x^2-17x-3}{(x+4)(x-4)}$
59. $\frac{x+1}{(x-2)(x-1)}$

EXERCISES 6.4, page 259

1. $\frac{8}{7}$
3. $\frac{10}{7}$
5. $\frac{1}{2}$
7. 3
9. $\frac{2x}{3y}$
11. $16xy$
13. $\frac{x-5}{x-4}$
15. $\frac{3x}{x-2}$
17. $\frac{2x(x+3)}{2x-1}$
19. $\frac{x+3}{x}$
21. $\frac{x+2}{4x}$
23. $\frac{2x}{3(x+6)}$
25. $\frac{x}{x-1}$
27. $\frac{x}{y+x}$
29. $\frac{2(x+1)}{x+2}$
31. $\frac{x+2}{x+3}$
33. $-\frac{5}{x+1}$
35. $\frac{29}{4(4x+5)}$
37. $\frac{x^2-3x-6}{x(x-1)}$
39. $\frac{x^2-4x-2}{(x+4)(x-4)}$

EXERCISES 6.5, page 265

1. $x = 21$ **3.** $y = -2$ **5.** $x = 48$ **7.** $x = \frac{108}{5}$ **9.** $x = \frac{49}{8}$

11. $x = \frac{80}{3}$ **13.** $x = 10$ **15.** $x = 20$ **17.** $x = \frac{37}{6}$ **19.** $x = 4$

21. $x = 7$ **23.** $x = -3$ **25.** $x = -\frac{34}{5}$ **27.** $x = -\frac{74}{9}$ **29.** $x = \frac{27}{4}$

31. \$8.50 **33.** 45 mi **35.** 144 bulbs **37.** \$3 **39.** 225 lb

41. $\frac{33}{4}$ in. **43.** 18 **45.** 21, 27 **47.** 20 **49.** 400, 250

EXERCISES 6.6, page 267

1. 384 rev per min **3.** 36 teeth **5.** $139\frac{7}{32}$ sq ft
7. 192 lb per sq ft **9.** 700 lb per sq ft **11.** 1.5 ml
13. 9 sacks, 27 cu ft sand, 45 cu ft gravel **15.** 900 lb
17. $1\frac{1}{3}$ ft from 960-lb weight **19.** $5\frac{1}{3}$ ohms

CHAPTER 6 REVIEW, page 271

1. $2x^2 + x - \frac{5}{3}$ **2.** $4x - 7 + \frac{3}{x}$ **3.** $\frac{3x}{5} + 2y + \frac{y^2}{x}$

4. $1 - 3y + \frac{8y^2}{7}$ **5.** $2x^2 - \frac{3xy}{2} + y^2$ **6.** $10\frac{8}{37}$

7. $12\frac{60}{61}$ **8.** $x - 2$ **9.** $x - 15 - \frac{1}{x + 1}$

10. $x + 2 + \frac{4}{4x - 3}$ **11.** $2x^2 - x + 3 - \frac{2}{x + 3}$ **12.** $4x + 13 + \frac{35}{x - 2}$

13. $x - 3 + \frac{18}{x + 3}$ **14.** $2x^2 - 3x + 2 - \frac{3}{3x + 2}$ **15.** $x^2 - 4x + 16$

16. $\frac{1}{x + 1}; x \neq 0, -1$ **17.** $-\frac{1}{3}; x \neq 4$ **18.** $\frac{2}{3}; x \neq -3$

19. $\frac{x}{x + 4}; x \neq -3, -4$ **20.** $\frac{x + 5}{2(x - 3)}; x = 3$ **21.** $-\frac{x - 5}{4 + x}; x \neq 4, -4$

22. $\frac{2x + 1}{x + 1}; x \neq -1, \frac{3}{4}$ **23.** $-\frac{x}{2}; x \neq 0, \frac{2}{5}$ **24.** $x - y$

25. 2 **26.** 3 **27.** $\frac{3}{4}$

28. $\frac{7}{3}$ **29.** $\frac{16x - 28}{x(x - 4)}$ **30.** $\frac{4}{(y + 2)(y + 3)}$

31. $\frac{x^2 - 2x}{3(x + 2)(x + 1)}$ **32.** $\frac{x + 4}{3x}$ **33.** $\frac{3x(x + 2)(x + 2)}{x + 3}$

34. $\frac{2x + 1}{x - 1}$ **35.** $\frac{4x + 25}{(x + 5)(x - 5)}$ **36.** $\frac{12x^2 + 52x - 24}{(x + 6)(x + 3)(x - 2)}$

37. $\frac{2x^2 + 14x - 8}{(x + 3)(x - 2)(x - 1)}$ **38.** $\frac{-4}{(x - 1)(x + 4)}$ **39.** $\frac{x - 2}{x + 3}$

40. $\frac{x - 4}{(x + 2)(x - 2)}$ **41.** $\frac{7 - x}{7x^2}$ **42.** $\frac{5x^2 - x - 47}{(x - 5)(x + 2)(x + 3)}$

43. $\frac{4}{13}$ **44.** $\frac{19}{14}$ **45.** $2(x + 1)$

46. $\frac{x - 1}{x + 1}$ **47.** $\frac{3x}{2 - x}$ **48.** $x = 52$

49. $x = 14\frac{2}{5}$ **50.** $x = \frac{9}{2}$ **51.** $x = \frac{13}{2}$

52. 24, 60 **53.** $4\frac{1}{2}$ in. by 6 in. **54.** 161

55. $18.00 **56.** 25,000 lb **57.** 3 ohms

CHAPTER 6 TEST, page 275

1. $2x + \frac{3}{2} - \frac{3}{x}$ **2.** $\frac{5}{3} + 2y + \frac{y^2}{x}$ **3.** $x + 9$

4. $x - 6 - \frac{2}{2x + 3}$ **5.** $x^2 + 3x + 9$ **6.** $\frac{x + 3}{x}; x \neq 0, \frac{1}{2}$

7. $\frac{2x + 1}{x + 1}$ **8.** $-\frac{x^2}{2}$ **9.** $\frac{7(4x + 3)}{x(x + 4)}$

10. $\frac{1}{2(2x + 1)}$ **11.** $\frac{10x - 13}{(x + 1)(x - 1)(x - 2)}$ **12.** $\frac{8x^2 - 19x + 15}{(x + 5)(x - 5)(x - 2)}$

13. $\frac{3x^2 - 8x - 27}{(x - 1)(x - 4)(x + 3)}$ **14.** $\frac{x(x - 3)}{2(x + 1)}$ **15.** $\frac{1}{x + 1}$

16. $\frac{6(x + 3)}{x + 18}$ **17.** $x = \frac{15}{26}$ **18.** $x = -\frac{5}{11}$

19. $3\frac{3}{5}$ ft **20.** 40 people

CHAPTER 7

EXERCISES 7.1, page 285

1. $x = -5$ **3.** $x = 5$ **5.** $x = 6$ **7.** $x = \frac{5}{2}$

9. $x = 2$ **11.** $x = 3$ **13.** $x = \frac{3}{2}$ **15.** $x = \frac{1}{2}$

17. $x = 3$ **19.** $x = 3$ **21.** $x = -7$ **23.** $x = -1$

25. $x = 21$ **27.** $x = -\frac{3}{2}$ **29.** $x = 20$ **31.** $x = -90$

33. $x = 17$ **35.** $x = \frac{13}{5}$ **37.** $x = -3$ **39.** $x = \frac{1}{9}$

41. $x = -13$ **43.** $x = \frac{10}{3}$ **45.** $x = \frac{1}{4}$ **47.** $x = -\frac{1}{27}$

49. $x = \frac{2}{3}$ **51.** $x = 0, x = 5$ **53.** $x = 0, x = -\frac{9}{5}$ **55.** $x = -1, x = 7$

57. $x = -5, x = 2$ **59.** $x = -1, x = \frac{6}{5}$ **61.** $x = -\frac{3}{5}, x = 2$ **63.** $x = \frac{1}{3}, x = 2$

65. $x = -2, x = \frac{5}{3}$ **67.** $x = 15$ **69.** $x = 0$ **71.** $x = -\frac{3}{2}$

73. $x = 1, x = 7$ **75.** $x = -3, x = \frac{2}{3}$ **77.** $x = 24$ **79.** $x \approx 12.41$

EXERCISES 7.2, page 292

1.

3.

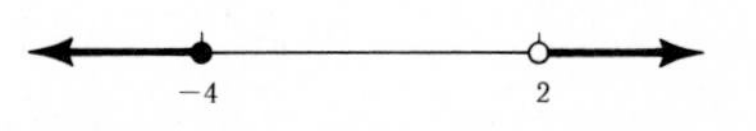

5.

7.

9.

11. $x > 4$

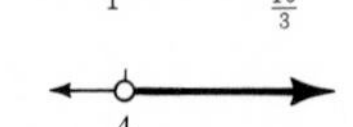

13. $x < -\frac{14}{3}$

15. $x \geq -\frac{11}{4}$

17. $x > -3$

19. $y \leq 5$

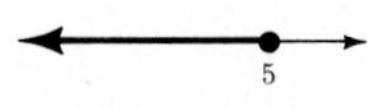

21. $x > 2$

23. $y < -\frac{8}{3}$

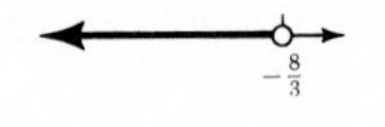

25. $x > -\frac{3}{2}$

27. $x \leq 8$

29. $x > \frac{9}{2}$

31. $x \geq -3$

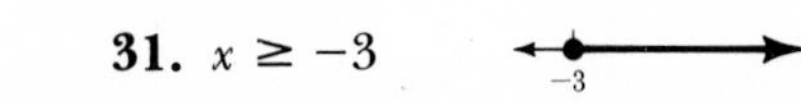

33. $x \leq \frac{1}{9}$

35. $x > -13$

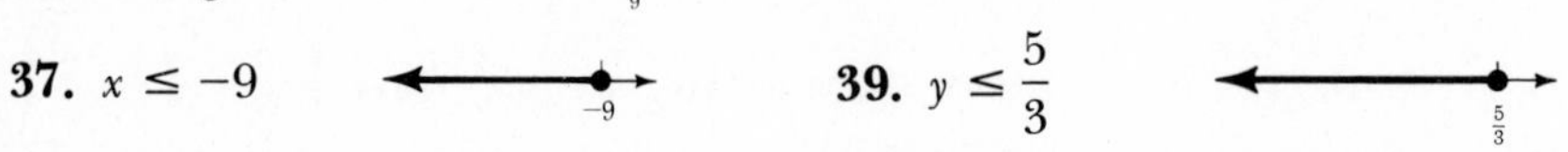

37. $x \leq -9$

39. $y \leq \frac{5}{3}$

41. $-\frac{3}{4} \le x \le 2$

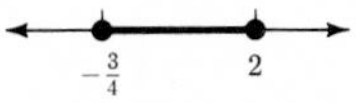

43. $-1 \le x < 2$

45. $3 < x \le 6$

47. $-\frac{3}{2} < x < \frac{3}{2}$

49. $-\frac{3}{2} < x < 2$

EXERCISES 7.3, page 295

1. 80 cm by 40 cm
3. 140 ft by 220 ft
5. 17 m by 35 m
7. 10 cm, 14 cm, 19 cm
9. 14 cm, 14 cm, 23 cm
11. 40 mph, 45 mph
13. June—25 mph; Sue—45 mph
15. 3 hr
17. 18 mi
19. 67.2 mph
21. 6 hr
23. 14 m, 8 m
25. 3 mph
27. 6 cm, 2 cm
29. $37\frac{1}{2}$ mph, $62\frac{1}{2}$ mph

EXERCISES 7.4, page 302

1. $1200
3. $1150
5. $800 @13%; $1500 @ 9%
7. $7000
9. $8500 @ 9%; $3500 @ 11%
11. $3800 @ 15%; $4200 @ 12%
13. $1\frac{1}{2}$ hr
15. $22\frac{1}{2}$ min
17. $2\frac{2}{9}$ hr
19. 6 hr
21. $4\frac{1}{2}$ days; 9 days
23. 2 hr; 6 hr
25. $6300 @ 6%; $3200 @ 3%
27. $7600 @ 9%; $12,400 @ 7%
29. 4 hr—Alice; 6 hr—Sharon

EXERCISES 7.5, page 307

1. 16 oz
3. 84 oz
5. 6.4 qt
7. 8 tons of 80%, 16 tons of 50%
9. $44 \le x \le 100$
11. $x \ge 67$, no
13. $-5° \le C \le 15°$
15. $6 < x < 13$
17. no
19. 50 lb of 30%, 40 lb of 12%
21. 8 qt of A, 32 qt of C
23. less than 31 ft
25. at least 171
27. 200 gal
29. 28 lb of $3.90, 12 lb of $3.50

CHAPTER 7 REVIEW, page 311

1. $x = 4$	**2.** $x = -6$	**3.** $x = -1$	**4.** $x = 6$
5. $x = -15$	**6.** $x = 4$	**7.** $x = 1$	**8.** $x = \frac{34}{3}$
9. $x = -31$	**10.** $x = \frac{100}{13}$	**11.** $x = -\frac{72}{7}$	**12.** $x = -\frac{3}{2}$
13. $x = \frac{5}{4}$	**14.** $x = 1$	**15.** $x = -9$	**16.** $x = 4$
17. $x = 0,\ x = -\frac{8}{3}$	**18.** $x = -\frac{3}{4},\ x = 2$	**19.** $x = 18$	**20.** $x = -9,\ x = 2$

21. −5 7

22. −1 4

23. −2 5

24. −2 $\frac{4}{5}$

25. −3 $\frac{7}{8}$

26. 2 14

27. −3 2.7

28. $x \leq 2$ 2

29. $x > 4$ 4

30. $x > 3$ 3

31. $x \leq \frac{1}{3}$ $\frac{1}{3}$

32. $x \leq -1$ −1

33. $x \geq \frac{5}{3}$ $\frac{5}{3}$

34. $x \geq 4$ 4

35. $x > -7$ −7

36. $x > -\frac{13}{3}$ $-\frac{13}{3}$

37. $x \leq -1$ −1

38. $x \leq 17$ 17

39. $-\frac{9}{4} \leq x \leq -1$ $-\frac{9}{4}$ −1

40. $\frac{3}{2} \leq x < 5$

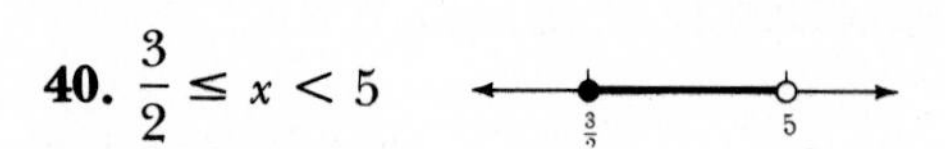

41. 15 m by 110 m

42. 12, 18

43. \$6500 @ 4%; \$19,500 @ 6%

44. 5 mph

45. between 12 and 17 yr, inclusive

46. 20 lb of 40% fat; 30 lb of 15% fat

47. 4 hr

48. 19 cm; 6 cm

49. Soo—42 mph; Lisa—57 mph

50. 7500 @ \$3.50; 5000 @ \$2.00

CHAPTER 7 TEST, page 314

1. $x = 1$ **2.** $x = 6$ **3.** $x = \frac{19}{10}$ **4.** $x = 20$

5. $x = 0, x = \frac{3}{2}$ **6.** $x = \frac{8}{5}$ **7.** 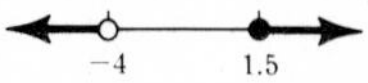 -4, 1.5 **8.** -1, $\frac{3}{2}$

9.  $-\frac{3}{4}$, 3 **10.** $x > \frac{17}{9}$ $\frac{17}{9}$

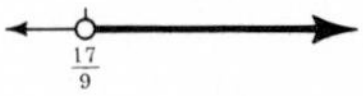

11. $-\frac{3}{5} < x < 1$ $-\frac{3}{5}$, 1 **12.** $x \leq \frac{9}{2}$ $\frac{9}{2}$

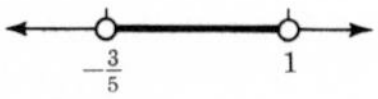

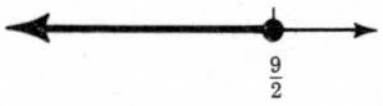

13. $x > \frac{19}{10}$ $\frac{19}{10}$  **14.** 8 ft by 25 ft

15. 52 mph, 65 mph **16.** \$5000 @ 18%; \$6200 @ 12%
17. 4 hr **18.** 1600 lb of 83%; 400 lb of 68%
19. $x \geq 86$ **20.** 22 @ \$95; 18 @ 120

CHAPTER 8

EXERCISES 8.1, page 324

1. $(-5, 1), (-3, 3), (-1, 1), (1, 2), (2, -2)$
3. $(-3, -2), (-1, 3), (-1, -3), (0, 0), (2, 1)$
5. $(-3, -4), (-3, 4), (0, 3), (0, -4), (4, 1)$
7. $(-5, 2), (-1, 6), (0, 0), (1, -6), (6, 4)$
9. $(-5, 0), (-2, 2), (-1, -4), (0, 6), (2, 0)$

11.
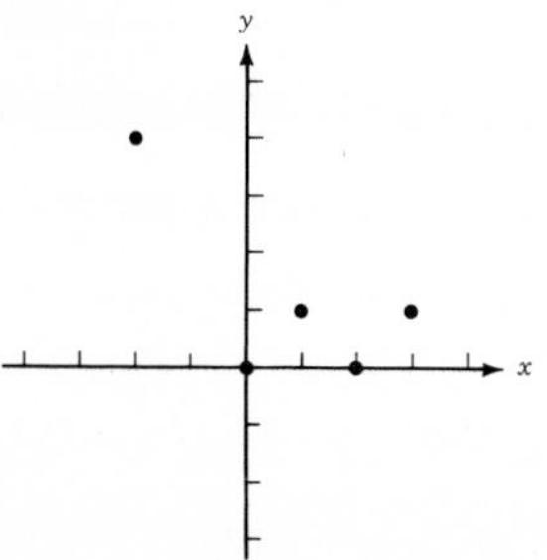

13.
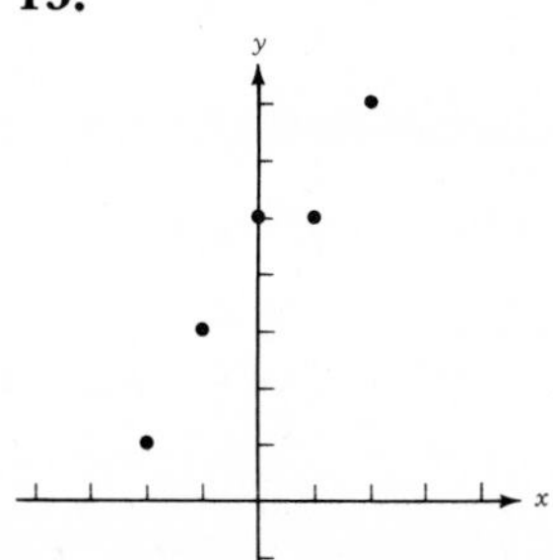

15.
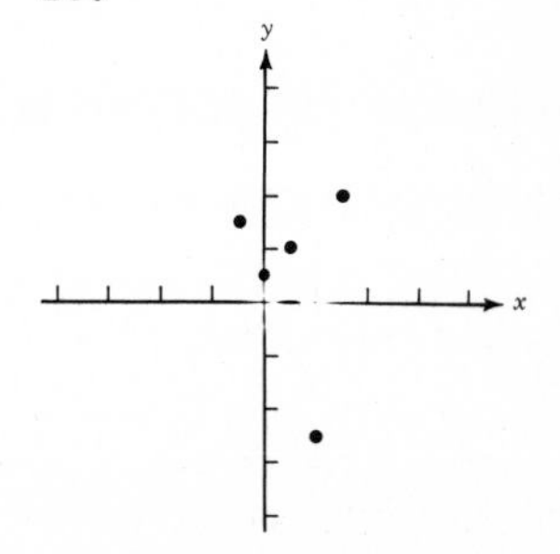

17.

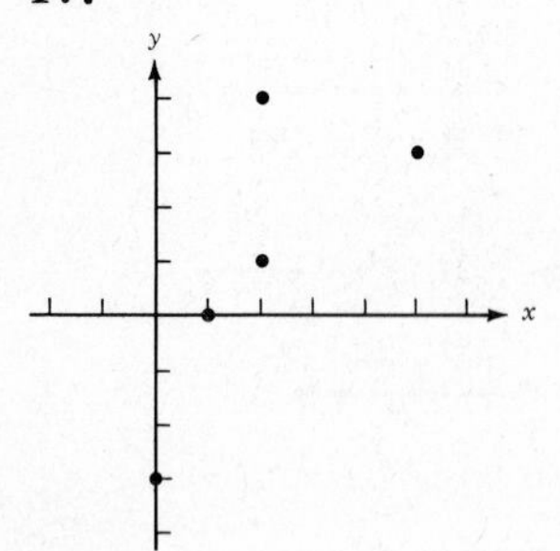

19.

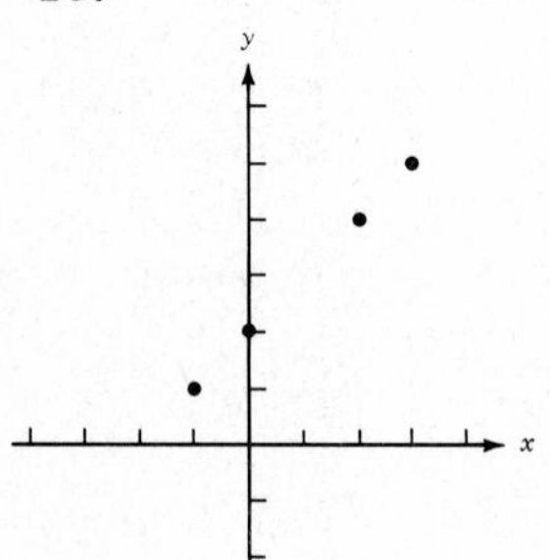

21.

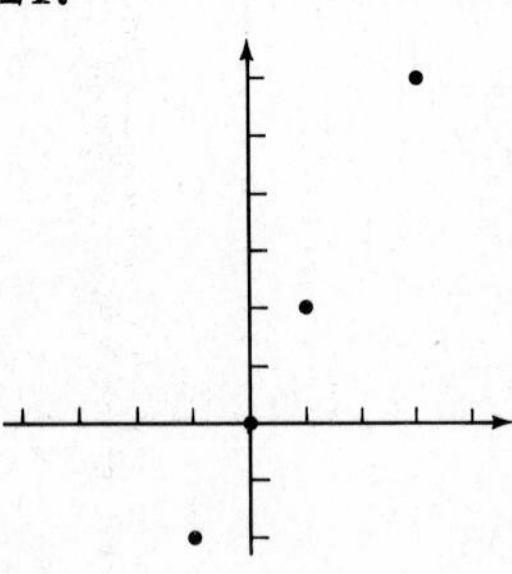

23.

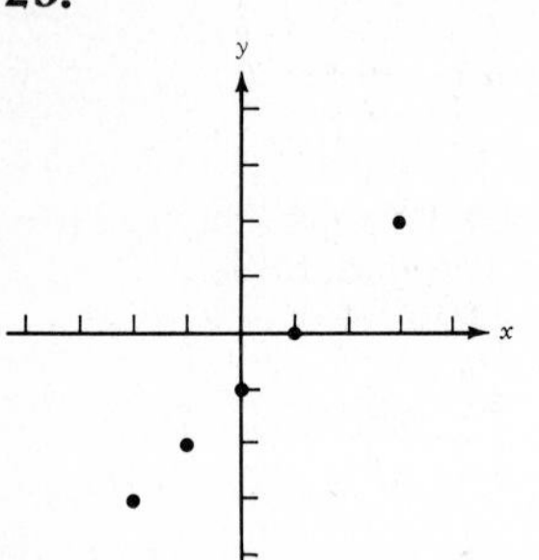

25.

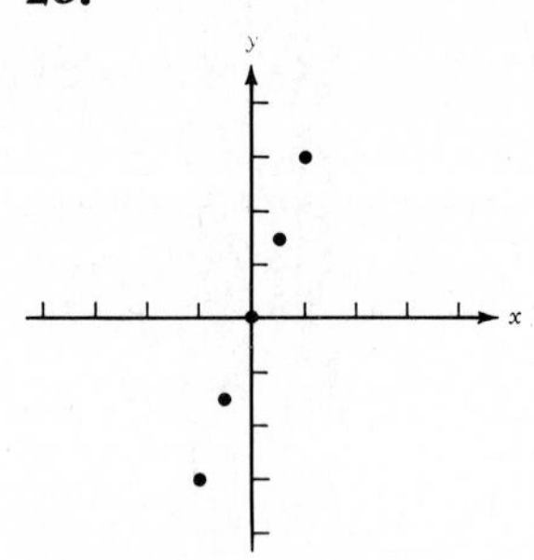

27.

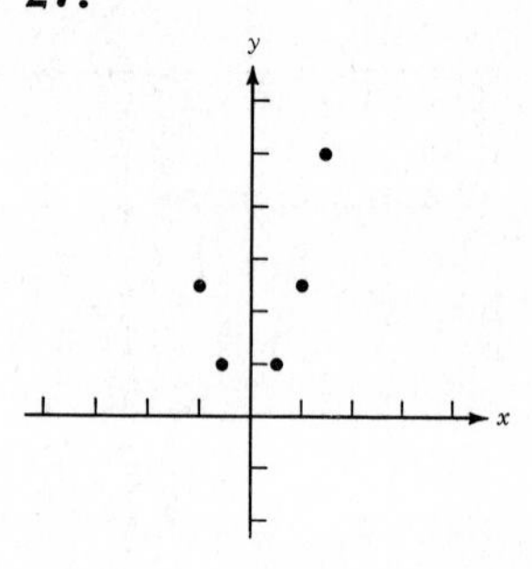

29.

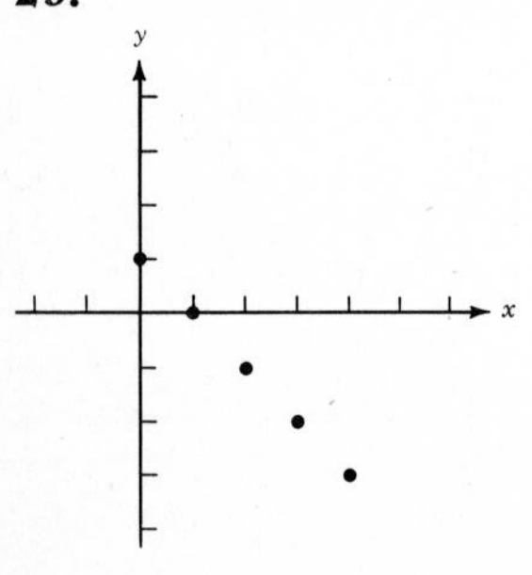

31.

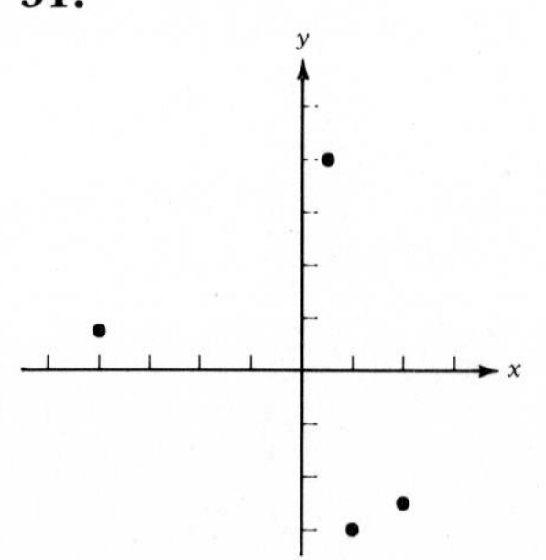

33.

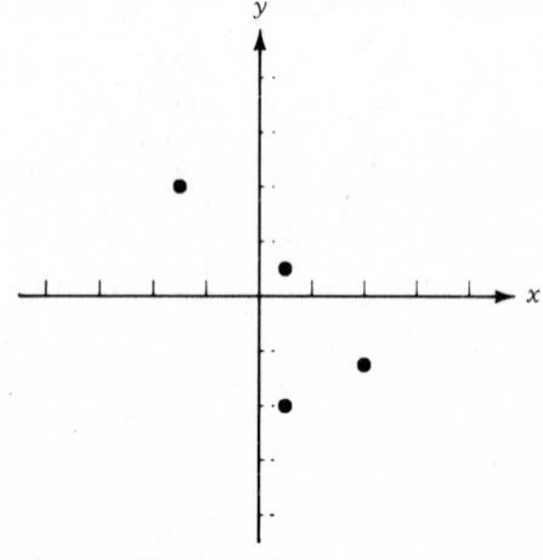

35.

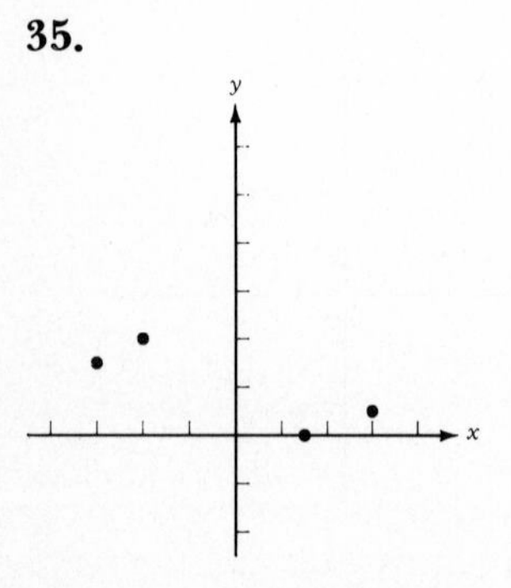

37. **a.** -5
b. -2
c. 1
d. 4
e. 7

39. **a.** 7
b. 2
c. -8
d. $\frac{11}{3}$
e. -3

41. **a.** 1
b. 0
c. −5
d. 3
e. 7

43. **a.** 2
b. 1
c. 2
d. 5
e. 5

45. **a.** 9
b. 1
c. 3
d. 19
e. 9

EXERCISES 8.2, page 330

1.

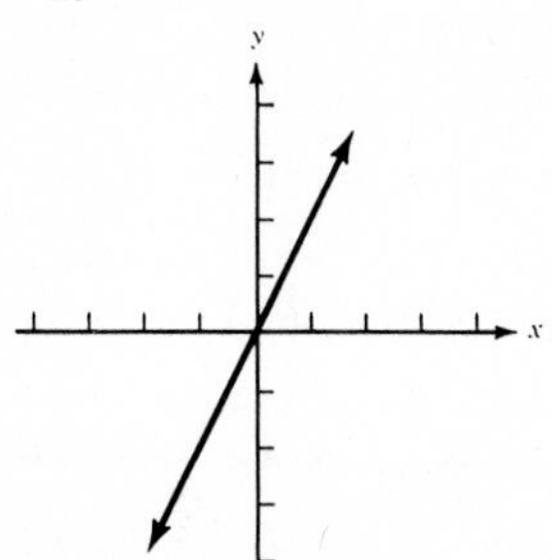

3.

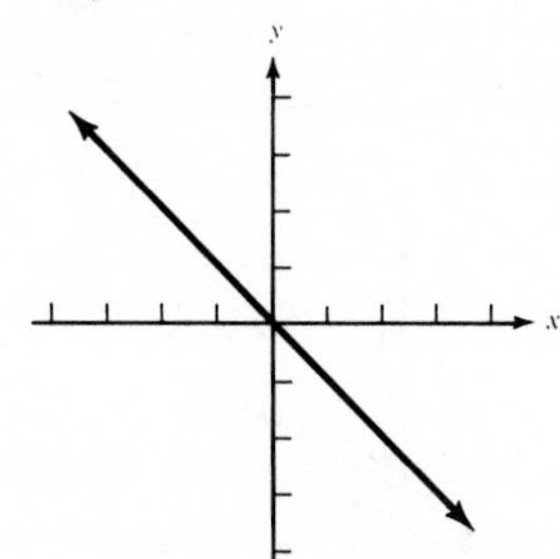

5.

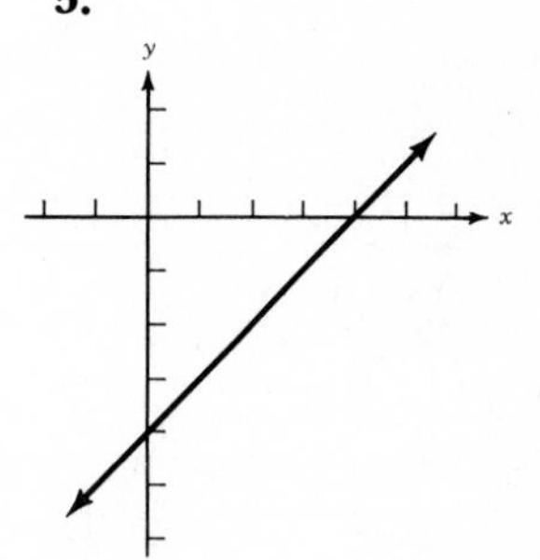

7.

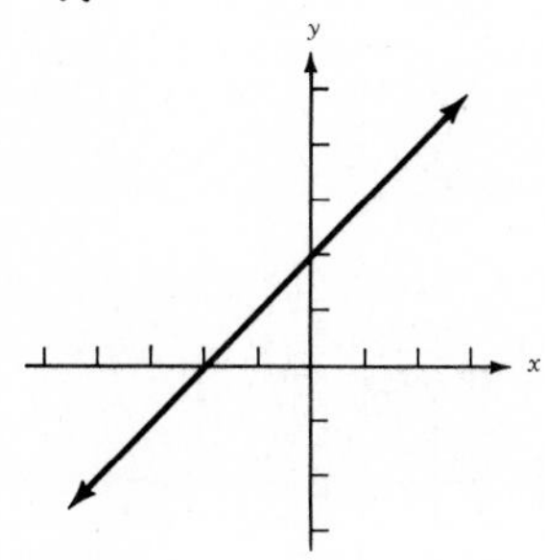

9.

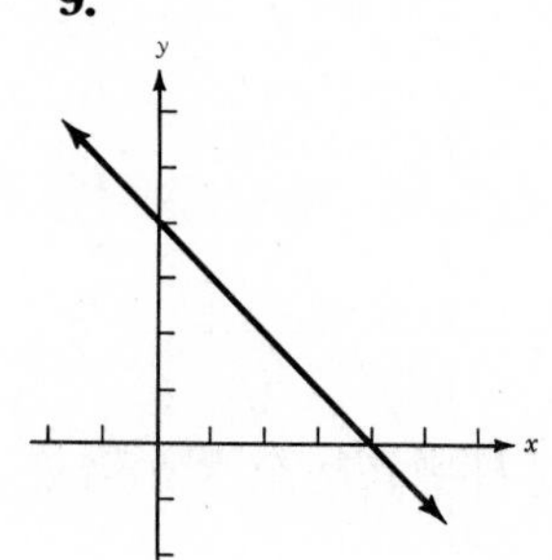

11.

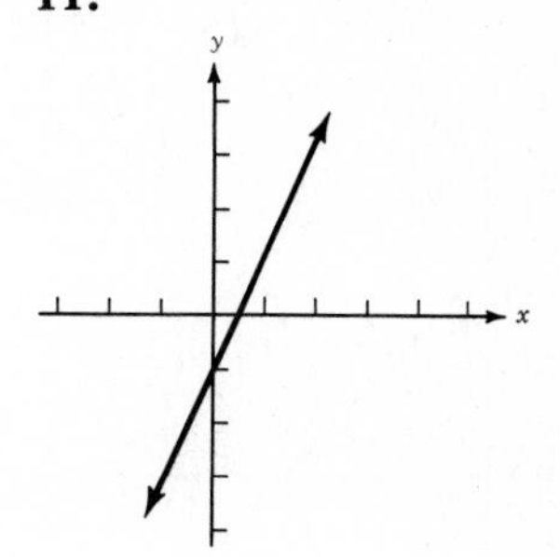

13.

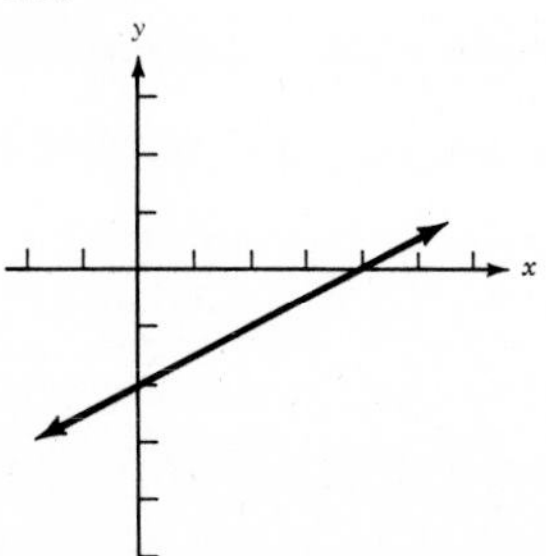

15.

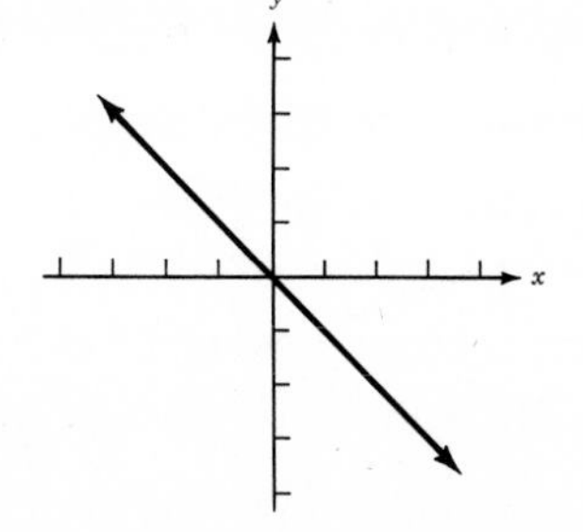

17.

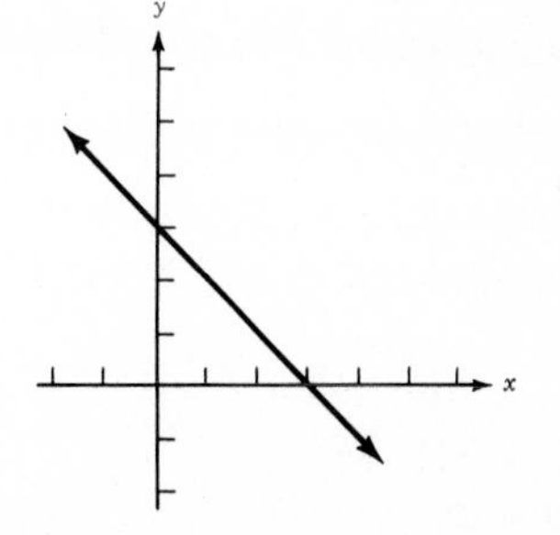

19.

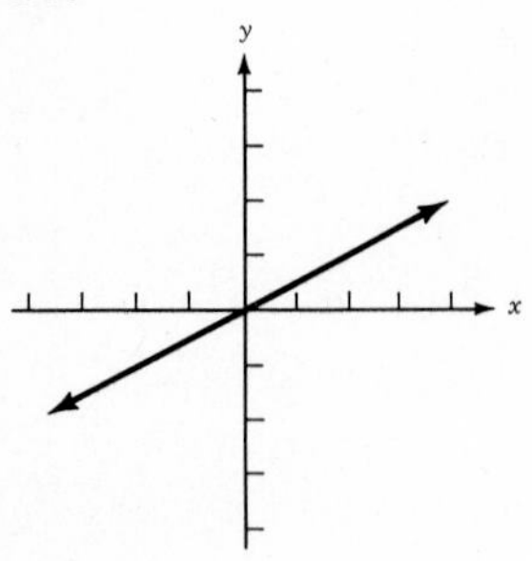

21.

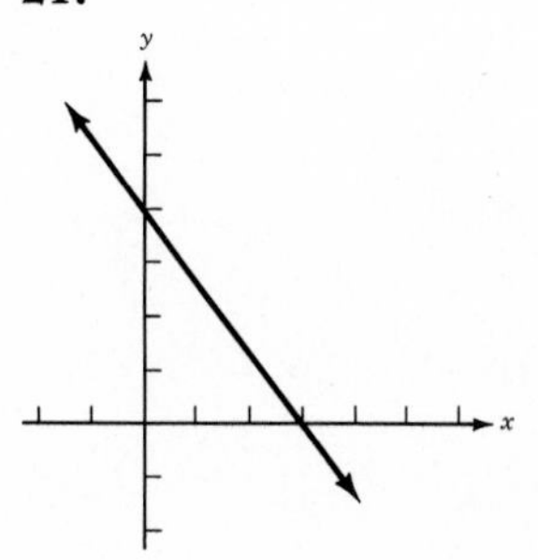

23.

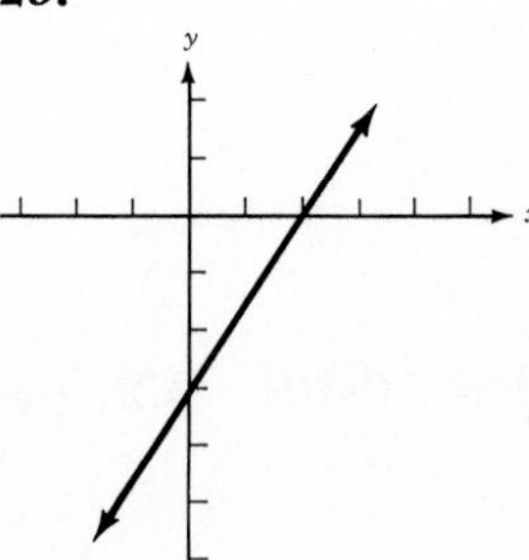

25.

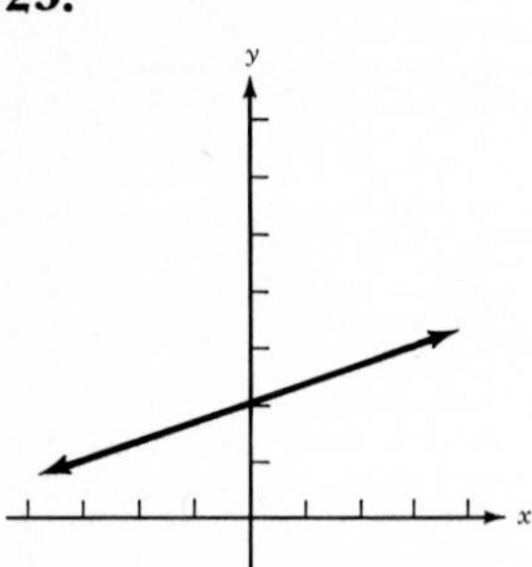

27.

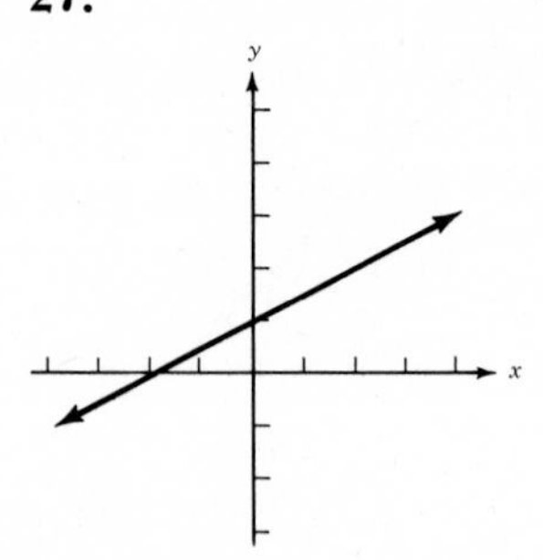

29.

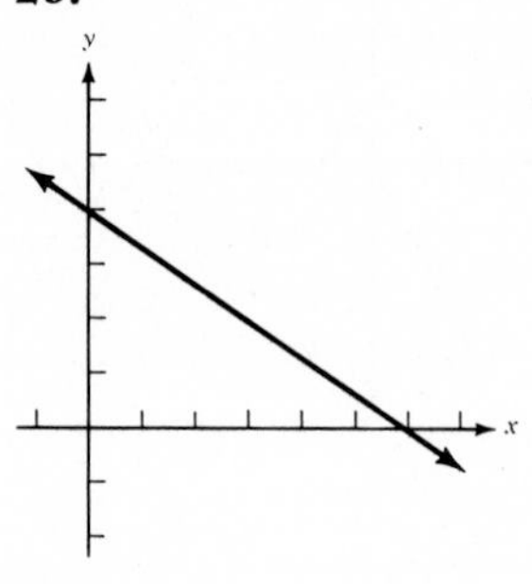

31.

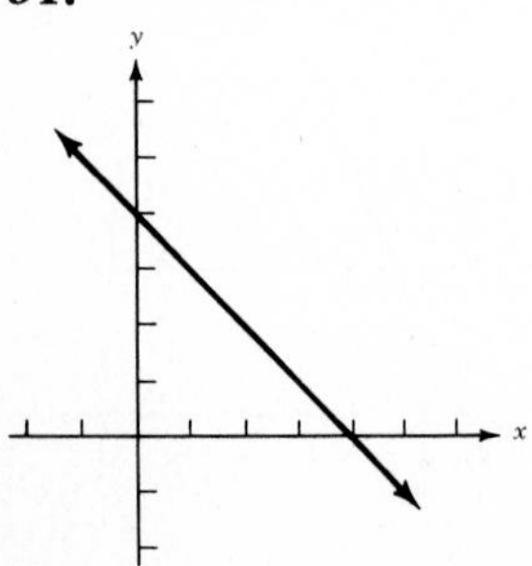

33.

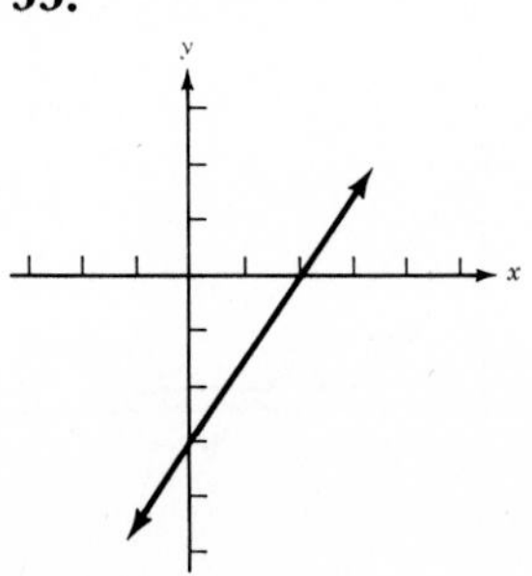

35.

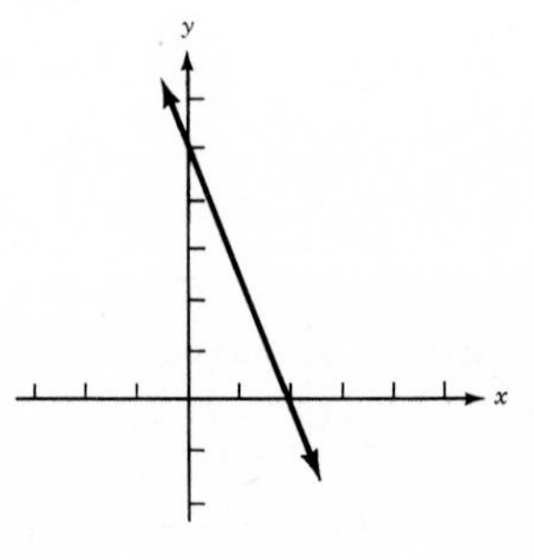

37.

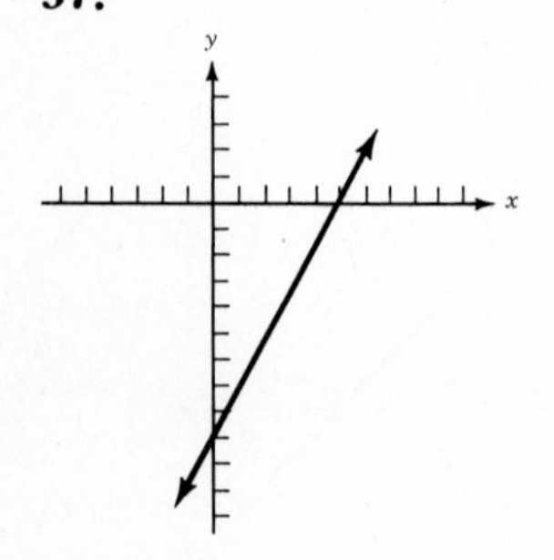

39.

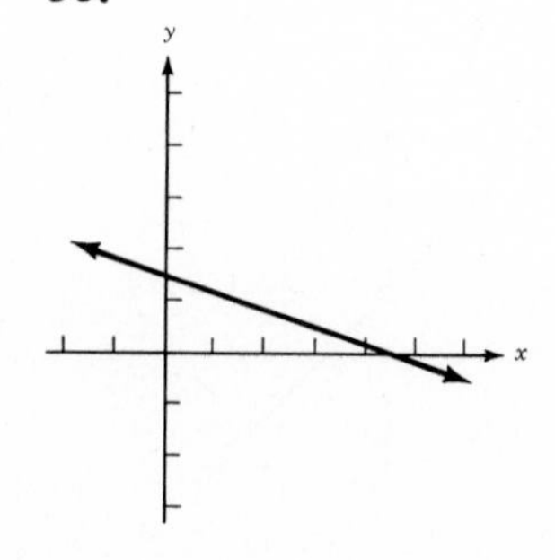

41.

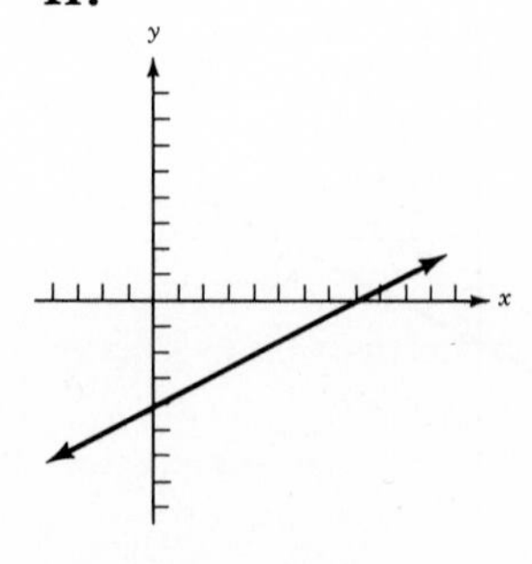

43.

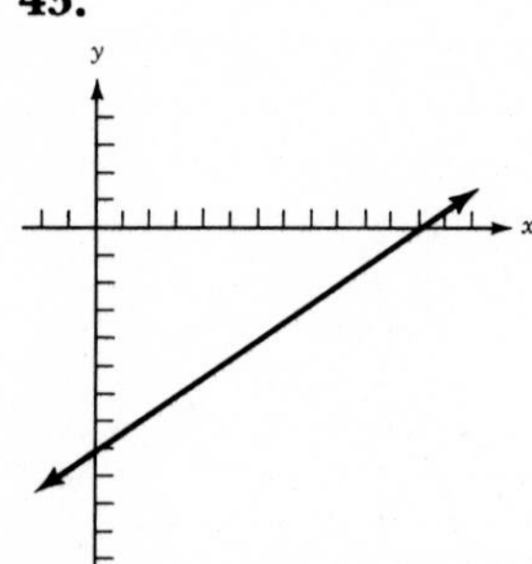

45.

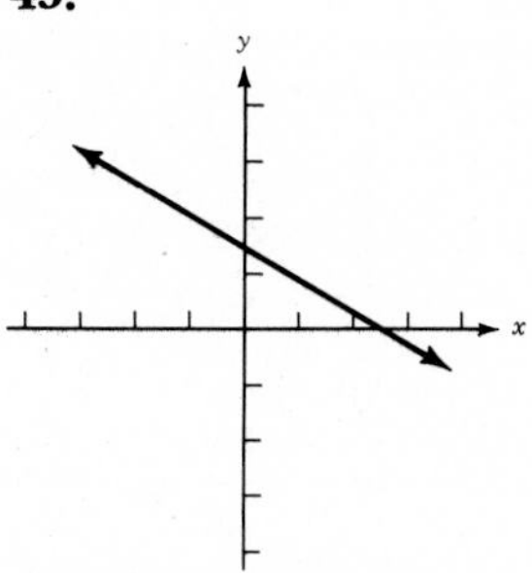

EXERCISES 8.3, page 340

1. $m = -5$

3. $m = \frac{8}{7}$

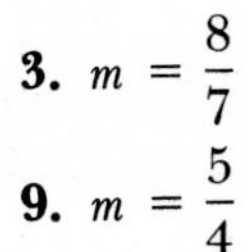

5. $m = \frac{1}{8}$

7. $m = \frac{3}{10}$

9. $m = \frac{5}{4}$

11. $y = \frac{2}{3}x$

13. $y = -\frac{3}{4}x - 3$

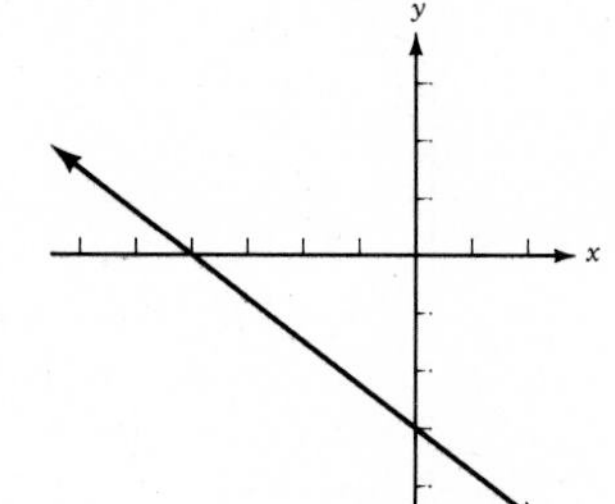

15. $y = -\frac{5}{3}x + 3$

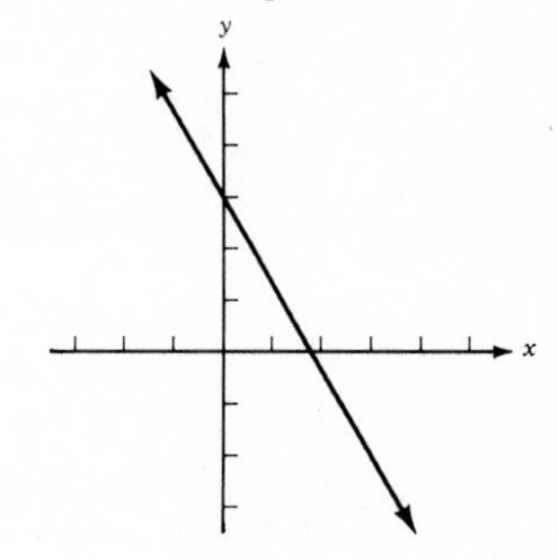

17. $y = 2x - 1$

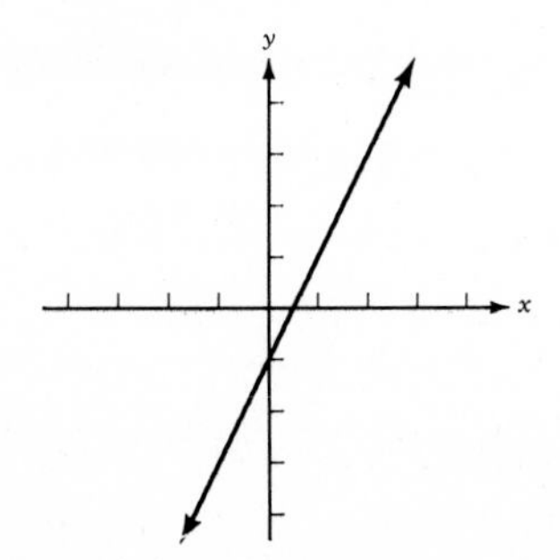

19. $y = -\frac{1}{4}x - 5$

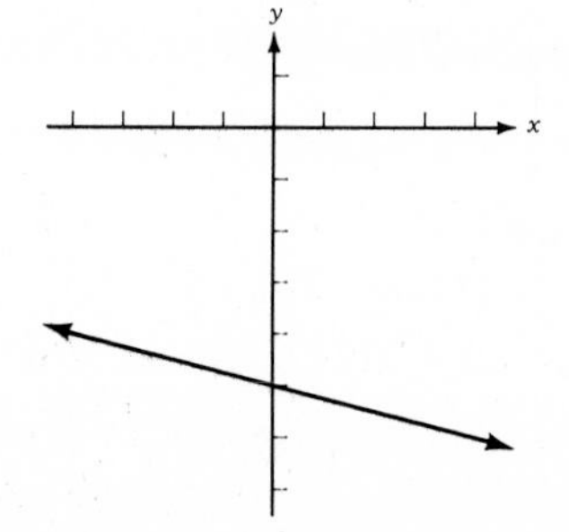

21. $y = \frac{3}{2}x - 4$

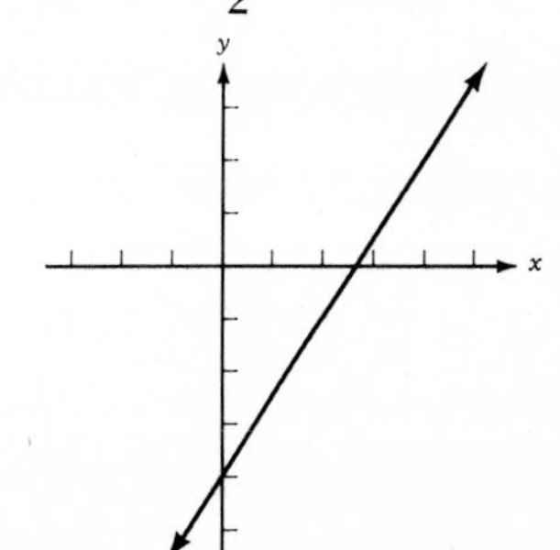

23. $y = x + 5$

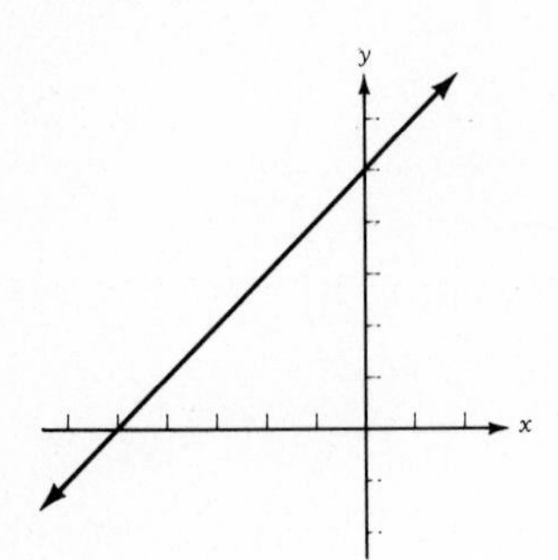

25. $y = \frac{2}{5}x - 6$

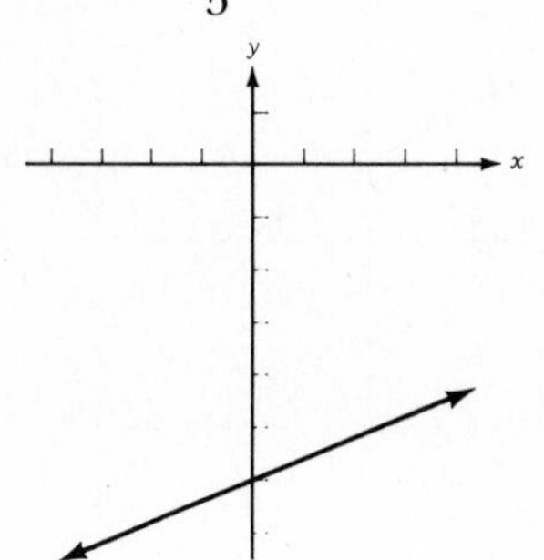

27. $m = 3,\ b = 4$

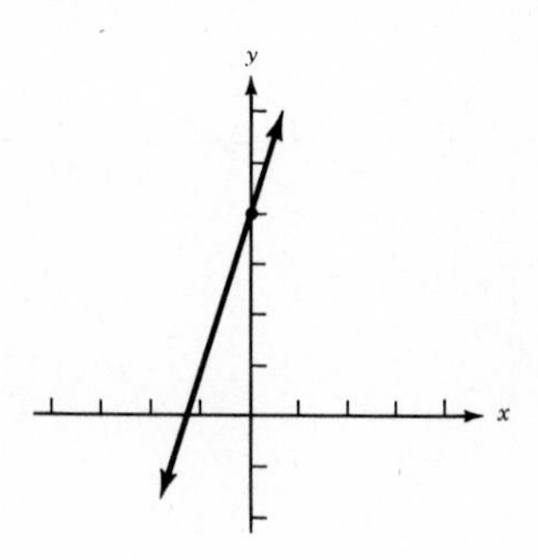

29. $m = -\frac{2}{3},\ b = 1$

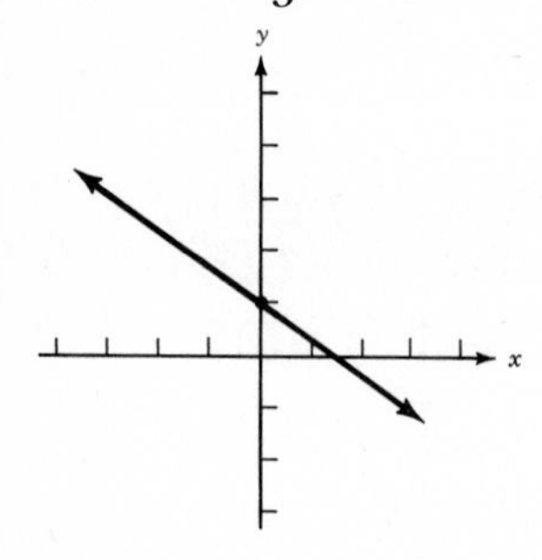

31. $m = 2,\ b = -3$

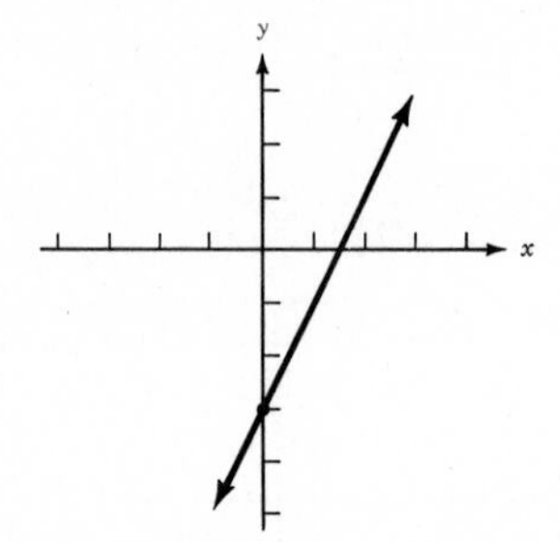

33. $m = -4,\ b = 0$

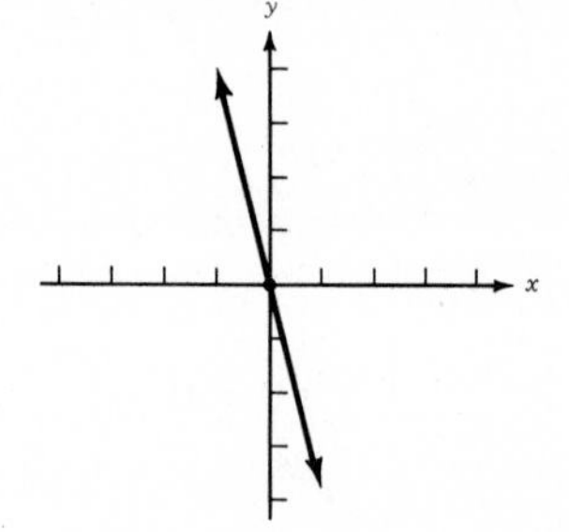

35. $m = -\frac{1}{4},\ b = -2$

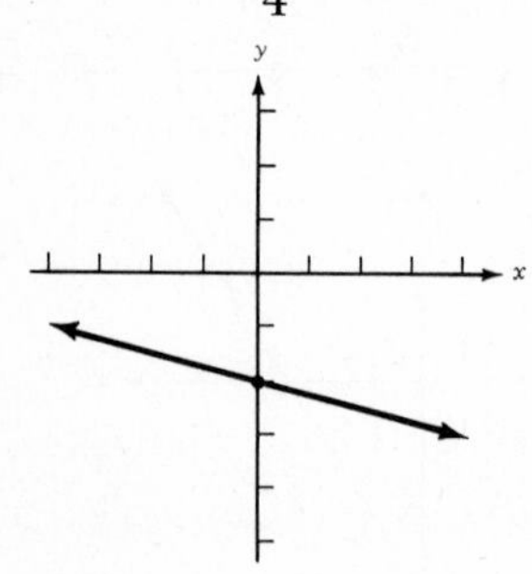

37. $m = \frac{2}{5},\ b = -2$

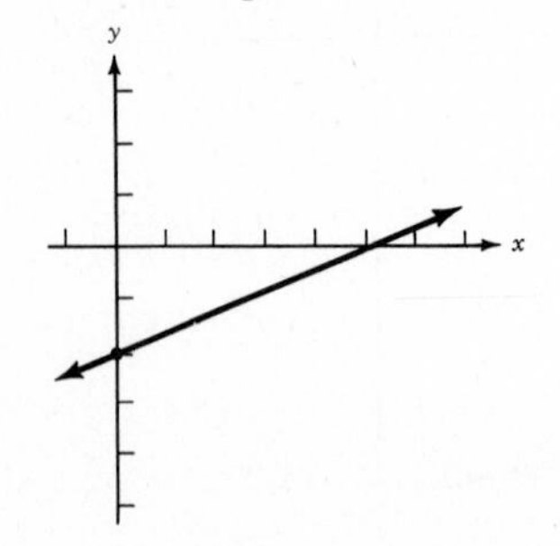

39. $m = -\frac{7}{2},\ b = 2$

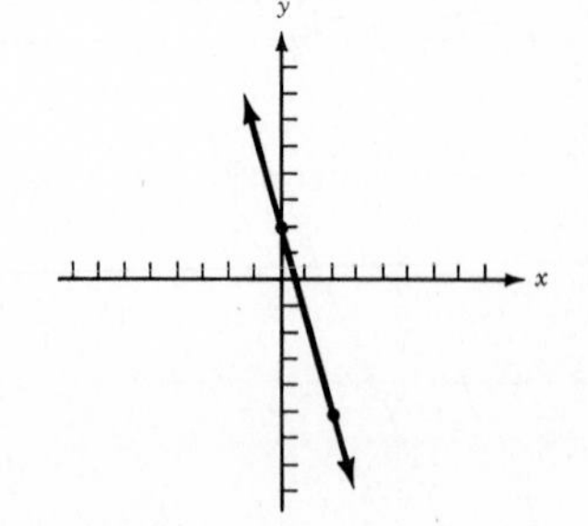

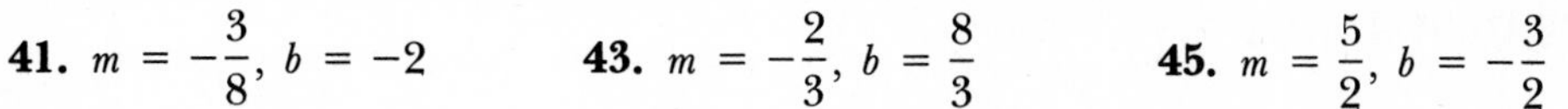
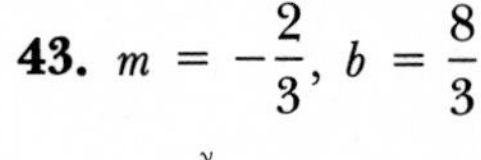

41. $m = -\frac{3}{8}$, $b = -2$ **43.** $m = -\frac{2}{3}$, $b = \frac{8}{3}$ **45.** $m = \frac{5}{2}$, $b = -\frac{3}{2}$

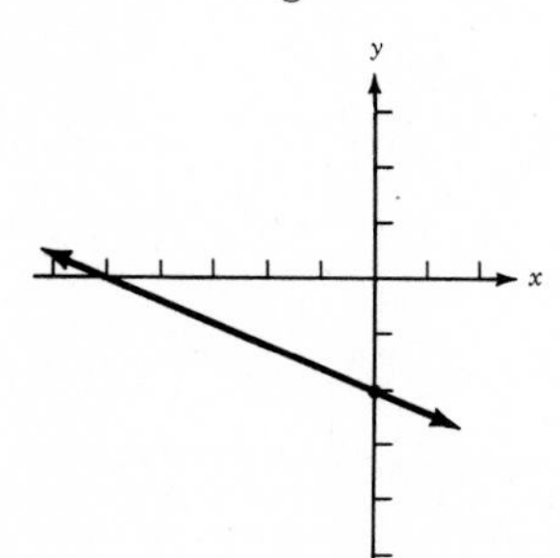

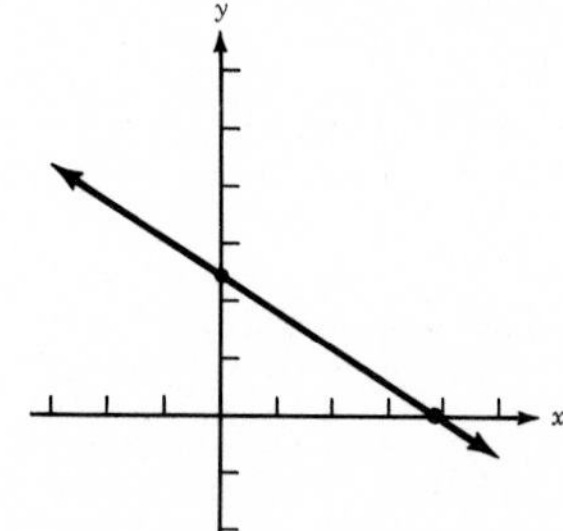

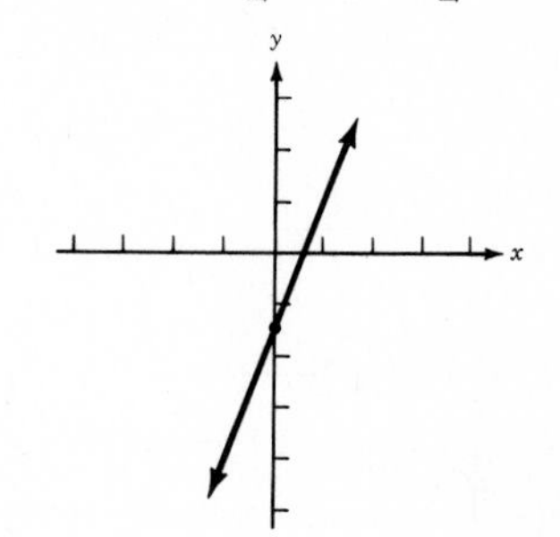

EXERCISES 8.4, page 345

1. $y = 2x + 3$ or $-2x + y = 3$

3. $y = -\frac{2}{5}x + \frac{13}{5}$ or $2x + 5y = 13$

5. $y = \frac{3}{4}x + \frac{17}{4}$ or $-3x + 4y = 17$

7. $y = -\frac{5}{6}x - \frac{14}{6}$ or $5x + 6y = -14$

9. $y = \frac{3}{2}x - \frac{5}{6}$ or $9x - 6y = 5$

11. $y = -\frac{1}{2}x - 1$ or $x + 2y = -2$

13. $y = -\frac{1}{3}x + \frac{7}{3}$ or $x + 3y = 7$

15. $y = -\frac{2}{3}x + 6$ or $2x + 3y = 18$

17. $y = \frac{3}{5}x - \frac{14}{5}$ or $3x - 5y = 14$

19. $y = \frac{1}{3}x + \frac{11}{3}$ or $-x + 3y = 11$

21. $y = -\frac{3}{10}x + \frac{4}{5}$ or $3x + 10y = 8$

23. $y = 2x + \frac{1}{2}$ or $-4x + 2y = 1$

25. $C = 0.15x + 30$

27. $S = 0.09x + 200$

29. $P = -\frac{1}{200}x + 12$

EXERCISES 8.5, page 349

1. $m = \frac{3}{4}$

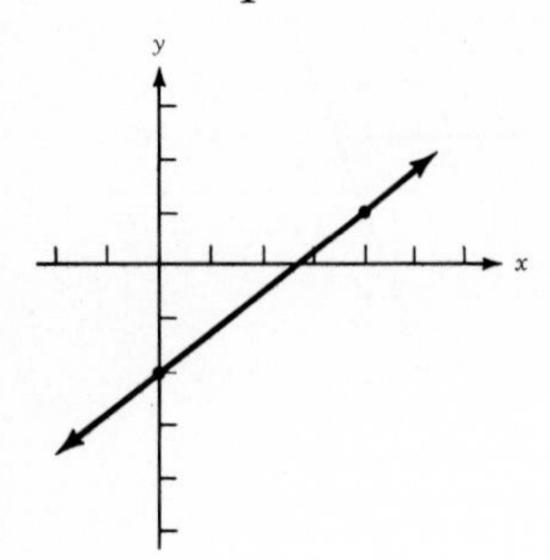

3. $m = 0$

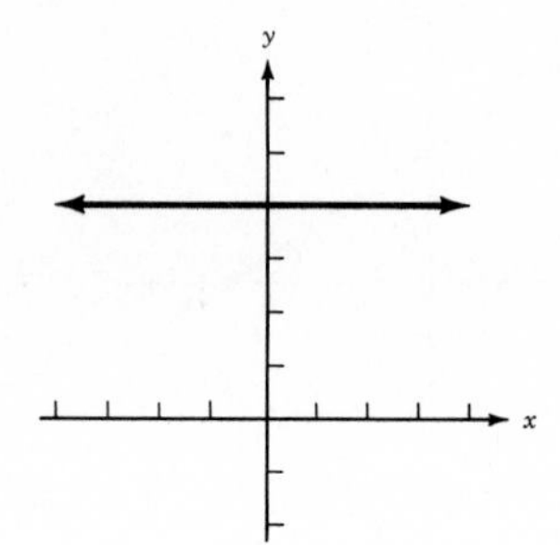

5. $m = \frac{3}{5}$

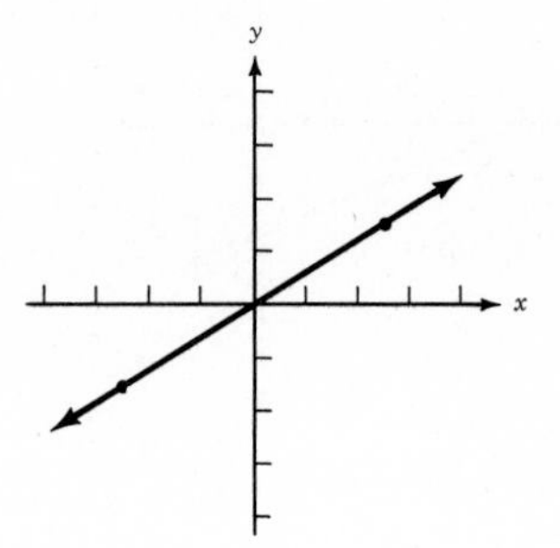

7. no slope

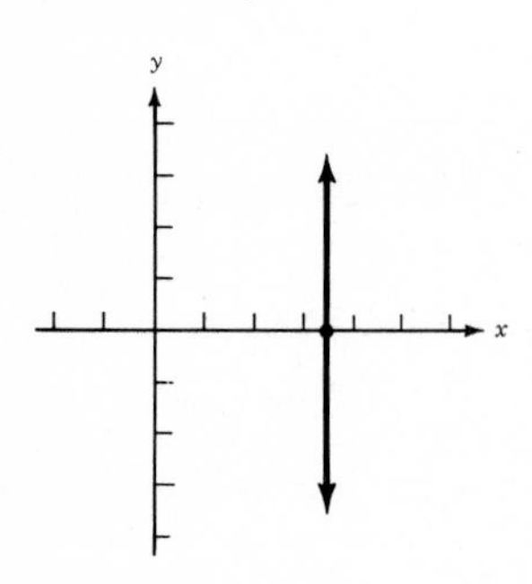

9. $m = 0$

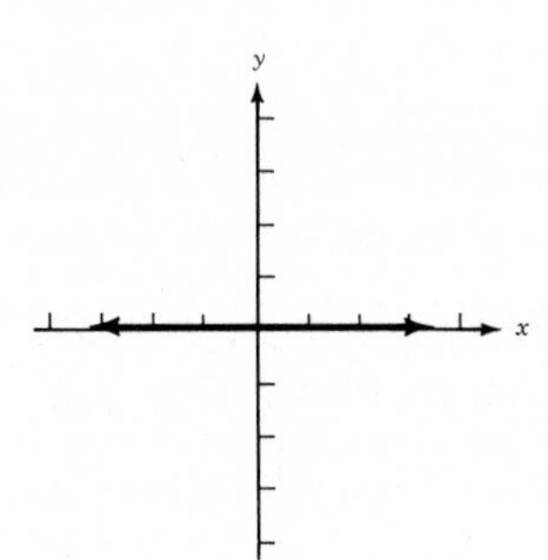

11. $m = \frac{1}{6}$

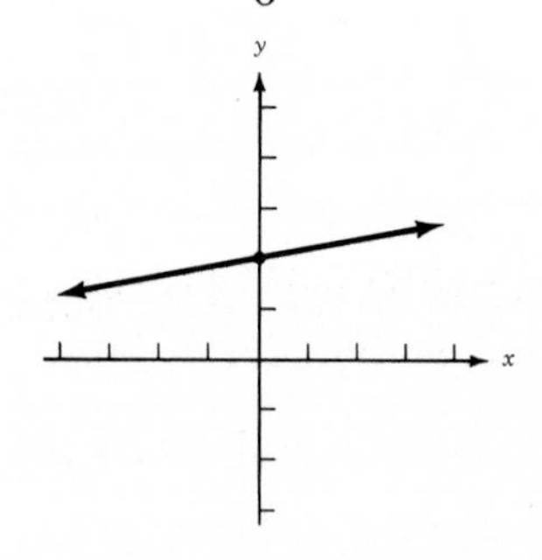

13. $m = 3$

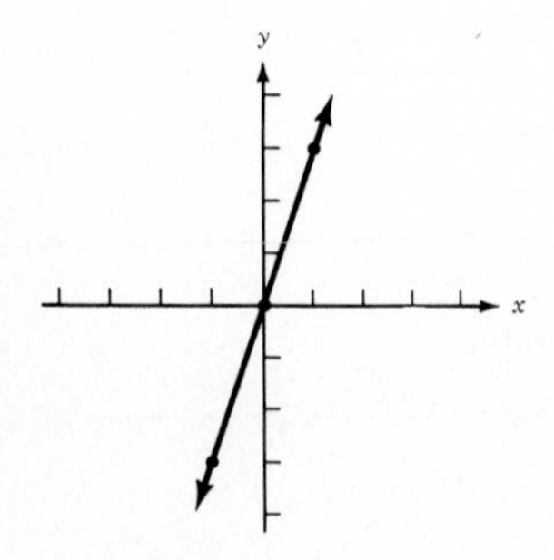

15. $m = -\frac{3}{5}$

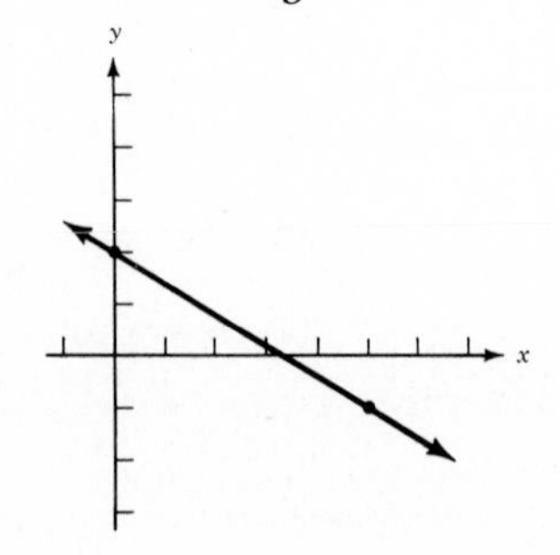

17. no slope

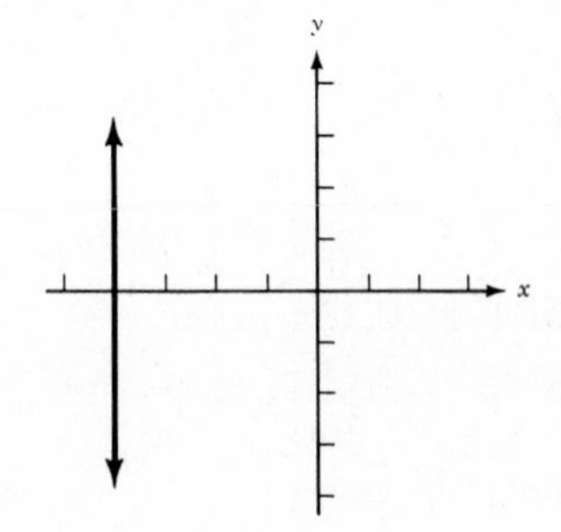

19. $m = -\dfrac{1}{3}$

21. $y = -3$

23. $3x - 4y = 3$

25. $2x + 3y = -14$

27. $x = -5$

29. $y = -1$

31. $y = 5$

33. $x = -1$

35. $3x - y = 0$

37. $y = 1$

39. $2x - y = 2$

EXERCISES 8.6, page 354

1.

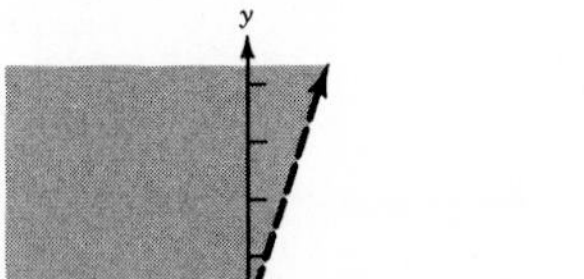

3.

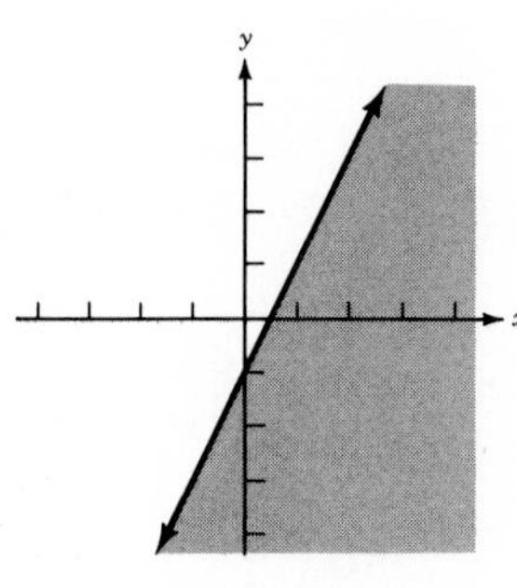

5.

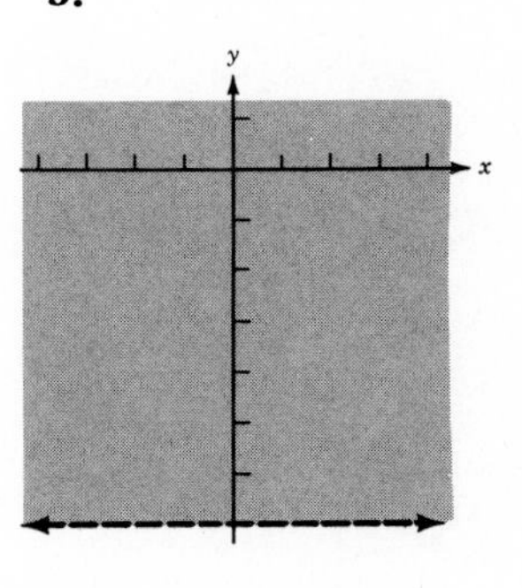

7.

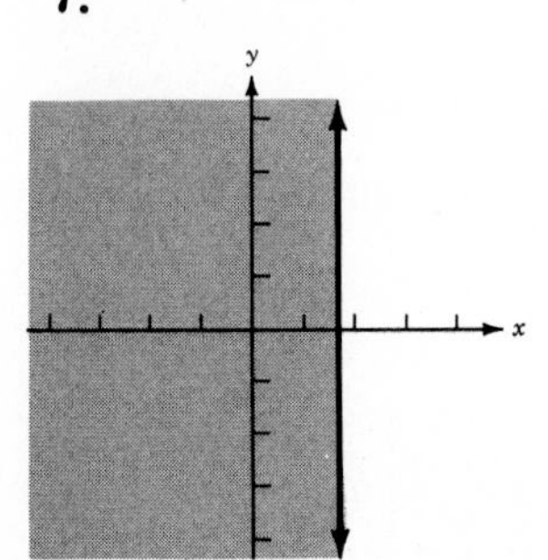

9.

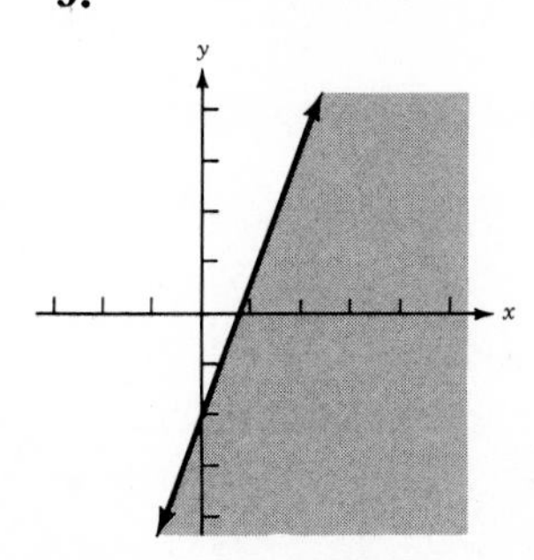

11.

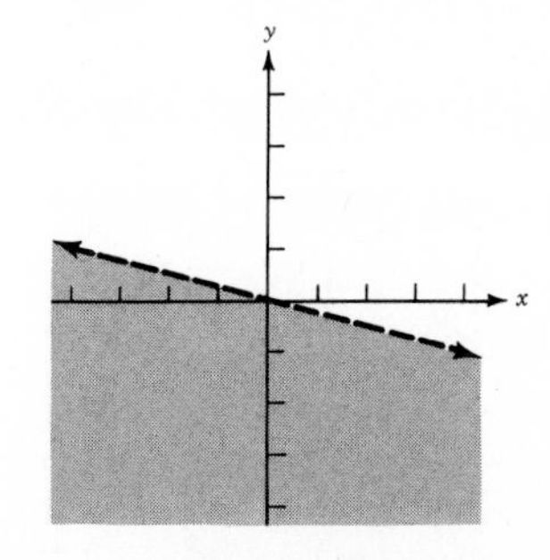

13.

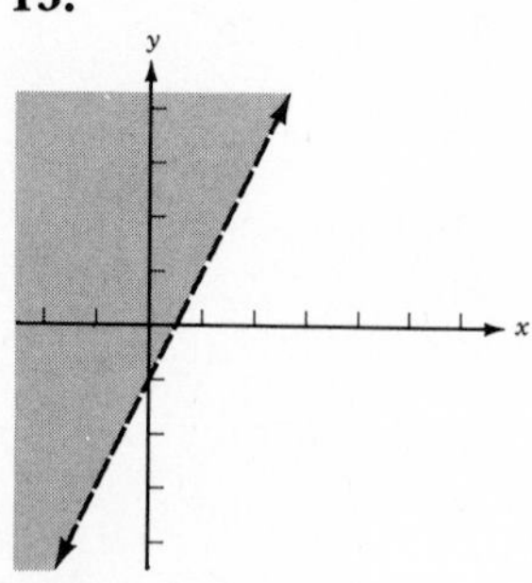

15.

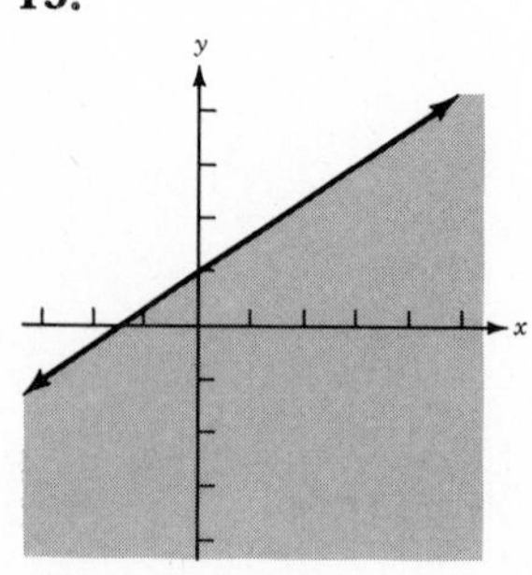

17.

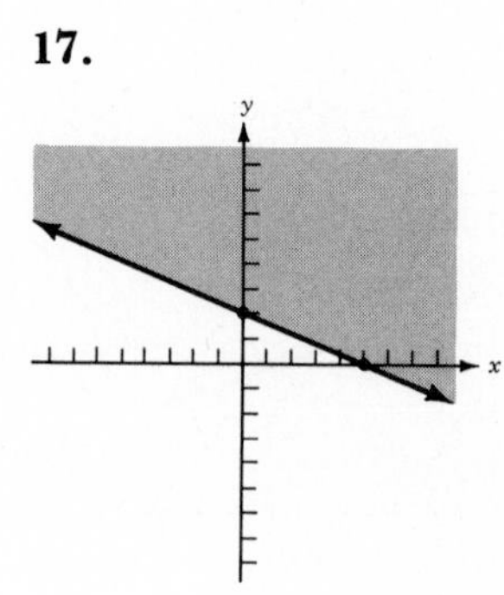

19.

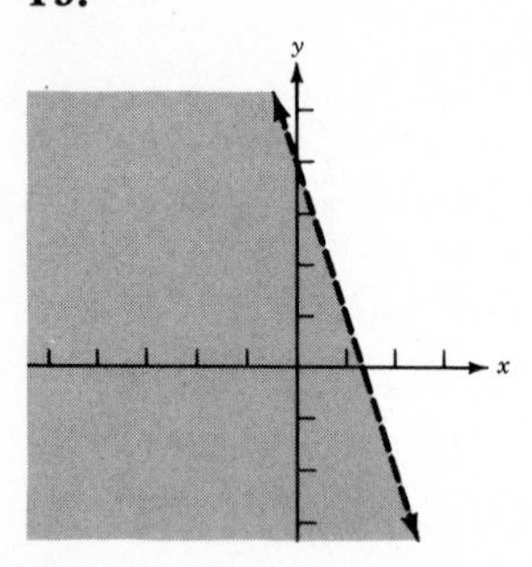

21.

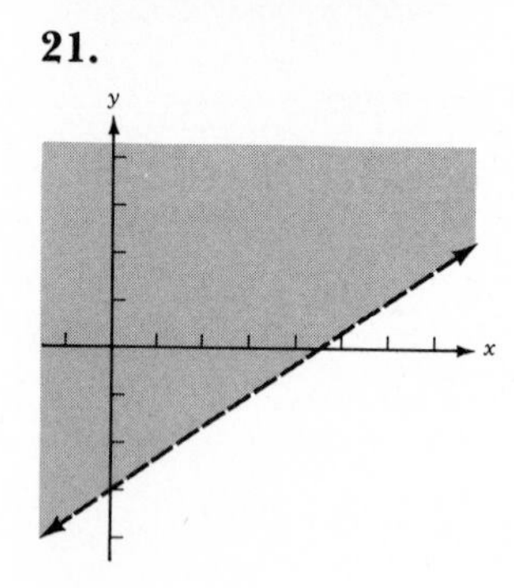

23.

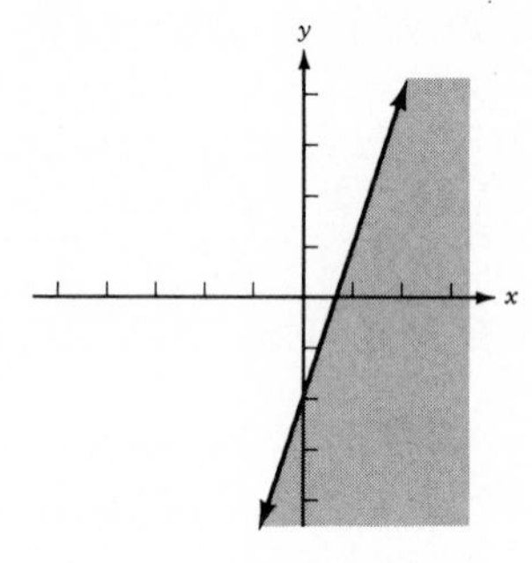

25.

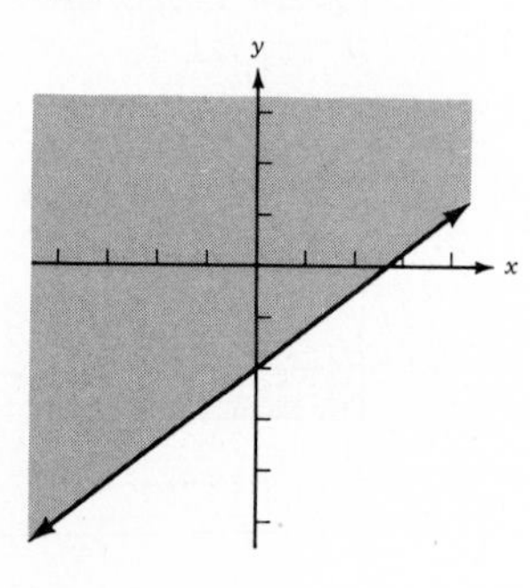

27.

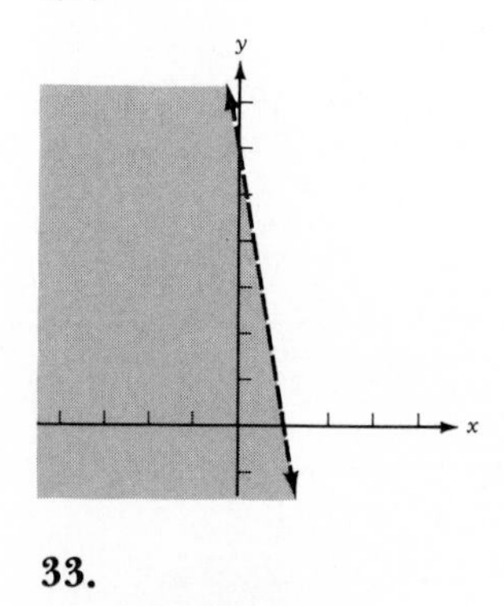

29.

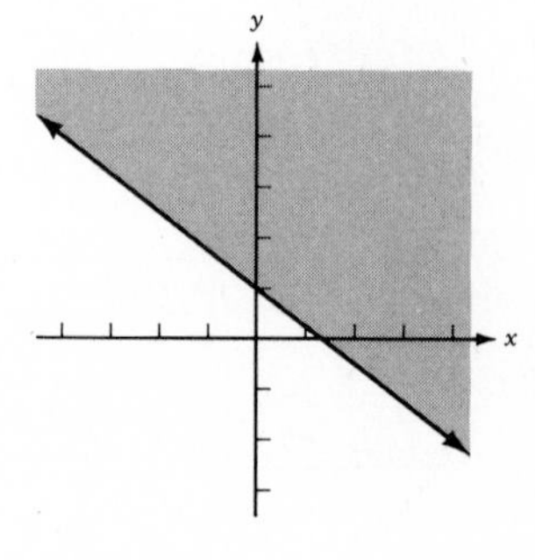

31.

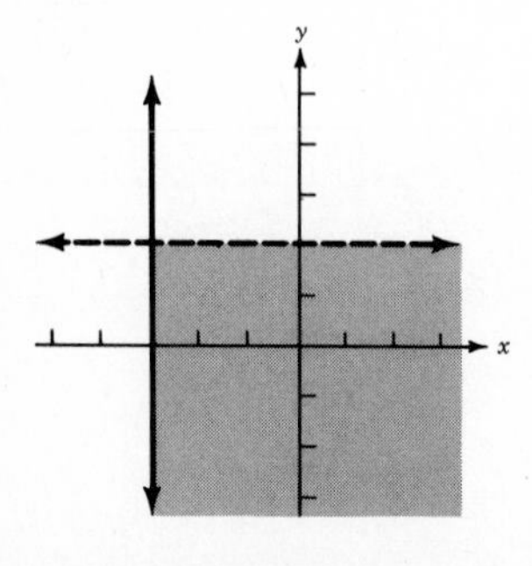

33.

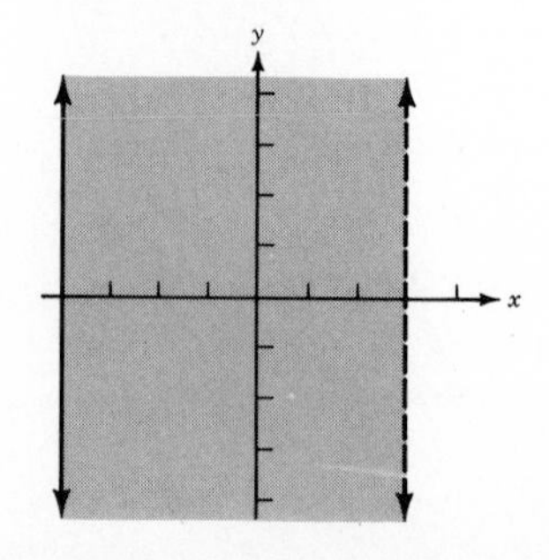

35.

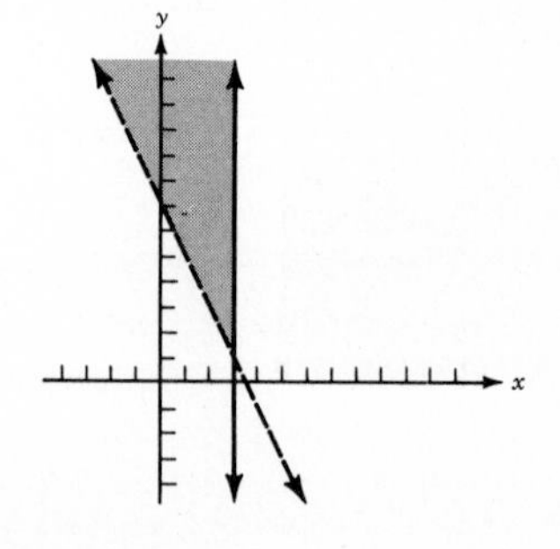

37.

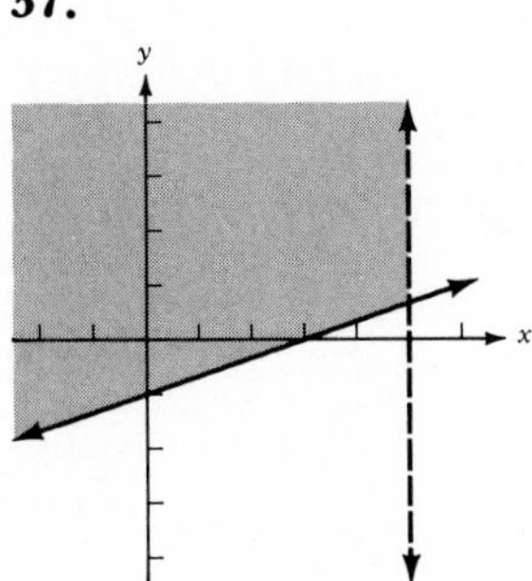

39.

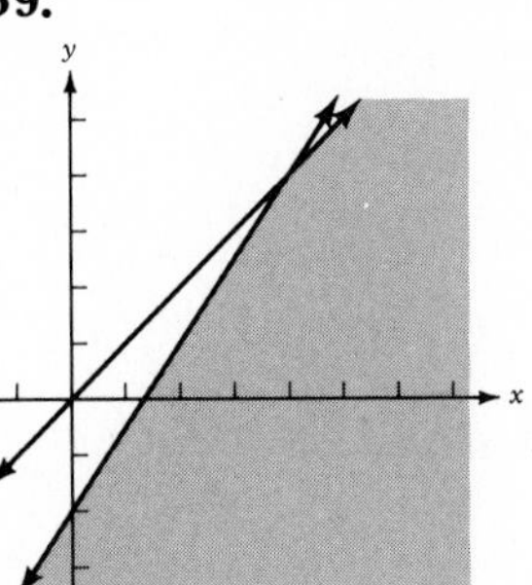

CHAPTER 8 REVIEW, page 357

1. $(-3, 1), (-2, 0), (-1, 1), (1, 1), (3, 3)$
2. $(-3, 1), (-1, 3), (-1, -1), (1, 1), (4, 3), (4, -2)$
3. $(-2, -1), (0, 0), (2, 1), (2, 2), (3, 3), (4, 4)$
4. $(-3, -1), (-1, -1), (0, -1), (1, 1), (2, 1), (3, 1)$
5. $(-5, 4), (-1, 1), \left(0, -\dfrac{9}{2}\right), (1, 3), (4, 0)$

6.

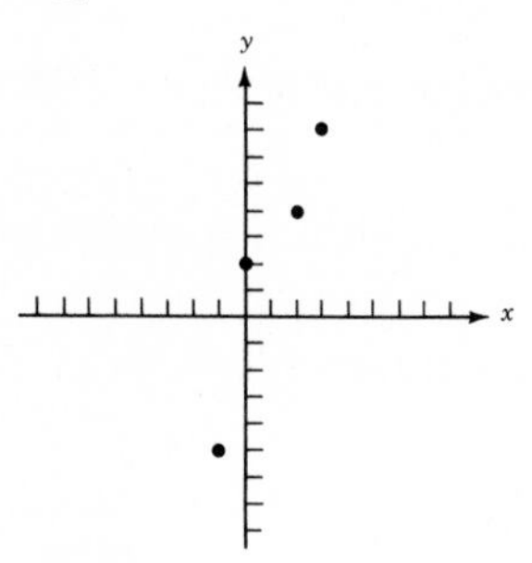

7.

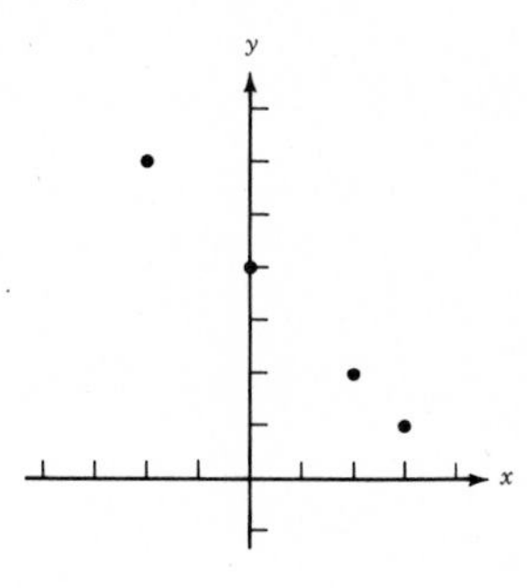

8.

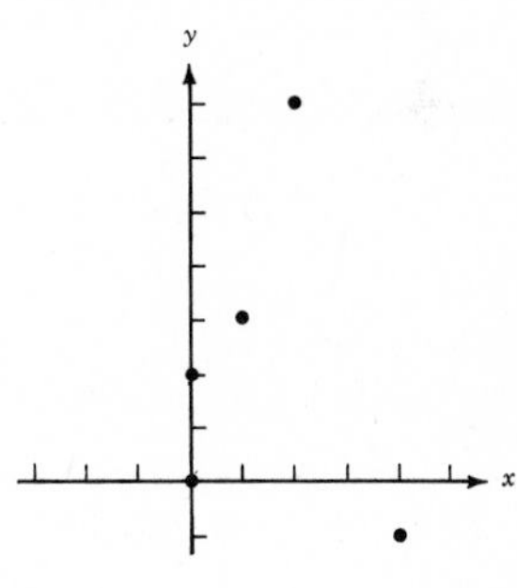

9. **10.**

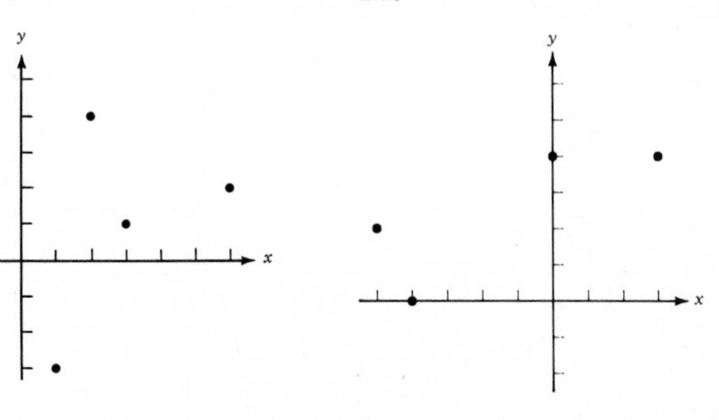

11.

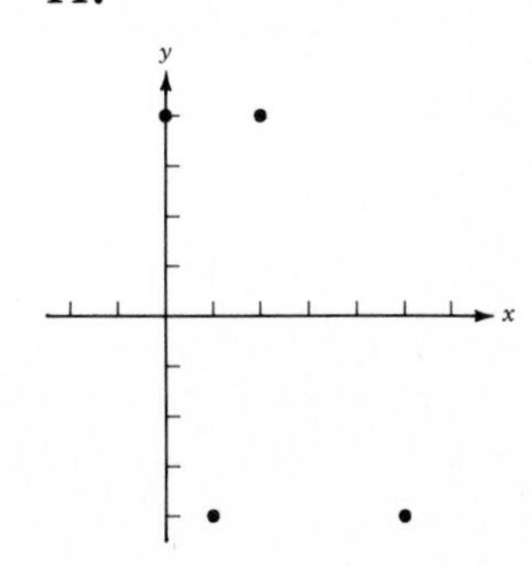

12.

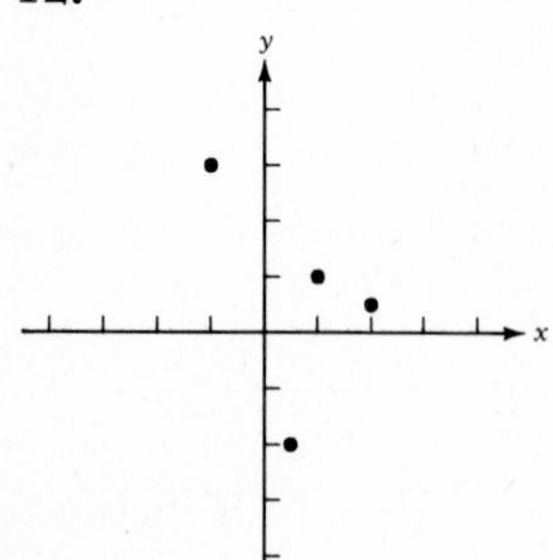

13.

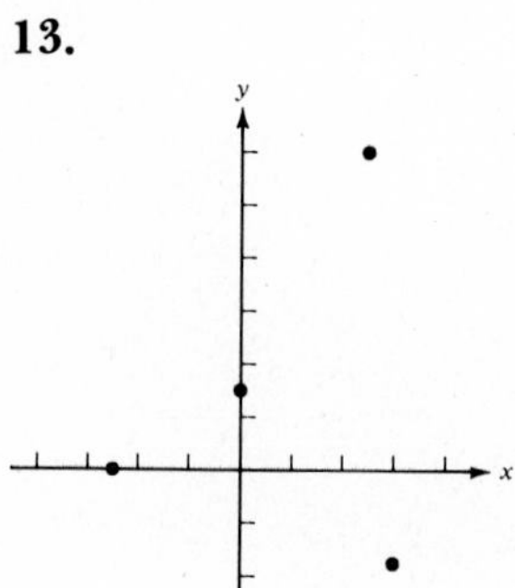

14. **a.** 5
b. −7
c. −3
d. 13

15. **a.** 6
b. 3
c. 5
d. $\frac{9}{2}$

16. **a.** −9
b. −8
c. 7
d. −5

17. **a.** 2
b. 11
c. 18
d. 3

18. **a.** 2
b. 1
c. 2
d. −8

19.

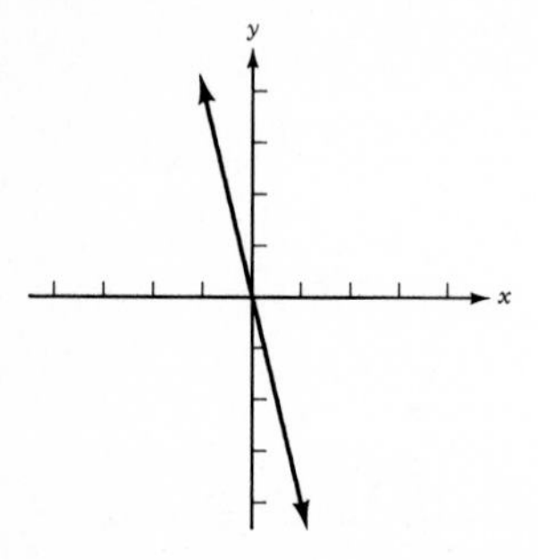

20.

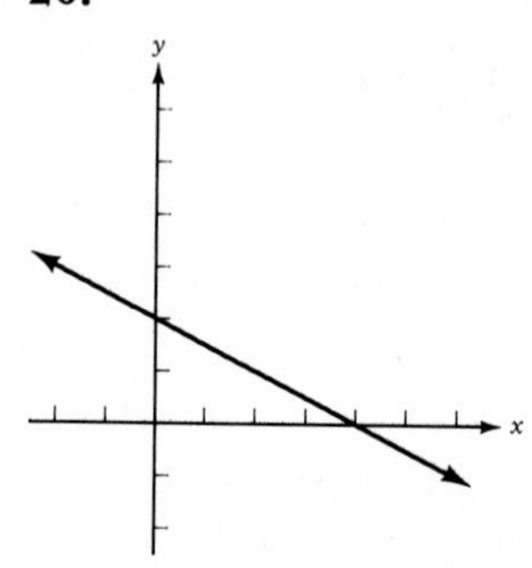

21.

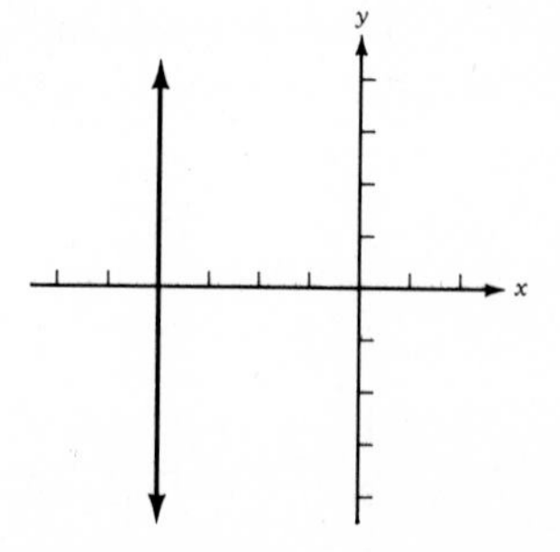

22.

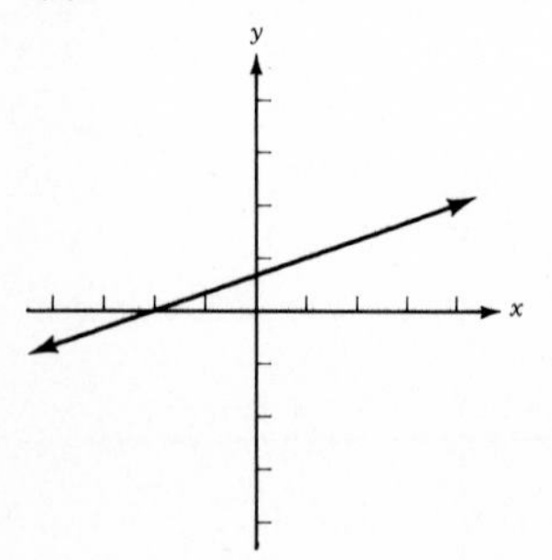

23.

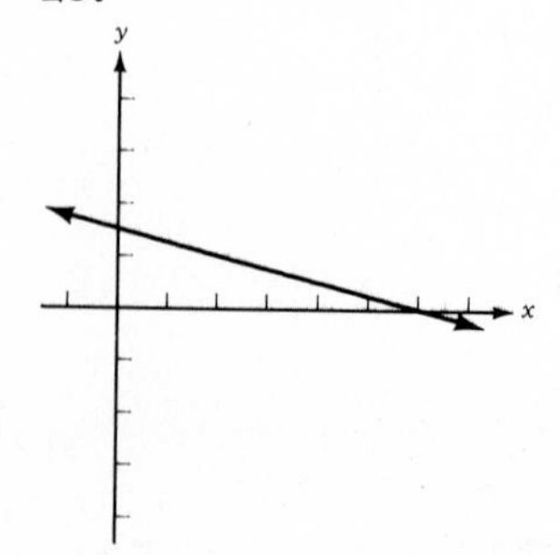

24.

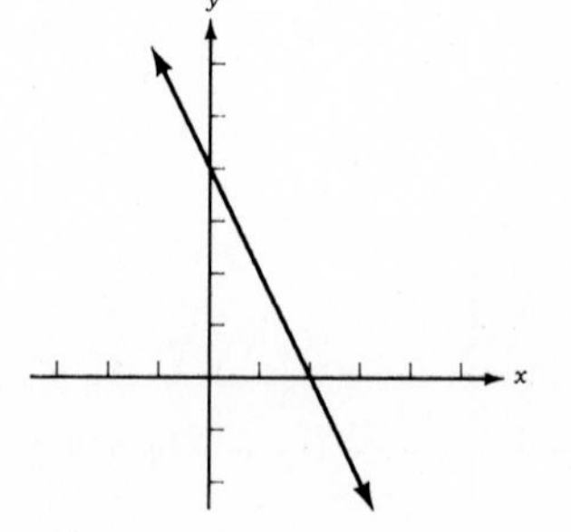

25.

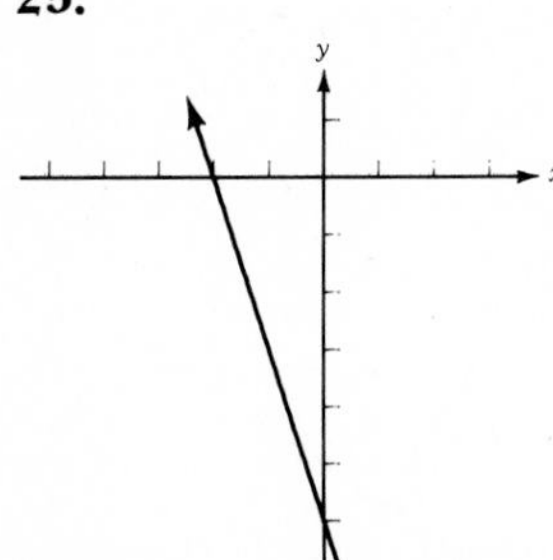

26.

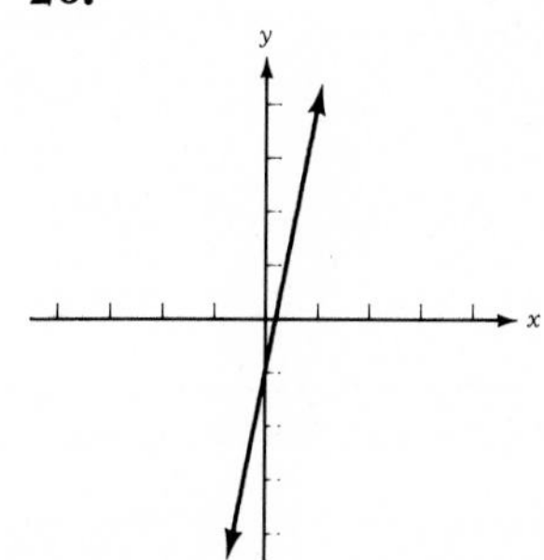

27.

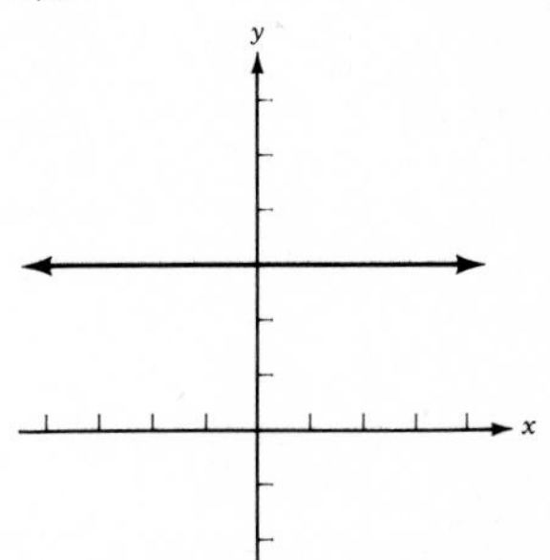

28.

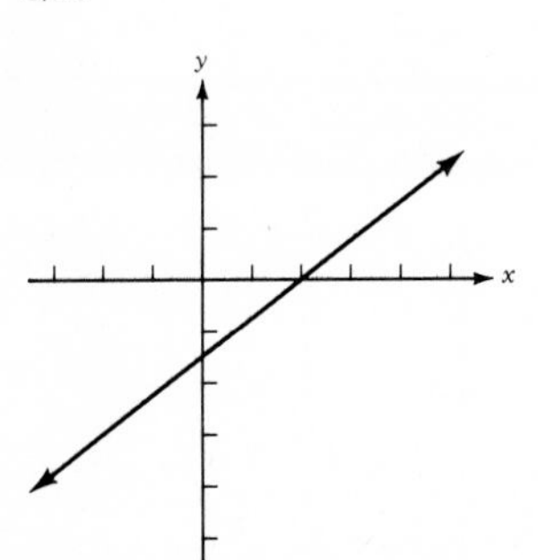

29. $m = 2,\ b = 3$

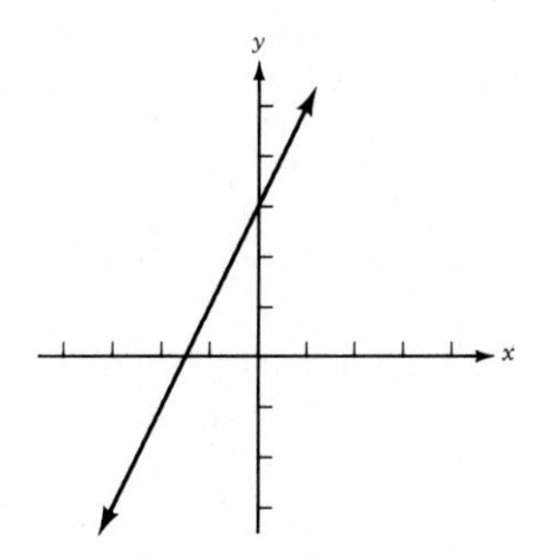

30. $m = -\dfrac{2}{5},\ b = 2$

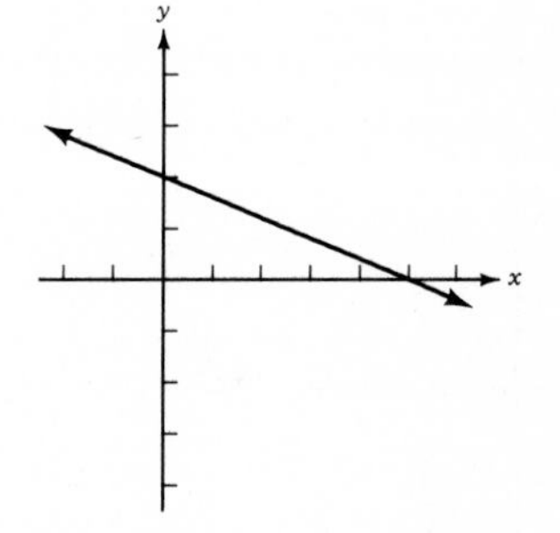

31. no slope, no y-intercept

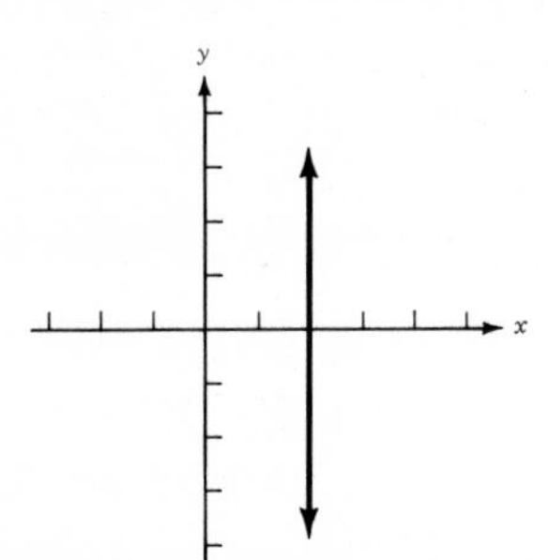

32. $m = \dfrac{4}{3},\ b = -\dfrac{7}{3}$

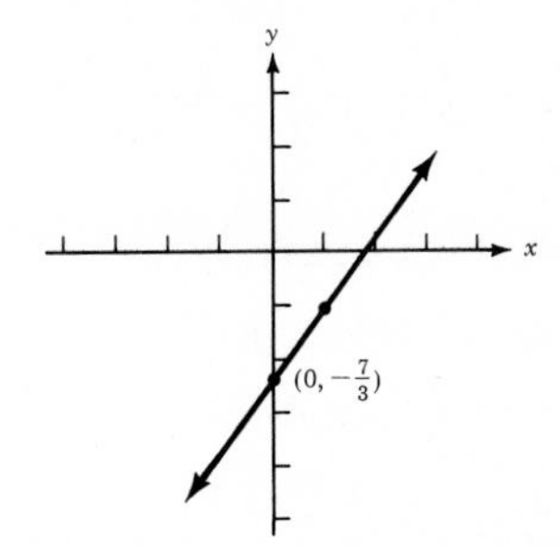

33. $m = -\dfrac{2}{3},\ b = \dfrac{5}{3}$

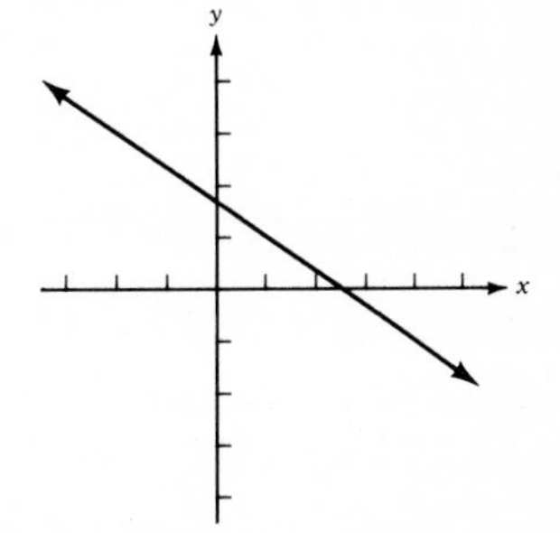

34. $m = -\dfrac{2}{3}$

35. $m = -\dfrac{4}{3}$

36. $m = -2$

37. $m = -\dfrac{8}{5}$

38. $y = -\frac{5}{2}x + \frac{29}{2}$
or $5x + 2y = 29$

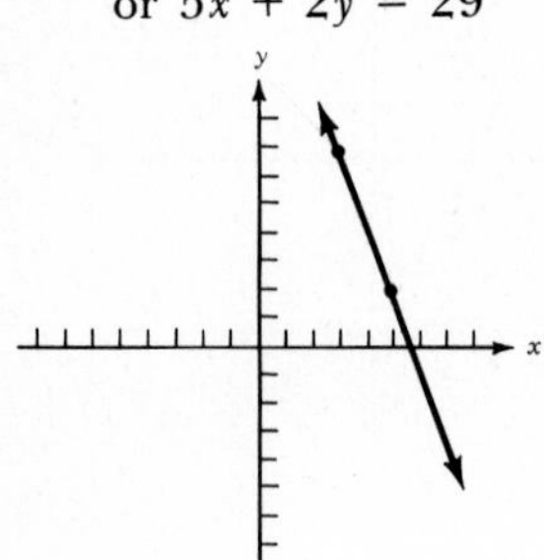

39. $y = 2$

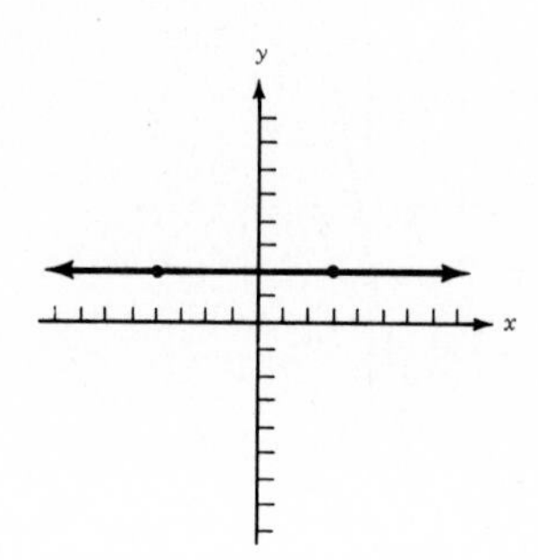

40. $y = \frac{2}{3}x - \frac{13}{3}$
or $2x - 3y = 13$

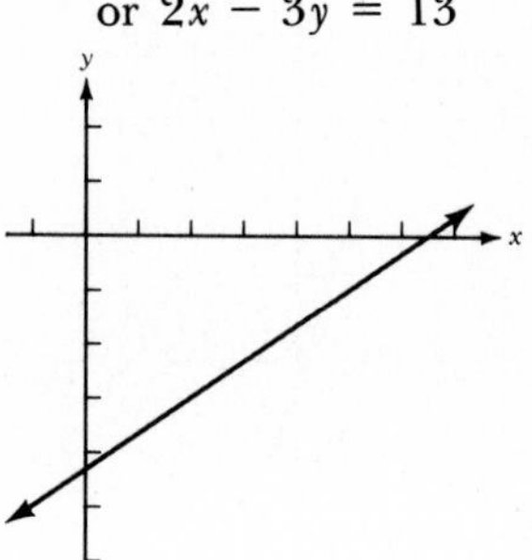

41. $y = -1$

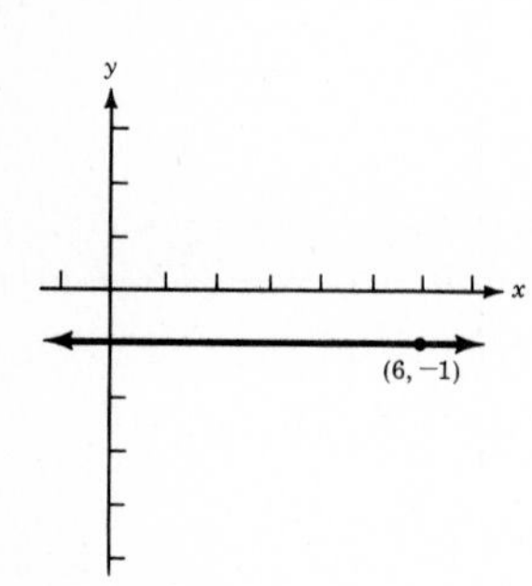

42. $y = -\frac{1}{4}x + \frac{13}{4}$
or $x + 4y = 13$

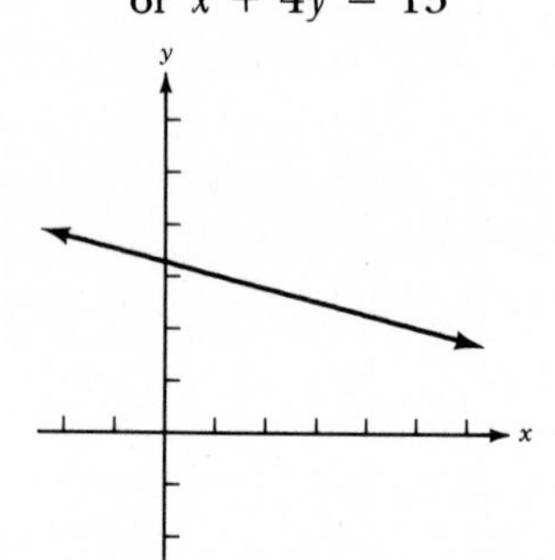

43. $x = -3$

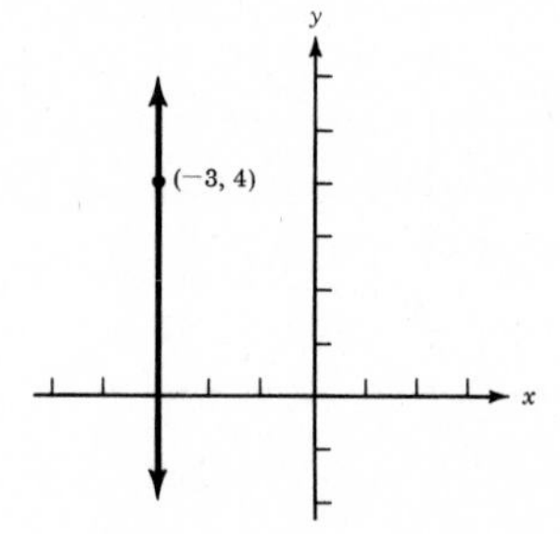

44. $y = -\frac{4}{3}x + \frac{20}{3}$
or $4x + 3y = 20$

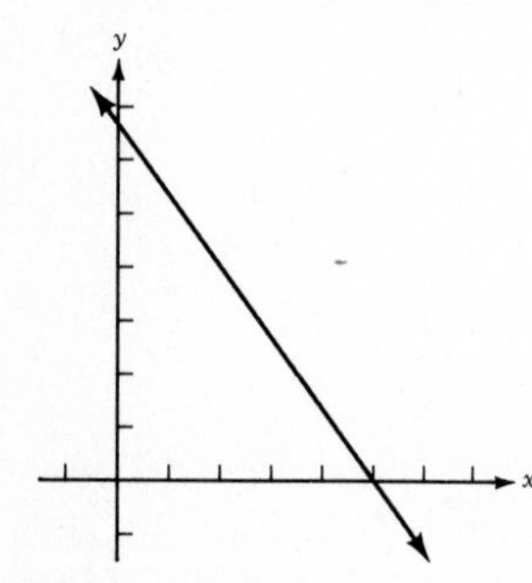

45. $y = \frac{3}{2}x - 11$
or $3x - 2y = 22$

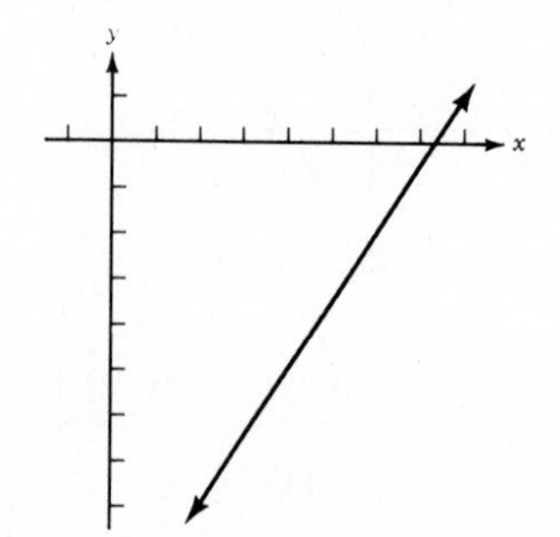

46. $x = -1$ **47.** $2x + y = 0$ **48.** $x - 2y = -6$ **49.** $N = 500t + 500$
50. $C = 17t + 20$

51.

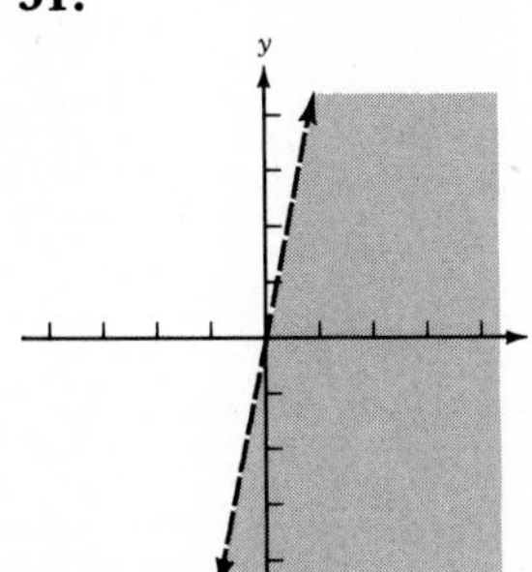

52.

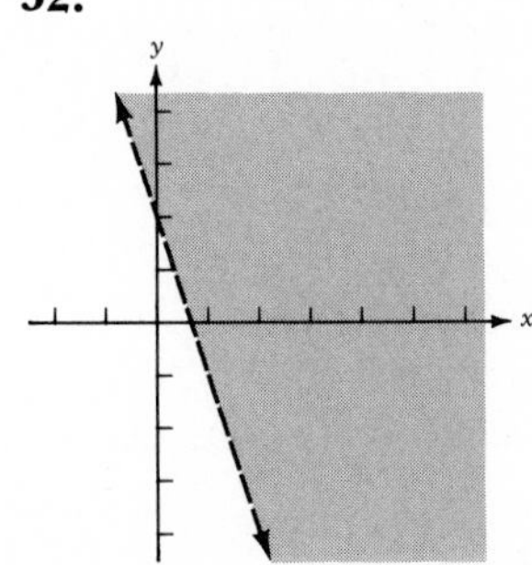

53.

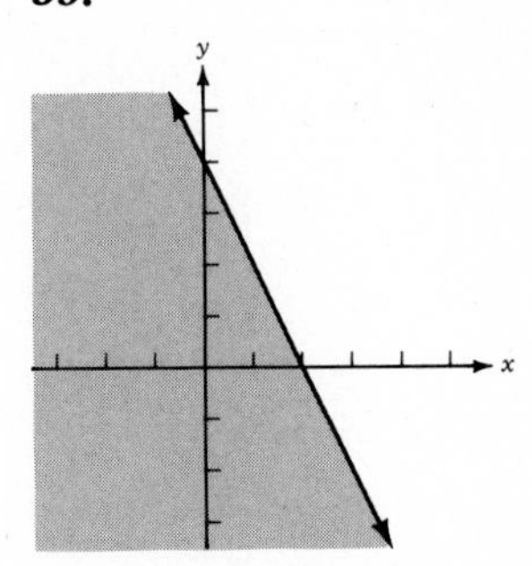

CHAPTER 8 TEST, page 360

1. $(-4, 2)$, $(-2, -1)$, $(0, -3)$, $(1, 2)$, $(3, 4)$, $(5, 0)$

2.

3. **a.** 7 **b.** 4 **c.** -2 **d.** $\frac{5}{2}$

4. **a.** -5 **b.** -6 **c.** -4 **d.** $-\frac{29}{6}$

5.

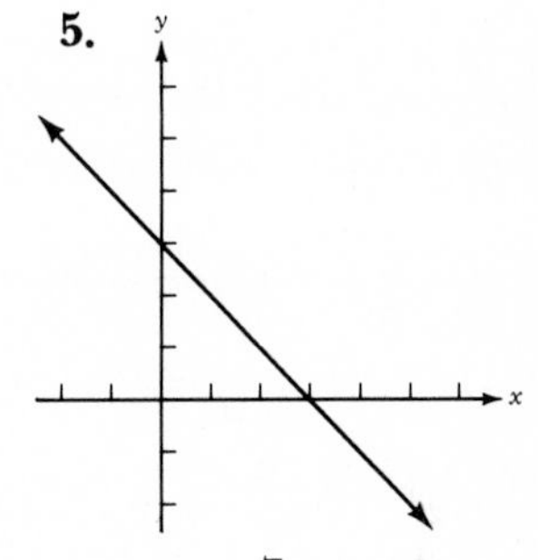

6.

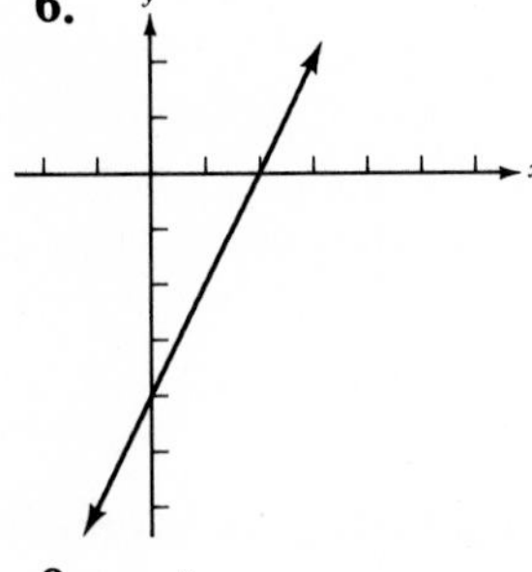

7.

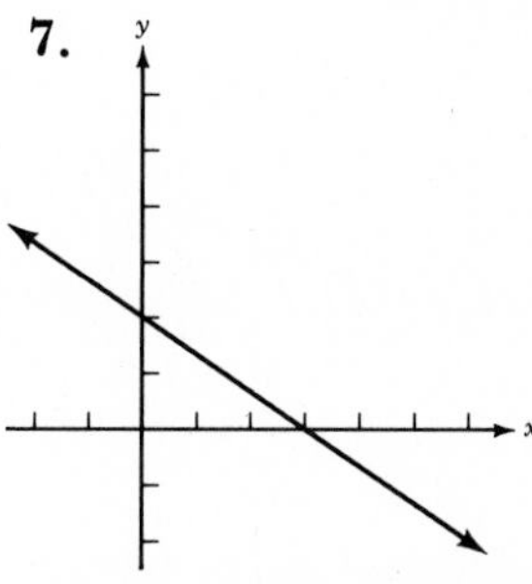

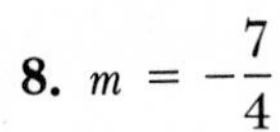

8. $m = -\frac{7}{4}$

9.

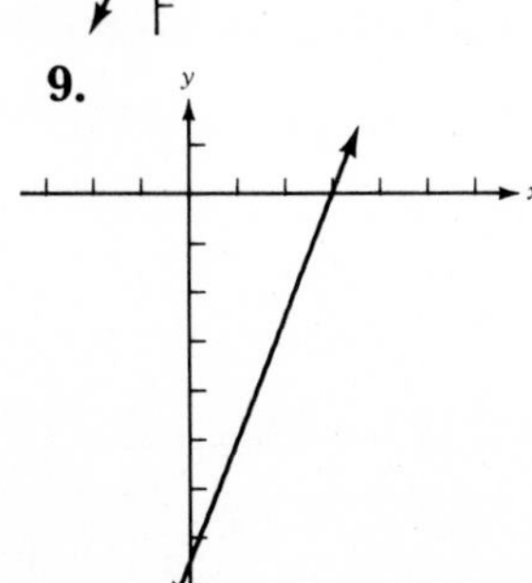

10. $m = -\frac{3}{5}, b = -3$

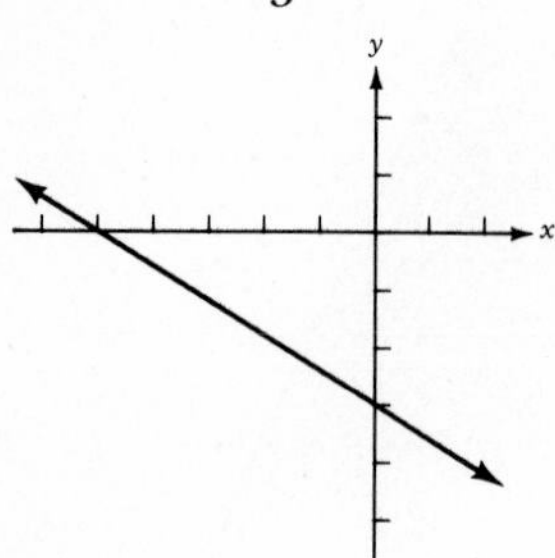

11. $m = -\frac{1}{2}, b = \frac{7}{4}$

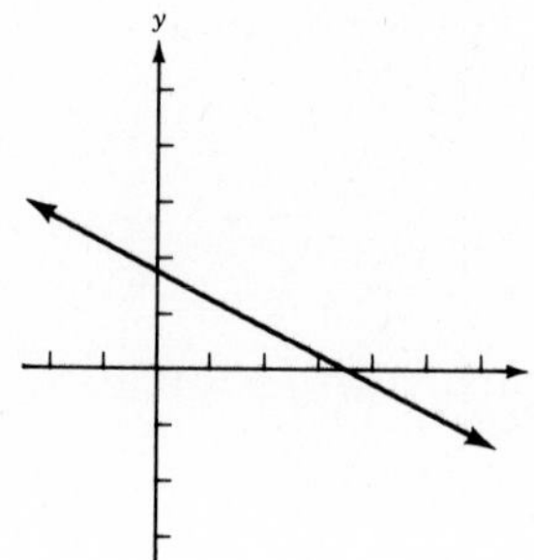

12.

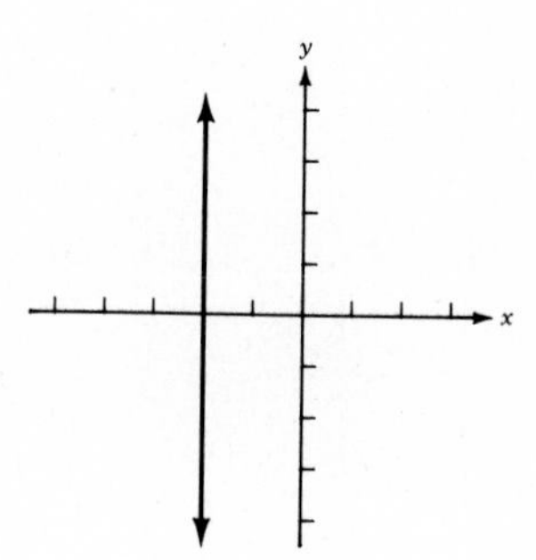

13. $y = \frac{1}{4}x + \frac{11}{4}$ or $-x + 4y = 11$

14. $x = 4$

15. $y = -x + 4$ or $x + y = 4$

16. $y = 5$

17. $y = -2$

18. $y = \frac{3}{2}x + 2$ or $3x - 2y = -4$

19. $y = \frac{1}{3}x + \frac{5}{3}$ or $x - 3y = -5$

20. $C = 4t + 9$

CHAPTER 9

EXERCISES 9.1, page 371

1. a. no
b. yes
c. yes
d. yes

3. a. yes
b. no
c. yes
d. no

5. a. no
b. no
c. yes
d. no

7. a. yes
b. no
c. yes
d. no

9. consistent

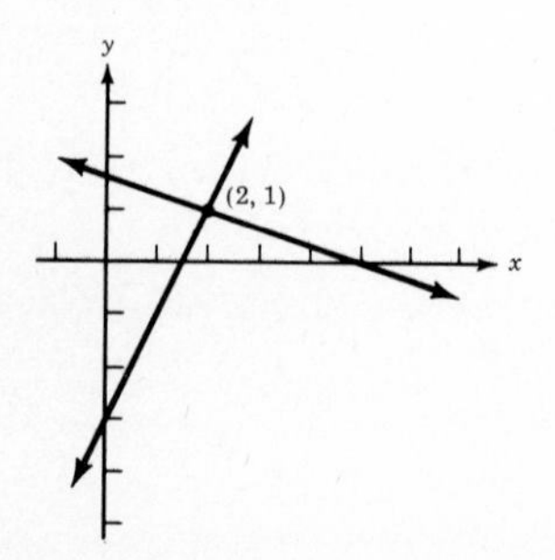

11. inconsistent

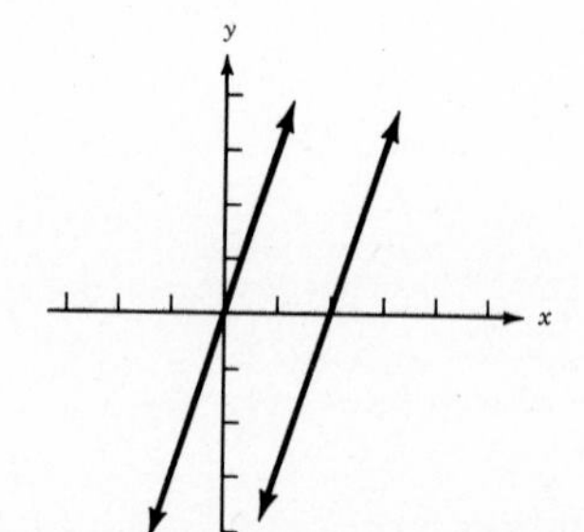

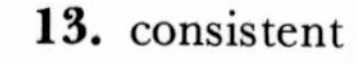

13. consistent

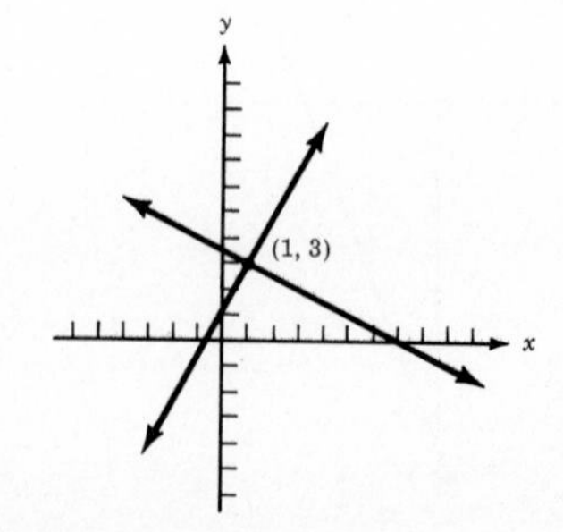

15. dependent

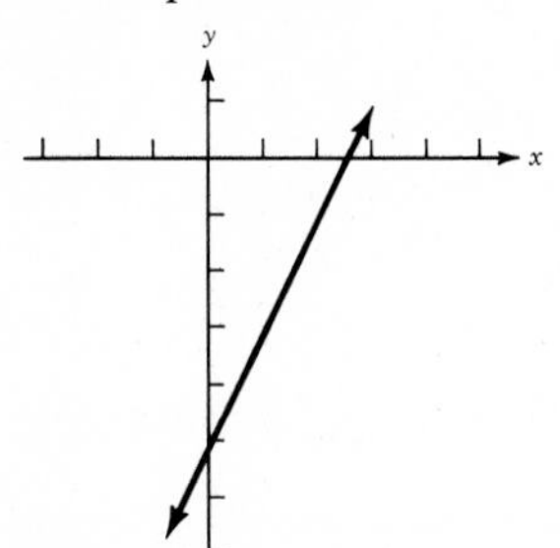

17. inconsistent

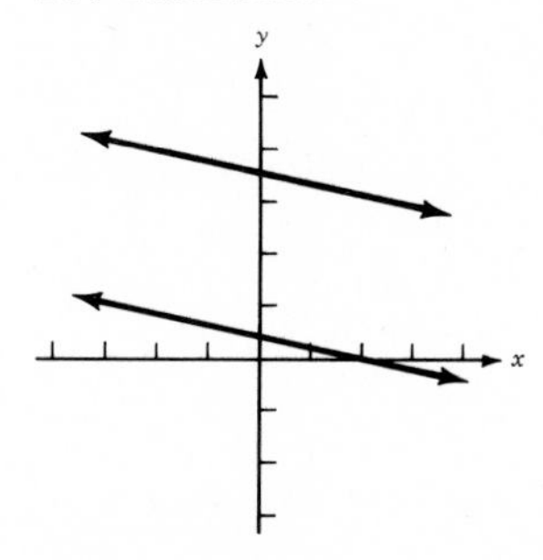

19. consistent

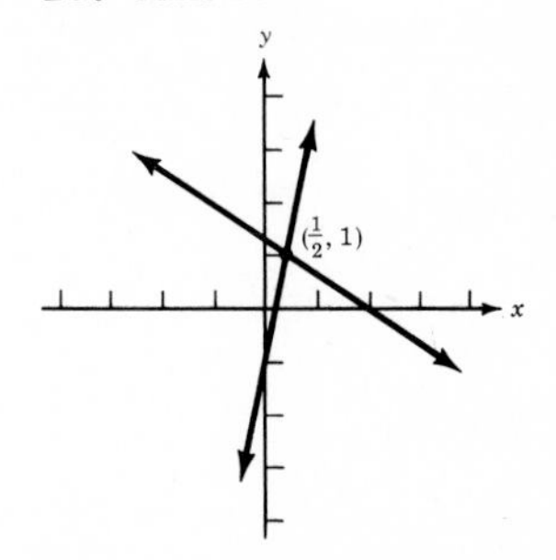

21. inconsistent

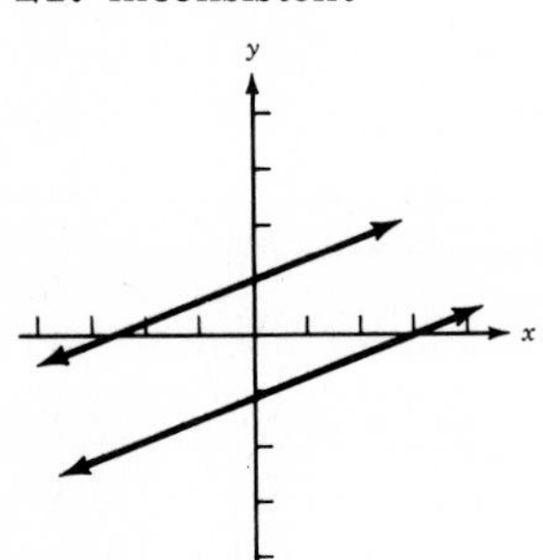

23. dependent

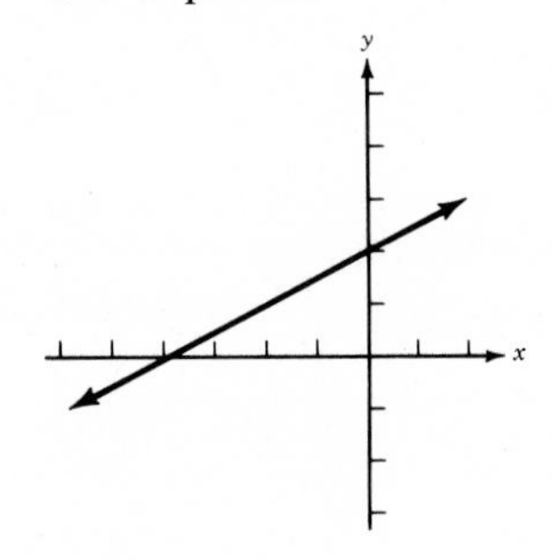

25. consistent

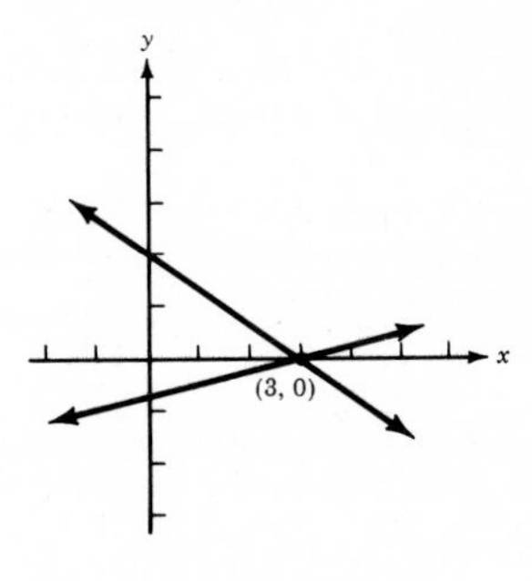

27. consistent

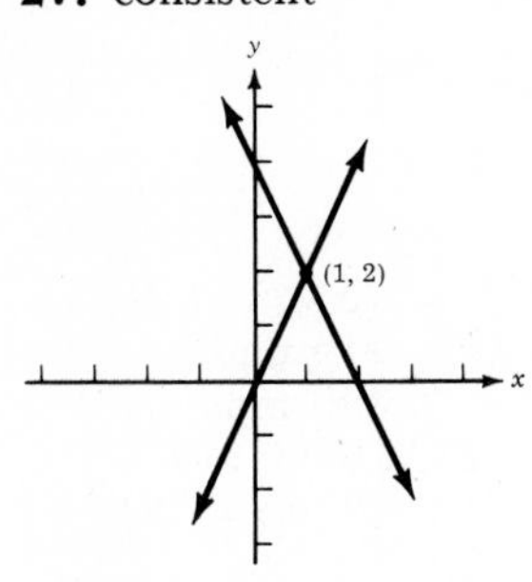

29. consistent

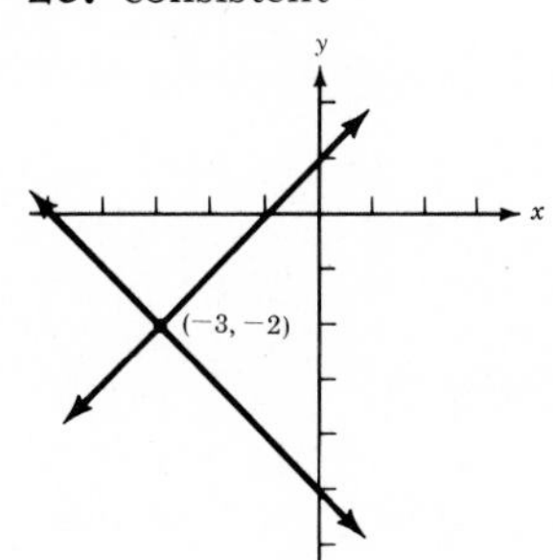

31. consistent

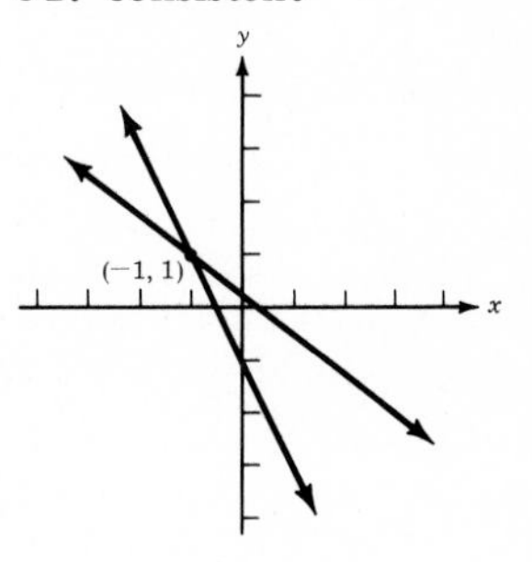

33. consistent

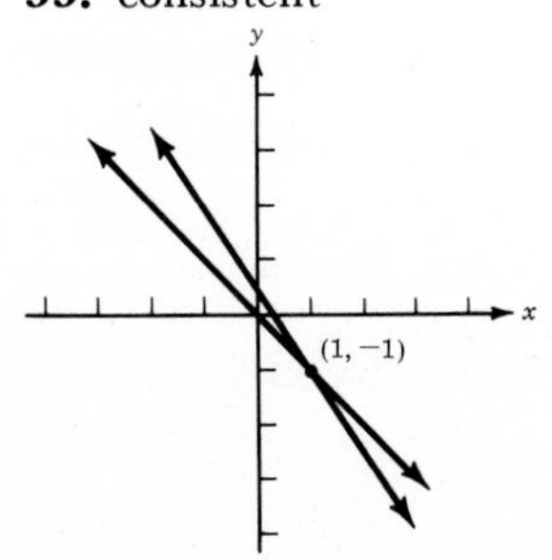

35. consistent

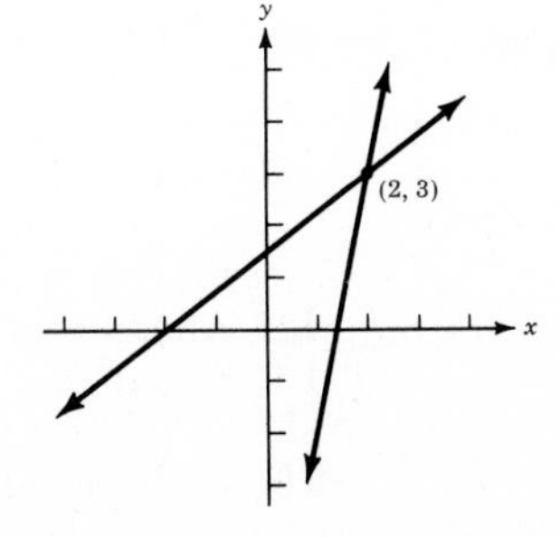

EXERCISES 9.2, page 375

1. $(2, 4)$
3. $(1, -2)$
5. $(-6, -2)$
7. $(4, 1)$
9. no solution, inconsistent
11. $(3, 2)$
13. dependent
15. $(2, 2)$
17. $(-8, 16)$
19. $\left(2, \frac{1}{3}\right)$
21. $\left(\frac{7}{2}, -\frac{1}{2}\right)$
23. dependent
25. $(2, -10)$
27. $(-5, 2)$
29. no solution, inconsistent
31. $\left(\frac{11}{7}, \frac{8}{7}\right)$
33. $\left(\frac{13}{5}, -\frac{39}{5}\right)$
35. $(10, 4)$
37. $(9, -3)$
39. $(6, -4)$

EXERCISES 9.3, page 381

1. $(3, -1)$
3. $\left(1, -\frac{3}{2}\right)$
5. inconsistent
7. $(1, -5)$
9. dependent
11. $(4, 3)$
13. $(-3, -1)$
15. $(2, -1)$
17. $(0, 4)$
19. $(-2, -3)$
21. dependent
23. $(3, -1)$
25. $(-1, -2)$
27. $(7, 0)$
29. $(5, 6)$
31. $(-3, 7)$
33. $(-4, -2)$
35. $(7, 5)$
37. $(20, 10)$
39. $\left(2, \frac{10}{9}\right)$
41. $y = 5x - 7$
43. $y = \frac{1}{9}x + \frac{13}{9}$
45. $y = \frac{11}{4}x - \frac{49}{4}$

EXERCISES 9.4, page 386

1. 23, 33
3. 8 ft, 10 ft
5. 81 ft, 40 ft
7. 23 won, 9 lost
9. 165 mph, 33 mph
11. 20 lb—20%; 30 lb—70%
13. \$1.40
15. 116 mph, 406 mph
17. \$450 at 5%
19. 8 tons—80%; 16 tons—50%
21. $37\frac{1}{2}$ mph, $62\frac{1}{2}$ mph
23. \$6300 at 6%, \$3200 at 3%
25. 15 stamps @ 22¢, 5 stamps @ 40¢
27. 12 m by 16 m
29. 4 hr

EXERCISES 9.5, page 390

1. \$20
3. \$20
5. **a.** $C = 2.40x + 9000$; $R = 4.20x$
b. 5000 items
7. **a.** $C = 30x + 8000$; $R = 80x$
b. 160 tables
9. 18 quarters, 9 dimes
11. Anna: 12 yrs, Beth: 22 yrs
13. 140 lb of newspapers, 40 lb of cans
15. $7\frac{1}{2}$ gal
17. 800 lb I, 875 lb II
19. 20 lb of 80%, 8 lb of 50%

CHAPTER 9 REVIEW, page 392

1. a. yes
b. yes
c. yes
d. no

2. a. yes
b. yes
c. yes
d. yes

3. $(-11, -9)$

4. $(6, 0), (2, -2), \left(3, -\frac{3}{2}\right)$

5. consistent

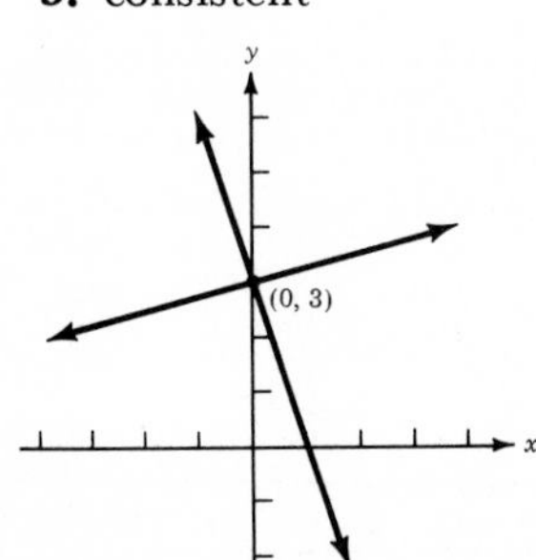

6. inconsistent

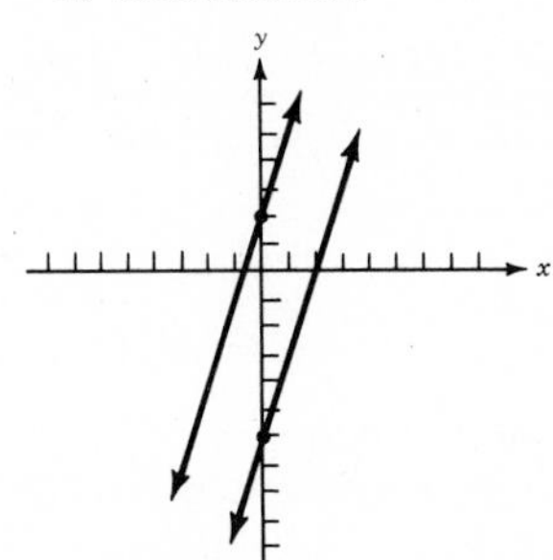

7. dependent

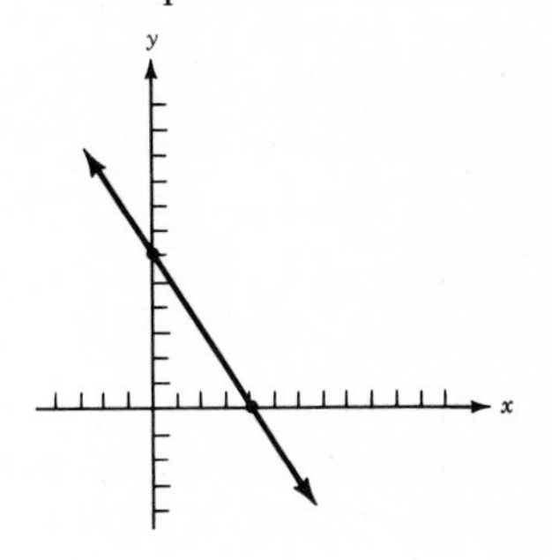

8. consistent

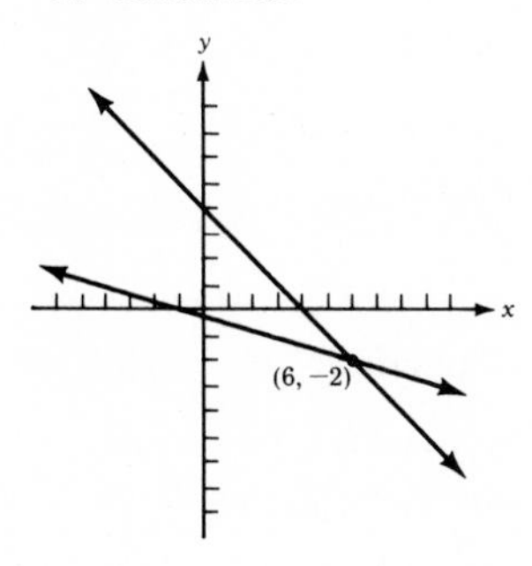

9. inconsistent

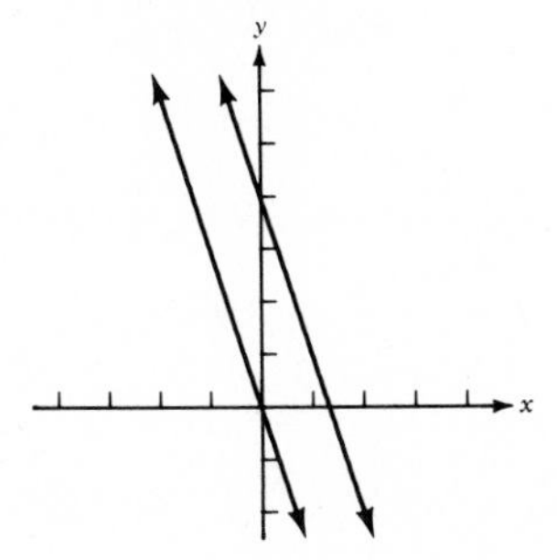

10. consistent

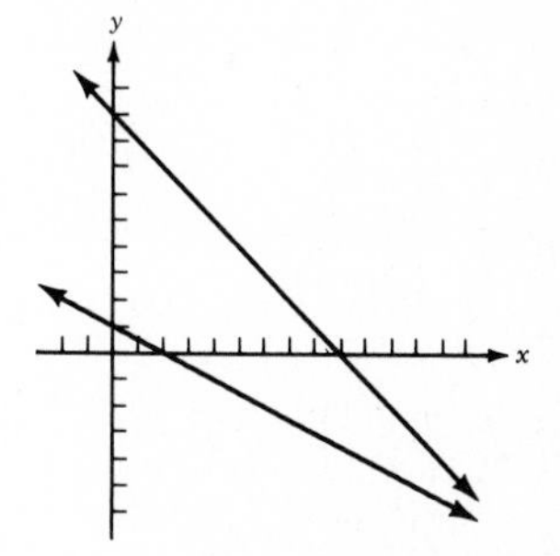

11. $(-6, 2)$
12. inconsistent
13. dependent
14. $(2, -4)$
15. $(0, 0)$
16. $\left(\frac{1}{3}, -1\right)$
17. $(2, 3)$
18. $(-1, -6)$
19. inconsistent
20. dependent
21. $(-2, -1)$
22. $(1, 1)$
23. $(1, 2)$
24. $\left(\frac{54}{11}, -\frac{5}{11}\right)$
25. dependent
26. $\left(-\frac{1}{3}, \frac{1}{2}\right)$
27. $\left(-\frac{21}{10}, -\frac{23}{10}\right)$
28. $\left(\frac{4}{5}, \frac{9}{5}\right)$
29. $y = -7x + 20$
30. $y = -\frac{4}{3}x + \frac{7}{3}$
31. 45 mph, 40 mph
32. 440 lb of 70%, 200 lb of 54%
33. 30 members
34. boat—14 mph, current—2 mph
35. 11 m by 14 m
36. $C = 10x + 2400$; $R = 22x$; $x = 200$ gloves
37. 4 of Model A; 18 of Model B

CHAPTER 9 TEST, page 395

1. **a.** no
b. yes
c. yes
d. no

2. $(-3, -2)$

3. consistent

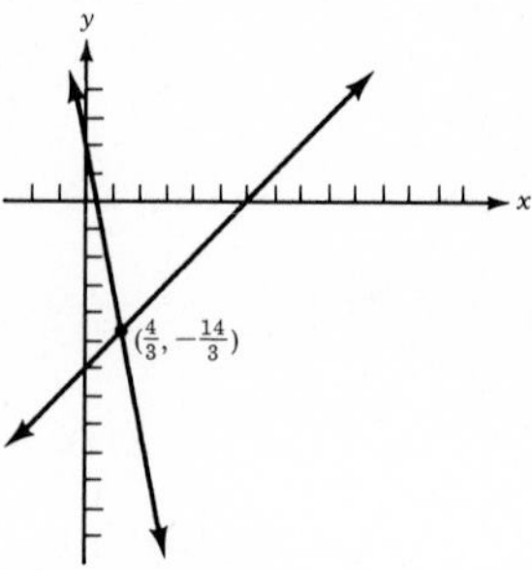

4. consistent

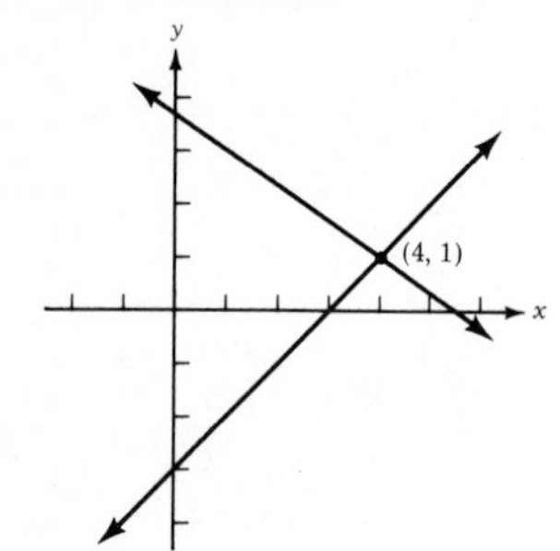

5. inconsistent

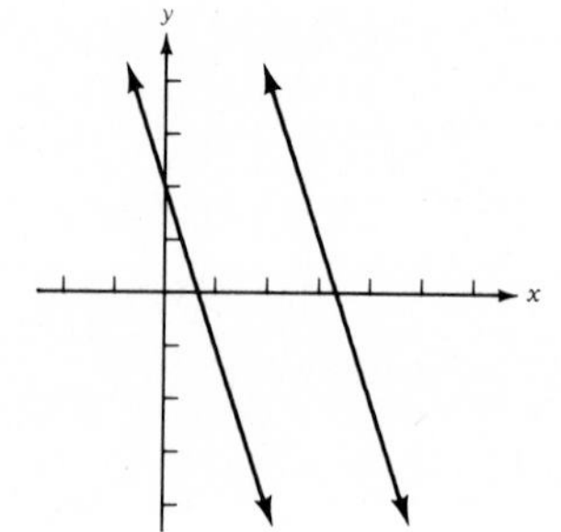

6. consistent

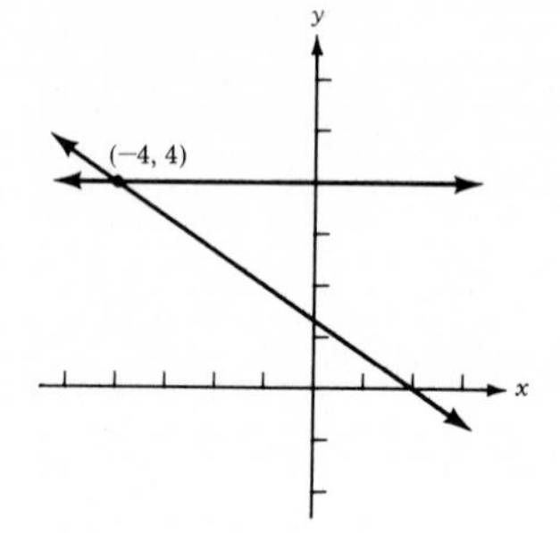

7. $(-8, -20)$

8. $(-2, 6)$

9. dependent

10. $\left(-\dfrac{3}{14}, \dfrac{13}{7}\right)$

11. $(-3, 5)$

12. inconsistent

13. $(-4, 1)$

14. $(3, -6)$

15. dependent

16. $\left(\dfrac{22}{21}, \dfrac{13}{21}\right)$

17. $y = \dfrac{7}{2}x - \dfrac{19}{2}$

18. boat—10 mph; current—2 mph

19. pencils—$0.08; pens—$0.79

20. 13 in. by 17 in.

CHAPTER 10

EXERCISES 10.1, page 406

1. rational
3. irrational
5. irrational
7. rational
9. irrational
11. rational
13. irrational
15. rational
17. rational
19. irrational
21. 6
23. $\dfrac{1}{2}$

25. $2\sqrt{2}$ **27.** $-2\sqrt{14}$ **29.** $\frac{\sqrt{5}}{3}$ **31.** $2\sqrt{3}$
33. $12\sqrt{2}$ **35.** $-6\sqrt{2}$ **37.** $-5\sqrt{5}$ **39.** $-\frac{\sqrt{11}}{8}$
41. $\frac{2\sqrt{7}}{5}$ **43.** $\frac{4\sqrt{2}}{9}$ **45.** $\frac{\sqrt{2}}{2}$ **47.** $-\frac{\sqrt{2}}{4}$
49. $\frac{1-\sqrt{3}}{2}$ **51.** $\frac{1+\sqrt{6}}{2}$ **53.** $\frac{4+\sqrt{5}}{5}$ **55.** $\frac{5-3\sqrt{3}}{2}$
57. $\frac{8-\sqrt{15}}{6}$ **59.** $\frac{3-2\sqrt{3}}{7}$

EXERCISES 10.2, page 408

1. $8\sqrt{2}$ **3.** $7\sqrt{5}$ **5.** $-3\sqrt{10}$ **7.** $-2\sqrt{5}$
9. $-\sqrt{11}$ **11.** $3\sqrt{a}$ **13.** $7\sqrt{x}$ **15.** $4\sqrt{2}+3\sqrt{3}$
17. $3\sqrt{2}-2\sqrt{5}$ **19.** $-4\sqrt{a}+8\sqrt{b}$ **21.** $5\sqrt{3}$ **23.** 0
25. $3\sqrt{3}$ **27.** $8\sqrt{2}-6\sqrt{3}$ **29.** $\sqrt{5}-6$ **31.** $\sqrt{3}-4\sqrt{2}$
33. $6\sqrt{7}-\sqrt{3}$ **35.** $4\sqrt{2x}$ **37.** $2y\sqrt{2y}$ **39.** $-4x\sqrt{3xy}$
41. $3\sqrt{2}-8$ **43.** 18 **45.** $-8\sqrt{3}$ **47.** $\sqrt{6}+12$
49. $\sqrt{xy}+2y$ **51.** $13-2\sqrt{2}$ **53.** $8-7\sqrt{6}$ **55.** $x-9$
57. -5 **59.** $x+2\sqrt{x}-15$

EXERCISES 10.3, page 412

1. $\frac{5\sqrt{2}}{2}$ **3.** $-\frac{3\sqrt{7}}{7}$ **5.** $2\sqrt{3}$ **7.** 3
9. 3 **11.** $\frac{1}{3}$ **13.** $\frac{2\sqrt{3}}{3}$ **15.** $\frac{3\sqrt{2}}{2}$
17. $\frac{\sqrt{x}}{x}$ **19.** $\frac{\sqrt{2xy}}{y}$ **21.** $\frac{\sqrt{2y}}{y}$ **23.** $\frac{3\sqrt{7}}{5}$
25. $\frac{-\sqrt{2y}}{5}$ **27.** $\sqrt{6}+2$ **29.** $\frac{\sqrt{5}+3}{-4}$ **31.** $\frac{-6(5+3\sqrt{2})}{7}$
33. $\frac{\sqrt{3}(\sqrt{2}-5)}{23}$ **35.** $\frac{\sqrt{7}(1+3\sqrt{5})}{-44}$ **37.** $\frac{\sqrt{3}+\sqrt{5}}{-2}$ **39.** $5(\sqrt{2}-\sqrt{3})$
41. $\frac{4(\sqrt{x}-1)}{x-1}$ **43.** $\frac{5(6-\sqrt{y})}{36-y}$ **45.** $\frac{8(2\sqrt{x}-3)}{4x-9}$ **47.** $\frac{2y\sqrt{5}+2\sqrt{3y}}{5y-3}$
49. $\frac{3(\sqrt{x}+\sqrt{y})}{x-y}$ **51.** $\frac{x(\sqrt{x}-2\sqrt{y})}{x-4y}$ **53.** $\frac{5+3\sqrt{3}}{-1}$

55. $\frac{11 - 5\sqrt{5}}{-4} = \frac{5\sqrt{5} - 11}{4}$

57. $\frac{x + 2\sqrt{x} + 1}{x - 1}$

59. $\frac{x\sqrt{3} + 2\sqrt{3x} - y\sqrt{x} - 2y}{3x - y^2}$ or $\frac{(\sqrt{x} + 2)(\sqrt{3x} - y)}{3x - y^2}$

EXERCISES 10.4, page 418

1. 4

3. 3

5. $\frac{7}{2}$

7. $\sqrt{2}$

9. $\sqrt{13}$

11. 13

13. $\frac{5}{7}$

15. 5

17. 13

19. 10

21. 5

23. 2

25. $|\overline{AB}|^2 = 45$
$|\overline{BC}|^2 = 20$
$|\overline{AC}|^2 = 65$
yes, since $65 = 45 + 20$

27. $|\overline{AB}| = \sqrt{80}$
$|\overline{BC}| = \sqrt{32}$
$|\overline{AC}| = \sqrt{80}$
yes, since $|\overline{AB}| = |\overline{AC}|$

29. $|\overline{AB}| = 4$
$|\overline{BC}| = 4$
$|\overline{AC}| = 4$

31. $|\overline{AC}| = \sqrt{61}$
$|\overline{BD}| = \sqrt{61}$

33. $|\overline{AB}| = \sqrt{80}$
$|\overline{BC}| = 5$
$|\overline{AC}| = 5$
$P = 10 + \sqrt{80}$

35. $|\overline{AB}| = \sqrt{41}$
$|\overline{BC}| = \sqrt{10}$
$|\overline{AC}| = \sqrt{65}$
$P = \sqrt{41} + \sqrt{10} + \sqrt{65}$

CHAPTER 10 REVIEW, page 421

1. rational **2.** irrational **3.** rational **4.** irrational

5. rational **6.** irrational **7.** irrational **8.** rational

9. 13 **10.** 14 **11.** $-4\sqrt{3}$ **12.** $3\sqrt{6}$

13. $2\sqrt{10}$ **14.** $5\sqrt{6}$ **15.** $\frac{\sqrt{7}}{14}$ **16.** $-\frac{\sqrt{5}}{11}$

17. $-\frac{\sqrt{5}}{4}$ **18.** $\frac{5\sqrt{3}}{8}$ **19.** $\frac{\sqrt{17}}{4}$ **20.** $\frac{4\sqrt{5}}{9}$

21. $4 + \sqrt{3}$ **22.** $\frac{5 - \sqrt{3}}{2}$ **23.** $\frac{6 + \sqrt{5}}{10}$ **24.** $\frac{3+\sqrt{2}}{2}$

25. $\frac{\sqrt{3} - \sqrt{5}}{2}$ **26.** $\sqrt{3} + \sqrt{2}$ **27.** $16\sqrt{2}$ **28.** $5\sqrt{5}$

29. $8\sqrt{2}$ **30.** $\sqrt{2x}$ **31.** $\sqrt{3}$ **32.** $9\sqrt{3} - 2\sqrt{2}$

33. $17 - 2\sqrt{14}$ **34.** $10\sqrt{2}$ **35.** $-5\sqrt{3y}$ **36.** 12
37. $20\sqrt{6}$ **38.** $8\sqrt{6}$ **39.** $7\sqrt{6}$ **40.** -22
41. $17 + 8\sqrt{2}$ **42.** $12 + 15\sqrt{3}$ **43.** $x - 2$ **44.** $32 - 2\sqrt{3}$
45. $2x - 13\sqrt{x} - 24$ **46.** $3\sqrt{3}$ **47.** $\frac{\sqrt{15}}{10}$ **48.** $\frac{\sqrt{15}}{9}$
49. $\frac{\sqrt{10}}{5}$ **50.** $\frac{2\sqrt{7x}}{x}$ **51.** $-\frac{3\sqrt{2y}}{2y}$ **52.** $\frac{2\sqrt{5y}}{5y}$
53. $-\frac{\sqrt{3} + 5}{11}$ **54.** $\frac{3 - \sqrt{2}}{7}$ **55.** $-2(\sqrt{6} + 3)$ **56.** $-\frac{\sqrt{5} + \sqrt{13}}{2}$
57. $\frac{\sqrt{3}(\sqrt{x} + 7)}{x - 49}$ **58.** $\frac{2(\sqrt{5} - \sqrt{2})}{3}$ **59.** $3(\sqrt{7} - \sqrt{3})$ **60.** $\frac{y - 4\sqrt{y} + 3}{y - 9}$
61. $\sqrt{58}$ **62.** $5\sqrt{2}$ **63.** $\sqrt{10}$ **64.** $2\sqrt{2}$
65. 2 **66.** 1
67. $|\overline{AB}| = \sqrt{18}$, $|\overline{BC}| = \sqrt{18}$, yes, since $|\overline{AB}| = |\overline{BC}|$
68. $|\overline{AC}| = 2\sqrt{10}$, $|\overline{BC}| = 2\sqrt{10}$, yes, since $|\overline{AC}| = |\overline{BC}|$
69. $P = 7 + \sqrt{41}$
70. $5 + \sqrt{34} + \sqrt{5}$

CHAPTER 10 TEST, page 424

1. irrational **2.** $5\sqrt{5}$ **3.** 12 **4.** $-\frac{1}{2}$
5. $-\frac{\sqrt{6}}{3}$ **6.** $\frac{\sqrt{6} + 3}{6}$ **7.** $\frac{3 - \sqrt{2}}{2}$ **8.** $13\sqrt{2}$
9. $22\sqrt{3}$ **10.** $-5\sqrt{2x}$ **11.** 75 **12.** $3\sqrt{5} + 5\sqrt{3}$
13. 14 **14.** $28 - \sqrt{2}$ **15.** $\frac{\sqrt{2}}{2}$ **16.** $\frac{2\sqrt{3}}{15}$
17. $\sqrt{3} + 1$ **18.** $\sqrt{6} - 2$ **19.** $3\sqrt{10}$ **20.** $8 + 3\sqrt{5} + \sqrt{13}$

CHAPTER 11

EXERCISES 11.1, page 433

1. $x = \pm 11$ **3.** $x = \pm\sqrt{42}$
5. $x = -1$, $x = 3$ **7.** $x = -\frac{3}{2}$, $x = -\frac{1}{2}$
9. $x = 3$, $x = 9$ **11.** $x = 7 \pm 2\sqrt{3}$
13. $x^2 + 12x + \underline{36} = (x + 6)^2$ **15.** $2x^2 - 16x + \underline{32} = 2(x - 4)^2$
17. $x^2 - 3x + \underline{\frac{9}{4}} = \left(x - \frac{3}{2}\right)^2$ **19.** $x^2 + x + \underline{\frac{1}{4}} = \left(x + \frac{1}{2}\right)^2$

21. $2x^2 + 4x + \underline{2} = 2(x + 1)^2$
23. $x = -7, x = 1$
25. $x = 9, x = -5$
27. $x = 8, x = -5$
29. $x = -\frac{4}{3}, x = 1$
31. $x = -\frac{1}{2}, x = \frac{3}{2}$
33. $x = -3 \pm \sqrt{6}$
35. $x = -1 \pm \sqrt{6}$
37. $x = -\frac{1}{2} \pm \frac{\sqrt{13}}{2} = \frac{-1 \pm \sqrt{13}}{2}$
39. $x = -\frac{3}{4} \pm \frac{\sqrt{17}}{4} = \frac{-3 \pm \sqrt{17}}{4}$
41. $x = \frac{9}{2} \pm \frac{\sqrt{73}}{2} = \frac{9 \pm \sqrt{73}}{2}$
43. $x = -\frac{7}{2} \pm \frac{\sqrt{105}}{2} = \frac{-7 \pm \sqrt{105}}{2}$
45. $x = 13, x = -2$
47. $x = -2, x = -\frac{1}{2}$
49. $x = 1$
51. $x = -\frac{5}{2} \pm \frac{\sqrt{33}}{2} = \frac{-5 \pm \sqrt{33}}{2}$
53. $x = \frac{2}{3} + \frac{\sqrt{10}}{6}$ or $x = \frac{2}{3} + \sqrt{\frac{5}{18}}$
55. $x = -\frac{7}{4} \pm \frac{\sqrt{17}}{4} = \frac{-7 \pm \sqrt{17}}{4}$
57. $x = \frac{1}{4} \pm \frac{\sqrt{13}}{4} = \frac{1 \pm \sqrt{13}}{4}$
59. $x = -\frac{5}{3}, x = -1$

EXERCISES 11.2, page 438

1. $a = 1, b = -3, c = -2$
3. $a = 2, b = -1, c = 6$
5. $a = 7, b = -4, c = -3$
7. $a = 3, b = -9, c = -4$
9. $a = 2, b = 5, c = -3$
11. $a = 2, b = 3, c = -3$
13. $x = -2 \pm \sqrt{5}$
15. $x = -1, x = 4$
17. $x = -\frac{1}{2}, x = 1$
19. $x = -1, x = \frac{2}{5}$
21. $x = \frac{3 \pm \sqrt{33}}{12}$
23. $x = \pm\sqrt{7}$
25. $x = 0, x = -1$
27. $x = 0, x = -\frac{4}{3}$
29. $x = \frac{-5 \pm \sqrt{65}}{4}$

EXERCISES 11.3, page 444

1. 11
3. 7 m by 9 m
5. 11 m by 18 m
7. 20 ft by 30 ft
9. 10, 11
11. 4
13. 5 cm
15. 5 m by 12 m
17. $6\sqrt{2}$ cm
19. 30 members
21. 50 mph
23. 3 mph
25. 20 members
27. 13 in. square
29. 40 min

CHAPTER 11 REVIEW, page 448

1. $x = \pm 7$

2. $x = \pm 5$

3. $x = \pm \frac{1}{2}$

4. $x = \pm 3\sqrt{2}$

5. $x = 0,\ x = -4$

6. $x = -2,\ x = 4$

7. $x = 5 \pm \sqrt{7}$

8. $x = 4 \pm 2\sqrt{2}$

9. $x^2 - 4x + \underline{4} = (x - 2)^2$

10. $3x^2 + 18x + \underline{27} = 3(x + 3)^2$

11. $x^2 - \frac{2}{3}x + \underline{\frac{1}{9}} = \left(x - \frac{1}{3}\right)^2$

12. $x^2 - x + \underline{\frac{1}{4}} = \left(x - \frac{1}{2}\right)^2$

13. $x^2 + 9x + \underline{\frac{81}{4}} = \left(x + \frac{9}{2}\right)^2$

14. $5x^2 - 10x + \underline{5} = 5(x - 1)^2$

15. $x = -1,\ x = 7$

16. $x = -2 \pm \sqrt{2}$

17. $x = -1 \pm \sqrt{5}$

18. $x = -2 \pm \frac{3\sqrt{2}}{2} = \frac{-4 \pm 3\sqrt{2}}{2}$

19. $x = -\frac{1}{2} \pm \frac{\sqrt{7}}{2} = \frac{-1 \pm \sqrt{7}}{2}$

20. $x = \frac{1}{3} \pm \frac{\sqrt{7}}{3} = \frac{1 \pm \sqrt{7}}{3}$

21. $x = -8,\ x = 2$

22. $x = -12,\ x = 4$

23. $x = -4,\ x = \frac{1}{2}$

24. $x = 0,\ x = \frac{6}{5}$

25. $x = 1 \pm \frac{\sqrt{10}}{5} = \frac{5 \pm \sqrt{10}}{5}$

26. $x = -\frac{1}{4} \pm \frac{\sqrt{13}}{4} = \frac{-1 \pm \sqrt{13}}{4}$

27. $x = -1 \pm \frac{2\sqrt{6}}{3} = \frac{-3 \pm 2\sqrt{6}}{3}$

28. $x = 0,\ x = -3$

29. $a = 2,\ b = 3,\ c = -2;\ x = -2,\ x = \frac{1}{2}$

30. $a = 3,\ b = 1,\ c = -4;\ x = -\frac{4}{3},\ x = 1$

31. $a = 2,\ b = -2,\ c = -1;\ x = \frac{1 \pm \sqrt{3}}{2}$

32. $a = 2,\ b = 5,\ c = -6;\ x = \frac{-5 \pm \sqrt{73}}{4}$

33. $a = 5,\ b = 1,\ c = -1;\ x = \frac{-1 \pm \sqrt{21}}{10}$

34. $a = \frac{3}{4},\ b = -2,\ c = \frac{1}{8};\ x = \frac{8 \pm \sqrt{58}}{6}$

35. $x = \pm 6$

36. $x = 2 \pm \sqrt{17}$

37. $x = 3 \pm \sqrt{5}$

38. $x = -1,\ x = \frac{7}{5}$

39. $x = -5,\ x = 12$

40. $x = \frac{3 \pm \sqrt{41}}{8}$

41. $x = -1,\ x = 4$

42. $x = \frac{2 \pm \sqrt{10}}{3}$

43. 12, 14

44. 25 cm, 11 cm

45. 12 m, 9 m
46. 6 ft
47. 6 mph
48. 12 hr, 6 hr
49. \$12, 25 shirts
50. 10 in. by 30 in.

CHAPTER 11 TEST, page 451

1. $x = -2, x = 12$

2. $x = \pm 2\sqrt{3}$

3. $x^2 - 24x + \underline{144} = (x - 12)^2$

4. $x^2 + 9x + \underline{\frac{81}{4}} = \left(x + \frac{9}{2}\right)^2$

5. $3x^2 + 9x + \underline{\frac{27}{4}} = 3\left(x + \frac{3}{2}\right)^2$

6. $x = 1, x = 2$

7. $x = 0, x = -\frac{12}{5}$

8. $x = 2, x = -\frac{5}{3}$

9. $x = -4, x = -2$

10. $x = 1 \pm \sqrt{6}$

11. $x = -\frac{3}{4} \pm \frac{\sqrt{33}}{4} = \frac{-3 \pm \sqrt{33}}{4}$

12. $x = \frac{-4 \pm \sqrt{10}}{3}$

13. $x = \frac{2 \pm \sqrt{10}}{2}$

14. $x = 1 \pm \sqrt{5}$

15. $x = -3 \pm \sqrt{11}$

16. $x = \frac{1}{3}, x = 2$

17. $x = \frac{3 \pm \sqrt{17}}{4}$

18. $x = 18\sqrt{2}$ in.

19. 4 mph

20. 20 in. by 23 in.

Index

C

D

E

I

K

L

M

N

O

P

Q

R

S

Powers, Roots, and Prime Factorizations

No.	Square	Square Root	Cube	Cube Root	Prime Factorization
1	1	1.0000	1	1.0000	
2	4	1.4142	8	1.2599	prime
3	9	1.7321	27	1.4423	prime
4	16	2.0000	64	1.5874	2•2
5	25	2.2361	125	1.7100	prime
6	36	2.4495	216	1.8171	2•3
7	49	2.6458	343	1.9129	prime
8	64	2.8284	512	2.0000	2•2•2
9	81	3.0000	729	2.0801	3•3
10	100	3.1623	1000	2.1544	2•5
11	121	3.3166	1331	2.2240	prime
12	144	3.4641	1728	2.2894	2•2•3
13	169	3.6056	2197	2.3513	prime
14	196	3.7417	2744	2.4101	2•7
15	225	3.8730	3375	2.4662	3•5
16	256	4.0000	4096	2.5198	2•2•2•2
17	289	4.1231	4913	2.5713	prime
18	324	4.2426	5832	2.6207	2•3•3
19	361	4.3589	6859	2.6684	prime
20	400	4.4721	8000	2.7144	2•2•5
21	441	4.5826	9261	2.7589	3•7
22	484	4.6904	10,648	2.8020	2•11
23	529	4.7958	12,167	2.8439	prime
24	576	4.8990	13,824	2.8845	2•2•2•3
25	625	5.0000	15,625	2.9240	5•5
26	676	5.0990	17,576	2.9625	2•13
27	729	5.1962	19,683	3.0000	3•3•3
28	784	5.2915	21,952	3.0366	2•2•7
29	841	5.3852	24,389	3.0723	prime
30	900	5.4772	27,000	3.1072	2•3•5
31	961	5.5678	29,791	3.1414	prime
32	1024	5.6569	32,768	3.1748	2•2•2•2•2
33	1089	5.7446	35,937	3.2075	3•11
34	1156	5.8310	39,304	3.2396	2•17
35	1225	5.9161	42,875	3.2711	5•7
36	1296	6.0000	46,656	3.3019	2•2•3•3
37	1369	6.0828	50,653	3.3322	prime
38	1444	6.1644	54,872	3.3620	2•19
39	1521	6.2450	59,319	3.3912	3•13
40	1600	6.3246	64,000	3.4200	2•2•2•5
41	1681	6.4031	68,921	3.4482	prime
42	1764	6.4807	74,088	3.4760	2•3•7
43	1849	6.5574	79,507	3.5034	prime
44	1936	6.6333	85,184	3.5303	2•2•11
45	2025	6.7082	91,125	3.5569	3•3•5
46	2116	6.7823	97,336	3.5830	2•23
47	2209	6.8557	103,823	3.6088	prime
48	2304	6.9282	110,592	3.6342	2•2•2•2•3
49	2401	7.0000	117,649	3.6593	7•7
50	2500	7.0711	125,000	3.6840	2•5•5